Worldwide Differential Calculus

David B. Massey

v.0727161038

ISBN 978-0-9842071-9-0

Contents

0.1 Preface . . . .

1 Rates of Change and the Derivative   1

1.1 Average Rates of Change   1

     1.1.1 Exercises 13

1.2 Prelude to IROC's 22

     1.2.1 Exercises 35

1.3 Limits and Continuity 41

     1.3.1 Limits 42

     1.3.2 Continuous Functions 50

     1.3.3 Limits involving Infinity 61

     1.3.4 Exercises 73

1.4 IROC's and the Derivative 81

     1.4.1 Exercises 100

1.5 Extrema and the Mean Value Theorem 108

     1.5.1 Exercises 120

1.6 Higher-Order Derivatives 126

     1.6.1 Exercises 134

Appendix 1.A Technical Matters 140

     1.A.1 Properties of Real Numbers and Extended Real Numbers 140

     1.A.2 Functions 142

     1.A.3 Proofs of Theorems on Limits and Continuity 145

1.A.4 Proofs of Theorems on Differentiability and Continuity 154

2 Basic Rules for Calculating Derivatives   **159**

2.1 The Power Rule and Linearity 159

    2.1.1 Exercises 165

2.2 The Product and Quotient Rules 171

    2.2.1 Exercises 179

2.3 The Chain Rule and Inverse Functions 186

    2.3.1 Exercises 199

2.4 The Exponential Function 205

    2.4.1 Exercises 221

2.5 The Natural Logarithm 228

    2.5.1 Exercises 238

2.6 General Exponential and Logarithmic Functions 243

    2.6.1 Exercises 251

2.7 Trigonometric Functions: Sine and Cosine 257

    2.7.1 Exercises 278

2.8 The Other Trigonometric Functions 286

    2.8.1 Exercises 292

2.9 Inverse Trig Functions 297

    2.9.1 Exercises 305

2.10 Implicit Functions 312

    2.10.1 Exercises 321

Appendix 2.A Technical Matters 330

3 Applications of Differentiation   **335**

3.1 Related Rates 335

    3.1.1 Exercises 340

3.2 Graphing 349

    3.2.1 Exercises 360

CONTENTS

3.3 Optimization 363

  3.3.1 Exercises 371

3.4 Linear Approximation 380

  3.4.1 Exercises 393

3.5 l'Hôpital's Rule 400

  3.5.1 Exercises 406

Appendix 3.A Technical Matters 413

4 Anti-differentiation & Differential Equations   419

4.1 What is a Differential Equation? 420

  4.1.1 Exercises 428

4.2 Anti-derivatives 435

  4.2.1 Exercises 452

4.3 Separable Differential Equations 461

  4.3.1 Exercises 472

4.4 Applications of Differential Equations 478

  4.4.1 Exercises 492

4.5 Approximating Solutions 501

  4.5.1 Exercises 517

Appendix A Parameterized Curves and Motion   523

A.1 Parameterized Curves 523

  A.1.1 Exercises 529

Appendix B Tables of Derivative Formulas   535

Appendix C Answers to Odd-Numbered Problems   537

Bibliography   555

Index   557

About the Author   565

0.1   Preface

Welcome to the Worldwide Differential Calculus textbook; the first textbook from the Worldwide Center of Mathematics.

Our goal with this textbook is, of course, to help you learn Differential Calculus– the Calculus of derivatives. But why publish a new textbook for this purpose when so many already exist? There are several reasons why we believe that our textbook is a vast improvement over those already in existence.

• Even if this textbook is used as a classic printed text, we believe that the exposition, explanations, examples, and layout are superior to every other Calculus textbook. We have tried to write the text as we would speak the material in class; though, of course, the book contains far more details than we would normally present in class. In the book, we emphasize intuitive ideas in conjunction with rigorous statements of theorems, and provide a large number of illustrative examples. Where we think it will be helpful to you, we include proofs, or sketches of proofs, but the more technical proofs are contained in the appendices (the "Technical Matters" sections) to chapters. This greatly improves the overall readability of our textbook, while still allowing us to give mathematically precise definitions, theorems, and proofs.

• Our textbook consists of Adobe pdf files, with linked/embedded/accompanying video content, annotations, and hyperlinks. With the videos contained in the supplementary files, you effectively possess not only a textbook, but also an online/electronic version of a course in Differential Calculus. Depending on the version of the files that you are using, clicking on the video frame to the right of each section title will either open an online, or an embedded, or a locally installed video lecture on that section. The annotations replace classic footnotes, without affecting the readability or formatting of the other text. The hyperlinks enable you to quickly jump to a reference elsewhere in the text, and then jump back to where you were.

• The pdf format of our textbook makes it incredibly portable. You can carry it on a laptop computer, on many handheld devices, e.g., an iPad, or can print any desired pages.

• Rather than force you to buy new editions of textbooks to obtain corrections and minor revisions, updates of this textbook are distributed free of cost.

• Because we have no print or dvd costs for the electronic version of this book or videos, we can make them available for download at an extremely low price. In addition, the printed, bound copies of this text and disks with the electronic files are priced as low as possible, to help reduce the burden of excessive textbook prices.

In this book, we assume you are familiar with high school algebra, analytic (Cartesian) geometry and graphing in the xy-plane, basic properties of trigonometric (trig) functions (in degrees), and basic properties of exponents and logarithms. We will define circular trig functions and radians in Section 2.7, define logarithms in Section 2.5, and define inverse trig functions in Section 2.9. However, in these sections, we will concentrate on the derivatives and graphs of the functions, not on algebraic properties.

We should remark on our approach to defining the exponential function in Section 2.4. There is always some difficulty in defining $\exp(x) = e^x$. If you already have integration, you can define the natural logarithm $\ln x$ via integration, and then define the exponential function as the inverse function of $\ln x$. This approach is elegant, but seems a bit "backwards" to most people, and would require developing integral Calculus before defining exp. This approach also makes it difficult to obtain bounds on the value of e. Then, there is the other calculator-based approach of "showing" that there is some number e, between 2 and 3, such that the limit, as h approaches zero, of $(e^h - 1)/h$ is equal to 1. The lack of rigor in this approach is worrisome and, once again, this approach makes it difficult to calculate bounds on the value of e.

We take a different approach from the two above. Our approach is via infinite series, a topic that is not covered in detail in this book. Consequently, we do not give a rigorous proof that our approach "works", but at least we can say that such a rigorous proof exists. We believe that there are several benefits to this series approach. First of all, we feel that students will have little trouble grasping that there is a sequence of polynomial functions such that the derivative of each element in the sequence is the previous element in the sequence (or zero), and that this sequence of functions can then be used to define a function which is its own derivative. Not only do we think that this approach poses no serious conceptual difficulty, but we hope that students will, in fact, find it "cool". Another advantage of using series to define $\exp(x)$ is that we can then show students how to calculate e, by hand, to any desired accuracy. A final advantage to our approach to exp is that we introduce students, briefly, to sequences, geometric series, and power series; we return to power series at the end of the section on sine and cosine, Section 2.7. We believe that these quick brushes with sequences and series will make students more comfortable when they look at these concepts in detail later.

We also wish to comment on our decision to include a chapter which contains a discussion of basic differential equations, Chapter 4. In any Calculus course which is preparing students for integration, the topic of anti-differentiation must be covered, and problems involving anti-differentiation frequently are phrased in terms of solving differential equations. However, once a student can anti-differentiate, he or she is fully prepared to solve separable differential equations. Therefore, it seemed reasonable to include both anti-differentiation and separable differential equations. We then decided there was a need for an introductory section, and for a section of applications of differential equations. We realize that many Calculus syllabi will include the anti-differentiation section, and omit the other sections on differential equations, but we nonetheless feel that the topics in the other sections of this chapter are natural to include.

The topics of definite integration and infinite series are not covered in this book, but will be covered in our second volume: Worldwide Integral Calculus, with infinite series. However, even this current volume contains some "warm-up" material on those topics. When discussing polynomial functions, in Section 2.1, we introduce the sigma notation for summations. Our approach to the exponential function in Section 2.4 requires us to briefly discuss sequences of numbers and functions, geometric series, and power series. In addition, while we use a traditional approach to defining, and working with, sine and cosine, we discuss their power series at the end of Section 2.7. It is our hope that these introductions to summations, sequences, and series will make those topics more comprehensible when they are dealt with in depth in Worldwide Integral Calculus, with Infinite Series.

This book is organized as follows:

Other than the Technical Matters sections, each section is accompanied by a video file, which is either a separate file, or an embedded video. Each video contains a classroom lecture of the essential contents of that section; if the student would prefer not to read the section, he or she can receive the same basic content from the video. Each non-technical section ends with exercises. The answers to all of the odd-numbered exercises are contained in Appendix C at the end of the book.

Important definitions are boxed in green, important theorems are boxed in blue. Remarks, especially warnings of common misconceptions or mistakes, are shaded in red. Important conventions, that will be used throughout the book, are boxed in black.

Very technical definitions and proofs from each section are contained in the Technical Matters at the end of the chapter containing the given section. This removal of the technicalities from the general exposition should make the presentation more clear, and more closely match what you normally experience in a classroom environment. Occasionally, we refer to external sources for results beyond the scope of this textbook; our favorite external technical source is the excellent textbook by William F. Trench, *Introduction to Real Analysis*, [3], which is available as a free pdf.

Internal references through the text are hyperlinked; simply click on the boxed-in link to go to the appropriate place in the textbook. If you have activated the "forward" and "back" buttons in your pdf-viewer software, clicking on the "back" button will return you to where you started before you clicked on the hyperlink.

Some terms or names are annotated; these are clearly marked in the margins by little blue "balloons". Comments will pop up when you click on such annotated items.

We sincerely hope that you find using our modern, multimedia textbook to be as enjoyable as using a mathematics textbook can be.

David B. Massey

February 2009

Chapter 1

Rates of Change and the Derivative

In this chapter, we discuss what the rate of change of one quantity, with respect to another, means. We use the intuitive notion of an *average rate of change* to lead us to a definition of the *instantaneous rate of change*: the *derivative*. The transition from the average rate of change to the instantaneous rate of change requires us to develop the idea of the *limit* of a function.

We also show that our mathematical definition of the instantaneous rate of change has many, or all, of the properties that you intuitively expect.

1.1 Average Rates of Change

The world around us is in a continual state of change. Positions of people or objects change with respect to time; this rate of change is called *velocity*. Velocities change with time; this rate of change is called *acceleration*. The radius of a balloon increases with respect to the volume of air blown into the balloon. Like many rates of change, this latter one has no name, and so we simply have to use the entire phrase "the rate of change of the radius of the balloon, with respect to volume". The price of a lobster changes, with respect to the weight of the lobster (usually with jumps at certain weights). The area of a flat television screen, either in "full-screen" 4:3 format or in "wide-screen" 16:9 format, changes, with respect to the diagonal length. The y-coordinate of the graph of a function $y = f(x)$ changes, with respect to the x-coordinate.

In this section, we will begin our mathematical discussion of how you calculate *average rates of change* (AROC's) when one quantity, such as position, velocity, radius, price, area, or the y-coordinate, depends on (i.e., is a function of) another quantity, such as time, volume, weight, length, or the x-coordinate. Our goal in the next section will be to use the notion of an AROC, developed in this section, to arrive at a reasonable definition of an *instantaneous rate of change* (IROC). The study of instantaneous rates of change is what Differential Calculus is all about.

Example 1.1.1. Suppose that a car is traveling down a straight road. At exactly noon, the driver notices that she passes a mile marker, mile marker 37 (measured from some important point 37 miles back). At exactly 12:02 pm, the driver notices that she passes mile marker 38. What was the velocity of the car during the two minutes from noon until 12:02 pm?

You should be asking "What do you mean by 'the velocity of the car'? Do you mean what would someone inside the car have seen on the speedometer at each moment during the two minutes, or do you simply mean that the car went exactly one mile in 1/30th of an hour, so that its velocity was

$$\frac{1 \text{ mile}}{1/30 \text{ hour}} = 30 \text{ miles/hour ?}"$$

The above example is intended to illuminate the difference between the instantaneous rate of change, the IROC, and the average rate of change, the AROC.

The velocity that you read on the speedometer is the IROC of the position, with respect to time; we shall discuss this concept in detail in Section 1.2, Section 1.4, and throughout much of the remainder of this book. This velocity is itself a function of time; at each time between $t = 0$ and $t = 1/30$, you can read the instantaneous velocity on the car's speedometer.

Suppose that we let $p(t)$ be the position of the car, measured in miles, as determined by the mile markers, at time t hours past noon. The 30 miles/hour that we calculated above is the AROC of the position, with respect to time, between times $t = 0$ and $t = 1/30$ hours. This is the average velocity of the car between times $t = 0$ and $t = 1/30$ hours. In terms of the position function, $p(t)$, the average velocity of the car between times $t = 0$ and $t = 1/30$ hours is

$$\frac{\text{change in } p(t)}{\text{change in } t} = \frac{p(1/30) - p(0)}{1/30 - 0} = \frac{38 - 37 \text{ miles}}{1/30 \text{ hours}} = 30 \text{ miles per hour.}$$

Note that knowing this average velocity does not, in any way, tell us what the speedometer of the car was reading at any time.

> **Remark 1.1.2.** The average velocity of the car between two given times is the AROC of the position with respect to time. It is **NOT** the average of the velocities at the two given times, that is, you do not add the velocities at the two different times and divide by 2. This fairly subtle difference in language leads to a huge difference in what you are calculating.

The phrase "the change in" that occurs when discussing various quantities in Calculus comes up so often that it is convenient to use one symbol to denote it. As is common, we shall use the Greek letter Δ for "the change in", so that the change in the position of the car in Example 1.1.1 would be denoted by Δp or $\Delta p(t)$. Of course, you cannot calculate Δp without being told the starting time and ending time, and without knowing the positions of the car at those times.

Using Example 1.1.1 as a guide, we would like to give the definition of the average rate of change for an "arbitrary" function. Of course, we want to use functions that you put real numbers into and from which you get real numbers back, that is, we want to use functions of the following type:

> **Definition 1.1.3.** *A real function f is a function whose domain and codomain are subsets of the set $\mathbb{R}$ of real numbers.*

(The term *codomain* may be unfamiliar to you. There is little harm done if you replace every occurrence of the word "codomain" with "range", which should be a familiar term. The range is the set consisting precisely of those values which are attained by the function. Technically, the codomain is merely some specified set in which we are told that the function takes its values; the codomain is allowed to contain extra elements that are not in the range. For instance, we may specify the function, whose domain and codomain are both the entire set of real numbers, given by $f(x) = 3x^6 - 4x^3 + 7x + 5$. It is difficult to determine the range of this f, but a codomain is easier to specify; as we wrote, the codomain is the set of all real numbers. Unlike the range, the codomain is merely required to be "big enough" to contain all of the possible values of f. For a technical discussion of functions, see Subsection 1.A.2.)

> All functions used throughout this book, for which the domain and codomain are not explicitly given, are assumed to be real functions.

In light of Example 1.1.1, we make the following definition for any (real) function $y = f(x)$. We use $y = f(x)$, since x, y, and f seem to be the favorite, generic variable and function names. You could just as easily use $z = p(t)$, and make the corresponding changes below.

Definition 1.1.4. *Suppose that a and b are in the domain of f, and $a < b$.* *Then, the* **average rate of change (the AROC) of f, with respect to x, between $x = a$ and $x = b$, or on the interval $[a, b]$,** *is*

$$\frac{\Delta y}{\Delta x} = \frac{\Delta f}{\Delta x} = \frac{f(b) - f(a)}{b - a} = \frac{f(a) - f(b)}{a - b}.$$

Note that the last equality above means that it doesn't really matter which is bigger, a or b, as far as calculating the AROC is concerned; we required $a < b$ simply so that the interval $[a, b]$ made sense, and to have that $a \neq b$, so that we did not divide by zero. What is important in calculating the AROC is to use the same order in the numerator as you use in the denominator in the quotient of differences (the *difference quotient*).

We should also remark that our definition of the AROC of f on $[a, b]$ requires that only a and b must be in the domain of f, and **not** that the entire interval $[a, b]$ has to be in the domain of f. However, we shall normally use the terminology "average rate of change" on intervals which are entirely contained in the domain of the function in question.

Remark 1.1.5. It is important to note that the units of the average rate of change of f, with respect to x, are the units of f divided by the units of x.

Example 1.1.6. Suppose that a car is moving (in a direction designated as positive) along a straight road, and that, at times $t = 0$, 1, and 5 hours (measured from some initial starting time), the car is moving at 30, 60, and 40 miles per hour, respectively. What are the average accelerations of the car on the intervals $[0, 1]$, $[0, 5]$, and $[1, 5]$?

Acceleration means the rate of change of the velocity, with respect to time. So, the average acceleration is the AROC of the velocity, with respect to time. If we let $v(t)$ denote the velocity of the car, in mph, at time t hours, then the average acceleration is $\Delta v / \Delta t$.

Hence, on the interval $[0, 1]$, the average acceleration is

$$\frac{v(1) - v(0)}{1 - 0} = \frac{60 - 30}{1 - 0} = 30 \text{ mph/hr (or mi/hr}^2).$$

On the interval $[0, 5]$, the average acceleration is

$$\frac{v(5) - v(0)}{5 - 0} = \frac{40 - 30}{5 - 0} = 2 \text{ mph/hr},$$

and, on the interval $[1, 5]$, the average acceleration is

$$\frac{v(5) - v(1)}{5 - 1} = \frac{40 - 60}{5 - 1} = -5 \text{ mph/hr}.$$

This negative average acceleration is an indication that the car decelerated.

Example 1.1.7. What is the average rate of change of the area A, in square inches, of a widescreen 16:9 television screen, with respect to the diagonal length d, between $d = 32$ inches and $d = 40$ inches? Between $d = 40$ inches and $d = 52$ inches?

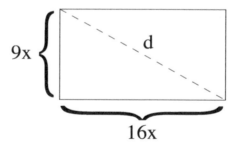

Figure 1.1: A widescreen 16:9 television.

The 16:9 ratio means that the width is 16/9 times the height. Suppose the height is $9x$. Then, the width is $16x$. The height, width, and diagonal measurements are related by the Pythagorean Theorem, so we have

$$(9x)^2 + (16x)^2 = d^2, \quad \text{and so} \quad 337x^2 = d^2.$$

Also, we know that the area $A = (9x)(16x) = 144x^2$. It follows that the area, as a function of d, is given by

$$A = A(d) = 144d^2/337 \text{ in}^2.$$

The AROC of the area, with respect to the diagonal length, on the interval $[32, 40]$, is

$$\frac{A(40) - A(32)}{40 - 32} = \frac{144}{337} \cdot \frac{40^2 - 32^2}{8} \approx 30.7656 \text{ in}^2/\text{in}.$$

The AROC of the area, with respect to the diagonal length, on the interval $[40, 52]$, is

$$\frac{A(52) - A(40)}{52 - 40} = \frac{144}{337} \cdot \frac{52^2 - 40^2}{12} \approx 39.3116 \text{ in}^2/\text{in}.$$

While it should not be totally surprising that the area increases more quickly per diagonal inch for a bigger television, you may still feel like saying "Wait a minute...doesn't adding a few diagonal inches to a television make a much bigger difference for smaller televisions?" This is true, except that we need to replace "bigger difference" with "bigger ratio" (something that is rarely, if ever, said). That is, if you start with two diagonal lengths $d_1 < d_2$ and add the same number of inches $h > 0$ to each one, then the ratios of the change in area over the original area satisfy the "reverse" inequality:

$$\frac{A(d_1 + h) - A(d_1)}{A(d_1)} > \frac{A(d_2 + h) - A(d_2)}{A(d_2)},$$

or, equivalently, dividing both sides by h (which is positive),

$$\frac{A(d_1 + h) - A(d_1)}{hA(d_1)} > \frac{A(d_2 + h) - A(d_2)}{hA(d_2)}. \tag{1.1}$$

To see this, we use that $A(d) = cd^2$, where c is a constant (namely, $144/337$). Thus, we find

$$\frac{A(d + h) - A(d)}{hA(d)} = \frac{c(d + h)^2 - cd^2}{hcd^2} = \frac{(d + h)^2 - d^2}{hd^2} =$$

$$\frac{d^2 + 2dh + h^2 - d^2}{hd^2} = \frac{2d + h}{d^2} = \frac{2}{d} + \frac{h}{d^2}.$$

Since d_1, d_2, and h are positive, and $d_1 < d_2$, it follows that $2/d_1 > 2/d_2$ and $h/d_1^2 > h/d_2^2$. Thus, we obtain Formula 1.1.

Note that Formula 1.1 tells us that the AROC of A on the interval $[d_1, d_1 + h]$, divided by $A(d_1)$, is greater than the AROC of A on the interval $[d_2, d_2 + h]$, divided by $A(d_2)$. Note also that the constant $c = 144/337$ cancelled out in the calculation, and so we would obtain the same result for a fullscreen 4:3 television. We shall revisit this example in Example 3.4.12 in Section 3.4.

Example 1.1.8. Consider the function $y = f(x) = 4 - x^2$. What is the AROC of y, with respect to x, on the intervals $[1, 2]$ and $[1, 1.5]$?

We need to calculate $\Delta y / \Delta x$. On the interval $[1, 2]$, we find

$$\frac{\Delta y}{\Delta x} = \frac{f(2) - f(1)}{2 - 1} = \frac{0 - 3}{1} = -3.$$

On the interval $[1, 1.5]$, we find

$$\frac{\Delta y}{\Delta x} = \frac{f(1.5) - f(1)}{1.5 - 1} = \frac{1.75 - 3}{0.5} = -2.5.$$

The $\Delta y / \Delta x$ in Example 1.1.8 and in Definition 1.1.4 should remind you of the slope of a line, the *rise* over the *run*. Can we, in fact, picture the AROC of an arbitrary (real) function in terms of the slope of some line? Certainly. The AROC will be the slope of the line defined by:

> **Definition 1.1.9.** *Given a function $y = f(x)$, and two x values a and b in the domain of f, with $a \neq b$, the **secant line** of f for $x = a$ and $x = b$ is the line through the two points $(a, f(a))$ and $(b, f(b))$.*

Clearly, we have

> **Proposition 1.1.10.** *Given a function $y = f(x)$, and $a < b$, where a and b are in the domain of f, the AROC of f on $[a, b]$ is equal to the slope of the secant line of f for $x = a$ and $x = b$.*

Example 1.1.11. Consider the function $y = f(x) = 4 - x^2$ from Example 1.1.8. The two AROC's on the intervals $[1, 2]$ and $[1, 1.5]$, calculated in Example 1.1.8, are the slopes of the two secant lines shown in Figure 1.2, on top of the graph of $y = 4 - x^2$.

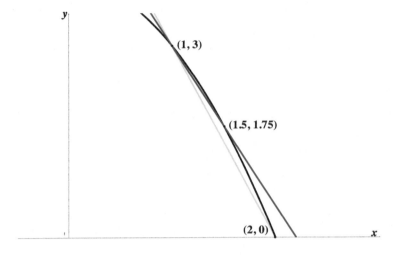

Figure 1.2: The graph of $y = 4 - x^2$ and two secant lines.

The green line has slope equal to the AROC of f on the interval $[1, 2]$, which we already found to be -3. Using the point-slope form, an equation for this secant line is $y - 3 = -3(x - 1)$. The blue line has slope equal to the AROC of f on the interval $[1, 1.5]$, which we already found to be -2.5. An equation for this secant line is $y - 3 = -2.5(x - 1)$.

Example 1.1.12. Assume that we have an ideal balloon, which stays perfectly spherical as it inflates. What is the AROC of the radius of the balloon, with respect to the volume of air inside the balloon, as the volume changes from 20 in^3 to 30 in^3?

The volume of the balloon V, in cubic inches, is related to the radius R, measured in inches, by $V = (4/3)\pi R^3$. Thus, the radius of the balloon can be considered as a function of the volume of air in the balloon:

$$R = R(V) = \left(\frac{3V}{4\pi}\right)^{1/3} = \left(\frac{3}{4\pi}\right)^{1/3} V^{1/3}.$$

The AROC of the radius with respect to volume on the interval $[20, 30]$ is

$$\frac{\Delta R}{\Delta V} = \frac{R(30) - R(20)}{30 - 20} = \left(\frac{3}{4\pi}\right)^{1/3} \frac{(30)^{1/3} - (20)^{1/3}}{30 - 20} \approx 0.0243683 \text{ in/in}^3.$$

Remark 1.1.13. In the next example, we will encounter a "problem" that will crop up from time to time: either the domain or codomain of the function being considered has units that do not naturally allow for arbitrarily small subdivisions.

For instance, the example below deals with the price, in dollars, of lobsters, as a function of the weight, in pounds, of the lobster. While you are probably comfortable assuming that the weight of a lobster could be any real number between some lower bound (1, in the example below) and some upper bound, you may question whether the price, in dollars, can be allowed to be a number that would use fractions of a dollar smaller than 0.01 (1 cent) or, worse, be an irrational number of dollars.

In another example, we may wish to look at a function which yields the number of people who would buy a certain product given the price, in dollars, of the product. Here, the smallest possible change in the price would be 0.01 dollars, while the smallest possible change in the number of people buying the product would be 1.

In such problems, we will usually ignore the fractional-units restrictions that would be placed on our mathematical functions, and assume that rounding up or down would yield a reasonable approximation, e.g., if our function tells us that the price of lobster "should" be 4π dollars, this is fine, and we assume that the cashier will ask for $12.56 or $12.57.

We could even take the philosophical position that our functions give the "right" answers, and it's not our fault that 4π dollars and 7.38 people don't exist in the physical world. There is also the physics question of whether any physical quantity can really vary by arbitrarily small amounts. We shall leave such discussions and debates for other books and other writers.

Example 1.1.14. The price of lobsters per pound typically "jumps" at certain weights, to take into account the fact that a larger lobster has a smaller percentage of its weight contained in the shell. In addition, lobsters which weigh less than one pound are not sold.

Suppose that the price, in dollars per pound, for a lobster is given by $p = p(w)$, where w is the weight of the lobster in pounds, and $w \geq 1$. Let's assume that $p(w)$ is $6/lb for $1 \leq w \leq 1.5$, $7/lb for $1.5 < w \leq 2$, $8/lb for $2 < w \leq 3$, and $9/lb for $w > 3$. The total cost $C(w)$, in dollars, of a lobster is then equal to the number of pounds that the lobster weighs times the price per pound, i.e., $C(w) = w \cdot p(w)$. The graph of $C(w)$ versus w is given in Figure 1.3.

If we take the secant line between two points that lie on the same line segment in the graph, then the secant line will simply be the line containing the given line segment, and so the average

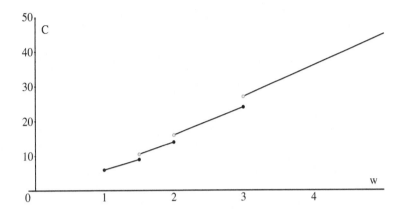

Figure 1.3: The cost C, in dollars, of a lobster of weight w, in pounds.

rate of change of C, with respect to w, on the interval determined by the w values will simply be the slope of the corresponding line segment. For instance, if $2 < a < b \leq 3$, then the AROC of C, with respect to w, on the interval $[a, b]$ is

$$\frac{\Delta C}{\Delta w} = \frac{C(b) - C(a)}{b - a} = \frac{8b - 8a}{b - a} = \$8/\text{lb.}$$

Let's look at the AROC of C, with respect to w, between $w = 2$ and $w = 2.1$. We find

$$\frac{\Delta C}{\Delta w} = \frac{C(2.1) - C(2)}{2.1 - 2} = \frac{(2.1)(8) - (2)(7)}{2.1 - 2} = \$28/\text{lb.}$$

This large AROC is caused by the "gap", the *discontinuity*, in the graph where $w = 2$, and it implies what many people know intuitively: it is **not** a good deal to buy a lobster that is just slightly bigger than one of the weights where the price per pound jumps.

In our final examples of this section, we wish to return to average velocity. We have already discussed velocity and acceleration in familiar contexts, where we did not believe confusion could arise. However, as we will use motion in a straight line in numerous examples throughout this book, we wish to clarify the standard set-up for such an example, and give precise definitions for terms that are frequently used interchangeably in everyday speech: "change in position" versus "distance traveled", and "velocity" versus "speed".

Definition 1.1.15. *When we discuss motion in a straight line (such as on a straight road), we assume that a coordinate axis has been laid out on the line, i.e., we assume that an origin, positive, and negative directions have been selected, and distances (with positive or negative signs) from the origin have been marked off. The* **position** *of an object on the line is simply its coordinate on the axis.*

The **average velocity** *of an object is the average rate of change of the position of the object, with respect to time.*

The **average acceleration** *of an object is the average rate of change of the velocity of the object, with respect to time.*

The **average speed** *of an object is the average rate of change of the distance traveled by the object, with respect to time.*

Remark 1.1.16. Note that the definitions above for average velocity, acceleration, and speed do not mention that they are only for motion in a straight line. The definitions also apply for motion in the plane or in space (or in higher dimensions), except that position is then a vector quantity. Consequently, average velocity and acceleration are, more generally, also vector quantities. Vectors have a magnitude and a direction. For motion in a straight line, the direction is simple to specify; the direction is given by a plus or minus sign. The average speed is always a non-negative number, not a vector quantity.

The treatment of Calculus involving vectors is the subject of Multi-Variable Calculus. In this book, we will generally restrict ourselves to motion in a straight line. However, occasionally, even in this book, we will consider more "complicated" motion. See Appendix A for a brief discussion of motion in a plane or in space.

We wish to give a quick example to demonstrate the difference between average velocity and average speed.

Example 1.1.17. Suppose that a car begins at some position, point A, and travels on a straight road to point B, and then turns around and comes back to point A. What is the average velocity of the car during this trip?

Without being given any distances or times (but assuming that **some** time elapsed during the trip), you can easily give the answer: 0, because the (net) change in **position** is 0. However, the average speed calculation requires more data, and will give a completely different result. Assuming that points A and B are 60 miles apart, and that the whole trip took 3 hours, the average speed of the car during the trip is $120/3 = 40$ mph.

Remark 1.1.18. Position can change from a larger coordinate to a smaller, but the distance traveled cannot decrease; hence, average velocity can be negative, while average speed must be non-negative. In general, instantaneous speed is the magnitude of the instantaneous velocity; for motion in a straight line, this means that instantaneous speed is the absolute value of the instantaneous velocity (see Definition 1.2.8). You must be very careful here; as Example 1.1.17 shows, if the direction of motion changes, the average speed need **not** be the absolute value of the average velocity.

Example 1.1.19. Suppose that the position, in meters, of a car (moving down a straight road, measured from some starting point, with a choice for a positive direction having been made), at time t seconds, is given by $p(t) = -1.5t^2 + 9t + 5$ for $0 \leq t \leq 20$.

What is the average velocity of the car between times $t = 0$ and 3 seconds? Between $t = 2$ and 3 seconds? Between $t = 3$ and 4 seconds?

The average velocity is the AROC of the position, with respect to time. We calculate, on the interval $[0, 3]$:

$$\frac{\Delta p}{\Delta t} = \frac{p(3) - p(0)}{3 - 0} = \frac{13.5}{3} = 4.5 \text{ m/s.}$$

On the interval $[2, 3]$, we find:

$$\frac{\Delta p}{\Delta t} = \frac{p(3) - p(2)}{3 - 2} = \frac{1.5}{1} = 1.5 \text{ m/s.}$$

Finally, on the interval $[3, 4]$, we have:

$$\frac{\Delta p}{\Delta t} = \frac{p(4) - p(3)}{4 - 3} = \frac{-1.5}{1} = -1.5 \text{ m/s}.$$

What does the minus sign in this last average velocity mean? It means that $p(4) < p(3)$, that is, the position of the car at time $t = 4$ seconds is less than the position of the car at time $t = 3$ seconds. In other words, the car "backed up" (moved in the negative direction) between times 3 and 4 seconds.

1.1.1 Exercises

In Exercises 1 through 7, find the average rates of change of the functions, with respect to their independent variables, on the given intervals.

1. $f(x) = x^2 - \dfrac{1}{x}$, from $x = 10$ to $x = 20$.

2. $z(t) = \dfrac{2t - 1}{t^2 - t + 3}$, from $t = 1$ to $t = 2$.

3. $g(s) = (s^2 + 1)^3$, from $s = 0$ to $s = 3$.

4. $t(y) = \dfrac{y + 2 - \sqrt{y}}{\sqrt{y}}$, from $y = 4$ to $y = 16$.

5. $y(x) = \sqrt{7x + 9}$, from $x = 0$ to $x = 1$.

6. $r(h) = 3h^4 - 2h^2 + 12h - 9$, from $h = 2$ to $h = 3$.

7. $k(q) = \dfrac{1}{2q + 3}$, from $q = 1$ to $q = 3$.

The graphs in Exercises 8 through 10 are graphs of functions of x. For each of these problems, use the slope of the secant line to approximate the average rates of change of the functions, with respect to x, on the given intervals.

8. $-1 \leq x \leq 2$

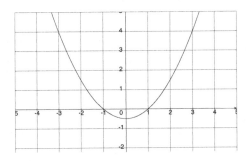

9. $-1 \leq x \leq 3$

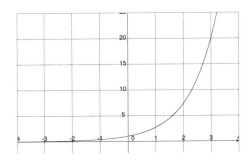

10. $-2 \leq x \leq 0$

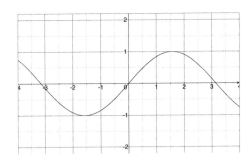

11. Explain how to determine the units of the AROC of a function over an interval.

12. Water drips into a circular puddle such that the radius of the puddle, in centimeters, at time t, in seconds, is given by the equation $r(t) = \sqrt{t}$.

     a. Write an equation for the area of the puddle as a function of t.

     b. What is the AROC of the area of the puddle with respect to time between $t = 0$ and $t = 16$?

     c. What is the AROC of the area of the puddle with respect to the radius between $t = 0$ and $t = 16$?

    d. What is the AROC of the area of the puddle with respect to the circumference of the puddle between $t = 0$ and $t = 16$?

In Exercises 13 through 18, determine whether or not the mapping is a real function. If not, explain why it is not.

13. This function takes an integer as its input. It outputs the number 0 if the input is even, and 1 if the input is odd.

14. This function takes an integer between 1 and 26 as its input and outputs the corresponding letter of the alphabet. For example, an input of 13 yields an output of m.

15. This function takes a polygon as its input and outputs the number of sides of the polygon. For example, given an input of square, the output is 4.

16.

$$g(x) = \begin{cases} \frac{1}{x^2} & \text{if } x \neq 0; \\ \infty & \text{if } x = 0 \end{cases}.$$

17. This function takes a fraction in the form $\frac{a}{b}$ where $a > 0$, $b > 0$, a and b are integers and share no common factors other than 1 as its input. The output is the denominator, b.

18. This function takes a fraction in the form $\frac{a}{b}$ where $a > 0$, $b > 0$, a and b are integers. The output is the denominator, b. Hint: how is this problem unlike the previous problem? Also, see the first paragraph of 1.A.2.

In Exercises 19 through 23, write an equation of the secant line of the function between the endpoints of the interval. Leave your answers in slope-intercept form, and use the same variable names in your equation that are used in giving the function.

19. $y = h(t) = t^3$, $[-1, 1]$.

20. $y = g(m) = \dfrac{m^2 - 1}{m^2 - 4}$, $[3, 4]$.

21. $y = j(z) = (z - 1)(z - 2)$, $[1, 2]$.

22. $y = v(x) = 2^x$, $[3, 5]$.

23. $w = f(y) = xy + xz + yz$, $[0, 1]$.

24. Let $h(x) = |x|$ and $k > 0$.

    a. Show that the AROC on $[0, k]$ is 1.

   b. Show that the AROC on $[-k, 0]$ is -1.

25. Suppose that $D(r)$ is a function which gives the population density, in hundreds of people per square mile, within r miles of the center of a very small town. Assume that there are 100 people per square mile within 5 miles of the town's center, and 350 people per square mile within 7 miles.

   a. Let $T(r)$ be the number of people (in hundreds) who live within r miles of the city. Write an expression for $T(r)$, in terms of r and $D(r)$.

   b. What is the average rate of change of the function $T(r)$ over the interval $[5, 7]$? Give your answer with units.

   c. Approximate the value of $T(6)$, using your answer to part (b). Do you suspect that this will be an overestimate, or an underestimate? Explain your answer.

26. All temperatures in this exercise are in degrees Fahrenheit.

   The internal temperature of a turkey, h hours after it is placed in a 350 degree oven, where $0 \leq h \leq 5.5$, can be modeled by the formula

$$T(h) = -3.5h^2 + 40.5h + 59.$$

   a. According to the FSIS (Food Safety and Inspection Service), a turkey should be cooked to an internal temperature of 165 degrees. How long should the turkey be cooked so that it is safe to eat?

   b. What is the average rate of change of internal temperature over the first hour that the turkey is in the oven? Over the second hour?

   c. Using some test points and your knowledge about quadratic functions, plot this temperature function on an appropriate graph, and draw the secant lines which correspond to the average rates of change from part (b).

27. Suppose spam, as a percentage of total email, is modeled by the equation

$$S(t) = 22.4\sqrt{t} + 20.3,$$

   where $t = 0$ corresponds to the year 2000 and t is measured in years. What is the AROC of S between the years 2001 and 2004?

28. Let

$$P(y) = \frac{71,800}{100y^2 + 718}$$

   model the percentage of people with y years of education who are unemployed.

    a. Without performing any calculations, do you expect this function to be an increasing or decreasing function of y?

    b. What is $P(0)$?

    c. Calculate the AROC of P between $y = 8$ and $y = 12$.

29. Assume that an ideal balloon stays perfectly spherical as it inflates.

    a. What is the AROC of the volume of the balloon, with respect to the radius, as the radius changes from 5 in to 6 in?

    b. What is the AROC of the volume of the balloon, with respect to the surface area, as the surface area changes from 4π in^2 to 9π in^2?

    c. What is the AROC of the surface area of the balloon, with respect to the volume, as the volume changes from $32\pi/3$ in^3 to 36π in^3?

30. Einstein's Special Theory of Relativity tells us that, relative to an observer on Earth, an astronaut traveling on a rocket at speeds close to the speed of light, c, will age more slowly than a human on Earth. Specifically, let r be the ratio of the aging rate of the astronaut to the aging rate of a human on Earth, as perceived by an observer on Earth. Then,

$$r(v) = \sqrt{1 - \frac{v^2}{c^2}}.$$

    a. Calculate $r(0)$, $r\left(\dfrac{c}{2}\right)$ and $r(c)$.

    b. Calculate the AROC of r between $v = c/2$ and $v = c/4$. Note that this is the average rate of change of the ratio of two rates.

31. Suppose that the projected number of citizens aged 65-85, in millions, is given by $E(t) = 5.2\sqrt{t} + 35.1$, where t is measured in years and $t = 0$ corresponds to the year 2010.

    a. Calculate the AROC between the years 2020 and 2030.

    b. Calculate the AROC between the years 2030 and 2040.

    c. Let

$$g(s) = \frac{E(s) - E(0)}{s}$$

    be the AROC between $t = 0$ and $t = s$. Is $g(s)$ an increasing or decreasing function?

32. Suppose that

$$\frac{[f(a)]^2 - [f(b)]^2}{a^2 - b^2} = 100,$$

that the average (mean) of $f(a)$ and $f(b)$ is 625, and the average of a and b is 25. What is the AROC of f on the interval $[a, b]$? ▶

33.   a. Let $f(x) = c$. What is the AROC of f on any interval $[a, b]$?

       b. Let $g(x) = mx + b$. What is the AROC of g on any interval $[c, d]$?

34. Suppose g and h both have domains equal to the set of all real numbers, and that f and j are defined as:

$$f(t) = tg(t) + (1 - t)h(t)$$
$$j(t) = (1 - t)g(t) + th(t).$$

   Let $AROC_I(y)$ be the AROC of a function y on the interval $[0, 1]$. Show that

$$AROC_I(f) + AROC_I(j) = AROC_I(g) + AROC_I(h).$$

35. Suppose $h(r) = ar^2 + 3r - b$ and that an equation of the secant line of h for $r = 2$ and $r = -3$ is $y = 0$. What are a and b?

36. Let f and g be functions defined on the domain $J = [a, b]$ and let c be any real number. Let $AROC_J(h)$ be the AROC of any function h on the interval J.

    a. Prove that $AROC_J(f + g) = AROC_J(f) + AROC_J(g)$.

    b. Prove that $AROC_J(cf) = c\,AROC_J(f)$. Note that cf is the function given by $(cf)(x) = cf(x)$.

37. Let

$$g(r) = \begin{cases} \frac{r}{2} & \text{if } 0 < r \le 2; \\ 3r + 2 & \text{if } 2 < r \le 4; \\ r^2 + 10.5 & \text{if } r > 4 \end{cases}.$$

   Solve for b, if the AROC of g between $r = 1$ and $r = b$ is 5.875.

38. Suppose an object is launched from an initial height of h_0, in meters, with initial velocity v_0, in meters per second. The height of the object after t seconds is given by the equation $h(t) = -\frac{1}{2}gt^2 + v_0 t + h_0$, where g is the acceleration downward produced by gravity. The velocity of the object at time t is $v(t) = -gt + v_0$.

    a. What is the average rate of change of the height of the object between $t = 0$ and $t = 1$?

    b. At what time t is the velocity zero?

    c. Let $t = b$ be the answer to part (b). What is the average rate of change of the velocity between $t = 0$ and $t = b$?

39. Consider the following statement: if $a < b < c$, then the AROC of a function on the interval $[a, c]$ is equal to the AROC of the function on the interval $[a, b]$ plus the AROC of the function on the interval $[b, c]$. Prove or disprove this statement.

40. A function f is *even* if it has the property that $f(-x) = f(x)$ for all x, and *odd* if $f(-x) = -f(x)$.

    a. What is the AROC of an even function on the interval $[-a, a]$?

    b. What is the AROC of an odd function on the interval $[-a, a]$?

    c. What is the AROC of $g(x) = 1/x$ on $[-a, a]$? Is g an even or odd function?

    d. What is the AROC of $h(x) = 1/x^2$ on $[-a, a]$? Is h an even or odd function?

41. The *present value* of a payment P made t years from now, where r is the interest rate, is given by $PV(P, r, t) = P(1 + r)^{-t}$.

    a. Suppose that a \$50,000 payment will be made three years from now. Calculate the AROC of the present value of this payment, with respect to the interest rate, between rates $r = 3\%$ and $r = 5\%$.

    b. Suppose that the interest rate is 4% and that a \$100,000 payment will be made in the future. Calculate the AROC of the present value of this payment, with respect to time, between $t = 2$ and $t = 5$.

    In both parts of the problem, be careful to correctly ascertain the units of the AROC.

42. Let f be a function of x, and assume that the AROC of f, with respect to x, on every interval $[0, 1]$, $[1, 2], \ldots, [10, 11]$ is 1. Draw three possible graphs of f on the interval $[0, 11]$.

43. A car is being driven on a straight road. Its velocity is measured every minute for 3 minutes, and the data are collected in the following table:

t (minutes)	0	1	2	3
v (miles per hour)	0	45	60	-10

    a. What was the car's average acceleration in the first two minutes? You may leave your answer with units of mph/min.

    b. Plot the points on a graph and draw the secant lines which represent the average acceleration of the car over the time intervals $[0, 1]$, $[1, 2]$, and $[2, 3]$.

    c. Judging from your picture from part (b), during what minute did the car experience an average acceleration with the largest absolute value? That is, during what minute was there the "most" average acceleration in either direction?

    d. Can you say whether the car stopped during this three minutes of observation?

44. A savings account is opened at noon on January 1, 2009, with an initial deposit of 500 dollars. The account offers a 4% annual interest rate, with interest compounded at the end of each quarter (of a year). No money, other than the interest, is deposited into, or withdrawn from, the account.

The amount of money, A, in dollars, in the account after m months, if m is an integer which is evenly divisible by 3, is given by

$$A \; = \; 500 \left(1 + \frac{0.01}{4}\right)^{m/3} .$$

However, as the bank deposits the interest only at the end of each 3-month period (at 12 am on the first day of April, July, October, and January), the amount of money in the account remains constant in-between these quarterly deposits. For instance, here is a small table of dates (in 2009) and corresponding amounts in the account at noon on the given day.

Date	January 1st	March 1st	May 1st	July 1st
Amount (in dollars)	500	500	505	510.05

a. Draw the graph representing the amount, in dollars, in the account at any given time from January 1, 2009 through September 30, 2009.

b. Calculate the average rate of change of the amount of money in the account, with respect to time (in months), from January 1st to February 1st. From January 1st to May 1st. From May 1st to July 1st. (All dates/times are in 2009 at noon.)

45. Alice travels 100 miles to a conference. The first 50 miles are along slow dirt roads; the second half of the trip is on a fast-moving interstate highway. Assume Alice travels at a constant speed on each of the two legs of the journey. ▶

a. If Alice's speed during the first 50 miles is 40 mph, what must her speed be during the second half of the journey so that her average speed over the entire 100 mile trip is 50 mph?

b. If Alice's speed during the first 50 miles is 25 mph, what must her speed be during the second half of the journey so that her average speed over the entire 100 mile trip is 50 mph?

46. At 6:00 am, Jenny starts driving from her house toward her coffee shop that is 20 miles away. She travels at a constant speed of 25 miles per hour for 10 minutes, and then stops suddenly (assume that her deceleration time is negligible), having realized that she may

have forgotten her keys to the shop. After 5 minutes idling on the side of the road, and realizing that she has, indeed, forgotten her keys, she speeds home at a constant rate of 35 miles per hour. Then, after 5 minutes of searching, she finds her keys and speeds off back toward the coffee shop at 40 miles per hour.

a. Suppose that Jenny makes no more stops. How long, in hours, does it take Jenny to reach her coffee shop? Call this value t_{final}.

b. Let t denote the number of hours after 6:00 am. Sketch the function $d(t)$ describing the distance that Jenny is from her house at time t, where $0 \leq t \leq t_{\text{final}}$.

c. Assume that the positive direction, from Jenny's house, is in the direction of the coffee shop. What is Jenny's average velocity, in miles per hour, over the first 5 minutes?

d. What is her average velocity over the first 15 minutes?

e. What was her average velocity for the entire trip (that is, by the time she reaches the coffee shop)?

f. What was her average **speed** for the entire trip?

1.2　Prelude to Instantaneous Rates of Change

In this section, we give a preliminary discussion of how you pass from average rates of change to instantaneous rates of change. We want to know, for instance, what the instantaneous velocity of an object means and how it's related to the average velocity. The fundamental idea is that an instantaneous rate of change is the *limit* of average rates of change, that is, the number that you approach if you look at average rates of change where the change in the independent variable gets closer and closer to zero. To make this rigorous, we must discuss limits in a serious way in the next section, but, here, we shall deal with them in an informal manner.

Let's return now to the situation of a car, moving along a straight road, on which we've placed a coordinate axis. Let $p(t)$ denote the position of the car, in miles, at time t hours, from some initial time. How do you determine the velocity of the car at time $t = 1$ hour?

In light of our discussions in Section 1.1, you may be asking "Do you mean the average velocity or the instantaneous velocity?" However, we wrote "the velocity at time $t = 1$ hour", not the velocity between two times or on some time interval. When we ask for the rate of change of one quantity p (or y, or any other variable name) with respect to another quantity t (or x, or any other variable) **at** or **when** t **equals** a particular value, that means that we are asking for the **instantaneous** rate of change, the IROC, whether the term "instantaneous" explicitly appears or not. Why? Simply because we can't mean the average rate of change, the AROC, if we don't specify two values, or an interval, for the independent variable.

Okay. So, how do you determine the instantaneous velocity of the car at time $t = 1$ hour? From inside the car, it's easy: look at the speedometer of the car when $t = 1$ hour (and also note whether you are traveling in the positive or negative direction). Our real question is: how does someone **outside the car** measure the instantaneous velocity of the car? Actually, our real question is: what does instantaneous velocity even mean?

We have some intuitive concept of velocity, and we believe that the speedometer of the car is measuring **something**. What is instantaneous velocity? We know what average velocity means; it's the average rate of change of the position, with respect to time. Can we use our definition of average velocity to arrive at a definition of instantaneous velocity? Let's think about it. Go back to Example 1.1.1, where the car was at mile marker 37 at exactly noon, and was at mile marker 38 at exactly 12:02 pm. Can we say what the velocity of the car was at noon?

Certainly not, at least not from this data. As we calculated in Example 1.1.1, the **average** velocity between noon and 12:02 pm would be 30 mph, but maybe the car was moving more slowly than this at noon and sped up by 12:02 pm, or was going faster at noon and slowed down by 12:02 pm, or sped up and slowed down multiple times in the intervening 2 minutes. The point is that 2 minutes of time is easily long enough for the velocity of the car to change appreciably, so that the average velocity of the car over a 2-minute period need not even be a good approximation of the actual (instantaneous) velocity at noon.

To get a good approximation of the velocity (as measured by the speedometer, and noting the direction of travel) at noon, we would need to calculate the average velocity between noon and some time that is so close to noon that we don't believe that the car's velocity could have changed much in the tiny time interval. Suppose, instead of looking at the car's position 2 minutes later, we looked at the car's position 10 seconds later, and calculated the average velocity between noon and 12:00:10 pm. This still doesn't seem like a small enough time interval to get a convincing, decent approximation of the actual velocity at noon. Maybe the driver braked severely between 12:00:01 pm and 12:00:10 pm, so that the average velocity between noon and 12:00:10 pm was far lower than the car's actual velocity at noon.

Certainly, 1/10th of a second seems like so small an interval of time that the car's velocity would be essentially unaffected by slamming on the brakes or stomping on the accelerator. Assuming that this is true, we could say that the instantaneous velocity of the car at noon is approximately the average velocity of the car between noon and the time 1/10 of a second later.

However, using the average velocity over a 1/10th of second interval to approximate the instantaneous velocity should feel unsatisfying for at least two reasons. First, why stop at 1/10th of a second? Surely we'd get an even better approximation if we used smaller intervals, like 1/100th of a second or 1/1000th of a second. Second, in order to decide that 1/10th of second was "good enough", we had to know something about the physical properties of a car, i.e., we needed to know that the car's velocity could not be changed any significant amount in 1/10th of a second. But, what if we were discussing the velocity of other objects, like bullets, atoms, or photons? How do we know, in all cases, when a time interval is small enough so that the average velocity yields a reasonable approximation of the instantaneous velocity?

An answer that may occur to you is to simply use a time interval of zero. Unfortunately, this doesn't work. If we try it, we get that the average rate of change of the position function $p(t)$ between $t = a$ and $t = a$ would be

$$\frac{p(a) - p(a)}{a - a} = \frac{0}{0} \, ,$$

which is undefined. Maybe we could take the time interval between t equals some number and the next biggest real number. Again, this doesn't work; there is no "next biggest real number".

Great. So, what do we do?

The answer is that we take the average velocity between times $t = a$ and $t = a + h$, where h is a variable, unequal to zero. We then see if this average velocity gets arbitrarily close to some number v, if h is "close enough" to 0. If there is such a number v, then we call that number the instantaneous velocity at $t = a$.

There is no reason to restrict ourselves here to velocity, which is the rate of change of position, with respect to time. If $y = f(x)$ is any function, we look at the average rate of change of f, with respect to x, between $x = a$ and $x = a + h$, where $h \neq 0$, and we see if this AROC gets arbitrarily close to some number L, as h gets close enough to 0. If there is such a number L, we say that the instantaneous rate of change of f, with respect to x, at $x = a$ exists and is equal to L.

To make this precise, we shall need the notion of a *limit*, which we discuss in Section 1.3 and in Section 1.A. However, in this section, we wish to give a number of preliminary examples. Note that, in order to calculate the AROC of f between $x = a$ and $x = a + h$, as h gets arbitrarily small, we must know the values of f for an infinite number of values of the independent variable. This means that a list or table of f values is not enough; we typically need a mathematical formula for f.

In these examples, the AROC that we are considering will, of course, be a function of h. It is very cumbersome to write over and over that some function of h, call it $q(h)$, gets arbitrarily close to some number L, as h gets close enough to 0. Thus, we will go ahead and adopt terminology and notation that we will not carefully explain until Section 1.3.

Definition 1.2.1. (Preliminary "Definition" of Limit, IROC, and Derivative) *If a function $q(h)$ gets arbitrarily close to some number L, as h gets close enough to 0, then we say that* **the limit of $q(h)$, as h approaches 0, exists and is equal to L**, *and we write* $\lim_{h \to 0} q(h) = L$.

Suppose we have $y = f(x)$ and we let $q(h)$ be the average rate of change of f, with respect to x, between $x = a$ and $x = a + h$, i.e.,

$$q(h) = \frac{f(a+h) - f(a)}{(a+h) - a} = \frac{f(a+h) - f(a)}{h}.$$

If, for this particular $q(h)$, $\lim_{h\to 0} q(h) = L$, then we say that the **instantaneous rate of change of f, with respect to x, at $x = a$**, *exists and equals L.*

This instantaneous rate of change of f, with respect to x, at $x = a$ is also called the **derivative of f at a and is denoted by $f'(a)$.**

It is tempting to look at Definition 1.2.1 and think "Ah - to calculate the limit as h approaches 0, and so to calculate the IROC, I simply have to plug in 0 for h."

However, this is clearly **not** what we want to do; if we were to put in 0 for h in the expression $(f(a+h) - f(a))/h$, then we would obtain the undefined quantity $0/0$. We must do some manipulations to somehow eliminate the division by h before we can "plug in" $h = 0$ and, even then, to know that plugging in $h = 0$ agrees with the limit as h approaches 0, we must use that the function under consideration is *continuous* everywhere that it is defined. We shall discuss this at length in Section 1.3

Example 1.2.2. Let's look again at Example 1.1.7, in which we had a widescreen television, which had area $A(d) = 144d^2/337$ in^2, where d is the diagonal length in inches. What is the **instantaneous** rate of change, the IROC, of A, with respect to d, when $d = 40$ in?

As before, let us write c for the constant $144/337$, simply to cut down on how much we have to write. So, $A = cd^2$. We wish to calculate the AROC of A, between $d = 40$ and $d = 40 + h$, where $h \neq 0$, and then see what happens to this AROC as h gets close to 0.

The AROC of A with respect to d between $d = 40$ and $d = 40 + h$ is

$$\frac{c(40+h)^2 - c(40)^2}{(40+h) - 40} = \frac{c\left[((40)^2 + 80h + h^2) - (40)^2\right]}{h} = c(80 + h) \text{ in}^2/\text{in}.$$

Does the limit of this AROC, as h approaches 0, exist, i.e., does the IROC at $d = 40$ exist? Yes. Namely,

$$\lim_{h\to 0} c(80 + h) = c \cdot 80 \approx 34.18398 \text{ in}^2/\text{in}.$$

Of course, we have not, at this point, proved any results about limits. We are simply appealing to your intuition that $c(80 + h)$ gets as close to $c \cdot 80$ as we want by taking h close enough to 0. How close is "close enough"? As we shall see in Section 1.3 and Section 1.A, that depends on how close we want $c(80 + h)$ to be to $c \cdot 80$.

What about the IROC at $d = 52$ inches? We do a similar calculation:

$$\frac{c(52+h)^2 - c(52)^2}{(52+h) - 52} = \frac{c\big[((52)^2 + 104h + h^2) - (52)^2\big]}{h} = c(104+h) \;\; \text{in}^2/\text{in.}$$

Does $c(104+h)$ get arbitrarily close to some number as h approaches 0? Certainly. It approaches $c(104) = (144/337)(104) \;\text{in}^2/\text{in}$. Therefore, we say that the instantaneous rate of change of the area of the television screen, with respect to the diagonal length, when $d = 52$ inches, exists and is equal to this number of square inches per inch.

Notice that the algebra that we had to do in the two calculations above was essentially the same in each case. We could have saved time and space, and calculated the IROC of A, with respect to d, for **every** possible d value, by simply leaving d as a variable in the calculation. We find that the IROC of A with respect to d, at each value of d, is

$$A'(d) = \lim_{h \to 0} \frac{A(d+h) - A(d)}{h} = \lim_{h \to 0} \frac{c(d+h)^2 - cd^2}{h} =$$

$$\lim_{h \to 0} c \cdot \frac{(d^2 + 2dh + h^2) - d^2}{h} = \lim_{h \to 0} c(2d+h) = 2cd =$$

$$\left(\frac{288}{337}\right) d \approx 0.85460\, d \;\; \text{in}^2/\text{in.}$$

As we saw in the example above, it was convenient to discuss the IROC of $A(d)$, with respect to d, at arbitrary values of d, i.e., it was convenient to just leave d as a variable in the derivative. Thus, we make the following definition, even before we have a rigorous definition of the limit. We restate this definition, a bit more carefully, in Definition 1.4.3, after we have investigated limits in Section 1.3. We should mention that it has become common practice to let the variable h denote Δx (or the change in whatever the independent variable is) in this definition.

Definition 1.2.3. *Suppose we have* $y = f(x)$. *Then, the new function* f', *given by*

$$f'(x) \;=\; \lim_{\Delta x \to 0} \frac{f(x + \Delta x) - f(x)}{\Delta x} \;=\; \lim_{h \to 0} \frac{f(x + h) - f(x)}{h},$$

is called the **derivative of** f, **with respect to** x, *and is the instantaneous rate of change of* f, *with respect to* x, *for any value of* x *for which the limit exists.*

Remark 1.2.4. You may look at Definition 1.2.3 and think "what's the difference between what's written for $f'(x)$ in Definition 1.2.3 and the definition of $f'(a)$ in Definition 1.2.1, other than that Definition 1.2.3 has an x where $f'(a)$ has an a?".

It is true that, in Definition 1.2.1, we defined $f'(a)$ by

$$f'(a) = \lim_{h \to 0} \frac{f(a + h) - f(a)}{h},$$

and a can be anything, just as x can be anything. So what's the point of putting in an a instead of x?

The point is that we frequently discuss functions in a convenient, but technically imprecise way, and replacing the variable x with a different letter helps avoid confusion.

Consider, for instance the function $f(x) = x^2$. The actual function is simply f, the squaring function. The x is what's referred to as a "dummy variable"; it's simply there as a named placeholder, but it doesn't matter what the name is. The function given by $f(t) = t^2$ is the same as the function given by $f(x) = x^2$. They are both the squaring function. The expression $f(x)$ is actually the value of the function at x; it is technically a real number, not a function. And yet, we frequently write $f(x)$ in place of the function f, or we write simply "the function x^2", assuming that the reader will know that we mean the function f defined by $f(x) = x^2$, where x is just a dummy variable.

But if we're going to use x^2 to denote the squaring function, then what do we write when we want to indicate simply a single value of f, **not the function** f, after we've plugged in a number that could be anything? The answer is that we write something like "consider the value of x^2, when $x = a$". This does exactly mean consider a^2, but the switch from our standard variable names, like x and t, is supposed to let the reader know that a^2 really means the value a^2, not the function $f(a) = a^2$, which would just be the squaring function again.

For more details on functions, and their domains and codomains, see Subsection 1.A.2.

Now that we have finished with that technical discussion, let's look at what happens to secant lines (Definition 1.1.9) as we take limits. Is there a graphical way to see the instantaneous rate of change, in addition to the average rates of change?

Example 1.2.5. Let's return to Example 1.2.2 above, where we considered the function $A = A(d) = (144/337)d^2$. The red lines in Figure 1.4 are the secant lines of A for the pairs $d = 20$ and 55, $d = 20$ and 40, and $d = 20$ and 27. That is, we have fixed one d-value, $d = 20$, and let the second d-value get closer and closer to $d = 20$. We cannot let the second d-value get **too** close to $d = 20$ and continue to see changes on the graph. If you are viewing this electronically, and can view videos, clicking on the graph in Figure 1.4 will produce an animation. Otherwise, you should be able to imagine the red lines approaching the fixed blue line as the second d-value approaches 20.

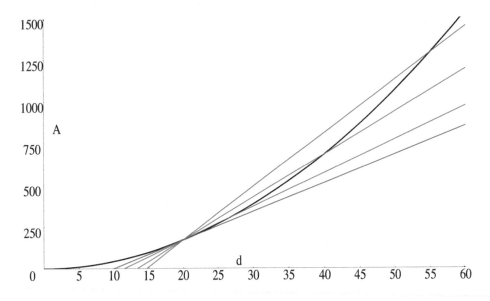

Figure 1.4: Limits of secant lines.

The blue line appears to glance off of the graph at the point $(20, A(20))$. Its slope is the limit of the slopes of the secant lines. But the slopes of the secant lines are the AROC's of A

between $d = 20$ and $d = 20 + h$, for values of h other than 0. Hence, the slope of the blue line is the limit of the AROC's, i.e., the blue line, the *tangent line to the graph of A, where d = 20*, is the unique line passing through the point $(20, A(20))$ with slope given by the **instantaneous** rate of change of A, with respect to d, at $d = 20$.

We calculated in Example 1.2.2 that the IROC of A, with respect to d, for any value of d, was given by $(288/337)d$. Therefore, the slope of the tangent line to the graph of A, where $d = 20$, is $(288/337)(20) \approx 17.09199$. In addition, $A(20) = (144/337)(20)^2 \approx 170.91988$. Using the point-slope form for an equation for a line, we find that the tangent line to the graph of A, where $d = 20$, is given by the equation

$$A - (57600/337) = (5760/337)(d - 20),$$

which would be approximated very closely by the line with equation

$$A - 170.91988 = 17.09199(d - 20).$$

In the example below, we will look at a function whose derivative does not exist at a particular point. However, you should keep in mind that our main focus in differential Calculus is on functions whose derivatives exist.

How can the instantaneous rate of change fail to exist at a point where the function exists? Consider the following example.

Example 1.2.6. Recall the lobster cost function $C(w)$ from Example 1.1.14. In that example, we calculated the AROC of C, with respect to w, on the interval $[2, 2.1]$, and found that it was large: \$28/lb. In fact, we claim that $C'(2)$, the instantaneous rate of change of C, with respect to w, at $w = 2$, does **not** exist.

We need to look at

$$C'(2) = \lim_{h \to 0} \frac{C(2 + h) - C(2)}{h}.$$

Thus, we are interested in the values of $C(w)$, when w is close to 2. Recall that $C(w) = 7w$ if $1.5 < w \leq 2$, and $C(w) = 8w$ if $2 < w \leq 3$. Therefore, if $0 < h < 1$, then $2 < 2 + h < 3$, and so the average rate of change of C, with respect to w, on the interval $[2, 2+h]$ is

$$\frac{C(2+h) - C(2)}{h} = \frac{8(2+h) - 7 \cdot 2}{h} = \frac{2 + 8h}{h} = \frac{2}{h} + 8 \ \ \$/\text{lb}. \tag{1.2}$$

Note that, by plugging in $h = 0.1$, we recover our earlier result that the AROC of C on the interval $[2, 2.1]$ is \$28/lb.

But what happens as we let h in Formula 1.2 approach 0 (but always choosing $h > 0$)? The $2/h$ portion gets arbitrarily large; we also say that it *increases without bound*. As a brief way of expressing this, we say that $2/h$ *approaches (positive) infinity as h approaches 0 from the right*, and write that $2/h \to \infty$ as $h \to 0^+$. The phrase "from the right" is used because we usually picture the positive direction on a coordinate axis to be to the right of the origin. We shall make all of this precise in Section 1.3.

Hence, the IROC of C, with respect to w, at $w = 2$ does not exist, for there is no real number that is being approached by the AROC between $w = 2$ and $w = 2 + h$, as h approaches 0 from the right.

Remark 1.2.7. In business and economics, derivatives are discussed all the time; the derivative of a function $f(x)$, with respect to x, is referred to as the *marginal value* of f (with respect to x). The marginal cost, marginal revenue, and marginal profit are all very important in the business world.

When we first gave Example 1.1.14, we had a price **per pound** for lobsters, and then the total price for for a lobster of weight w; in order to avoid confusion over which "price" we were discussing, we referred to the total price as the "total cost", and denoted this by $C(w)$. In business terminology, this is very bad. "Price" is what you sell something for, "cost" is how much the item costs you, the seller, to produce or acquire. (The revenue, per item, would be the same as the price, after the sale is made, assuming the sale is made at the given price.) So, really, we should refer to $C(w)$ in Example 1.1.14 and Example 1.2.6 as the **price** of a lobster of weight w pounds and, using the "marginal" terminology, what we showed in Example 1.2.6 would be described as the marginal price per pound of a lobster of weight 2 is infinite. Such infinite marginal quantities are **not** something that are discussed often; as we mentioned above, we typically deal with cases where the derivatives actually exist.

Before we look at two final examples in this section, we want to clearly define instantaneous velocity, speed, acceleration, using our definitions of the average velocity, speed, and acceleration, together with our preliminary definitions of limit and derivative in Definition 1.2.1. Of course, while these definitions are the actual definitions, they won't technically have rigorous mathematical meaning until after the next section, in which we give the formal definition of the limit.

> **Definition 1.2.8.** *The* **instantaneous velocity** *of an object is the instantaneous rate of change of the position of the object, with respect to time, as defined in Definition 1.2.3. Thus, if $p(t)$ is the position of the object as a function of the time t, then the instantaneous velocity of the object at time t is $p'(t)$.*

> *The* **instantaneous acceleration** *of an object is the instantaneous rate of change of the velocity of the object, with respect to time. Hence, if $v(t)$ is the velocity of the object as a function of the time t, then the instantaneous acceleration of the object at time t is $v'(t)$.*

> *The* **instantaneous speed** *of an object is the instantaneous rate of change of the distance traveled by the object, with respect to time. Therefore, if $d(t)$ is the distance that the object has traveled, as a function of the time t, then the instantaneous speed of the object at time t is $d'(t)$. This is equivalent to defining instantaneous speed to be the magnitude of the instantaneous velocity; for motion in a straight line, this means that the instantaneous speed is the absolute value of the instantaneous velocity.*

Remark 1.1.18 was important enough that we repeat some of it here.

> **Remark 1.2.9.** You must be careful. As Example 1.1.17 shows, and as we discussed in Remark 1.1.18, even if an object is traveling in a straight line, if the direction of motion changes, the **average** speed need **not** be the absolute value of the **average** velocity.

We will give two more examples of using our intuitive notion of limits to calculate IROC's, one fairly easy example and one hard one. The point of the hard example is to show you that the algebra involved in calculating limits can be quite involved.

Example 1.2.10. Suppose that a particle is moving along a coordinate axis in such a way that its position $p(t)$, in meters, at time t seconds, is given by

$$p(t) = 5t^2 - 40t.$$

What is the (instantaneous) velocity of the particle at an arbitrary time t seconds? When does the particle "stop" for an instant of time?

Solution:

The instantaneous velocity of the particle is the instantaneous rate of change of the position, with respect to time. This is the limit of the average velocities of the particle, as the interval of time approaches zero, i.e., the velocity $v(t)$, in m/s, at time t seconds, provided it exists, is given by

$$v(t) = p'(t) = \lim_{h \to 0} \frac{p(t+h) - p(t)}{h} = \lim_{h \to 0} \frac{\left(5(t+h)^2 - 40(t+h)\right) - \left(5t^2 - 40t\right)}{h} =$$

$$\lim_{h \to 0} \frac{5t^2 + 10th + 5h^2 - 40t - 40h - 5t^2 + 40t}{h} = \lim_{h \to 0} \frac{10th + 5h^2 - 40h}{h}$$

$$\lim_{h \to 0} \frac{(10t + 5h - 40)h}{h} = \lim_{h \to 0} (10t + 5h - 40) = 10t - 40 \text{ m/s.}$$

We have found that the velocity $v(t) = 10t - 40$ m/s. When does the particle stop? When its velocity is zero. We set $10t - 40$ equal to 0, and solve for t to find that the particle is stopped at the instant $t = 4$ seconds.

Example 1.2.11. Suppose that a particle is moving along a coordinate axis in such a way that its position $p(t)$, in meters, at time t seconds, where $t > 0$, is given by

$$p(t) = \frac{1}{\sqrt{t^3 + 1}}.$$

What is the (instantaneous) velocity of the particle at an arbitrary time $t > 0$ seconds?

Solution:

Note that the restriction that $t > 0$ is present just as a convenient way of avoiding the division by 0 if we were to let $t = -1$.

As in the previous example, but now using our new $p(t)$, we need to calculate

$$v(t) = p'(t) = \lim_{h \to 0} \frac{p(t+h) - p(t)}{h} = \lim_{h \to 0} \frac{\frac{1}{\sqrt{(t+h)^3+1}} - \frac{1}{\sqrt{t^3+1}}}{h} =$$

$$\lim_{h \to 0} \frac{\sqrt{t^3+1} - \sqrt{(t+h)^3+1}}{h \sqrt{(t+h)^3+1} \sqrt{t^3+1}}, \tag{1.3}$$

where we shall omit the units of m/s until we write our final answer.

It is still unclear at this point that the ugly fraction in Formula 1.3 approaches anything as h approaches 0; if we simply "plug in" $h = 0$, we get the meaningless 0/0. Our goal is to eliminate the division by h, so that we can calculate the limit by just plugging in $h = 0$. (We remind you that calculating limits simply by plugging in requires that the function being considered is continuous. This is something that we shall just assume below. See Section 1.3.)

To eliminate the division by h, we multiply the numerator and denominator of Formula 1.3 by the "conjugate" of the numerator:

$$\sqrt{t^3+1} + \sqrt{(t+h)^3+1}$$

to obtain that

$$v(t) = \lim_{h \to 0} \frac{(t^3+1) - ((t+h)^3+1)}{h \sqrt{(t+h)^3+1} \sqrt{t^3+1} \left[\sqrt{t^3+1} + \sqrt{(t+h)^3+1}\right]} =$$

$$\lim_{h \to 0} \frac{(t^3+1) - (t^3 + 3t^2h + 3th^2 + h^3 + 1)}{h \sqrt{(t+h)^3+1} \sqrt{t^3+1} \left[\sqrt{t^3+1} + \sqrt{(t+h)^3+1}\right]} =$$

$$\lim_{h \to 0} \frac{-3t^2h - 3th^2 - h^3}{h \sqrt{(t+h)^3+1} \sqrt{t^3+1} \left[\sqrt{t^3+1} + \sqrt{(t+h)^3+1}\right]}.$$

Now, at last, we can cancel the h factor in the denominator with an h factor in the numerator to obtain

$$v(t) = \lim_{h \to 0} \frac{-3t^2 - 3th - h^2}{\sqrt{(t+h)^3 + 1} \, \sqrt{t^3 + 1} \, \left[\sqrt{t^3 + 1} + \sqrt{(t+h)^3 + 1}\right]},$$

and, finally, we can calculate this limit (of this continuous function) by letting h actually equal 0; we find that the velocity of the particle at any time $t > 0$ is

$$v(t) = \frac{-3t^2}{\sqrt{t^3 + 1} \, \sqrt{t^3 + 1} \, \left[\sqrt{t^3 + 1} + \sqrt{t^3 + 1}\right]} = \frac{-3t^2}{2(t^3 + 1)^{3/2}} \quad \text{m/s}.$$

In order to make precise all that we have written in this section, we must give a **real** definition of limit, a mathematically rigorous definition, and discuss theorems on limits. We do these things in the next section, Section 1.3.

To calculate instantaneous rates of change, i.e., derivatives, we do not want to have to perform horrendous algebra or trigonometric manipulations, such as in Example 1.2.11 over and over again. Hence, we will prove "rules" for calculating derivatives in Chapter 2. Once you memorize these relatively few rules for calculating derivatives, the calculation of instantaneous rates of change becomes very simple and quick for many, many functions.

Later, when we have so many rules for calculating derivatives, you may feel that you can safely forget the definition of the derivative as a limit of average rates of change.

Try to always remember: the rules that we shall develop will help us calculate derivatives easily, but **the reason that the derivative is something that we want to calculate is because it is the instantaneous rate of change, and the reason it is the instantaneous rate of change is because it is the limit of the average rates of change, as the change in the independent variable approaches 0.**

The point being that, even after we have the rules for calculating derivatives, you should never forget the definition of the derivative. The definition of the derivative is where all of its applications come from.

1.2.1   Exercises

For the functions in Exercises 1 through 9, find the average rates of change on the intervals $[x, x+h]$, for x given in the problem, and $h = 1$, 0.1, 0.01, 0.001, and 0.0001. Use this data to estimate the IROC at the point x.

1. $f(x) = 7x + 3$, at $x = 9$.

2. $f(x) = 11$, at $x = 2$.

3. $f(x) = x^2 - 9$, at $x = 3$.

4. $g(x) = \dfrac{1}{\sqrt{x}}$, at $x = 9$.

5. $k(x) = (2x - 4)^3$, at $x = 1$.

6. $l(x) = \begin{cases} 0 & \text{if } -\infty < x \leq 0.5; \\ x + 5 & \text{if } 0.5 < x < \infty \end{cases}$, at $x = 0.25$.

7. $z(x) = \dfrac{x^2 - 1}{x - 1}$, at $x = 0$.

8. $w(x) = \sqrt{25 - x}$, at $x = 16$.

9. $j(x) = ax^2 + bx + c$, $a \neq 0$, at $x = 1$.   ▶

In Exercises 10 through 13, you are given a position function $x(t)$, in feet, for a particle traveling along the x-axis, where t is measured in seconds. In each case, calculate the average velocity of the particle between times t and $t + h$. "Simplify" your answer to the point where you obtain a function of h (and t) which is defined and continuous at $h = 0$. Then, "plug in" $h = 0$ to find the instantaneous velocity $v = v(t)$ of the particle at time $t > 0$. Include units in your answers.

10. $x(t) = \dfrac{1}{t}$.

11. $x(t) = \dfrac{1}{t^2}$.   ▶

12. $x(t) = \sqrt{t}$.

13. $x(t) = mt + b$, $m \neq 0$.

14. Find the acceleration $a = a(t)$ of the particle from Exercise 10. Include units in your answer.

15. Find the acceleration $a = a(t)$ of the particle from Exercise 11. Include units in your answer.

16. Find the acceleration $a = a(t)$ of the particle from Exercise 12. Include units in your answer.

17. Find the acceleration $a = a(t)$ of the particle from Exercise 13. Include units in your answer.

In Exercises 18 through 20, solve the equation for x and calculate the IROC of x with respect to y at the given y value. Assume $x > 0$

18. $y = 3x + 7$, $y = 22$.

19. $y = x^2 + 6x + 9$, $y = 49$.

20. $y = \sqrt{100 - x^2}$, $y = 6$.

21. What is the limit as h gets close to 0 of the function $f(h) = \begin{cases} 3 & \text{if } h \neq 0; \\ 5 & \text{if } h = 0 \end{cases}$?

    Use Definition 1.2.1.

22. Consider a car traveling at constant speed around a circular track.

    a. Is the car's velocity changing?

    b. Is the car accelerating?

23. Consider the function:
$$f(x) = \begin{cases} 3 & \text{if } x \neq 0; \\ 5 & \text{if } x = 0. \end{cases}$$

    a. What is the instantaneous rate of change of f with respect to x at $x = a$ if $a \neq 0$?

    b. Does the instantaneous rate of change of f with respect to x at $x = 0$ exist? If so, what is it? If not, why not?

The height of a ball dropped from a platform 50 meters off the ground, with initial velocity 0, at time t seconds, is given by: $h(t) = -4.9t^2 + 50$. Use this information in Exercises 24 through 26.

24. a. Calculate the average rate of change of the height with respect to time on the intervals: $[2, 2.5]$, $[2, 2.25]$, $[2, 2.05]$, $[2, 2.001]$.

     b. Based on part (a), what do you suspect is the IROC of the height with respect to time at $t = 2$?

25. As long as the platform is high enough so that the ball does not reach the ground before $t = 2$ seconds, the IROC at $t = 2$ is independent of the height of the platform. Justify this statement.

26. Show that the acceleration at any time t is -9.8 meters/sec^2.

The statements in Exercises 27 through 29 are all false. Write correct versions of each statement.

27. The instantaneous velocity of an object is the instantaneous rate of change of the distance traveled by an object with respect to time.

28. The instantaneous acceleration of an object is the instantaneous rate of change of the speed of the object with respect to time.

29. The instantaneous speed of an object is the magnitude of the instantaneous acceleration.

30. The graph of the function $y = \sqrt{9 - x^2}$ is the upper half of the circle of radius 3 centered at the origin.

    a. Draw this semicircle and label the point $A = (0, 3)$.

    b. On your graph, draw secant lines between the point A and other points on the graph near A.

    c. Use the geometric notion of the IROC at a point to find the IROC of y, with respect to x, for this function at the point $(0, 3)$.

    d. Now, find the formula for the IROC of y, with respect to x, at the point $x = 0$.

31. Einstein's Special Theory of Relativity states that the mass m of a particle, to an outside observer, is a function of the particle's velocity v. If m_0 is the mass of the particle at rest and c is the speed of light, then

$$m(v) = \frac{m_0}{\sqrt{1 - \dfrac{v^2}{c^2}}}.$$

Assume $m_0 = 1000$ kg and $c = 299{,}792.458$ km/sec. Calculate the AROC of m, with respect to v, on the intervals $[a, a + h]$ where $a = \frac{\sqrt{3}}{2}c$ and $h = 0.1c, 0.01c, 0.001c$ to estimate the IROC at a.

32. Suppose \$1,000 is invested in an account where the interest rate is 10%, compounded annually. The table below shows the amount of money in the account after each of the first five years.

Year	1	2	3	4	5
Balance ($)	1100	1210	1331	1464.10	1610.51

    a. Calculate the AROC of the balance for each consecutive pair of years.

    b. Does the AROC appear to be increasing, decreasing, or staying the same?

    c. If $B(n)$ is the balance at the end of the nth year, let $P(n) = \dfrac{B(n) - B(n-1)}{B(n-1)}$.
       Calculate $P(n)$ for $n = 2, 3, 4, 5$.

    d. Does $P(n)$ appear to be increasing, decreasing, or staying the same?

33. The height of a tree is often modeled as a function of its diameter. The table below shows the diameter in centimeters of a tree, and the corresponding predicted height in meters.

Diameter (cm)	15	45	75	120
Height (m)	12	20	25	30

    a. Calculate the slopes of the secant lines between the points $(15, 12)$ and each of the other three points.

    b. Calculate the slopes of the secant lines between the points $(45, 20)$ and each of the other three points.

    c. Calculate the slopes of the secant lines between the points $(75, 120)$ and each of the other three points.

    d. Based on your answers to the first three parts, does the IROC appear to be increasing, decreasing or staying the same?

34. Suppose a function has the property that $f(x + y) = f(x) + f(y)$, for all x and y. Show that $f'(a) = f'(b)$ for all a and b. This is the same as saying the derivative or IROC is the same at every point.

In Exercises 35 through 37, show that the limit as $h \to 0$ of the slope of the secant line of the function on the interval $[a, a+h]$ is equal to the limit as $h \to 0$ of the slope of the secant line on the interval $[a - h, a + h]$.

35. $f(x) = x^2$, $a = 0$.

36. $f(x) = \frac{1}{\sqrt{x^3+1}}$, $a = 1$.

37. $f(x) = 3x + 1$, $a = 4$.

In Exercises 38 through 40, show that the limit as $h \to 0$ of the slope of the secant line of the function on the interval $[a, a + h]$ is not equal to the limit as $h \to 0$ of the slope of the secant on the interval $[a - h, a + h]$.

38. $h(x) = |x|$, $a = 0$.

39. $j(x) = 1/x^2$, $a = 0$.

40. $p(x) = (\sqrt[3]{x})^2$, $a = 0$.

41. Viscosity of a liquid is a measure of how resilient a liquid is to pressure exerted in a lateral direction, measured in kilogram meters per second. The viscosity of water at different temperatures is given in the table below:

Temperature (°C)	10	20	30	40	50
Viscosity (kg·m/s)	0.0013	0.0010	0.0008	0.0007	0.0005

   a. Plot these points on an appropriately scaled graph.

   b. Calculate the slopes of the secant lines which pass through the point $(20, 0.0010)$, and through each of the other points on your graph (points that come from the table).

   c. Use these slopes to estimate the instantaneous rate of viscosity, with respect to temperature, when the temperature is 20°C.

42. Demand for a product, $D(s)$, is the number of consumers who will buy a product at a certain fixed price s dollars. Suppose that the demand for this book fits the model $D(s) = s^2 - 600s + 90000$, for $0 \leq s < 301$.

   a. According to the model, how many books are sold when the price is 50 cents?

   b. According to the model, at what price will no one buy the book?

   c. Find and interpret, in words, the instantaneous rate of change of demand, with respect to price, when the price is \$10.

43. Suppose that a heater can increase the temperature in a small room from 40°F to 80°F in 20 minutes. Suppose that the function $T(m)$ representing the temperature, in °F, in the room m minutes after the heater is turned on is given by the function

$$T(m) = -\frac{1}{10}m^2 + 4m + 40,$$

for $0 \leq m \leq 20$.

   a. What is the average rate of change of temperature, with respect to time, over the entire 20 minutes? Over the first 10 minutes? Over the second 10 minutes?

   b. What is the instantaneous rate of change of the temperature, with respect to time, in the room 5 minutes, 10 minutes, and 15 minutes from the moment that the heater is turned on?

In physics, it's common to describe the horizontal and vertical coordinates of a particle as two separate functions of time. In this case, we write $p(t) = (x(t), y(t))$. The horizontal velocity is just the **IROC** of the function $x(t)$. Similarly, the vertical velocity is just the **IROC** of the function $y(t)$. Calculate the instantaneous horizontal and vertical velocities of the particles in Exercises 44 through 46.

44. $p(t) = (t, 3t + 5)$.

45. $p(t) = (t, \sqrt{9 - t^2})$ (defined when $0 < t < 3$).

46. $p(t) = (1/t, 1/t^2)$.

1.3 Limits and Continuity

Despite the fact that we have placed all of the proofs of the theorems from this section in the Technical Matters section, Section 1.A, at the end of the chapter, the material presented here is, nonetheless, necessarily technical. We suggest that you spend a significant amount of time "digesting" the definition of *limit* in Definition 1.3.2.

After you understand what it means to write $\lim_{x \to b} f(x) = L$, then you should understand one main point: essentially every function $f(x)$ that you have ever seen, which was not explicitly defined in cases or pieces, is a *continuous function* (Definition 1.3.20), which means that, if b is in the domain of f, then $\lim_{x \to b} f(x)$ simply equals $f(b)$, i.e., to calculate the limit, you simply plug in $x = b$. (Here, we have assumed that an open interval around b is contained in the domain of f; the preceding statement needs to be modified a bit otherwise. See Theorem 1.3.23.)

What do we mean by "essentially every function $f(x)$ that you have ever seen"? We mean any *elementary function*: a function which is a constant function, a power function (with an arbitrary real exponent), a polynomial function, an exponential function, a logarithmic function, a trigonometric function, or inverse trigonometric function, or any finite combination of such functions using addition, subtraction, multiplication, division, or composition.

As we shall see, this means that, if you want to calculate $\lim_{x \to b} g(x)$, and $g(x)$ is equal to an elementary $f(x)$, for all x in some open interval around b, except possibly at b itself, and b is in the domain of f, then $\lim_{x \to b} g(x)$ exists and is equal to $f(b)$. In practice, this means that you typically proceed as follows to calculate $\lim_{x \to b} g(x)$: you assume that $x \neq b$, and manipulate or simplify $g(x)$ until it is reduced to an elementary function which is, in fact, defined at b; at this point, you simply plug in b to obtain the limit of $g(x)$ as x approaches b.

> It is important for you to understand what we have, and have not, written above. We have **not** claimed that somehow the definition of limit means that you calculate the limit of an arbitrary function $f(x)$, as x approaches b, simply by manipulating $f(x)$ until you can plug in $x = b$ and get something defined. We **have** claimed that it is an important **theorem** that this "method" does, in fact, work for **elementary functions**, i.e., it is extremely important that all elementary functions are continuous.

1.3.1 Limits

In the previous section, we gave a "definition" of limit; this definition used phrases like "arbitrarily close" and "close enough". You should have realized that such a definition is no real definition at all, merely a colloquially phrased intuitive idea of what the term "limit" should mean.

The actual, mathematical definition of limit seems very technical, especially since it is traditional to use Greek letters in the definition. We are tempted to put this technical definition in the Technical Matters section, Section 1.A, and yet, it is the definition of limit that forms the basis for all of the theorems in Calculus. Therefore, we will give the rigorous definition of limit here, and state the theorems on limits that we shall need throughout the remainder of the book. However, we shall put the proofs of the theorems on limits in Section 1.A.

We want to emphasize that all functions $y = f(x)$ discussed in this section are assumed to be real functions (see Subsection 1.A.2).

If b is a real number, possibly not even in the domain of f, what should it mean to say/write that "the limit as x approaches b of $f(x)$ is equal to (the real number) L", i.e., to write $\lim_{x \to b} f(x) = L$?

We want it to mean that we can make $f(x)$ get as close to L as we want (except, possibly, equalling L) by picking x close enough, but not equal, to b. In what sense do we mean "close"? We mean in terms of the distance between the numbers, and this is most easily stated in terms of absolute value; the distance between any two real numbers p and q is $|p - q| = |q - p|$.

Rewriting what we wrote before, but now phrasing things in terms of absolute values, $\lim_{x \to b} f(x) = L$ should mean that we can make $|f(x) - L|$ as close to 0 as we want, except, possibly, equal to 0, by making $|x - b|$ sufficiently close to 0, without being 0. Saying that these absolute values are close to 0 is the same as saying that they are less than "small" positive numbers. But this just pushes the question back to "what does a small positive number mean?". The answer almost seems like cheating; we simply use **all** positive numbers, which will certainly include anything that you would consider "small". Hence, we arrive at the definition of limit in Definition 1.3.2, in which f is a real function, and all of the numbers involved are real.

However, before we can proceed to that definition, we must make another definition.

> **Definition 1.3.1.** *Let b be a real number. A **deleted open interval around** b is any subset of the real numbers formed by taking an open interval containing b and then removing b.*

Now, we can give the definition of limit.

Definition 1.3.2. (**Definition of Limit**) *Suppose that L is a real number. We say that* **the limit as x approaches b of $f(x)$ exists and is equal to L,** *and write $\lim_{x \to b} f(x) = L$, if and only if the domain of f contains a deleted open interval around b and, for all real numbers $\epsilon > 0$, there exists a real number $\delta > 0$ such that, if x is in the domain of f and $0 < |x - b| < \delta$, then $|f(x) - L| < \epsilon$.*

If there is no such L or if the domain of f does not contain a deleted open interval around b, then we say **the limit as x approaches b of $f(x)$ does not exist.**

Remark 1.3.3. It is important to note that $\lim_{x \to b} f(x)$ does **not** depend at all on the value of f at b. In fact, the limit may exist even though b is not in the domain of f; this case is what **always** occurs when calculating derivatives. It is also important to note that $\lim_{x \to b} f(x)$ depends only on the values of f in any deleted open interval around b; thus, two functions which are identical when restricted to a deleted open interval around b will have the same limit as x approaches b.

You should think of ϵ and δ in Definition 1.3.2 as being small, and think of the definition of the limit as saying "if you specify how close (this is a choice of ϵ) you want $f(x)$ to be to L, other than equal to L, you can specify how close (this is giving a δ) to take x to b to make $f(x)$ as close to L as you specified, except, possibly, when $x = b$."

Note that, while you may not "pick" ϵ to be 0 in Definition 1.3.2, we did not write "...then $0 < |f(x) - L| \ldots$". Thus, it is allowable for $f(x)$ to equal L for x values close to b. It is also allowable for b to be in the domain of f and for $f(b)$ to equal L, as we shall see when we discuss *continuous functions*.

It is important that we get to produce δ AFTER ϵ is specified; if we choose a new, smaller $\epsilon > 0$, then we will usually need to pick a smaller $\delta > 0$.

Example 1.3.4. Consider the function

$$y = f(x) = \frac{x^2 - 1}{x - 1}.$$

We use the "natural" domain for f; that is, the domain of f is taken to be the set of all real numbers other than 1. The graph of this function is

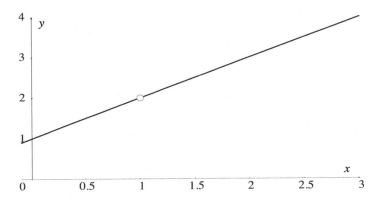

Figure 1.5: The graph of $y = f(x) = (x^2 - 1)/(x - 1)$.

What is $\lim_{x \to 1} f(x)$?

We may factor the numerator as the difference of squares to obtain $x^2 - 1 = (x - 1)(x + 1)$, and so $f(x) = x + 1$, but with a domain that excludes 1. Of course, this explains why the graph looks the way it does. While we have no theorems on limits yet to help us calculate, we can determine $\lim_{x \to 1}(x + 1)$ "barehandedly".

We suspect that $\lim_{x \to 1}(x + 1) = 2$, but can we prove it? Suppose that we are given an $\epsilon > 0$. Can we find, possibly in terms of ϵ, a number $\delta > 0$ such that, if $0 < |x - 1| < \delta$, it follows that $|(x + 1) - 2| < \epsilon$? Certainly, simply pick δ to be ϵ, because it is certainly true that $|x - 1|$ being less than ϵ implies that $|(x + 1) - 2| = |x - 1|$ is less than ϵ.

We can think of $\lim_{x \to 1} f(x)$ as being the "correct value" to fill in the hole in the graph in Figure 1.5. Sometimes people say that "f wants to equal 2 when $x = 1$."

In the example above, we showed that $\lim_{x \to 1} f(x) = 2$. But, how do we know that it is not also true that $\lim_{x \to 1} f(x) = 17$? That is, how do we know that an L as specified in Definition 1.3.2 is unique, if such an L exists? We have the following theorem.

Theorem 1.3.5. (**Uniqueness of Limits**) *Suppose that* $\lim_{x \to b} f(x) = L_1$ *and* $\lim_{x \to b} f(x) = L_2$. *Then,* $L_1 = L_2$. *In other words,* $\lim_{x \to b} f(x)$, *if it exists, is unique.*

The following theorem will be of great use to us later. It goes by various names, such as the *Pinching Theorem* or the *Squeeze Theorem*, and it tells us that if one function is in-between two others, the limit of the function in the middle can get "trapped", provided that the two bounding functions approach the same limit.

> **Theorem 1.3.6. (The Pinching Theorem)** *Suppose that f, g and h are defined on some deleted open interval U around b, and that, for all $x \in U$, $f(x) \leq g(x) \leq h(x)$. Finally, suppose that $\lim_{x \to b} f(x) = L = \lim_{x \to b} h(x)$. Then, $\lim_{x \to b} g(x) = L$.*

> **Corollary 1.3.7.** *Suppose that A and B are constants, that f and g are defined on some deleted open interval U around b, and that, for all $x \in U$, $A \leq g(x) \leq B$. Finally, suppose that $\lim_{x \to b} f(x) = 0$. Then, $\lim_{x \to b} f(x)g(x) = 0$.*

Example 1.3.8. In order to give a classic example of how you use the Pinching Theorem or its corollary above, we have to assume that you have some familiarity with the sine function, $\sin(x)$, which we won't really discuss in depth until Section 2.7.

The domain of the function $y = \sin(x)$ is the entire real line, and the range is the closed interval $[-1, 1]$. Thus, for all x, $-1 \leq \sin(x) \leq 1$.

Consider now the function $g(x) = \sin\left(\frac{1}{x}\right)$, which is defined for all $x \neq 0$. Then, for all $x \neq 0$, $-1 \leq g(x) \leq 1$, and the limit, as x approaches 0 of $g(x)$ does not exist. Nonetheless, Corollary 1.3.7 tells us that

$$\lim_{x \to 0} \left[x \sin\left(\frac{1}{x}\right) \right] = 0.$$

Some functions, such as our lobster price-function from Example 1.1.14, exhibit one type of behavior as the independent variable approaches a given value b through values greater than b, and a different type of behavior as the independent variable approaches b through values less than b. These "one-sided" limits will be of importance to us in various situations.

As when we defined the "ordinary" limit, before we can define the one-sided limits, we need to define some special types of sets near b.

Definition 1.3.9. *Let b be a real number. A **deleted open right interval of** b is a subset of the real numbers of the form $(b, b + p)$ for some $p > 0$.*

*A **deleted open left interval of** b is a subset of the real numbers of the form $(b - q, b)$ for some $q > 0$.*

Now we can define the one-sided limits.

Definition 1.3.10. *We say that **the limit as x approaches b, from the right, of** $f(x)$ **exists and is equal to** L, and write $\lim_{x \to b^+} f(x) = L$ if and only if the domain of f contains a deleted open right interval of b and, for all real numbers $\epsilon > 0$, there exists a real number $\delta > 0$ such that, if x is in the domain of f and $0 < x - b < \delta$ (i.e., $b < x < b + \delta$), then $|f(x) - L| < \epsilon$.*

*If there is no such L or if the domain of f does not contain a deleted open right interval of b, then we say **the limit as x approaches b, from the right, of** $f(x)$ **does not exist**.*

*We say that **the limit as x approaches b, from the left, of** $f(x)$ **exists and is equal to** L, and write $\lim_{x \to b^-} f(x) = L$ if and only if the domain of f contains a deleted open left interval of b and, for all real numbers $\epsilon > 0$, there exists a real number $\delta > 0$ such that, if x is in the domain of f and $0 < b - x < \delta$ (i.e., $b - \delta < x < b$), then $|f(x) - L| < \epsilon$.*

*If there is no such L or if the domain of f does not contain a deleted open left interval of b, then we say **the limit as x approaches b, from the left, of** $f(x)$ **does not exist**.*

*The limits from the left and right are referred to as **one-sided limits**. In order to distinguish the limit from the one-sided limits, the regular limit is sometimes referred to as the **two-sided limit**.*

The following theorem relates the ordinary, two-sided limit, to the one-sided limits.

Theorem 1.3.11. $\lim_{x \to b} f(x) = L$ *if and only if* $\lim_{x \to b^+} f(x) = L$ *and* $\lim_{x \to b^-} f(x) = L$.

Remark 1.3.12. Thus, the two-sided limit exists if and only if the two one-sided limits exist and are equal, in which case the two-sided limit equals the common value of the two one-sided limits.

Example 1.3.13. Let's look back at Example 1.1.14, in which we looked at the price of lobsters.

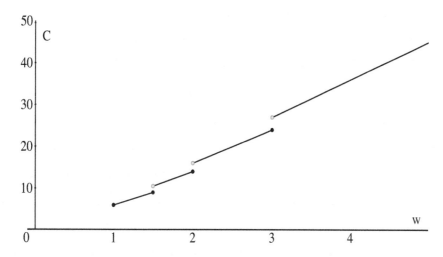

Figure 1.6: The cost C, in dollars, of a lobster of weight w, in pounds.

For lobsters whose weight w in pounds satisfies $1.5 < w \leq 2$, the cost of the lobster, in dollars, is $7w$. If $2 < w \leq 3$, $C(w) = 8w$. While we cannot calculate the limits rigorously until we have some theorems, what you are supposed to see in the graph is that $\lim_{x \to 2^-} C(w) = \lim_{x \to 2^-} 7w = 14$, while $\lim_{x \to 2^+} C(w) = \lim_{x \to 2^+} 8w = 16$.

Therefore, the two one-sided limits exist, but, since they have different values, the two-sided limit $\lim_{x \to 2} C(w)$ does not exist.

In order to calculate limits rigorously, we need a number of results, all of which are proved in Subsection 1.A.3.

We begin with two basic limits from which we will derive more complicated ones.

Theorem 1.3.14.

1. *Suppose that c is a real number and, for all x, $f(x) = c$, i.e., f is a constant function, with value c. Then, $\lim_{x \to b} f(x) = c$. In other words, $\lim_{x \to b} c = c$.*

2. *$\lim_{x \to b} x = b$.*

The next four results on limits tell us that we can "do algebra with limits" as we would expect. These properties enable us to break up complicated-looking limit calculations into a collection of smaller, easier calculations.

Theorem 1.3.15. *Suppose that* $\lim_{x \to b} f(x) = L_1$, $\lim_{x \to b} g(x) = L_2$, *and that c is a real number. Then,*

1. $\lim_{x \to b} c f(x) = c L_1$;

2. $\lim_{x \to b} [f(x) + g(x)] = L_1 + L_2$;

3. $\lim_{x \to b} [f(x) - g(x)] = L_1 - L_2$;

4. $\lim_{x \to b} [f(x) \cdot g(x)] = L_1 \cdot L_2$; *and*

5. *if* $L_2 \neq 0$, $\lim_{x \to b} [f(x)/g(x)] = L_1/L_2$.

In addition, the theorem remains true if every limit, in the hypotheses and the formulas, is replaced by a one-sided limit, from the left or right.

Remark 1.3.16. When applying Theorem 1.3.15, it is standard to apply the algebraic results first, and then show that they were valid by demonstrating the existence of the limits $\lim_{x \to b} f(x)$ and $\lim_{x \to b} g(x)$ near the end of a calculation. That is, you typically write something like

$$\lim_{x \to 1}(3x + 5) \;=\; \lim_{x \to 1}(3x) + \lim_{x \to 1} 5 \;=\; 3 \lim_{x \to 1} x + \lim_{x \to 1} 5 \;=\; 3 \cdot 1 + 5, \qquad (1.4)$$

knowing that you have to know/show that $\lim_{x \to 1} x$ and $\lim_{x \to 1} 5$ exist and know their values (which we get from Theorem 1.3.14). Thus, you typically apply Theorem 1.3.15 by working "backwards".

The proof in the "correct direction" would be: Theorem 1.3.14 tells us that $\lim_{x \to 1} x = 1$ and $\lim_{x \to 1} 5 = 5$. By Item 1 of Theorem 1.3.15,

$$\lim_{x \to 1}(3x) = 3 \cdot 1 = 3$$

and now, by Item 2 of Theorem 1.3.15,

$$\lim_{x \to 1}(3x + 5) = 3 + 5.$$

It is cumbersome to write the calculation/proof is this direction; we shall stick with the usual practice of calculating backwards (or, forwards, depending on your point of view), as in Formula 1.4.

You should understand the point: to know that you may algebraically decompose more complicated limits into smaller limit pieces, you must first know that the smaller pieces have limits that exist. Theorem 1.3.15 says **nothing** about what happens if either $\lim_{x \to b} f(x)$ or $\lim_{x \to b} g(x)$ fails to exist.

Example 1.3.17. Let's calculate the limit of the polynomial function

$$g(x) = 4x^3 - 5x + 7,$$

as $x \to 2$. Using Theorem 1.3.14 and Theorem 1.3.15, we find

$$\lim_{x \to 2}(4x^3 - 5x + 7) = 4\lim_{x \to 2}(x^3) - 5\lim_{x \to 2}x + \lim_{x \to 2}7 =$$

$$4(\lim_{x \to 2}x) \cdot (\lim_{x \to 2}x) \cdot (\lim_{x \to 2}x) - 5\lim_{x \to 2}x + \lim_{x \to 2}7 = 4 \cdot 2^3 - 5 \cdot 2 + 7 = 29.$$

In the example above, we see that the limit simply equals what you would get by substituting $x = 2$ into $g(x)$. In fact, this is true much more generally. By using Theorem 1.3.14, and iterating the formulas in Theorem 1.3.15 (technically, by using *induction*), we arrive at:

Corollary 1.3.18. *Suppose that $f(x)$ and $g(x)$ are polynomial functions. Then,*

1. $\lim_{x \to b} f(x) = f(b)$ and $\lim_{x \to b} g(x) = g(b)$;

2. if $g(b) \neq 0$, then $\lim_{x \to b} [f(x)/g(x)] = f(b)/g(b)$.

A quotient of polynomial functions, such as that appearing in Item 2 above, is referred to as a *rational function*. Note that Item 2 above actually implies Item 1, since any polynomial function is a rational function with the constant function 1 in the denominator.

Example 1.3.19. To calculate the limit

$$\lim_{t \to 3} \frac{4t^6 + t^2}{\sqrt{5}\, t^3 + 2t - 7},$$

Corollary 1.3.18 tells us that we just stick in $t = 3$, once we make sure that the denominator is not zero there. We obtain

$$\lim_{t \to 3} \frac{4t^6 + t^2}{\sqrt{5}\, t^3 + 2t - 7} = \frac{4(3^6) + 3^2}{\sqrt{5}(3^3) + 2 \cdot 3 - 7} = \frac{2925}{27\sqrt{5} - 1}.$$

It would be completely understandable at this point if you were saying to yourself: "You mean we have this complicated definition of limit, and all it amounts to is that you plug the given value into the function? What a waste of time!"

You should keep in mind that we are looking at limits in order to calculate the instantaneous rate of change, the derivative, and that you cannot calculate the limit which defines the derivative,

$$\lim_{h \to 0} \frac{f(x + h) - f(x)}{h},$$

by just "plugging in $h = 0$". However, what Corollary 1.3.18 **does** mean is that, if you can perform some algebraic (or other) manipulations and show, for h in a deleted open interval around 0, that $[f(x + h) - f(x)]/h$ is equal to a rational function $q(h) = f(h)/g(h)$, in which $g(h) \neq 0$ (for instance, if $g(h) = 1$, so that $q(h)$ is a polynomial), then

$$\lim_{h \to 0} \frac{f(x + h) - f(x)}{h} = q(0).$$

In other words, once you "simplify" $[f(x + h) - f(x)]/h$ to a rational function of h, which is defined when $h = 0$, you can calculate the limit simply by letting h be equal to 0.

1.3.2   Continuous Functions

Functions $f(x)$ for which we can calculate all of the limits $\lim_{x \to b} f(x)$ simply by plugging in $x = b$ (assuming $f(b)$ exists) are so common that we give them a name: *continuous functions.*

The definition below doesn't mention limits explicitly, though it is clearly related. The precise connection between the more general definition below and limits is given in Theorem 1.3.23 .

Definition 1.3.20. *The function f is **continuous** at b if and only if b is a point in the domain of f and, for all $\epsilon > 0$, there exists $\delta > 0$ such that, if x is in the domain of f and $|x - b| < \delta$, then $|f(x) - f(b)| < \epsilon$.*

*The function f is **discontinuous** at b if and only if b is a point in the domain of f and f is not continuous at b.*

*We say that f **is continuous** (without reference to a point) if and only if f is continuous at each point in its domain.*

Remark 1.3.21. If b is a real number which is not in the domain of f, f is neither continuous nor discontinuous at b; **there is no function f to discuss at b.** It would be like asking "Is the real function f continuous or discontinuous at the planet Venus?". The function f has no meaning at Venus; Venus is not in the domain of f.

Consider the function $y = f(x) = 1/x$, with its "natural" domain of all $x \neq 0$. We shall see shortly that this function is continuous at each point in its domain. It is, therefore, a continuous function. On the other hand, it disagrees with what some people want to call "continuous"; they want continuous to mean that the graph is connected, i.e., in one piece. Connectedness of the graph was, perhaps, the initial motivation for defining continuous functions. However, in modern mathematics, there is no disagreement: $f(x) = 1/x$ is a continuous function, which has a disconnected domain.

What the term "discontinuous" means, even in present-day mathematics, is not so clear-cut. Some authors take it to mean precisely "not continuous", and so a function would be discontinuous at any point which is not in its domain. The down-side to using this as a definition is that it would mean that we would need to say that continuous functions, such as $f(x) = 1/x$, can be discontinuous at some points. (It would also mean that all continuous real functions are discontinuous at Venus.) We choose not to adopt this terminology.

Understand the main point: a continuous real function need not be defined everywhere on the real line. In particular, the domain of a continuous real function is allowed to be the union of disjoint (non-intersecting) intervals.

We should mention two other pieces of terminology which are used when discussing continuity.

Suppose that $\lim_{x \to b^-} f(x)$ and $\lim_{x \to b^+} f(x)$ both exist and are equal; call the common value L. Then, there is one, and only one, value for $f(b)$ that would make f continuous at b; namely, we need to have $f(b) = L$. Thus, if f is **not** continuous at b (still assuming that the one-sided limits exists and are equal), then either b is not in the domain of f, or $f(b)$ is defined but is not equal to L. In either of these cases, some books would say that f has a *removable discontinuity* at b, for you can remove the lack of continuity by defining, or redefining, $f(b)$ to equal L.

If $f(b)$ is defined, but unequal to L, then we too would say that f has a removable discontinuity at b; for f has a discontinuity at $x = b$, which can be removed by redefining $f(b)$.

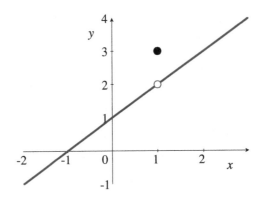

Figure 1.7: $f(x)$ has a removable discontinuity at 1.

However, if $f(b)$ is originally undefined, then, as we said earlier, we would **not** say that f is discontinuous at b, and so we would also not say, in this case, that f has a removable discontinuity at b. We would need to use the more cumbersome phrase: *f has a removable lack of continuity at b.*.

Suppose, on the other hand, that $\lim_{x \to b^-} f(x)$ and $\lim_{x \to b^+} f(x)$ both exist but are **unequal**. Then, regardless of how you try to define/redefine $f(b)$, you will not produce a function that is continuous at b. Some textbooks say, in this case, that f has a *jump discontinuity* at b. We, ourselves, would also use the phrase "jump discontinuity" if $f(b)$ is, in fact, defined, so that f is discontinuous at b. For instance, we saw in Example 1.3.13 that $C(w)$ has a jump discontinuity at 2.

It is trivial to prove, but nonetheless important, that:

Proposition 1.3.22. *Suppose that f is continuous. Then, any function obtained from f by restricting its domain, or by restricting its codomain to a set which contains the range of f, is also continuous.*

Looking at the definitions of limit and continuity, it is easy to show that we have the following characterizations of continuity.

Theorem 1.3.23. *Suppose that the domain of f contains an open interval around b. Then, f is continuous at b if and only if $\lim_{x \to b} f(x) = f(b)$ or, equivalently, $\lim_{h \to 0} f(b+h) = f(b)$.*

If the domain of f is an interval of the form $[b, \infty)$, $[b, c]$, or $[b, c)$, where $c > b$, then f is continuous at b if and only if $\lim_{x \to b^+} f(x) = f(b)$ or, equivalently, $\lim_{h \to 0^+} f(b+h) = f(b)$.

If the domain of f is an interval of the form $(-\infty, b]$, $[a, b]$, or $(a, b]$, where $a < b$, then f is continuous at b if and only if $\lim_{x \to b^-} f(x) = f(b)$ or, equivalently, $\lim_{h \to 0^-} f(b+h) = f(b)$.

Our next theorem is very useful in dealing with limits of compositions of functions. In particular, it is frequently used to simplify a limit calculation by making a "substitution". That is, suppose we want to calculate $\lim_{x \to b} f(g(x))$. You should think "if $g(x)$, the function on the inside, approaches M as x approaches b, then $f(g(x))$ should approach $f(M)$". Or, at least, we would hope, letting $y = g(x)$, that $\lim_{x \to b} f(g(x)) = \lim_{y \to M} f(y)$. This is true under the right hypotheses, and we refer to it as *making the substitution $y = g(x)$ in the original limit.*

Before we state the actual theorem, lets look at a simple example.

Example 1.3.24. Consider

$$\lim_{x \to \frac{1}{2}} \frac{(2x)^2 - 1}{2x - 1}.$$

As x approaches $1/2$, $2x$ approaches 1, so we expect (and Theorem 1.3.25 guarantees) that

$$\lim_{x \to \frac{1}{2}} \frac{(2x)^2 - 1}{2x - 1} = \lim_{y \to 1} \frac{y^2 - 1}{y - 1} = \lim_{y \to 1} (y + 1) = 2.$$

Theorem 1.3.25. (Limit Substitution) *Suppose that* $\lim_{x \to b} g(x) = M$.

Then,

$$\lim_{x \to b} f(g(x)) = \lim_{y \to M} f(y),$$

provided that $\lim_{y \to M} f(y)$ *exists, and that either f is continuous at M, or that $g(x)$ does not obtain the value M infinitely often in every open interval around b.*

In particular, if f is a continuous function and $\lim_{x \to b} g(x)$ exists and is in the domain of f, then

$$\lim_{x \to b} f(g(x)) = f\left(\lim_{x \to b} g(x)\right).$$

Note that $g(x)$ would have to be a pretty strange function in order for $g(x)$ to hit the value M infinitely often in every open interval around b; in other words, we may substitute as described in Theorem 1.3.25 essentially all of the time.

Example 1.3.26. Consider the limit

$$\lim_{x \to 1} \frac{\left(\dfrac{2x+1}{x}\right)^2 - 9}{\dfrac{2x+1}{x} - 3}.$$

We want to make the substitution $y = \dfrac{2x+1}{x}$, where we quickly see from Corollary 1.3.18 that $\lim_{x \to 1} y = 3$. Thus, we want to say that

$$\lim_{x \to 1} \frac{\left(\dfrac{2x+1}{x}\right)^2 - 9}{\dfrac{2x+1}{x} - 3} = \lim_{y \to 3} \frac{y^2 - 9}{y - 3}.$$

Is this correct? Yes, assuming that the limit on the right exists, and that $(2x+1)/x$ does not equal 3 an infinite number of times in every open interval around 1. We leave it to you to show that the only x value for which $(2x+1)/x = 3$ is $x = 1$. Thus, we may use limit substitution.

Factoring the numerator as the difference of squares, we find

$$\lim_{y \to 3} \frac{y^2 - 9}{y - 3} = \lim_{y \to 3} \frac{(y-3)(y+3)}{y-3} = \lim_{y \to 3} (y+3) = 6.$$

We state the following theorem for arbitrary continuous functions, without assuming that "nice" intervals are contained in the domain. Thus, technically, the following theorem does not follow immediately from the corresponding theorems on limits. Nonetheless, the proofs are so similar that we omit them, even in the Technical Matters section, Section 1.A.

Theorem 1.3.27. *Constant functions and the identity function are continuous.*
Sums, differences, products, quotients (with their new domains), and compositions of continuous functions are continuous.

In particular, we can now restate Corollary 1.3.18, noting again that the domain of a rational function is the open set where the denominator is not zero.

Corollary 1.3.28. *Rational functions (which include polynomials) are continuous.*

The following two theorems will be extremely important for us later, and describe fundamental properties of continuous functions.

Theorem 1.3.29. (Intermediate Value Theorem) *Suppose that f is continuous, and that the domain, I, of f is an interval. Suppose that $a, b \in I$, $a < b$, and y is a real number such that $f(a) < y < f(b)$ or $f(b) < y < f(a)$. Then, there exists c such that $a < c < b$ (and, hence, c is in the interval I) such that $f(c) = y$.*

Theorem 1.3.30. (Extreme Value Theorem) *If the closed interval $[a, b]$ is contained in the domain of the a continuous function f, then $f([a, b])$ is a closed, bounded interval, i.e., an interval of the form $[m, M]$. Therefore, f attains a minimum value m and a maximum value M on the interval $[a, b]$.*

We wish to consider taking n-th roots or, what's the same thing, raising to the power $1/n$ for non-zero integers n. Note that, for odd integers n, the domain of $x^{1/n}$ is all real numbers. For even, non-zero integers n, the domain of $x^{1/n}$ is $[0, \infty)$.

This is a special case of a more general result that applies to *inverse functions*. See Subsection 1.A.2) for a longer discussion of general properties of functions.

> **Definition 1.3.31.** *Suppose that a function $f : A \to B$ is 1) one-to-one and 2) onto. This means that: 1) for all x_1 and x_2 in A, if $f(x_1) = f(x_2)$, then $x_1 = x_2$, and 2) B is the range of f, i.e., for all y in B, there exists x in A such that $f(x) = y$.*
>
> *Then, the* **inverse function of** f, $f^{-1} : B \to A$, *is defined by: for all y in B, $f^{-1}(y)$ is equal to the unique x in A such that $f(x) = y$. Thus, for all x in A, $f^{-1}(f(x)) = x$, and, for all y in B, $f(f^{-1}(y)) = y$.*
>
> *Furthermore, if we do not initially assume that $f : A \to B$ is one-to-one and onto, but we have a function $g : B \to A$ such that, for all x in A, $g(f(x)) = x$, and, for all y in B, $f(g(y)) = y$, then f is, in fact, one-to-one and onto, and $g = f^{-1}$.*

Remark 1.3.32. In practice, you frequently determine that f is one-to-one and onto, and find a formula for its inverse, as follows: Suppose that y is in B. Solve $y = f(x)$ for x, in terms of y, and show that your solution is unique, i.e., write $x = g(y)$, for some function $g : B \to A$. The fact that you can find **some** solution x to $y = f(x)$ means that f is onto. The fact that the solution x is **unique** tells you that f is one-to-one. Then, the fact that $x = g(y)$ tells you that $g = f^{-1}$.

Example 1.3.33. Graphically, it is frequently easy to see whether or not a given function f is one-to-one.

Fix a number b. If two different points x_1 and x_2 in the domain of f are such that $b = f(x_1) = f(x_2)$, then, if we look at the horizontal line $y = b$, we will see that this horizontal line intersects the graph of f in (at least) two points: (x_1, b) and (x_2, b). Thus, if f is **not** one-to-one, there will be a horizontal line that intersects the graph of f in more than one point. For instance, $f(x) = x^2$ is not one-to-one, and we see, in Figure 1.8, a horizontal line which intersects the graph twice.

Conversely, if every horizontal line intersects the graph of f in exactly one point, or not at all, then f is one-to-one. This observation is usually referred to as the *horizontal line test*.

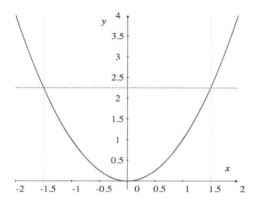

Figure 1.8: The graph of $y = x^2$.

Remark 1.3.34. If f is one-to-one, it is fairly common to say that f^{-1} exists. When we, and other authors, write this, we mean that f is actually replaced by f with its codomain restricted to be the range of the original f, i.e., f is assumed to be onto. This means that the domain of f^{-1} is the range of the original f.

> **Theorem 1.3.35.** *If $f : I \to J$ is a continuous, one-to-one, and onto real function, and I is an interval, then $f^{-1} : J \to I$ is continuous.*

In fact more can be said under the hypotheses of Theorem 1.3.35; see the full statement in Theorem 1.A.19.

Remark 1.3.36. When n is an odd natural number, x^n is one-to-one, the domain of x^n is the interval $(-\infty, \infty)$, and the range is the interval $(-\infty, \infty)$. The n-th root function, for odd n, is the inverse of this function

If n is an even natural number, then x^n is **not** one-to-one, since $a^n = (-a)^n$, for all a. However, we may restrict the domain of x^n to obtain a one-to-one function. We define a function p_n, whose domain is the interval $[0, \infty)$, whose codomain and range are $[0, \infty)$, and such that, for all $x \geq 0$, $p_n(x) = x^n$. This function **is** one-to-one. The n-th root function, for even n, is defined to be the inverse of the function p_n. As p_n is obtained from a continuous function, x^n, by restricting its domain, p_n is continuous.

Corollary 1.3.37. *For all non-zero integers n, $x^{1/n} = \sqrt[n]{x}$ is continuous.*

By applying the composition statement in Theorem 1.3.27 to x^m and $x^{1/n}$, for integers m and n, where $n \neq 0$, and by taking reciprocals, we obtain:

Theorem 1.3.38. *If r is a positive rational number, then x^r and x^{-r} are continuous functions.*

Remark 1.3.39. It is worthwhile to discuss carefully exactly what $x^{m/n}$ means, for positive integers m and n. You may have been told such things as "to calculate $x^{m/n}$, you may either first take the n-th root of x and then raise that to the m-th power, or you may first raise x to the m-th power and then take the n-th root of that." If $x \geq 0$, then it's true that either order of the operations will yield the same result.

But what about the domains of the functions $(x^{1/n})^m$ and $(x^m)^{1/n}$? If m and n share an even factor, then the "natural" domains of these compositions are different; $(x^m)^{1/n}$ is defined for all real x, while $(x^{1/n})^m$ is not defined for $x < 0$. There is a further issue. We have $(-1)^{6/2} = (-1)^3 = -1$, and yet $\left((-1)^6\right)^{1/2} = 1$.

The point is, if we are allowing x to be negative, and r is a positive rational number, then we must be slightly careful in defining the function x^r. We write r in its reduced form $r = p/q$, where p and q are positive integers with no common factor, and then $x^r = (x^p)^{1/q} = (x^{1/q})^p$, where the domain is all real x in the case where q is odd, and $x \geq 0$ in the case where q is even.

We will postpone our serious discussions about exponential functions, logarithmic functions, trigonometric functions, and inverse trigonometric functions until Section 2.4, Section 2.5, Section 2.7, Section 2.8, and Section 2.9, respectively. However, looking ahead, we wish to restate what we wrote at the beginning of the chapter:

Definition 1.3.40. *An* **elementary function** *is any function which is a constant function, a power function (with an arbitrary real exponent), a polynomial function, an exponential function, a logarithmic function, a trigonometric function, or inverse trigonometric function, or any finite combination of such functions using addition, subtraction, multiplication, division, or composition, and functions obtained from any of these by restricting domains or codomains.*

Remark 1.3.41. It is worth noting, as a special case, that the absolute value function, $|x|$, is an elementary function, since it is the composition $\sqrt{x^2}$.

The following theorem summarizes many results.

Theorem 1.3.42. *All of the elementary functions are continuous.*

Remark 1.3.43. Theorem 1.3.42 forms a big part of what you will use to calculate almost all limits (except for functions explicitly defined in cases/pieces - like the lobster price function).

You will typically proceed as follows:

You want to calculate $\lim_{x \to b} f(x)$, where f is an elementary function, and the domain of f contains a deleted open interval around b.

If b itself is in the domain of f, then Theorem 1.3.42 tells you to just plug in $x = b$. That is $\lim_{x \to b} f(x) = f(b)$.

If b is not in the domain of f, then you manipulate $f(x)$, algebraically or by some other means, to show that, for all x in some deleted open interval U around b, $f(x)$ equals another elementary function $g(x)$ (which we usually just think of as a simplified version of f), and what we want is that b **is** in the domain of g. Then, since $f(x)$ and $g(x)$ are equal for all x in U, we have

$$\lim_{x \to b} f(x) = \lim_{x \to b} g(x) = g(b),$$

where the last equality once again follows from the fact that all elementary functions are continuous.

Example 1.3.44. Let's calculate

$$\lim_{x \to 4} \frac{\sqrt{x} - 2}{x - 4}.$$

We cannot simply substitute 4 for x, for the function that we're taking the limit of is not defined when $x = 4$.

We want to perform algebraic operations, which don't change the function anywhere close to $x = 4$, except, possibly, exactly at $x = 4$, and end up with an elementary function which is defined at $x = 4$; hence, the new elementary function will be continuous at $x = 4$, so that the limit we're after can then be found by plugging in 4 for x.

How do we know what algebra to do? We don't know immediately. We have to think. The problem is the $x - 4$ in the denominator, and we'd like to somehow get rid of it.

After some thought, hopefully you come up with multiplying the numerator and denominator of the fraction by the "conjugate" of the numerator, that is, we consider

$$\lim_{x \to 4} \left[\frac{\sqrt{x} - 2}{x - 4} \cdot \frac{\sqrt{x} + 2}{\sqrt{x} + 2} \right].$$

Why would you do this??? Because you thought really hard and realized that it leads to something good, for now the numerator is the factorization of $x - 4 = (\sqrt{x})^2 - 2^2$ as the difference of squares. We find

$$\lim_{x \to 4} \frac{\sqrt{x} - 2}{x - 4} = \lim_{x \to 4} \left[\frac{\sqrt{x} - 2}{x - 4} \cdot \frac{\sqrt{x} + 2}{\sqrt{x} + 2} \right] = \lim_{x \to 4} \frac{x - 4}{(x - 4)(\sqrt{x} + 2)} =$$

$$\lim_{x \to 4} \frac{1}{\sqrt{x} + 2} = \frac{1}{4},$$

where, to obtain the last equality, we used that the elementary function $1/(\sqrt{x} + 2)$ is defined and, hence, continuous at $x = 4$.

1.3.3 Limits involving Infinity

You have seen the symbols ∞ and $-\infty$ before, for instance, in interval notation, such as $(0, \infty)$. These symbols are referred to as *positive* and *negative infinity*, respectively. The set of *extended real numbers* is the set of real numbers, together with ∞ and $-\infty$, where we extend the standard ordering on the real numbers by declaring that, for all real numbers x, $-\infty < x < \infty$.

We wish to make sense of our intuitive feeling for certain limits involving $\pm\infty$ (plus or minus infinity). For instance, as x gets "really big", $1/x$ gets "really close to" 0; so we feel that

$$\lim_{x \to \infty} \frac{1}{x}$$

should be defined and equal to 0. Also, as x get "really close" to 0, but using only x's that are greater than 0, $1/x$ gets "really big"; so we feel that

$$\lim_{x \to 0^+} \frac{1}{x}$$

should equal ∞ (which exists as an extended real number, but not as a real number).

Such limits do not fit in to our earlier definition because $\pm\infty$ are not real numbers, and expressions such as $0 < |x - \infty| < \delta$ are not what we want. The real problem is that, when we say "x approaches ∞" or the "limit equals ∞", we don't really mean that x or $f(x)$ are approaching some **number**; we mean that x or $f(x)$ gets arbitrarily large, or that they *increase without bound*. Similarly, "approaching $-\infty$" means *decreasing without bound*, that is, getting more negative than any prescribed negative number.

To handle limits involving $\pm\infty$, we must make some new definitions, definitions which fit with our discussion above. Below, when we write x and $f(x)$, we are assuming they are real numbers, i.e., not $\pm\infty$.

Definition 1.3.45. *Let L be a real number.*

Then, $\lim_{x \to \infty} f(x) = L$ if and only if the domain of f contains an interval of the form (a, ∞) and, for all real $\epsilon > 0$, there exists a real number $M > 0$ such that, if x is in the domain of f and $x > M$, then $|f(x) - L| < \epsilon$.

Similarly, $\lim_{x \to -\infty} f(x) = L$ if and only if the domain of f contains an interval of the form $(-\infty, a)$ and, for all real $\epsilon > 0$, there exists a real number $M < 0$ such that, if x is in the domain of f and $x < M$, then $|f(x) - L| < \epsilon$.

You should think about the definitions above. You should see, for instance, that the definition of $\lim_{x \to \infty} f(x) = L$ means precisely that, if we pick x big enough, we can make $f(x)$ get arbitrarily close to L.

Definition 1.3.46. *Let b be a real number.*

Then, $\lim_{x \to b} f(x) = \infty$ if and only if the domain of f contains a deleted interval around b and, for all real $M > 0$, there exists a real number $\delta > 0$ such that, if x is in the domain of f and $0 < |x - b| < \delta$, then $f(x) > M$.

Similarly, $\lim_{x \to b} f(x) = -\infty$ if and only if the domain of f contains a deleted interval around b and, for all real $M < 0$, there exists a real number $\delta > 0$ such that, if x is in the domain of f and $0 < |x - b| < \delta$, then $f(x) < M$.

We leave it to you to modify the definitions above in order to obtain definitions of $f(x)$ approaching $\pm \infty$, as x approaches b from the right or from the left.

Finally, we need to define limits where both b and L are infinite. After you have understood the definitions above, when b and L are separately infinite, the following definitions should come as no surprise.

Definition 1.3.47. *We write $\lim_{x \to \infty} f(x) = \infty$ if and only if the domain of f contains an interval of the form (a, ∞) and, for all real $N > 0$, there exists a real number $M > 0$ such that, if x is in the domain of f and $x > M$, then $f(x) > N$.*

We write $\lim_{x \to \infty} f(x) = -\infty$ if and only if the domain of f contains an interval of the form (a, ∞) and, for all real $N < 0$, there exists a real number $M > 0$ such that, if x is in the domain of f and $x > M$, then $f(x) < N$.

We write $\lim_{x \to -\infty} f(x) = \infty$ if and only if the domain of f contains an interval of the form $(-\infty, a)$ and, for all real $N > 0$, there exists a real number $M < 0$ such that, if x is in the domain of f and $x < M$, then $f(x) > N$.

We write $\lim_{x \to -\infty} f(x) = -\infty$ if and only if the domain of f contains an interval of the form $(-\infty, a)$ and, for all real $N < 0$, there exists a real number $M < 0$ such that, if x is in the domain of f and $x < M$, then $f(x) < N$.

We wish to give a few basic limits involving infinities.

Theorem 1.3.48. *Suppose that c is a real number, and b is an extended real number. If b is a real number, assume $f(x) \neq 0$ for all x in some deleted open interval around b. If $b = \infty$ (respectively, $b = -\infty$), assume that there exists a real number c such that, for all x in (c, ∞) (respectively, $(-\infty, c)$), $f(x) \neq 0$.*

Then,

1. $\lim\limits_{x \to \pm\infty} c = c$;   *2. $\lim\limits_{x \to \pm\infty} x = \pm\infty$; and*   *3. $\lim\limits_{x \to b} f(x) = 0$ if and only if $\lim\limits_{x \to b} \dfrac{1}{|f(x)|} = \infty$.*

We want to know that we can apply algebraic rules to limits involving infinity. First, we need to define algebraic operations on the extended real numbers.

Definition 1.3.49. *Let x be an extended real number. We make the following algebraic definitions.*

1. If $x \neq -\infty$, then $\infty + x = \infty$;

2. If $x \neq \infty$, then $(-\infty) + x = -\infty$;

3. if $x \neq \pm\infty$, then $\dfrac{x}{\pm\infty} = 0$;

4. If $x > 0$, then $\infty \cdot x = \infty$ and $(-\infty) \cdot x = -\infty$;

5. If $x < 0$, then $\infty \cdot x = -\infty$ and $(-\infty) \cdot x = \infty$.

In addition, the sums and products above are commutative.

Definition 1.3.50. *The following operations, and those obtained by commuting the arguments, on the extended real numbers are undefined and are referred to as* **indeterminate forms***.*

1. $\infty + (-\infty)$ *or* $\infty - \infty$;   *2.* $0 \cdot (\pm\infty)$;   *3.* $\dfrac{\pm\infty}{\pm\infty}$;   *4.* $\dfrac{0}{0}$.

Now that we have defined some algebraic operations involving infinities, we can state the following theorem.

Theorem 1.3.51. *Theorem 1.3.15 holds for* b, L_1, *and* L_2 *in the extended real numbers, whenever the resulting algebraic operations on* L_1 *and* L_2 *do not yield indeterminate forms.*

That is, we may add, subtract, multiply, and divide limits involving infinities, providing that the result is defined in the extended real numbers.

In addition, Theorem 1.3.11 holds for $L = \pm\infty$, *and Theorem 1.3.6 and Corollary 1.3.7 hold for* $b = \pm\infty$.

Limit Substitution, Theorem 1.3.25, holds for $L = \pm\infty$, $b = \pm\infty$, *and* $M = \pm\infty$, *with the understanding that a "deleted open interval around* ∞ *(respectively,* $-\infty$*)" means an open interval of the form* (a, ∞) *(respectively,* $(-\infty, a)$*), and that Condition 3b of Theorem 1.3.25 is automatically satisfied if* $M = \pm\infty$.

It follows from Theorem 1.3.48 that:

Corollary 1.3.52. *Suppose that* p *and* q *are positive integers with no common divisor, other than 1. Then,*

1. $\displaystyle\lim_{x \to \infty} \frac{1}{x^{p/q}} = 0$, *and* $\displaystyle\lim_{x \to 0^+} \frac{1}{x^{p/q}} = \infty$;

2. if q *is odd,* $\displaystyle\lim_{x \to -\infty} \frac{1}{x^{p/q}} = 0$, *and* $\displaystyle\lim_{x \to 0^-} \frac{1}{x^{p/q}} = (-1)^p \infty$.

Remark 1.3.53. As special cases of Corollary 1.3.52, we have $\lim_{x \to 0^+}(1/x) = \infty$ and $\lim_{x \to 0^-}(1/x) = -\infty$. Consequently, by the last line of Theorem 1.3.51, the two-sided limit $\lim_{x \to 0}(1/x)$ does not exist even as an extended real number.

However, Corollary 1.3.52 also tells us that, if p is even (and, hence, q is odd), then $\lim\limits_{x\to 0} \dfrac{1}{x^{p/q}} = \infty$.

Note that the case when $q = 1$ is important in Corollary 1.3.52; by combining this case with other extended algebraic operations, via Theorem 1.3.48, we can quickly conclude such things as

$$\lim_{x\to\pm\infty} \left(\frac{100}{x} + \frac{9987}{x^2} - \frac{\pi}{x^5} \right) = 0.$$

Of course, you shouldn't have to memorize theorems to evaluate this last limit; the theorems are just rigorous forms of what hopefully seems intuitively clear: if x is really big (or negatively big), then any positive power of x is big in absolute value, and so 1 over a positive power of x would be close to 0. In the limit, you obtain 0's, and these 0's can be multiplied and added.

Remark 1.3.54. The extended real numbers are convenient to define and use in limits. However, $\pm\infty$ are not real numbers, and if a limit equals one of these infinities, then that limit **does not exist**. It is true, however, that saying that a limit is $\pm\infty$ is a particularly nice way of saying that a limit fails to exist; it tells us that the limit fails to exist because the function gets unboundedly large or unboundedly negative. In addition, the algebraic operations on the extended real numbers mean that we can frequently work with infinite limits in very useful ways.

Rational Functions and Infinite Limits

We wish to end this subsection on infinite limits by giving a classic collection of examples: limits of rational functions $r(x) = p(x)/q(x)$, where p and q are polynomial functions (and $q(x)$ is not the zero function, i.e., $q(x)$ is not the polynomial function which is **always** zero).

There are two quick results which tell us how to deal with rational functions $p(x)/q(x)$ as $x \to \pm\infty$ or as x approaches a root of $q(x)$. We wrote that these results are "quick". However, when written out in full generality, the results can look a bit overwhelming. Read the discussions/examples preceding and following; they should make it clear what's going on.

Consider the rational function

$$r(x) = \frac{5x^m - 2x^3 + 3x + 12}{-8x^n + 7x^2 - 100},$$

where m and n are integers that are greater than, or equal to, 4.

When x is large in absolute value, that is, as x heads to plus or minus infinity, the largest degree terms in the numerator and denominator of $r(x)$ "overwhelm" the smaller degree terms; in other words, the smaller degree terms in the numerator and denominator become negligible compared to the highest degree terms. Thus, the limit of $r(x)$, as $x \to \pm\infty$, is the same as the limit

$$\lim_{x \to \pm\infty} \frac{5x^m}{-8x^n},$$

and it's easy to see that this last limit depends on whether m is bigger than n, m is less than n, or m equals n.

If $m = n$,

$$\lim_{x \to \pm\infty} \frac{5x^m}{8x^n} = \lim_{x \to \pm\infty} \frac{5}{-8} = -\frac{5}{8}.$$

If $n > m$, then $n - m > 0$ and

$$\lim_{x \to \pm\infty} \frac{5x^m}{-8x^n} = \lim_{x \to \pm\infty} \frac{5}{-8x^{n-m}} = \frac{5}{\pm\infty} = 0.$$

If $m > n$, then $m - n > 0$ and

$$\lim_{x \to \pm\infty} \frac{5x^m}{-8x^n} = \lim_{x \to \pm\infty} \frac{5x^{m-n}}{-8},$$

and now there's a slight additional complication. Certainly, as $x \to \infty$, we find

$$\lim_{x \to \infty} \frac{5x^{m-n}}{-8} = -\frac{5}{8} \cdot \lim_{x \to \infty} x^{m-n} = -\frac{5}{8} \cdot \infty = -\infty.$$

As $x \to -\infty$, we once again have

$$\lim_{x \to -\infty} \frac{5x^{m-n}}{-8} = -\frac{5}{8} \cdot \lim_{x \to -\infty} x^{m-n},$$

but, now, it's important whether $m - n$ is even or odd; if $m - n$ is even, $x^{m-n} \to \infty$, so that the entire limit is $-\infty$, and if $m - n$ is odd, $x^{m-n} \to -\infty$, so that the entire limit is ∞.

Should you memorize all of the cases that we just discussed, which are also stated in the proposition below? **NO**. Just remember that, as x approaches $\pm\infty$, it's the largest degree terms that matter in a rational function, and so what you're left with a limit of the form

$$\lim_{x \to \pm\infty} \frac{a_m x^m}{b_n x^n},$$

where a_m and b_n are constants. Then just do the algebra and calculations that come naturally.

However, we will go ahead and state the general, technical, result.

Proposition 1.3.55. *Let m be the degree of $p(x)$ and let a_m be the coefficient of x^m in $p(x)$. Let n be the degree of $q(x)$ and let b_n be the coefficient of x^n in $q(x)$. Then, as limits in the extended real numbers, with extended algebraic operations,*

$$\lim_{x \to \infty} \frac{p(x)}{q(x)} = \frac{a_m}{b_n} \cdot \lim_{x \to \infty} x^{m-n},$$

and the same result is true replacing both ∞'s with $-\infty$'s. In particular, the limit of a rational function as $x \to \pm\infty$ is completely determined by the highest-degree terms in the numerator and denominator.

Thus,

    a. *if $n > m$, then $\lim_{x \to \pm\infty} p(x)/q(x) = 0$;*

    b. *if $n = m$, then $\lim_{x \to \pm\infty} p(x)/q(x) = a_m/b_n$;*

    c. *if $m > n$, then $\lim_{x \to \infty} p(x)/q(x) = \pm\infty$, where the $\pm$ sign in the result agrees with the sign of a_m/b_n when $x \to \infty$; as $x \to -\infty$, the sign of the result is the same as the sign of $(-1)^{m-n} a_m/b_n$, and so depends on whether $m - n$ is even or odd.*

Proof. In all cases, this is proved by dividing the numerator and denominator of the rational function by x^m or x^n, which does not change the function (except, possibly, by excluding zero from the domain, which does not affect the limits as x approaches $\pm\infty$).

For example, when $n > m$, we divide the numerator and denominator by x^n to obtain, for $x \neq 0$, that

$$\frac{p(x)}{q(x)} = \frac{a_m x^m + a_{m-1} x^{m-1} + a_{m-2} x^{m-2} + \cdots + a_1 x + a_0}{b_n x^n + b_{n-1} x^{n-1} + b_{n-2} x^{n-2} + \cdots + b_1 x + b_0} =$$

$$\frac{\frac{a_m}{x^{n-m}} + \frac{a_{m-1}}{x^{n-m+1}} + \frac{a_{m-2}}{x^{n-m+2}} + \cdots + \frac{a_1}{x^{n-1}} + \frac{a_0}{x^n}}{b_n + \frac{b_{n-1}}{x} + \frac{b_{n-2}}{x^2} + \cdots + \frac{b_1}{x^{n-1}} + \frac{b_0}{x^n}}.$$

By Theorem 1.3.15 and Corollary 1.3.52, all of the summands in the numerator and denominator of this final fraction approach 0 as $x \to \pm\infty$, with the exception of b_n. Thus, as $x \to \pm\infty$, $p(x)/q(x) \to 0/(b_n + 0) = 0$.

We leave the other cases as exercises. $\square$

Example 1.3.56. Consider the rational functions

$$f(x) = \frac{(x+2)^2(x-2)}{x^2+3}, \quad g(x) = \frac{(x+2)(x-2)}{x^2+3}, \quad \text{and} \quad h(x) = \frac{(x+2)(x-2)}{(x^2+3)(x-2)}.$$

The natural domain of a rational function is all real numbers which do not make the denominator equal 0. Thus, the domains of f and g are all real numbers, while the domain of h is all real numbers other than 2. The graphs of these functions appear below in Figure 1.9, Figure 1.10, and Figure 1.11.

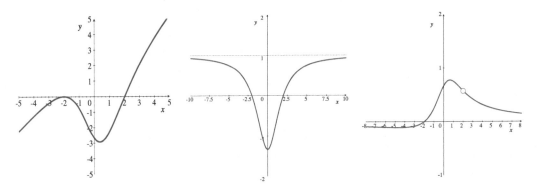

Figure 1.9: $y = f(x)$.   Figure 1.10: $y = g(x)$.   Figure 1.11: $y = h(x)$

Though the numerators and denominators of f, g, and h are factored, you can still see their degrees; f is a degree 3 polynomial over a degree 2 polynomial, g is a degree 2 over a degree 2, and h is a degree 2 polynomial over a degree 3 polynomial. Thus, in order, we see graphs of rational functions as described in cases c), b), and a) in Proposition 1.3.55.

From the graph, you may be able to tell that, as guaranteed by Proposition 1.3.55, the value of $y = g(x)$ is approaching 1 as x approaches $\pm\infty$ and, in the graph, we have included a dotted line at $y = 1$. The line given by $y = 1$ is called a *horizontal asymptote* of the graph of $y = g(x)$; it is a horizontal line which the graph of g approaches as the graph goes out arbitrarily far. We discuss asymptotes and graphs of more general functions in Section 3.2.

Similarly, the graph of $y = h(x)$ approaches the horizontal asymptote given by $y = 0$.

Understand that Proposition 1.3.55 describes what happens when x gets very large in absolute value. It does not tell us what happens for x's which are "small" in absolute value. In particular, while 2 is clearly not in the domain of h, the fact that the graph has a "hole" in it above $x = 2$ is **not** explained by Proposition 1.3.55. For that explanation, we need the next proposition.

We now want to look at limits of rational functions $r(x)$ as x approaches a root b of the polynomial in the denominator; in fact, we want to consider the limit as x approaches b from the left and from the right, i.e., as $x \to b^{\pm}$.

Consider, for example,

$$\lim_{x \to 1^-} \frac{(x-5)(x-1)^m}{(x^2+7)(x-1)^n},$$

where m and n are positive integers. As in the calculation of the limit when x approached $\pm\infty$, what happens in this one-sided limit is a question of whether $m = n$, $m > n$, or $n > m$ and, in one of the cases, we need to know whether the difference in the exponents is even or odd.

How do you see this? Once again, you basically do the algebra and calculations that come naturally.

If $m = n$, then

$$\lim_{x \to 1^-} \frac{(x-5)(x-1)^m}{(x^2+7)(x-1)^n} = \lim_{x \to 1^-} \frac{x-5}{x^2+7} = \frac{-4}{8} = -\frac{1}{2}.$$

If $m > n$, then $m - n > 0$ and

$$\lim_{x \to 1^-} \frac{(x-5)(x-1)^m}{(x^2+7)(x-1)^n} = \lim_{x \to 1^-} \frac{(x-5)(x-1)^{m-n}}{(x^2+7)} = \frac{-4 \cdot 0}{8} = 0.$$

If $n > m$, then $n - m > 0$ and

$$\lim_{x \to 1^-} \frac{(x-5)(x-1)^m}{(x^2+7)(x-1)^n} = \lim_{x \to 1^-} \frac{x-5}{(x^2+7)(x-1)^{n-m}} = \frac{-4}{8} \cdot \lim_{x \to 1^-} \frac{1}{(x-1)^{n-m}}.$$

Now, as $x \to 1^-$, $(x-1)^{n-m} \to 0$, and so the limit is definitely $\pm\infty$, but we need to know if

$(x-1)^{n-m}$ approaches 0 through positive numbers or negative numbers, i.e., if $(x-1)^{n-m} \to 0^+$ or $(x-1)^{n-m} \to 0^-$. Why do we need to know this? Because, if $(x-1)^{n-m} \to 0^+$, then $1/(x-1)^{n-m} \to \infty$, while, if $(x-1)^{n-m} \to 0^-$, then $1/(x-1)^{n-m} \to -\infty$. Of course, multiplying by $-4/8$ will negate the results in the final limit.

Okay. So, how do you tell if $(x-1)^{n-m}$ approaches 0 from the left or right, as x approaches 1 from the left? Well... as $x \to 1^-$, certainly $x-1 \to 0^-$; this is true, because all it really says is that, when x is slightly less than 1, $x-1$ is slightly less than 0. Now we need to know whether $n-m$ is even or odd. For, if $n-m$ is even, then $(x-1)^{n-m} \geq 0$ and so $(x-1)^{n-m} \to 0^+$. On the other hand, if $n-m$ is odd, and $x-1 \leq 0$, then $(x-1)^{n-m} \leq 0$ and so $(x-1)^{n-m} \to 0^-$.

Once again, you should **NOT** try to memorize all of the cases discussed above, and stated in full generality in the proposition below. If you're taking a left or right limit of a rational function as x approaches a root b of the denominator, just cancel as many powers of $(x-b)$ as you can in the numerator and denominator, and then think about the easier limit that you obtain.

Of course, factoring out all of the powers of $(x-b)$ in the numerator and denominator can be a substantial algebra problem!

The general, technical, result is:

Proposition 1.3.57. *Let b be a real root of $q(x)$, so that $q(b) = 0$, $x-b$ divides $q(x)$, and $p(x)/q(x)$ is undefined at b.*

Let m be the largest power of $x-b$ which divides $p(x)$ (m could be 0) and let n be the largest power of $x-b$ which divides $q(x)$. Let $\hat{p}(x)$ and $\hat{q}(x)$ be the resulting quotient polynomials, i.e., let $\hat{p}(x)$ and $\hat{q}(x)$ be the unique polynomials such that $p(x) = \hat{p}(x)(x-b)^m$ and $q(x) = \hat{q}(x)(x-b)^n$.

Then, neither $\hat{p}(b)$ nor $\hat{q}(b)$ is zero, and, if we let $c = \hat{p}(b)/\hat{q}(b)$, then

$$\lim_{x \to b^+} \frac{p(x)}{q(x)} = c \cdot \lim_{x \to b^+} (x-b)^{m-n},$$

and the same result is true if we replace both b^+'s with b^-'s.

Thus,

    a. if $m > n$, then $\lim_{x \to b^{\pm}} p(x)/q(x) = 0$;

    b. if $m = n$, then $\lim_{x \to b^{\pm}} p(x)/q(x) = c$;

    c. if $n > m$, then $\lim_{x \to b^{\pm}} p(x)/q(x) = \pm\infty$, where the $\pm$ sign in the result agrees with the sign of c when $x \to b^{+}$; as $x \to b^{-}$, the sign of the result is the same as the sign of $(-1)^{m-n}c$, and so depends on whether $m - n$ is even or odd.

Example 1.3.58. Consider the rational functions

$$h(x) = \frac{(x+2)(x-2)}{(x^2+3)(x-2)}, \quad i(x) = \frac{(x+2)(x-2)}{(x^2+3)(x-2)^2}, \quad \text{and} \quad j(x) = \frac{(x+2)(x-2)}{(x^2+3)(x-2)^3}.$$

The graphs of these functions appear below in Figure 1.12, Figure 1.13, and Figure 1.14.

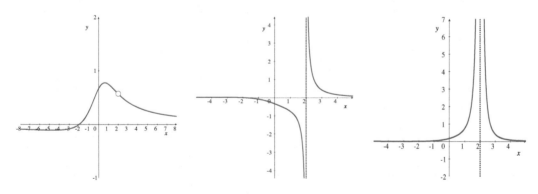

Figure 1.12: $y = h(x)$ Figure 1.13: $y = i(x)$.   Figure 1.14: $y = j(x)$

Looking at the graph of

$$h(x) = \frac{(x+2)(x-2)}{(x^2+3)(x-2)}$$

in Figure 1.11, you should immediately notice two things: $y = 0$ is a horizontal asymptote of the graph (as we discussed above) and there's a **hole** in the graph near (or at) the point where $x = 2$.

Why is there a hole in the graph of $h(x)$? Because h is not defined $x = 2$, for the denominator is 0 there, and yet the $(x - 2)$ factor in the denominator can be "cancelled" with the $(x - 2)$

factor in the numerator. That is, when $x \neq 2$, we have

$$h(x) \;=\; \frac{(x+2)(x-2)}{(x^2+3)(x-2)} \;=\; \frac{x+2}{x^2+3},$$

and so

$$\lim_{x \to 2} g(x) \;=\; \lim_{x \to 2} \frac{x+2}{x^2+3} \;=\; \frac{4}{7};$$

this is the content of case b) in Proposition 1.3.57. Thus, as x gets arbitrarily close to 2, the corresponding point on the graph gets arbitrarily close to $(2, 4/7)$, and yet, since $h(x)$ is undefined at $x = 2$, the point $(2, 4/7)$ is not on the graph; hence, we see a hole in the graph of $y = h(x)$ at $(2, 4/7)$.

Recall our discussion in Remark 1.3.21: some references would say that "h has a removable discontinuity at $x = 2$". However, according to our definition, Definition 1.3.20, a function must be defined at a point to be called discontinuous (or continuous) there, and h is not defined at 2. Hence, we prefer to say that "h has a *removable lack of continuity* at $x = 2$" or that "h extends to a continuous function at $x = 2$", even though these phrases don't exactly roll off the tongue like "removable discontinuity".

At $x = 2$, $i(x)$ and $j(x)$ are examples of case c) of Proposition 1.3.57; $i(x)$ has $m - n = -1$, which is odd, while $j(x)$ has $m - n = -2$, which is even. As promised by Proposition 1.3.57, as x approaches 2 from the right or left, $i(x)$ and $j(x)$ head towards $\pm\infty$. We say that the graphs have *vertical asymptotes* at $x = 2$, because, from each side, the graphs approach the vertical line given by $x = 2$. Since, at $x = 2$, $m - n$ is even for $j(x)$, we see that, from the left or right, the graph heads in the same direction – here, in the direction of $+\infty$. On the other hand, at $x = 2$, $m - n$ is odd for $i(x)$, and we see that, from the left and right, the graph heads in opposite directions, to $-\infty$ from the left and $+\infty$ from the right.

It is, of course, possible to produce rational functions whose graphs have any (finite) number of vertical asymptotes and/or "holes"; you simply have to have denominators with the appropriate roots, and the appropriate cancellation/non-cancellation of factors in the numerator and denominator. Note, however, there can be at most one **horizontal** asymptote.

1.3.4 Exercises

In Exercises 1 and 2, estimate and/or calculate the limits by using the graphs and/or by using the definition of the limit, Definition 1.3.2, or explain why the limit does not exist:

1. $\lim_{x \to 1} \dfrac{x^2 + 3x + 2}{x + 1}$

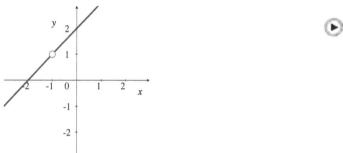

2. $\lim_{x \to 0} f(x)$, where $f(x) = \begin{cases} 3x + 1 & \text{if } x \le 0; \\ 2 & \text{if } x > 0. \end{cases}$

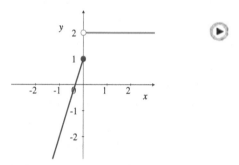

In Exercises 3 through 5, find the given limits (if they exist) by using any methods from this section. If a given limit does not exist, explain why.

3. $\lim_{x \to -1} \dfrac{x^2}{\sqrt{x + 2}}$

4. $\lim_{x \to 10} \left[\dfrac{1}{x} \cdot (x^2 - 200)^2 - x \right]$

5. $\lim_{x \to \infty} \dfrac{\frac{2}{9} - \frac{2}{x}}{\sqrt{x^2 - 10}}$

In Exercises 6 - 8 you are given a continuous function, an ϵ, and a point. Find a corresponding δ that is guaranteed to exist by the definition of continuity.

6. $f(x) = 3x + 7$, $x = 2$, $\epsilon = 0.05$.

7. $g(x) = x^2 + 4$, $x = 0$, $\epsilon = 0.04$.

8. $h(x) = \dfrac{1}{x}$, $x = 3$, $\epsilon = 1$.

In Exercises 9 and 10 you are given a function, an extended real number b such that $\lim_{x \to b} f(x) = \infty$ or $\lim_{x \to b} f(x) = -\infty$, and a number N as in Definition 1.3.45. Find the corresponding M value.

9. $f(x) = x^3$, $b = \infty$, $N = 64{,}000$.

10. $f(x) = 5x + 7$, $b = -\infty$, $N = -4500$.

11. Construct a rational function $f(x)$ such that $\lim_{x \to -2} f(x) = -6$ but where $f(-2)$ is undefined (there's more than one correct answer).

12. Suppose $\lim_{y \to 7} f(y) = 5$, that $f(7) = 5$, that $g(5)$ is undefined, that f is continuous and that $\lim_{x \to 5} g(x) = 7$. What is $\lim_{x \to 5} f(g(x))$?

13. Newton's 1st Law of Motion, also known as the Law of Inertia, states that a body will continue moving in a straight line at a constant speed unless made to change that state by a force acting on it. This means that a force must be acting on the Earth to prevent it from floating out of orbit. This force is the gravitational attraction of the Sun. In the general case where a body with mass m is traveling at constant speed v around a circle of radius r, the magnitude of the force needed to hold a mass in a circular orbit is given by the equation: $F = \dfrac{mv^2}{r}$.

    a. Calculate $\lim\limits_{r \to \infty} F(r)$.

    b. Calculate $\lim\limits_{r \to 0^+} F(r)$.

14. Let $f(x) = \dfrac{1}{2^n}$ for $n \le x < n + 1$, $n = 0, 1, 2, \ldots$. The domain of f is $[0, \infty)$.

    a. Calculate $\lim_{x \to n^+}$ for $n = 1, 2, 3, \ldots$

    b. Calculate $\lim_{x \to n^-}$ for $n = 1, 2, 3, \ldots$

    c. Let $g(z) = \lim_{x \to z^-} f(x) - \lim_{x \to z^+} f(x)$. Calculate $\lim_{z \to \infty} g(z)$

15. Why doesn't $\lim_{v \to 0} \sqrt{v}$ exist? What is $\lim_{v \to 0^+} \sqrt{v}$?

16. Is the function $f(v) = \sqrt{v}$ continuous at $v = 0$? Why or why not?

17. Does $\lim_{x \to 1} \sqrt{1 - x^2}$ exist? What about the one-sided limits?

18. Show the function $\sqrt{1 - x^2}$ is continuous on the interval $[-1, 1]$.

Use facts about the continuity of rational functions and the square-root function to determine the maximal domains on which the functions in Exercises 19 - 21 are continuous.

19. $h(x) = \dfrac{3x^9 + 5x + 11}{x^2 - 4}$

20. $j(x) = \dfrac{5}{x+2} + \dfrac{7x^3 - 9x^5}{(x-1)^2} + \dfrac{x^3(3x^5 + 7x^2 + 1)}{(x+2)(x-1)}$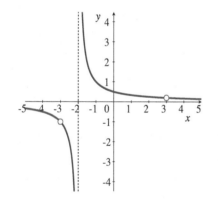

21. $k(x) = \sqrt{x+1} + \dfrac{x^5 + 3x + 1}{(x+3)(x-3)}$

For each function in Exercises **22** and **23**, find all of the points P at which the continuity of elementary functions does not guarantee that the function is continuous, including points P not in the domain of the function. At each such point, P, estimate and/or calculate the limits as x approaches P from the left and from the right, or explain why the limits do not exist, decide if the function is continuous at P, and decide if the function has a removable lack of continuity at P (see Remark 1.3.21).

22. $f(x) = \dfrac{x^2 - 9}{x^3 + 2x^2 - 9x - 18} = \dfrac{(x-3)(x+3)}{(x-3)(x+2)(x+3)}$. See Figure 1.15.

Figure 1.15: The graph for Exercise 22.　　Figure 1.16: The graph for Exercise 23.

23. $g(x) = \begin{cases} x & \text{if } x < -10; \\ 0 & \text{if } x = -10; \\ -x - 20 & \text{if } -10 < x \leq 0; \\ 40/(x-2) & \text{if } x > 0. \end{cases}$　　See Figure 1.16.

24. Solve for b if $g(x)$ is continuous on $(-\infty, \infty)$, where $g(x) = \begin{cases} 2x + 3 & \text{if } x < 4; \\ x^2 + bx - 13 & \text{if } x \geq 4. \end{cases}$

  ▶

The Intermediate Value Theorem is often used to show that equations have solutions. Show that the following equations have solutions in the specified intervals.

25. $x^6 + 5x^3 = 1$, $I = (0, 1)$.   ▶

26. $\dfrac{(x+3)^2}{x+2} = -1$, $I = (-3, -2.1)$.

27. $\sqrt{x^2 + 2x + 2} - \sqrt{x + 3} = 0$, $I = (-2, 0)$.

28. Suppose $f(x)$ is always positive, $g(x)$ is always negative, and both functions are continuous on the reals and elementary. Show that $h(x) = xf(x) + (1 - x)g(x) = 0$ has a solution in $(0, 1)$.

29. Let $f(x) = ax^2 + bx + c$, $g(x) = mx^2 + nx + p$, and $h(x) = \dfrac{f(x)}{g(x)}$. Assume none of the six coefficients are zero.

    a. Can Theorem 1.3.49 be used to calculate $\lim_{x \to \infty} h(x)$? If so, what is the limit? If not, why not?

    b. Divide the numerator and denominator by x^2. Calculate $\lim_{x \to \infty} h(x)$, being careful to support each step with a result from this chapter.

    c. Generalize the result. If $f(x) = a_n x^n + a_{n-1} x^{n-1} + \dots + a_1 + a_0$ and $g(x) = b_n x^n + b_{n-1} x^{n-1} + \dots + b_1 + b_0$, calculate $\lim_{x \to \infty} \dfrac{f(x)}{g(x)}$. Assume neither a_n nor b_n is zero.

30. The total number of degrees in a regular polygon with n sides is $D(n) = 180(n - 2)$. The number of degrees per angle, DA, is obtained by dividing D by the number of sides: $DA(n) = \dfrac{180(n - 2)}{n}$. For example, a regular pentagon has 108 degrees per angle for atotal of 540 degrees. Calculate $\lim_{n \to \infty} DA(n)$.

31. Suppose $f : I \to J$ is a continuous function where the domain is the interval $I = [a, b]$. Suppose that the m guaranteed by the Extreme Value Theorem is positive. Prove that the area A under the graph of f and above the x-axis satisfies $m(b - a) \leq A \leq M(b - a)$.

32. The Extreme Value Theorem requires the domain of the function to be a closed interval. Construct a continuous, unbounded function on an arbitrary open interval (a, b).

33. Prove that the Extreme Value Theorem is still true for a continuous function with a domain equal to the union of finitely many closed intervals. What if the domain is the union of infinitely many closed intervals?

34. Suppose a pendulum is constructed by hanging a mass from a string. The motion of the pendulum begins when the mass is released from some elevated position. The *period* of a pendulum is the time it takes for the pendulum to complete one cycle and return, approximately, to its starting position. The period, T, of a pendulum is approximated by the equation $T = 2\pi\sqrt{\dfrac{L}{g}}$ where L is the length of the string and g is the acceleration due to gravity.

   a. The acceleration due to gravity decreases as a pendulum moves into deep space. Calculate $\lim_{g \to 0^+} T$.

   b. The acceleration due to gravity increases as a pendulum moves towards a massive object. Calculate $\lim_{g \to \infty} T$

   c. What happens to the period as the string gets shorter? Set up a limit and evaluate it.

   d. What happens to the period as string gets longer? Set up a limit and evaluate it.

35. Wholesalers often offer a discount for large orders. Suppose a supplier charges \$0.80 per CD if you purchase 500 or fewer CD's. If you purchase between 501 and 1000 CD's, the price drops to \$0.72 each. The price bottoms out at a price of \$0.65 per CD for orders in excess of 1000. See Figure 1.17.

   a. Write an equation for the price per CD, $P(x)$, in terms of the number of CD's purchased, x.

   b. Write an equation for the total cost, $C(x)$, of purchasing x CD's.

   c. Estimate from the graph, and/or calculate, $\lim_{x \to 1000^+} C(x)$ and $\lim_{x \to 1000^-} C(x)$.

36. Every real number x can be written uniquely as $\pm(n + f)$, where $n \geq 0$ is an integer and $0 \leq f < 1$; this $\pm f$ is called the *fractional part* of x. For instance, the fractional part of 3 is 0, the fractional part of 3.2 is 0.2, and the fractional part of -3.2 is -0.2.

   Define the function $\text{Frac}(x)$ to be the fractional part of x. Those familiar with programming in C could note that this is the operation $x\%1$.

   a. Graph $\text{Frac}(x)$ for $-3.5 \leq x \leq 3.5$.

   b. Where is $\text{Frac}(x)$ continuous?

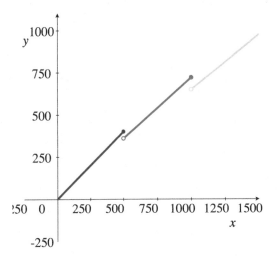

Figure 1.17: The graph of $y = C(x)$.

    c. For every integer n, find $\lim_{x \to n^+} \mathrm{Frac}(x)$ and $\lim_{x \to n^-} \mathrm{Frac}(x)$.

37. Critique the following argument: Let $f(x) = (6x^2 - 4x - 10)/(x+1)$. Notice that $f(-2) = -22$ and $f(0) = -10$. Since $-22 = f(-2) < -16 < f(0) = -10$, by the Intermediate Value Theorem (1.3.27), there exists a c such that $-2 < c < 0$ and $f(c) = -16$.

38. Consider the cubic polynomial $f(x) = x^3 - 4x$

    a. Graph the function by hand. What is the domain of this function?

    b. Calculate $f(-4)$ and $f(0)$.

    c. State the conditions and consequence of the Extreme Value Theorem (Theorem 1.3.30) with regards to this function, on the interval $[-4, 0]$.

    d. Estimate the minimum and maximum values of this function, on the interval $[-4, 0]$, as guaranteed by this theorem (these will be precisely calculable in Section 1.5).

39. Consider the rational function $f(x) = 1/(x^2 + 1)$, whose domain is the interval $[0, \infty)$ and whose codomain (for now) is the interval $(-\infty, \infty)$. We seek to apply Theorem 1.3.35 (concerning inverse functions) to this function.

    a. First, find the range, J, of f, so that the restriction $f : [0, \infty) \to J$ is the type of function required in the theorem, i.e., such that f restricted to this codomain is one-to-one and onto. For the remainder of this exercise, let f denote this restricted function.

    b. Find a formula for $f^{-1}(y)$.

c. What does Theorem 1.3.35 tell you about f^{-1}?

40. A frequent source of counterexamples is the function $g(x) = \begin{cases} 0 & \text{if } x \text{ is irrational} \\ 1 & \text{if } x \text{ is rational} \end{cases}$ Show that $g(x)$ is discontinuous at every real number.

41. Let $f(x) = \begin{cases} 0 & \text{if } x \text{ is irrational} \\ x & \text{if } x \text{ is rational} \end{cases}$
    Show that $f(x)$ is continuous at one and only one point.

42. A function $f : I \to J$ between two intervals is a *homeomorphism* if f is continuous, one-to-one, onto and has a continuous inverse f^{-1}.

    a. Show that $f(x) = 2x + 7$ is a homeomorphism from $(0, 1)$ to $(7, 9)$

    b. More generally, show that $g(x) = (b - a)x + a$ is a homeomorphism from $(0, 1)$ to (a, b).

    c. Find a homeomorphism between the intervals (a, b) and (c, d).

43. Suppose the domain of a function $f(x)$ is an interval I. f is said to be *uniformly continuous* on I if for every $\epsilon > 0$ there exists $\delta > 0$ such that, if x and y are in I and $|x - y| < \delta$, then $|f(x) - f(y)| < \epsilon$. Uniformly continuous functions are a subclass of continuous functions where δ depends only on ϵ. The δ for a continuous function may depend on both ϵ and x.

    a. Show that $f(x) = 2x + 1$ is uniformly continuous on its natural domain, $(-\infty, \infty)$.

    b. Show that $f(x) = \dfrac{1}{x}$ is not uniformly continuous on the domain $(0, \infty)$.

44. Use the graph of $y = f(x)$ in Figure 1.18 to determine/estimate "values" for the given quantities, including, possibly, $\pm\infty$, or to determine that the quantity does not exist (DNE) for some reason other than being $\pm\infty$. The unit distance, indicated on the x- and y-axes, is 1.

    a. $\displaystyle\lim_{x \to 0^-} f(x)$ b. $\displaystyle\lim_{x \to 0^+} f(x)$ c. $\displaystyle\lim_{x \to 0} f(x)$ d. $f(0)$

    e. $\displaystyle\lim_{x \to 2^-} f(x)$ f. $\displaystyle\lim_{x \to 2^+} f(x)$ g. $\displaystyle\lim_{x \to 2} f(x)$ h. $f(2)$

    i. $\displaystyle\lim_{x \to -3^-} f(x)$ j. $\displaystyle\lim_{x \to -3^+} f(x)$ k. $\displaystyle\lim_{x \to -3} f(x)$ l. $f(-3)$

    m. $\displaystyle\lim_{x \to -6^-} f(x)$ n. $\displaystyle\lim_{x \to -6^+} f(x)$ p. $\displaystyle\lim_{x \to -6} f(x)$ q. $f(-6)$

    r. $\displaystyle\lim_{x \to -\infty} f(x)$ s. $\displaystyle\lim_{x \to \infty} f(x)$

45. $\displaystyle\lim_{x \to -3} \frac{x^2 - 9}{x + 3} = ?$

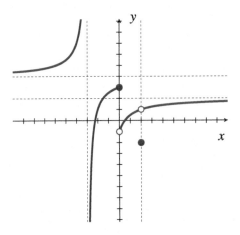

Figure 1.18: The graph of $y = f(x)$.

46. $\lim\limits_{x \to 1} \dfrac{\frac{1}{\sqrt{x}} - 1}{x - 1} = ?$

47. $\lim\limits_{t \to 2} \dfrac{\frac{1}{t} - \frac{1}{2}}{t - 2} = ?$

48. $\lim\limits_{h \to 0} \dfrac{(5 + h)^2 - 25}{h} = ?$

49. $\lim\limits_{h \to 0} \dfrac{\frac{1}{(5+h)^2} - \frac{1}{25}}{h} = ?$

50. $\lim\limits_{x \to a} \dfrac{x^2 - a^2}{x - a} = ?$

1.4 Instantaneous Rates of Change: The Derivative

Now that we have a rigorous mathematical definition, Definition 1.3.2, of the limit, we can define the instantaneous rate of change of a function $f(x)$, with respect to x, when x is equal to a specific value a (as opposed to between two values), by taking the limit of the average rates of change of f, with respect to x, as the change in x approaches 0, where the x values are close to a. This is what we stated in Definition 1.2.1, where we used an intuitive "definition" of the limit.

Throughout this section, we will use properties of limits that we stated in the previous section, usually without referring to the theorems being used. You may wish to reread Remark 1.2.4 before proceeding, in order to understand why we switch between $f(x)$ and $f(a)$, or something similar, in various places.

We need a preliminary definition of the types of points in the domain of a real function at which the instantaneous rate of change, the IROC, will be defined.

Definition 1.4.1. *If E is a subset of the real numbers (e.g., an interval), an* **interior point of** *E is a point a in E such that there is an open interval around a which is also contained in E.*

Example 1.4.2. The most recurring example of interiors of sets that we shall consider are the interiors of intervals. If I is an interval, the interior of I is the interval obtained by deleting the endpoints of the interval, if they were present. So, the interiors of intervals of the form $[a, b]$, $[a, b)$, $(a, b]$, and (a, b) are all the open interval (a, b).

The fundamental object of study throughout this book is:

Definition 1.4.3. (The Derivative) *Suppose that a is in the interior of the domain of f. Then, the* **instantaneous rate of change of f, with respect to x, at $x = a$** *is*

$$\lim_{\Delta x \to 0} \frac{f(a + \Delta x) - f(a)}{\Delta x} \;=\; \lim_{h \to 0} \frac{f(a + h) - f(a)}{h},$$

provided this limit exists. If the limit exists, it is denoted by $f'(a)$, (read: f prime of a) and we say that $f'(a)$ **exists**. *The instantaneous rate of change $f'(a)$ is called the* **derivative of f at a**.

If a is not in the interior of the domain of f, or if the limit above fails to exist, then we say that **the derivative $f'(a)$ is undefined**.

The **derivative of f** *is the real function f', which has as its domain the set of all x values such that $f'(x)$ exists, has as its codomain the set of all real numbers, and is given by the rule*

$$f'(x) \;=\; \lim_{\Delta x \to 0} \frac{f(x + \Delta x) - f(x)}{\Delta x} \;=\; \lim_{h \to 0} \frac{f(x + h) - f(x)}{h}.$$

Thus, $f'(x)$ is the instantaneous rate of change of f, with respect to x, for any x value whatsoever (as long as x is in the domain of f').

The process of producing f', given f, is called **differentiation**. *Thus, we* **differentiate** *f, with respect to x, to produce f'. If a is an x value such that $f'(a)$ exists, then we say that f* **is differentiable at** *a. A function which is differentiable at each point in its domain is simply said to be* **differentiable**.

If $y = f(x)$, then other notations for the derivative function f' are y', $\dfrac{dy}{dx}$, or $\dfrac{df}{dx}$; in these notations, $f'(a)$ would be denoted by $y'(a)$, or any of

$$\frac{dy}{dx}\Big|_{x=a}, \qquad \frac{dy}{dx}\Big|_{a}, \qquad \frac{df}{dx}\Big|_{x=a} \qquad \text{or} \qquad \frac{df}{dx}\Big|_{a}.$$

We also refer to the symbol $\dfrac{d}{dx}$ as the **"differentiation with respect to x" operator** *and write $\dfrac{d}{dx}(f(x))$ for $f'(x)$.*

It is possible to define one-sided derivatives, by taking the limit, as h approaches 0, from the left or right. See, for instance, [3] or [2]. We shall not use this notion.

Remark 1.4.4. If $y = f(x)$, then the definition of the derivative is frequently written as the very simple

$$\frac{dy}{dx} = \lim_{\Delta x \to 0} \frac{\Delta y}{\Delta x},$$

where Δy, of course, denotes the change in y which corresponds to the change in x given by Δx, i.e., $\Delta y = f(x + \Delta x) - f(x)$. This is a good way to think about the derivative, but you need to be careful not to think of the limit, itself, as being a fraction.

> Despite the fact that dy/dx and df/dx **look** like fractions, they are not; they are simply notations for the derivative, which is a limit of fractions. There are not two quantities dy and dx, or df and dx, that are being divided to yield dy/dx or df/dx. To emphasize this fact, dy/dx and df/dx are **not** read as "dy divided by dx" or "df divided by dx", instead you read these as "$dydx$" and "$dfdx$". The operator d/dx is read as "ddx".

The notation $f'(x)$ is due to Lagrange (1736-1813). The notation dy/dx and its variants are due to Leibniz (1646-1716), who, along with Newton (1643-1727), was one of the founders of Calculus.

The notation used by Newton was $\dot{y}$, which is still used fairly often by some authors (not us) when the independent variable represents time. The notation used by Euler (1707-1783) was Df, Dy, $D_x f$, and $D_x y$.

We shall restrict ourselves to Lagrange's and Leibniz's notation throughout this book.

We should make two final remarks before looking at some examples.

We will frequently refer to the **instantaneous** rate of change, the IROC, to distinguish it clearly from the **average** rate of change, the AROC. However, it is actually unnecessary to include the word "instantaneous"; if we speak (or write) of the rate of change of $f(x)$, with respect to x, when x equals a specific value, **instead of between two values** of x or **on an interval**, then we must mean the IROC, not the AROC. For instance, if we refer to the velocity of a car at noon, this must be a reference to the instantaneous velocity, for we do not give two times between which to calculate the average velocity.

The rate of change of one quantity y, with respect to another quantity x, is frequently referred to as the **rate of increase of y, with respect to x**. For instance, you might say the temperature, in $°F$, is increasing, with respect to time t seconds, at a rate of 0.05 $°F/s$.

If we were to say that the rate of change of y, or the rate of increase, with respect to x, were negative, that would mean, in fact, that y is decreasing (as x increases). It is also true that a positive rate of **decrease** means a corresponding negative rate of increase, and a negative rate of decrease would, in fact, describe a quantity which is increasing.

Now, let's go through some examples, and use our new notation and terminology.

Example 1.4.5. Fix a constant c. Define three functions

$$w = f(x) = c, \qquad y = g(x) = x, \qquad \text{and} \qquad z = m(x) = x^2.$$

Let's find the derivatives of all three of these functions, and evaluate those derivatives at $x = 1$.

First, we find

$$\frac{dw}{dx} = f'(x) = \lim_{h \to 0} \frac{f(x+h) - f(x)}{h} = \lim_{h \to 0} \frac{c - c}{h} = \lim_{h \to 0} 0 = 0.$$

When $x = 1$, we have

$$\frac{dw}{dx}\bigg|_1 = f'(1) = 0.$$

If you think about this in terms of rates of change, it sounds completely silly; we just showed that the instantaneous rate of change of a function which doesn't change – a constant function – is zero. Not terribly surprising.

Now, for the next function, $y = g(x) = x$, we find

$$\frac{dy}{dx} = g'(x) = \lim_{h \to 0} \frac{g(x+h) - g(x)}{h} = \lim_{h \to 0} \frac{x + h - x}{h} =$$

$$\lim_{h \to 0} \frac{h}{h} = \lim_{h \to 0} 1 = 1.$$

When $x = 1$, we have

$$\frac{dy}{dx}\bigg|_1 = g'(1) = 1.$$

Finally, for our third function, $z = h(x) = x^2$, we have

$$\frac{dz}{dx} = m'(x) = \lim_{h \to 0} \frac{m(x+h) - m(x)}{h} = \lim_{h \to 0} \frac{(x+h)^2 - x^2}{h} =$$

$$\lim_{h \to 0} \frac{x^2 + 2xh + h^2 - x^2}{h} = \lim_{h \to 0} (2x + h) = 2x.$$

When $x = 1$, we have

$$\frac{dz}{dx}\Big|_1 = m'(1) = 2.$$

It is worth recording the results of the example above in a proposition.

Proposition 1.4.6. *Suppose that c is a constant.*

1. $(c)' = 0$;   *2. $(x)' = 1$;*   *3. $(x^2)' = 2x$;*

where the primes denote derivatives with respect to x.

Remark 1.4.7. Note that we did not explicitly state that the functions in the above proposition were, in fact, differentiable. Instead, we simply gave formulas for the derivatives which are defined for all values of x. This is a standard practice, which we shall follow throughout this book. When we give formulas for derivatives, we mean that the original function was differentiable at each x value for which the given formula is defined, unless we explicitly state otherwise.

It is important for you to remember our comments from Remark 1.2.4. When we write $(x^2)' = 2x$, we are using convenient, though technically incorrect, notation. What we actually mean is that, if f is the function defined by $f(x) = x^2$, then $f'(x) = 2x$. Sentences such as this latter one are simply too cumbersome to write over and over again.

You must be **very** careful here. If $(x^2)' = 2x$, then can you plug in $x = 5$ and conclude that $(25)' = 10$? Certainly not. As we stated in Proposition 1.4.6, the derivative of a constant is zero. What is true is that if $f(x) = x^2$, then $f'(5) = 10$. To indicate this without giving the squaring function a name, we write $(x^2)'\big|_5 = 10$ or $\frac{d}{dx}(x^2)\big|_5 = 10$.

In Example 1.2.5, we described the tangent line to a graph as a limit of secant lines and, hence, the slope of the tangent line was the limit of the slopes of the secant lines. Therefore, we now make the following definition.

Definition 1.4.8. *Suppose that $f(x)$ is differentiable at $x = a$. Then, the **tangent line to the graph of f at** $(a, f(a))$ or, simply, the **tangent line of f where** $x = a$ is the line with slope $f'(a)$, which passes through the point $(a, f(a))$, i.e., the line given by the point-slope equation*

$$y - f(a) = f'(a)(x - a).$$

*We also refer to $f'(a)$ as the **slope of the graph of f where** $x = a$ or, simply, the **slope of f where** $x = a$.*

Example 1.4.9. Consider the three functions from the previous example:

$$w = f(x) = c, \qquad y = g(x) = x, \qquad \text{and} \qquad z = m(x) = x^2.$$

Let's look at their tangent lines. Again, we will use various notations for the derivative throughout this example, in order to help you become more familiar with the different things that people write.

We know that $f'(x) = 0$, which means that the slope of the tangent line of f is always zero. Of course, this is correct; the graph of $f(x)$ is a horizontal line at w-coordinate c, i.e., the line given by $w = c$. Hence, the tangent line of f, where x is any coordinate, is also just $w = c$, i.e., where $x = a$, an equation for the tangent line is

$$w - c = f'(a)(x - a) = 0(x - a) = 0.$$

What about the tangent lines to the graph of $y = g(x) = x$? We know that, for all x, $\dfrac{dy}{dx} = 1$. Thus, at any point on the graph of $y = x$, where $x = a$, an equation for the tangent line is $y - g(a) = 1(x - a)$, which gives us $y - a = x - a$, i.e., $y = x$.

Thus, as in the case of $f(x) = c$, the tangent line at each point to the graph of $y = x$ is the same as the graph of the original function. This should not be surprising; it should be intuitively clear that the tangent line at each point on a (non-vertical) line is the original line itself. We shall show this rigorously in Section 1.4 by calculating the derivative of $y = ax + b$.

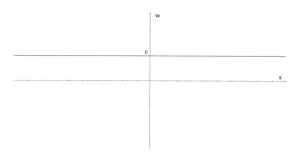

Figure 1.19: The graph of $w = c$.

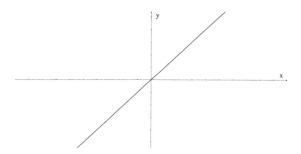

Figure 1.20: The graph of $y = x$.

Finally, let's look at the tangent lines to the graph of $z = m(x) = x^2$. From Proposition 1.4.6, we know that $\dfrac{dm}{dx} = 2x$. Thus, at any point on the graph of $z = x^2$, where $x = a$, an equation for the tangent line is $z - m(a) = 2a(x - a)$, which gives us $z - a^2 = 2ax - 2a^2$, i.e., $z = 2ax - a^2$. In Figure 1.21, we have drawn the tangent lines to the graph at the points $(0.5, 0.25)$ and $(1, 1)$; these lines have equations $z = x - 0.25$ and $z = 2x - 1$, respectively.

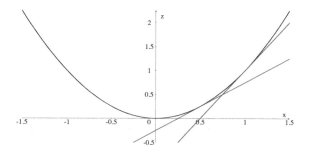

Figure 1.21: The graph of $z = x^2$.

Remark 1.4.10. In the above example, for $z = x^2$, we used that the slope of the tangent line at any x-coordinate is $2x$. Note, however, that when we gave the general equation for the tangent line at an arbitrary point on the graph of $z = x^2$, we did **not** use (x, x^2) as our arbitrary point, but, rather, used (a, a^2). It is actually crucial to introduce a new variable here.

Why? We wish to use x and z as our variables in the tangent line equations, but then those (x, z) pairs will describe points on lines, not points on the graph of $z = x^2$. We **could** introduce some sort of new x and z variables, such as $\hat{x}$ and $\hat{z}$, and then write an equation for the tangent line in the $\hat{x}\hat{z}$-plane, but we prefer to graph the tangent line in the same coordinate plane as the original graph. Thus, we pick a new variable, such as a, to represent the x-coordinate at which we find an equation for the tangent line.

> The primary danger here is that you may be tempted to use (x, x^2) for the point that appears in the point-slope form for the tangent line equations, and still use x and z for the variables describing the tangent lines. This would make you write something awful for the tangent line equations, something like $z - x^2 = 2x(x - x)$, i.e., $z - x^2 = 0$, which would just define the parabola again, not its tangent lines.

Example 1.4.11. Suppose that a particle is moving along the x-axis in such a way that its position at time t seconds is $x = x(t) = \sqrt{t}$ meters, for all $t \geq 0$. What is the velocity of the particle as a function of time? What is the velocity of the particle at time $t = 4$ seconds?

Solution:

The velocity $v = v(t)$ is the IROC of the position, with respect to time, i.e., in meters per second, the velocity is

$$v = \frac{dx}{dt} = \lim_{h \to 0} \frac{p(t+h) - p(t)}{h} = \lim_{h \to 0} \frac{\sqrt{t+h} - \sqrt{t}}{h},$$

for t values in the interior of the domain of $p(t)$, i.e., where $t > 0$, and at points where the limit exists. We remark that, if $t > 0$ and h is close enough to zero (e.g., $|h| < t$), then $t + h > 0$ and $\sqrt{t+h}$ is defined.

Now, for $t > 0$, we find that

$$v = \lim_{h \to 0} \frac{\sqrt{t+h} - \sqrt{t}}{h} = \lim_{h \to 0} \frac{\left(\sqrt{t+h} - \sqrt{t}\right)\left(\sqrt{t+h} + \sqrt{t}\right)}{h\left(\sqrt{t+h} + \sqrt{t}\right)},$$

where we have multiplied the numerator and denominator by the same thing in order not to change the fraction. That "same thing" that we multiplied by is the "conjugate" of $\sqrt{t+h} - \sqrt{t}$, and so we have in our new numerator the standard factorization of a difference of squares. Therefore, we find

$$v = \lim_{h \to 0} \frac{\left(\sqrt{t+h}\right)^2 - \left(\sqrt{t}\right)^2}{h\left(\sqrt{t+h} + \sqrt{t}\right)} = \lim_{h \to 0} \frac{(t+h) - t}{h\left(\sqrt{t+h} + \sqrt{t}\right)} = \lim_{h \to 0} \frac{h}{h\left(\sqrt{t+h} + \sqrt{t}\right)} =$$

$$\lim_{h \to 0} \frac{1}{\sqrt{t+h} + \sqrt{t}} = \frac{1}{2\sqrt{t}} \ \text{m/s},$$

which exists for all $t > 0$.

You may be thinking to yourself "How in the world would I think to multiply the numerator and denominator of our original limit by the conjugate of $\sqrt{t+h} - \sqrt{t}$?".

The answer to this is not as clear-cut as you might like. The real answer is: We know that we want to eliminate the division by h in order to arrive at a continuous function which is defined at $h = 0$. Hence, you think of all of the algebra that you've ever learned that leaves unchanged the function of h given by $(\sqrt{t+h} - \sqrt{t})/h$, for $h \neq 0$ (or in some deleted open interval around 0), and which allows us to get rid of the division by 0 when we plug in $h = 0$. There is no set of instructions, no recipe, to tell you how to proceed in every case; different limits require different manipulations.

The velocity of the particle at $t = 4$ seconds is

$$v(4) \ = \ \left.\frac{dx}{dt}\right|_4 \ = \ \frac{1}{2\sqrt{4}} \ = \ \frac{1}{4} \ \text{m/s}.$$

Let's look at the graph of $p(t) = \sqrt{t}$. If you look at the slope of the graph in Figure 1.22, you should be able to see that, as t approaches 0 from the right, i.e., through positive values, the slope of the graph gets unboundedly large, i.e., the slope of the graph approaches ∞. The slope of the graph at $(t, \sqrt{t})$ is, of course, the instantaneous rate of change of the position, with respect to time, i.e., $p'(t)$, which equals $v(t)$.

So, we can also see from our formula $v(t) = 1/(2\sqrt{t})$, for $t > 0$, that the slope of the graph of $p(t)$ approaches ∞ as $t \to 0^+$, for we know that

$$\lim_{t \to 0^+} \frac{1}{2\sqrt{t}} \ = \ \frac{1}{2} \cdot \lim_{t \to 0^+} \frac{1}{\sqrt{t}} \ = \ \infty,$$

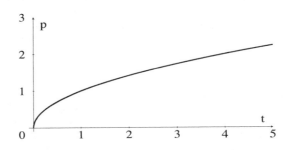

Figure 1.22: The graph of $p(t) = \sqrt{t}$.

by Corollary 1.3.52.

As a final comment on this example, we should mention that, in the real world, the velocity of a particle cannot exceed the speed of light. Hence, there is no particle whose position can be given by the particular function $p(t)$ that we used, since the velocity gets arbitrarily large as t approaches 0 from the right. We could avoid this "difficulty" by considering only t values bigger than something reasonable; for example, we could say that the position of the particle, for $t > 0.0001$, is given by $p(t) = \sqrt{t}$.

However, it is frequently useful when making mathematical points to ignore some physical limitations, even in the midst of a problem which is supposed to address questions which arise from physics. We shall take such liberties from time to time (but not **too** often).

Before giving another example, it will be convenient to derive another differentiation formula. We would like to know, for instance, that since $(x^2)' = 2x$, it must be true that $(5x^2)' = 5 \cdot 2x = 10x$.

More generally, suppose that c is a constant, and that f is a real function. The function cf is the function with the same domain as f such that, for all x in the domain of f, the value of cf at x is $cf(x)$.

We will show the following:

Proposition 1.4.12. *Suppose that c is a constant, and that f is differentiable at x. Then, cf is also differentiable at x, and $(cf)'(x) = cf'(x)$; this is normally written as $(cf(x))' = cf'(x)$, or*

$$\frac{d}{dx}(cf) = c\,\frac{df}{dx}.$$

Proof. Given our properties of limits, this is very easy. We have

$$(cf)'(x) \;=\; \lim_{h\to 0} \frac{cf(x+h) - cf(x)}{h} \;=\; \lim_{h\to 0} c\cdot\frac{f(x+h) - f(x)}{h} \;=\; cf'(x),$$

where we used Item 1 of Theorem 1.3.15, together with the fact that f is differentiable at x, to obtain the last equality above. $\qquad\square$

Now let's look at another application of derivatives in a problem related to motion.

Example 1.4.13. Let's look again at the particle from Example 1.4.11. Its position at time t seconds was $x = x(t) = \sqrt{t}$ meters, for all $t \geq 0$, and we determined that its velocity, in m/s, was $v = v(t) = 1/(2\sqrt{t})$, for all $t > 0$.

What is the acceleration of this particle as a function of t? What is its acceleration at time $t = 4$ seconds? If the particle has a mass of 1.6726×10^{-27} *kg* (approximately that of a proton), how much force must be acting on the mass at time $t = 4$ seconds?

Solution:

Acceleration is the instantaneous rate of change of the velocity, with respect to time. Hence, the acceleration, $a = a(t)$ in meters per second per second, at time $t > 0$, seconds is given by

$$a \;=\; \frac{dv}{dt} \;=\; \lim_{h\to 0} \frac{v(t+h) - v(t)}{h} \;=\; \lim_{h\to 0} \frac{1/(2\sqrt{t+h}) - 1/(2\sqrt{t})}{h} \;=$$

$$\frac{1}{2}\cdot\lim_{h\to 0} \frac{\frac{\sqrt{t}}{\sqrt{t}\cdot\sqrt{t+h}} - \frac{\sqrt{t+h}}{\sqrt{t}\cdot\sqrt{t+h}}}{h} \;=\; \frac{1}{2}\cdot\lim_{h\to 0} \frac{\sqrt{t} - \sqrt{t+h}}{h\sqrt{t}\sqrt{t+h}}.$$

As in our previous example, we once again multiply the numerator and denominator by the "conjugate" of the numerator to obtain:

$$a = \frac{1}{2} \cdot \lim_{h \to 0} \frac{t - (t+h)}{h\sqrt{t}\sqrt{t+h}(\sqrt{t} + \sqrt{t+h})} = \frac{1}{2} \cdot \lim_{h \to 0} \frac{-h}{h\sqrt{t}\sqrt{t+h}(\sqrt{t} + \sqrt{t+h})}.$$

Now, we can cancel the h's in the numerator and denominator, and let h equal 0 in the resulting continuous function to obtain

$$a = \frac{-1}{2\sqrt{t}\sqrt{t}(\sqrt{t} + \sqrt{t})} = \frac{-1}{4t^{3/2}} \text{ (m/s)/s},$$

for all $t > 0$.

At $t = 4$ seconds, we find that $a(4) = -1/32$ (m/s)/s. The negative sign means that the IROC of the velocity is negative, i.e., the particle is decelerating.

How much force must be acting on the particle at $t = 4$ seconds? We must use **Newton's 2nd Law of Motion** for an object of constant mass: the sum of the forces (i.e., the net force), F, acting on an object is equal to the (instantaneous) rate of change of the momentum of the object, with respect to time. The *momentum* of an object is its mass, m, times its velocity, v. Thus, we can summarize Newton's 2nd Law, for objects with constant mass, by

$$F = \frac{d}{dt}(mv).$$

In our problem, the mass, m, the object/particle is not changing; this is the standard situation. However, in some applications, such as the motion of rockets or hail, the fact that the mass of the object is changing is **very** important, and you can still use that $F = d(mv)/dt$. This is correct when the added or subtracted mass is moving with zero velocity, instantaneously, as the mass is added/subtracted. Here, we mean the velocity of the added/subtracted mass is zero relative to an observer that measures the object as moving with velocity v. For instance, hail moving with velocity v gains mass by smacking into essentially motionless water molecules.

However, when the mass is constant, Proposition 1.4.12 tells us that Newton's 2nd Law becomes the more commonly known formula:

$$F = m\frac{dv}{dt} = ma.$$

Therefore, in our problem, we find that the net force acting on the particle, at $t = 4$ seconds, is

$$m \cdot a(4) \; = \; (1.6726 \times 10^{-27} \; kg) \left(-\frac{1}{32} \; (m/s)/s \right) \; = \; -5.226875 \times 10^{-29} \; \text{Newtons.}$$

The negative sign indicates that the force is "pushing" the particle in the negative direction.

Before giving our next example, it will once again be convenient to derive another differentiation formula. We would like to know, for instance, that since $(x^2)' = 2x$ and $(x)' = 1$, it must be true that $(x^2 + x)' = 2x + 1$.

More generally, suppose that f and g are real functions. The function $f + g$ is the function which has as its domain all of those x values that are in both the domain of f and the domain of g (i.e., the domain of $f + g$ is the intersection of the domains of f and g) such that, for all x in the domain of $f + g$, the value of $f + g$ at x is $f(x) + g(x)$.

We will show the following:

Proposition 1.4.14. *Suppose that f and g are differentiable at x. Then, $f + g$ is also differentiable at x, and $(f+g)'(x) = f'(x)+g'(x)$; this is normally written as $(f(x)+g(x))' = f'(x) + g'(x)$, or*

$$\frac{d}{dx}(f + g) = \frac{df}{dx} + \frac{dg}{dx}.$$

Proof. Given our properties of limits, this is also very easy. We have

$$(f + g)'(x) \; = \; \lim_{h \to 0} \frac{(f(x + h) + g(x + h)) - (f(x) + g(x))}{h} \; =$$

$$\lim_{h \to 0} \frac{(f(x + h) - f(x)) + (g(x + h) - g(x))}{h} \; =$$

$$\lim_{h \to 0} \left[\frac{f(x + h) - f(x)}{h} + \frac{g(x + h) - g(x)}{h} \right] = f'(x) + g'(x),$$

where we used Item 2 of Theorem 1.3.15, together with the facts that f and g are differentiable at x, to obtain the final equality. $\qquad\square$

We now wish to look at a problem about mass and density.

Example 1.4.15. Suppose we have a straight rod of length 1 meter, with a circular cross section of constant radius 4 cm (so, 0.04 m). An x-axis is marked off along the rod with the origin at one of the rod and the other end at $x = 1$ meter. We are interested in the case where the density of the rod varies as x varies.

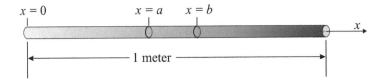

Figure 1.23: A rod of varying density.

Assume that the mass of the rod between 0 and any x value less than or equal to 1 is given by

$$m(x) = 5x^3 + x \quad \text{kg},$$

and assume that the density, the mass per volume, of the rod is constant at any fixed x value, i.e., assume that the density at each point in a cross section perpendicular to the rod is constant.

Determine the density $\rho(x)$ of the rod at each point in the perpendicular cross section of the rod at coordinate x, for $0 < x < 1$.

Solution:

We need to think about this one. What is the **average** density of the rod between $x = a$ and $x = b$ meters?

It's the total mass of the rod between $x = a$ and $x = b$, divided by the volume of that portion of the rod. The total mass of the rod between $x = a$ and $x = b$ is simply $m(b) - m(a)$ kg. The volume of the rod between $x = a$ and $x = b$ is the length of that part of the rod times the cross-sectional area, i.e., $(b - a)\pi(0.04)^2$ cubic meters.

Therefore, the average density, in kg/m^3, of the rod between $x = a$ and $x = b$ meters is

$$\frac{m(b) - m(a)}{(b - a)\pi(0.04)^2} = \frac{(5b^3 + b) - (5a^3 + a)}{(b - a)\pi(0.04)^2}. \tag{1.5}$$

If we want a good approximation for the density at each point in the rod where $x = a$, what should we do? If you think about it for a few minutes, you should realize that you want to take b "close to a". How close? Arbitrarily close, just as we did when we wanted to obtain the instantaneous velocity from the average velocities.

Therefore, what we want to calculate, for each x between 0 and 1, is the limit as h approaches 0, of the average density of the rod between x and $x + h$ meters. That is, the density, in kg/m^3, at each point in the rod in the cross section at x is

$$\rho(x) = \lim_{h \to 0} \frac{(5(x+h)^3 + (x+h)) - (5x^3 + x)}{h\pi(0.04)^2} =$$

$$\frac{1}{\pi(0.04)^2} \cdot \lim_{h \to 0} \frac{(5(x+h)^3 + (x+h)) - (5x^3 + x)}{h} =$$

$$\frac{1}{\pi(0.04)^2} \cdot m'(x).$$

Of course, we still wish to calculate $m'(x)$.

We find

$$m'(x) = \lim_{h \to 0} \frac{[5(x+h)^3 + (x+h)] - (5x^3 + x)}{h} =$$

$$\lim_{h \to 0} \frac{5(x+h)^3 - 5x^3 + h}{h} = \lim_{h \to 0} \frac{5x^3 + 15x^2h + 15xh^2 + 5h^3 - 5x^3 + h}{h} =$$

$$\lim_{h \to 0} \frac{15x^2h + 15xh^2 + 5h^3 + h}{h} = \lim_{h \to 0} \left(15x^2 + 15xh + 5h^2 + 1\right) = 15x^2 + 1.$$

Hence, we find that

$$\rho(x) = \frac{1}{\pi(0.04)^2} \cdot m'(x) = \frac{15x^2 + 1}{\pi(0.04)^2} \ \text{kg/m}^3.$$

In the calculation above of $\rho(x)$, as part of our work, we calculated $m'(x)$, the instantaneous rate of change of the mass, with respect to the x-coordinate, that is, with respect to distance along the rod, as measured from the end at 0. However, the original ratio that we considered in Formula 1.5 was a change in mass over a change in **volume**, not a change in distance. Why didn't we just calculate dm/dV, where V is the volume?

The reason was that we had the mass given to us in terms of x, not V. However, we could let $V(x) = \pi(0.04)^2 x$ cubic meters, determine m as a function of V, and then calculate dm/dV. But then there would still be a further problem; our density calculation would give us the density as a function of V, not x, as was desired.

We will return to this discussion in Example 2.3.12 in Section 2.3, when we will have another method for dealing with such problems.

As our final topic for this section, we would like to discuss how functions can fail to be differentiable at interior points of their domains.

Example 1.4.16. Let us return once more to Example 1.1.14, where we looked at the cost of lobsters, in dollars, as a function, $C(w)$, of their weight, w, in pounds. In Example 1.2.6, we found that, for $h > 0$,

$$\frac{C(2 + h) - C(2)}{h} = \frac{8(2 + h) - 7 \cdot 2}{h} = \frac{2 + 8h}{h} = \frac{2}{h} + 8 \;\; \$/\text{lb},$$

and concluded that $C'(2)$ does not exist, since $\displaystyle\lim_{h \to 0^+} \left(\frac{2}{h} + 8\right)$ is infinite, and so fails to exist.

In the example above, we found that the function C was not differentiable at 2. We also saw, informally, back in Example 1.3.13, that C is not continuous at 2. Is it true more generally that functions are not differentiable at points where they are discontinuous? Yes.

Theorem 1.4.17. *If f is differentiable at x, then f is continuous at x.*

    Equivalently, if f is not continuous at x, then f is not differentiable at x.

Proof. See 1.A.24.   □

Theorem 1.4.17 tells us one way that a function can fail to be differentiable at an interior point of its domain, but it leaves us asking: if x is an interior point of the domain of f **and** f is continuous at x, then must f be differentiable at x? The answer to this is "no", as we see in the following example.

Example 1.4.18. Let $f(x) = |x|$. Recall that, if $x \geq 0$, then $|x| = x$, but if $x \leq 0$, then $|x| = -x$. The graph of $y = |x|$ appears in Figure 1.24. From Remark 1.3.41 (or, informally, by looking at the graph), f is continuous. We will show that f is not differentiable.

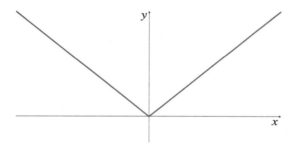

Figure 1.24: The graph of $f(x) = |x|$.

For any x-coordinate b greater than 0, the slope of the secant line of f for $x = 0$ and $x = b$ is clearly equal to 1. For any x-coordinate b less than 0, the slope of the secant line of f for $x = 0$ and $x = b$ is clearly equal to -1. Thus, we see a sharp point on the graph at $(0, 0)$. Does this mean that f is not differentiable at 0? Yes.

Consider the two one-sided limits of

$$\frac{f(0 + h) - f(0)}{h} = \frac{|h|}{h},$$

as h approaches 0 from the left and right. What we wrote above is the graphical argument for seeing that the two one-sided limits are different. The actual calculations are:

$$\lim_{h \to 0^-} \frac{|h|}{h} = \lim_{h \to 0^-} \frac{-h}{h} = -1$$

and

$$\lim_{h \to 0^+} \frac{|h|}{h} = \lim_{h \to 0^-} \frac{h}{h} = 1.$$

As the two one-sided limits are different, Theorem 1.3.11 tells us that the two-sided limit

$$\lim_{h \to 0} \frac{f(0+h) - f(0)}{h}$$

does not exist, i.e., $|x|$ is not differentiable at 0.

Example 1.4.19. There are other ways that the derivative can fail to exist at interior points of the domain of a function, even if the function is continuous at the point.

Consider the functions $f(x) = x^{1/3}$ and $g(x) = x^{2/3}$. The domain of each of these functions is the entire set of real numbers, each function is elementary and, hence, continuous everywhere. Let's look at the limits which you would calculate to find $f'(0)$ and $g'(0)$, if they were to exist.

We find

$$\lim_{h \to 0} \frac{f(h) - f(0)}{h} \;=\; \lim_{h \to 0} \frac{h^{1/3}}{h} \;=\; \lim_{h \to 0} \frac{1}{h^{2/3}} = \infty,$$

by Remark 1.3.53. Therefore, f is not differentiable at 0.

For g, we find

$$\frac{g(h) - g(0)}{h} \;=\; \frac{h^{2/3}}{h} \;=\; \frac{1}{h^{1/3}}.$$

By Corollary 1.3.52, as $h \to 0^-$, this quantity approaches $-\infty$, while, as $h \to 0^+$, this quantity approaches ∞. Therefore, g is not differentiable at 0.

It is interesting to look at the graphs of these two functions. They are given in the two figures below.

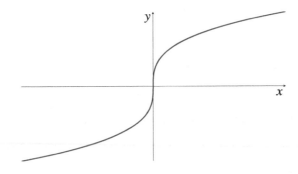

Figure 1.25: The graph of $f(x) = x^{1/3}$.

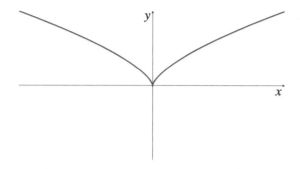

Figure 1.26: The graph of $g(x) = x^{2/3}$.

Note that the graph of $x^{1/3}$ does not have a sharp point at the origin, while the graph of $x^{2/3}$ does. The sharp point in the graph of $g(x) = x^{2/3}$ results from the fact that the slopes of the secant lines of g for $x = 0$ and $x = h$ approach $-\infty$ as $h \to 0^-$, and approach ∞ as $h \to 0^+$ (as we saw above). On the other hand, as $h \to 0$, both one-sided limits of the slopes of the secant lines of $x^{1/3}$, for $x = 0$ and $x = h$, are equal to ∞.

We shall revisit this type of phenomenon in Section 2.3, where we will see the relationship between the derivative of $f(x) = x^{1/3}$ and the derivative of its *inverse function* $f^{-1}(x) = x^3$. At that point, we will see that it is reasonable to define the vertical line given by $x = 0$ (i.e., the y-axis) as the tangent line to the graph of $f(x)$ at $(0,0)$.

Remark 1.4.20. As our final remark for this section, we should mention that you can encounter, in other sources, a "different" limit defining the derivative. You may see:

$$f'(x) = \lim_{t \to x} \frac{f(t) - f(x)}{t - x}.$$

That this yields the same derivative and, in fact, exists for the same x values as in the definition in Definition 1.4.3 follows from the limit substitution result in Theorem 1.3.25; you let $h = t - x$, so that $t = x + h$, and notice that $h \to 0$ if and only if $t \to x$.

We shall occasionally use this equivalent characterization of the derivative.

1.4.1   Exercises

In each of Exercises 1 through 8, find the derivative of the given function at the given point.

1. $f(x) = \dfrac{1}{\sqrt{x}}$; $x = 4$.

2. $g(z) = z^{3/2}$; $z = 9$.

3. $z(w) = \pi$; $w = \pi^2$.

4. $n(t) = \dfrac{1}{t} - 4t$; $t = 10$.

5. $l(r) = r\sqrt{r+1}$; $r = 0$.

6. $h(\theta) = 2\theta^2 + 4\theta$; $\theta = 1$.

7. $d(y) = x^2 + y^2 + z^2$; $y = 3$.

8. $b(x) = \dfrac{2x}{x^2-1}$; $x = 0$. Hint: you may find it useful to first show $\dfrac{2x}{x^2-1} = \dfrac{1}{x+1} + \dfrac{1}{x-1}$.

9. Solve for a and b, if f is differentiable, where $f(x) = \begin{cases} 2x + 3, & \text{if } x \leq 4; \\ x^2 + ax + b, & \text{if } x > 4. \end{cases}$

10. Let $f(x) = |x|$, $g(x) = x^2$, and $h(x) = g(f(x))$.

    a. Calculate $h'(0)$.

    b. Based on your answer to part (a), comment on the statement: if $h(x) = g(f(x))$ is differentiable, then f and g must also be differentiable.

11. Suppose that $y = 3z^2 + 4z - 7\pi$. Find $\dfrac{dy}{dz}$.

12. Suppose that $r = 4 + 8\theta^2$. Find $\dfrac{dr}{d\theta}$.

13. Suppose that $z = \sqrt{2}\,u + 7u^2$. Find $\dfrac{dz}{du}$.

14. Suppose that $\gamma = 1/(\alpha + 4)$. Find $\dfrac{d\gamma}{d\alpha}\Big|_1$.

15. Suppose that $m = \sqrt{4x + 9}$. Find $\dfrac{dm}{dx}\Big|_0$.

In each of Exercises 16 through 19, find an equation of each tangent line to the graph of the given function at the given point(s), then graph each function, together with the tangent line(s) that you found.

16. $t(x) = \dfrac{4}{\sqrt{x}}; \quad x = 16.$

17. $f(x) = x^2 + x - 20; \quad x = -5, 4.$

18. $h(x) = \begin{cases} 2 & \text{if } x \le 1 \\ -x + 3 & \text{if } x > 1 \end{cases}; \quad x = 0, 2.$

19. $r(x) = \dfrac{1}{1+x}; \quad x = 0, 1, 2.$

20. Rewrite the point-slope equation for the tangent line to the graph of a function $f(x)$ at the point $x = a$ in slope-intercept form.

For the functions in Exercises 21 - 23, find a general equation for the tangent line, i.e., the tangent line at any point $x = a$. If there are restrictions on those a for which the tangent line exists, note them.

21. $v(x) = 7x + 3$

22. $f(x) = x\sqrt{x+1}.$

23. $y(x) = \dfrac{x+9}{x-4}.$

24. Let $f(x) = \begin{cases} (x+3)^{\frac{2}{3}} & \text{if } x < -1 \\ |x| & \text{if } -1 \le x < 1 \\ -x^2 + 3x - 1 & \text{if } x \ge 1 \end{cases}$

    a. At what points is f discontinuous?

    b. At what points is f not differentiable?

25. Two particles leave the origin at the same time. The first particle travels along the y-axis and its distance, in meters, from the origin at time t seconds is $Y(t) = 3t$. The second particle travels along the x-axis and its distance from the origin, in meters, at time t seconds is $X(t) = 4t$.

    a. What is the distance between the two particles at time t?

    b. At what rate is the distance between the two particles changing at time t?

26. A rectangular piece of pizza dough is placed in an oven, where the dough expands. The length of the pizza is exactly twice the width throughout the warming process.

    a. Show that if P is the perimeter of the dough and A the area, then
       $A = P^2/18.$

    b. What is the derivative of A with respect to P?

27. Suppose $f(x) = x^3 + 2x^2 + 1$.

    a. Calculate the average rate of change of the function on the following intervals:

        i. $[0, 2]$

        ii. $[0, 1]$

        iii. $[0, 0.5]$

        iv. $[0, 0.25]$

    b. Calculate the instantaneous rate of change of the function, if it exists, at $x = 0$.

28. Suppose $f(x) = x^{4/3}$.

    a. Calculate the average rate of change of the function on the following intervals:

        i. $[0, 2]$

        ii. $[0, 1]$

        iii. $[0, 0.5]$

        iv. $[0, 0.25]$

    b. Calculate the instantaneous rate of change of the function, if it exists, at $x = 0$.

In Exercises 29 through 34, you are given a function $p = p(t)$, which gives the position, in meters, of an object whose mass is 5 kg, at time t seconds. Calculate the velocity $v = dp/dt$, and the acceleration $a = dv/dt$ of the object. Apply Newton's 2nd Law of Motion to determine the net force which is acting on the object. Your answers should be (possibly constant) functions of time, with the appropriate units.

29. $p = 5t - 7$.

30. $p = 5t^2 - 7$.

31. $p = \sqrt{t^2 + 1}$.

32. $p = 1/(t^2 + 1)$.

33. $p = 5t^2 - 7 + 4\sqrt{t^2 + 1}$.

34. $p = (t + 1)/t^2$.

The Two-Sided Derivative is an alternative definition of the derivative and is given by the equation $f'(x) = \lim\limits_{h \to 0} \dfrac{f(x + h) - f(x - h)}{2h}$. Use this new definition to calculate the derivative in Exercises 35 - 37

35. $g(x) = x^2$.

36. $j(x) = \dfrac{1}{x}$.

37. $m(x) = \sqrt{x}$.

38. Show that f is differentiable at $x = 0$, if $f(x) = \begin{cases} x^2 & \text{if } x \text{ is rational;} \\ 0 & \text{if } x \text{ is irrational.} \end{cases}$

39. Let $g(x) = x^3 + ax^2 + bx + c$. Prove that the tangent line to the graph of g is horizontal at exactly two points when $a^2 > 3b$, exactly one point when $a^2 = 3b$ and at no points when $a^2 < 3b$.

40. Let $f(x) = 3yx^2 + y^2$ where y is some constant. Calculate $f'(x)$.

41. Let $g(y) = 3yx^2 + y^2$ where x is some constant. Calculate $g'(y)$.

In Exercises 42-44, determine the x values where the tangent line to the graph of the function is horizontal. Note that this may occur at one or more points, or at zero points.

42. $u(x) = \sqrt{x}$.

43. $v(x) = \sqrt{x} - 0.25x$.

44. $w(x) = x^3 - 12x$.

45. $h(x) = ax^2 + bx + c$.

    a. Show that $h(0) = h'(0)$ when $b = c$.

    b. Show that $h(1) = h'(1)$ when $a = c$.

    c. Based on the results in parts (a) and (b), when is it true that $h(0) = h'(0)$ and $h(1) = h'(1)$?

46. When a current of $i = i(t)$ amperes (where t is the time, in seconds) passes through a capacitor, a charge of $q = q(t)$ Coulombs builds up across the capacitor. The current is equal to the rate of change of the charge across the capacitor, with respect to time, i.e., $i = dq/dt$.

    a. Suppose that the charge across the capacitor is constant. What does this tell you about the current passing through the capacitor?

    b. Suppose that the charge across the capacitor grows linearly with time, i.e., $q = at + b$, for some constants a and b. Then, what can you say about the current passing through the capacitor?

c. Suppose that the charge across the capacitor is given by $q = t^{1/3}$. What would you say is happening with the current at $t = 0$?

47. Given molecules A and B, a second order chemical reaction occurs when one molecule of A and one molecule of B collide to give a single new molecule C. Suppose we have a mixture of molecules of A, B, and C. Let $a > 0$ and $b > 0$ be the initial concentrations (fractions of the total number of molecules in the mixture) of A and B, respectively, and let $c(t)$ be the concentration of C at time t. Suppose that $c'(t) = 4(a - c(t))(b - c(t))$.

   a. If the initial concentration of C is 0, i.e., $c(0) = 0$, is $c'(0)$ zero, negative or positive?

   b. If the concentration of C at time t_1 is greater than the initial amount of A and less than the initial concentration of B, is $c'(t_1)$ zero, negative or positive?

   c. Describe the chemical conditions necessary at time t for $c'(t) = 0$.

48. The *weight* of a mass is the force due to gravity acting on the body. Let R be the radius of the Earth. The weight of a body $x \geq 0$ meters from the surface of the Earth is given by

$$w(x) = \frac{mgR^2}{(R + x)^2}.$$

Recall that $g \approx 9.8$ m/s^2 is the acceleration due to gravity.

   a. What is the weight of an object on the surface of the Earth?

   b. Calculate $w'(x)$.

49. Let $k(x) = 2x^3 - 9x^2 - 22x$. Show that there are two points where the tangent lines to the graph of $k(x)$ are perpendicular to the line $y = -\frac{1}{2}x + 3$.

50. Show by example that the following statement is *incorrect*:

$$\frac{d}{dx}\left[f(x)g(x)\right] = \frac{d}{dx}f(x) \cdot \frac{d}{dx}g(x).$$

51. The payoff of a call option at maturity depends on two quantities: the price of the stock, S, and the strike price, K, according to the equation

$$P(S) = \max(S - K, 0)$$

For example, if the price per share of company XYZ is \$40 and the strike price is \$30, the holder of the option receives a payment of \$10. If the price per share is only \$20, the investor receives no payment. Given that K is fixed, show that $P(S)$ is continuous but not differentiable.

52. The height of an object, in meters, in free-fall at time t seconds after being dropped, ignoring wind resistance, is given by $h(t) = h_0 - 4.9t^2$ where h_0 is the initial height of the object.

    a. How long does it take for the object to reach the ground?

    b. What is the velocity of the object at time t?

    c. What is the speed of the object as it hits the ground?

53. The kinetic energy of a body with mass m with velocity v is given by the equation $E = \frac{1}{2}mv^2$. Show that dE/dv is the momentum of the body.

54. When sodium is placed in a beaker containing a solution of hydrochloric acid, the resulting reaction produces a solution of sodium chloride and lets off hydrogen, H_2, in the form of a gas. Suppose that the number of grams of sodium chloride $S(t)$, which have been produced t seconds after the start of the reaction, is given by the function

$$S(t) = t^{\frac{2}{3}} + 0.1t,$$

for $0 \le t \le 30$, after which time the reaction ceases.

    a. How much sodium chloride is synthesized by the end of the reaction?

    b. At what rate is sodium chloride being synthesized 10 seconds after the start of the reaction?

55. A gas canister with a variable tap is used to fill a balloon in such a way that the volume of the balloon at time t seconds after the balloon starts to fill is given by

$$V(t) = -t^2 + 9t$$

cubic centimeters.

    a. At what time does the balloon pop, if its maximum volume is 20 cubic centimeters?

    b. At what rate was the volume in the balloon increasing just before it popped?

    c. At what rate was the volume increasing when the canister started filling the balloon?

56. Suppose that it costs a company $C(n)$ dollars to produce n cans of chili, and that a model for the cost of production is given by

$$C(n) = -0.00056n^2 + 1.12n$$

dollars, for $n \leq 1000$.

   a. What is the average cost per can of chili, if the company produces 1000 cans of chili?

   b. How much does it cost the company to produce 100 cans of chili?

   c. Recall from Remark 1.2.7 that the marginal cost is the derivative of the cost function, with respect to the number of items produced (using a differentiable approximation to the cost function). The marginal cost, calculated at n, is used as an estimate of the extra cost incurred by producing $n + 1$ items. Find the marginal cost when $n = 100$, and combine this with your answer to part (b), in order to approximate the cost of producing 101 cans of chili.

   d. Approximate the marginal cost of producing one more can of chili, when the company is producing 505 cans of chili.

57. Suppose that a particle travels along the x-axis, with velocity function

$$v(t) = 2t^2 - 42t + 180 \quad \text{meters per minute,}$$

measured t minutes from some starting time. Recall that (instantaneous) speed is the absolute value of (instantaneous) velocity.

   a. Find the speed function $s(t)$ for the particle, written in cases (i.e., without using absolute value signs).

   b. Determine the instantaneous rate of change of speed, with respect to time, one minute after the initial observation.

   c. What are the two points at which the speed function fails to be differentiable?

58. Coulomb's Law describes the interaction between charged objects. If object one has charge q_1, object two has charge q_2 (both in Coulombs), and the two objects are separated by r meters, then Coulomb's Law states that $F(r) = k\dfrac{q_1 q_2}{r^2}$, where $k \approx 9.0 \times 10^9$ Nm^2/C^2 is a proportionality constant.

   a. Two protons (each with charge $\approx 1.602 \times 10^{-19}$ Coulombs) are placed 0.002 meters away from one electron (with precisely the same magnitude of charge as the protons, but measured as a negative value). According to Coulomb's Law, what force is exerted on the electron by the pair of protons?

b. Find the instantaneous rate of change of the force, with respect to the distance between the charges. What does this quantity mean?

1.5   Extrema and the Mean Value Theorem

The derivative is defined in such a way that it is the instantaneous rate of change. Intuitively, then, there are a number of things which we expect to be true.

We expect that if the derivative is positive, the function increases as the variable increases, and if the derivative is negative, the function decreases as the variable increases. For instance, if the rate of change, with respect to time, of the money that you have is positive, you should have more money as time increases.

Suppose you throw a ball up in the air. As the ball is going up, the height of the ball is increasing, and so the derivative of the height, with respect to time, should be positive. When the ball is falling back down, the height is decreasing, and so the derivative of the height, with respect to time, should be negative. Thus, the derivative switches from being positive to being negative when the ball is at its highest point. Therefore, we would expect the derivative to be zero when the ball is at its highest point.

Suppose that a car travels 60 miles in exactly 1 hour, so that the average velocity of the car during the trip is 60 mph. Was there ever a time when the instantaneous velocity of the car was exactly 60 mph?

If the car went exactly 60 mph for the whole trip, then certainly the answer is "yes". Suppose the car did not go exactly 60 mph for the entire trip. If the car always went under 60 mph, then the average velocity would have been under 60 mph. If the car always went over 60 mph, then the average velocity would have been over 60 mph. Therefore, if the car did not go exactly 60 mph for the entire trip, then there must have been a time when the instantaneous velocity was under 60 mph and a time when the instantaneous velocity was over 60 mph; in-between these two times, we would expect that the velocity of the car had to pass through exactly 60 mph.

Finally, and most obviously, shouldn't it be true that, if the rate of change of a quantity is zero, then that quantity is constant? Note that what we already know is that the derivative of a constant function is zero. However, we don't have any theorem (yet) telling us that, if the derivative is zero, then the function had to be constant.

In this section, we shall prove that all of the above expectations are correct, under the right technical hypotheses.

First, we need to give rigorous definitions of a few intuitive concepts.

Definition 1.5.1. *A real function* $f : A \to B$ **attains a maximum value of** M **at a point** a *in* A *if and only if* $M = f(a)$ *and, for all* $x \in A$, $f(x) \leq M$.

Analogously, $f : A \to B$ **attains a minimum value of** m **at a point** a *in* A *if and only if* $m = f(a)$ *and, for all* $x \in A$, $m \leq f(x)$.

The function f **attains a local (or, relative) maximum (resp., minimum) value at a point** a *in* A *if and only if there exists an open interval* I *around* a *such that the function* f, *restricted to points that are in both* A *and* I *(that is,* $f_{|A \cap I}$*), attains a maximum (resp., minimum) value at* a.

Remark 1.5.2. Of course, if f attains a maximum (resp., minimum) value at a, then f obtain a local maximum (resp., minimum) at a, but the converse is not true. See Example 1.5.3.

To emphasize that maximum and minimum values of functions are not (necessarily) local (or, relative), we shall usually write that they are **global** (or, **absolute**) maximum or minimum values.

Instead of writing "local maximum values", we usually write simply **local maxima**. Naturally, we write **local minima** in place of "local minimum values", and the collective term for both maxima and minima is **extreme values** or **extrema**.

Example 1.5.3. Consider the function $f : [-1, \infty) \to \mathbb{R}$ given by $y = f(x) = x^2$.

This function attains a global minimum value of 0 at $x = 0$. However, f does not attain a global maximum value, since $f(x)$ gets arbitrarily large as x gets large. Still, f does attain a local maximum value of 1 at $x = -1$; 1 is the largest value that f attains for any x value close to -1 (and in the domain of f).

Note that we created the local maximum value f at $x = -1$ by "artificially" chopping off the domain of $y = x^2$.

The function f above is continuous, but is "allowed" to not obtain a global maximum because its domain is $[-1, \infty)$, which is not a closed, bounded interval $[a, b]$. Had the domain of f been

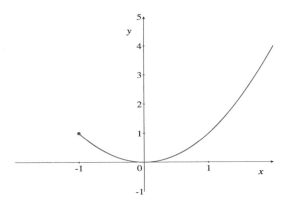

Figure 1.27: An example of local and global extrema.

a closed, bounded interval, we would have been able to apply the Extreme Value Theorem, Theorem 1.3.30, to conclude that f must obtain a global minimum and global maximum value.

Note that, in Example 1.5.3, f attains a minimum value at $x = 0$, and $f'(0) = 0$, since $f'(x) = 2x$ (for $x > -1$). This is similar to the phenomenon that we discussed for the height of a ball that is tossed upward; when the ball reaches its maximum height, we expect that the derivative of the height, with respect to time, should be 0.

This type of phenomenon is explained by:

Theorem 1.5.4. *If f is differentiable at x, and obtains a local extreme value at x, then $f'(x) = 0$.*

Proof. We shall prove this assuming that f attains a local maximum at x. The proof in the case where f attains a local minimum at x is essentially identical.

So, assume that f attains a local maximum value at x and that f is differentiable at x. Then, x is in the interior of the domain of f and, by definition of a local maximum value, there exists an open interval $(x - \delta, x + \delta)$ containing x, and contained in the domain of f, such that the restriction of f to

$$(x - \delta, x + \delta)$$

attains a maximum value at x.

Therefore, if h is in $(-\delta, \delta)$, then $f(x) \geq f(x + h)$, i.e., $f(x + h) - f(x) \leq 0$. Now consider the limit defining the derivative

$$\lim_{h \to 0} \frac{f(x+h) - f(x)}{h} = f'(x).$$

The left- and right-hand limits as $h \to 0$ must both equal $f'(x)$. However, the limit, as h approaches 0 from the left, uses negative values of h, and the quotient $\left(f(x+h) - f(x)\right)/h$ would be greater than or equal to 0 for h values close to 0. Hence, $f'(x) \geq 0$. On the other hand, the limit, as h approaches 0 from the right, uses positive values of h, and the quotient $\left(f(x+h) - f(x)\right)/h$ would be less than or equal to 0 for h values close to 0. Hence, $f'(x) \leq 0$. Therefore $f'(x) = 0$. $\qquad\square$

Remark 1.5.5. Note that Theorem 1.5.4 does **not** say that if $f'(x) = 0$, then f attains a local extreme value at x. This is, in general, false. For instance, if $f(x) = x^3$, then

$$f'(0) = \lim_{h \to 0} \frac{h^3}{h} = \lim_{h \to 0} h^2 = 0,$$

but f does not attain a local extreme value at 0, since x values smaller than 0 give f values smaller than 0, while x values greater than 0 give f values greater than 0.

What Theorem 1.5.4 does say is that the only x values that you need to consider when looking for places where a function f attains local extreme values are x values in the domain of f at which f is not differentiable or at which $f'(x) = 0$. This is important enough to warrant a new definition.

Definition 1.5.6. *A **critical point** of a function f is a point x in the domain of f at which $f'(x) = 0$ or at which f is not differentiable. A **critical value** of f is the value of f at any critical point.*

The following is a rephrasing of Theorem 1.5.4, using our new terminology.

Theorem 1.5.7. *If f attains a local extreme value at x, then x must be a critical point of f.*

Remark 1.5.8. You may think "Wait a minute. In Example 1.5.3, we had the function $y = x^2$, which is differentiable everywhere, and we found a local maximum value occurred at $x = -1$, even though the derivative at $x = -1$ is $2(-1) = -2 \neq 0$."

However, this ignores that the function f in Example 1.5.3 had a restricted domain, and $x = -1$ was an endpoint of that domain. Thus, $x = -1$ was not an interior point of the domain of f, and so, f was not differentiable there, i.e., $x = -1$ was a critical point of f.

Understand - in this book, when we discuss a function f on an interval I, we mean that we are restricting the domain of f to the interval I, even if f was initially defined on a bigger set than I. If an endpoint of I is included in the interval, i.e., I is of the form $[a, b]$, $[a, b)$, or $(a, b]$, then such an endpoint is a critical point of the restricted function, since the derivative does not exist there (since we can no longer take a limit from one side of the endpoint).

Be aware of the fact that other books may not always refer to included endpoints of intervals as critical points. In such books, they are not restricting their function to the given interval, but rather are looking at the function defined on a larger set. For instance, when looking at the function $y = f(x) = x^2$ on the interval $[-1, \infty)$, some books would not refer to $x = -1$ as a critical point of f. This makes the statement of where functions can attain local extreme values more complicated, since included endpoints of intervals would have to be allowed as a special case.

Looking back at our discussion at the beginning of this section, we would now like to address the case of the car which averages a velocity of 60 mph during a a trip. Must the car have gone exactly 60 mph at some time?

The answer to this is provided by the **Mean Value Theorem**, Theorem 1.5.10, below. First, however, we need to prove a lemma.

Lemma 1.5.9. (Rolle's Theorem) *Suppose that $a < b$, and that f is a continuous function on the closed interval $[a, b]$, which is differentiable on the open interval (a, b). Suppose also, that $f(a) = f(b)$. Then, there exists c in (a, b) such that $f'(c) = 0$.*

Proof. By Theorem 1.3.30, f restricted to $[a, b]$ attains a maximum and minimum value. If the maximum value M equals the minimum value m, then f is constant on $[a, b]$, and, hence, its derivative is zero at each point in (a, b).

Thus, we need to deal with the case where $M \neq m$. In this case, at least one of M or m is not equal to $f(a) = f(b)$, and so must be attained at some c in the open interval (a, b). By Theorem 1.5.4, $f'(c)$ must equal 0. $\square$

Now we can prove the Mean Value Theorem, which tells us that, under the right conditions, there must be a point in an interval at which the instantaneous rate of change of a function is equal to the average rate of change of the function over the interval.

Theorem 1.5.10. (Mean Value Theorem). *Suppose that $a < b$, and that f is a continuous function on the closed interval $[a, b]$, which is differentiable on the open interval (a, b). Then, there exists c in (a, b) such that*

$$f'(c) = \frac{f(b) - f(a)}{b - a}.$$

Proof. This follows quickly from Rolle's Theorem. Let

$$g(x) = \left(\frac{f(b) - f(a)}{b - a} \right)(x - a) - f(x) + f(a).$$

Then, g is continuous on $[a, b]$ and differentiable on (a, b). In addition, $g(a) = g(b) = 0$. By Rolle's Theorem, there exists c in (a, b) such that $g'(c) = 0$. However, for x in (a, b),

$$g'(x) = \frac{f(b) - f(a)}{b - a} - f'(x),$$

and, hence, $g'(c) = 0$ tells us that

$$f'(c) = \frac{f(b) - f(a)}{b - a}.$$

$\square$

Now we can prove the "obvious" theorem, which says that, if the rate of change of a function is zero, on an open interval, then the function had to be constant on that interval.

Theorem 1.5.11. *Suppose f is continuous on an interval I and that, for all x in the interior of I, $f'(x) = 0$. Then, f is constant on I, i.e., there exists a real number C such that, for all x in I, $f(x) = C$.*

Proof. Take any two distinct points in I; let x_1 denote the smaller number and x_2 the larger. We need to show that $f(x_1) = f(x_2)$.

We may apply the Mean Value Theorem to conclude that there exists c in (x_1, x_2) such that $f'(c) = (f(x_2) - f(x_1))/(x_2 - x_1)$. Now, $f'(c) = 0$ by assumption, so $f(x_1) = f(x_2)$. $\square$

A very important corollary is:

Corollary 1.5.12. *Suppose that f and g are continuous on an interval I and that, for all x in the interior of I, $f'(x) = g'(x)$. Then, there exists a constant C such that, for all x in I, $f(x) = g(x) + C$, i.e., f and g differ by a constant on I.*

Proof. Apply Theorem 1.5.11 to $f - g$. $\square$

Example 1.5.13. Suppose that we know that a function f, which is differentiable on the entire real line, is such that

$$\frac{df}{dx} = x,$$

for all x. Can we say what the function f is? Yes and no.

We know that $(x^2)' = 2x$. Dividing both sides of this equality by 2, we find

$$\frac{1}{2}(x^2)' = x.$$

By Proposition 1.4.12, we may move the 1/2 inside the parentheses to obtain

$$\left(\frac{x^2}{2}\right)' = x.$$

Now, Corollary 1.5.12 tells us that, since the f that we're looking for and $x^2/2$ both have the same derivative on the entire real line, then they must differ by a constant, i.e., there exists a constant C such that $f(x) = (x^2/2) + C$.

Can we possibly say what the constant C is? No – not without more data; for every choice of the number C,

$$\left(\frac{x^2}{2} + C\right)' = x.$$

To determine C, we would need more data about f, such as knowing the value of f at one x-coordinate. For instance, if we know that $f(2) = 7$, then we have $7 = f(2) = [(2^2)/2] + C$. From which we conclude that $C = 5$, and so $f(x) = (x^2/2) + 5$.

If we fix a function g, then those functions f, such that $f' = g$, are called **anti-derivatives of** g. If we have one particular anti-derivative, f_0, of g, then the expression $f_0 + C$, where C denotes an arbitrary constant, represents what **every** anti-derivative of g looks like (on an open interval); thus, $f_0 + C$ is called the **general anti-derivative of** g.

We also say that we are **solving the differential equation** $df/dx = g(x)$. If we are given both the differential equation, and a value of f at a particular value of x – so that we can solve for the constant C – then determining the function f is referred to as **solving an initial value problem**.

We shall look at such problems in more depth in Chapter 4.

We wish now to investigate the relationship between the sign (positive or negative) of the derivative of a function, and where the function is increasing or decreasing. First, we need to make some careful definitions.

Definition 1.5.14. *Let I be an interval, and f be a function whose domain contains I.*
  *f is **increasing on** I if and only if, for all a and b in I, if $a \leq b$, then $f(a) \leq f(b)$.*
  *f is **strictly increasing on** I if and only if, for all a and b in I, if $a < b$, then $f(a) < f(b)$.*
  *f is **decreasing on** I if and only if, for all a and b in I, if $a \leq b$, then $f(b) \leq f(a)$.*
  *f is **strictly decreasing on** I if and only if, for all a and b in I, if $a < b$, then $f(b) < f(a)$.*

---

  *We say that f is **monotonic on** I provided that f is increasing on I or decreasing on I.*

  *We say that f is **strictly monotonic on** I provided that f is strictly increasing on I or strictly decreasing on I.*

> The terminology that we have adopted for increasing and decreasing varies from source to source. You need to be aware of this if you look at other books.

Now, as another corollary of the Mean Value Theorem, we have:

Corollary 1.5.15. *Suppose that f is continuous on the interval I and differentiable at each interior point of I. If, for all interior points x of I,*

1. *$f'(x) \geq 0$, then f is increasing on I;*

2. *$f'(x) > 0$, then f is strictly increasing on I;*

3. *$f'(x) \leq 0$, then f is decreasing on I;*

4. *$f'(x) < 0$, then f is strictly decreasing on I.*

Proof. Suppose that a and b are in I and $a < b$. Then, f is continuous on $[a, b]$ and differentiable on (a, b). By the Mean Value Theorem, there exists c in (a, b) such that

$$f'(c) \;=\; \frac{f(b) - f(a)}{b - a}.$$

The four cases follow immediately.   □

Example 1.5.16. Functions are frequently increasing on some intervals, and decreasing on others. The easiest example of this is $f(x) = x^2$.

You can see where the function is decreasing; you look for where the graph roughly goes from the upper left to the lower right (or stays horizontal), i.e., where the slope is negative or zero. The corresponding x-coordinates yield the interval on which f is decreasing.

In the same way, the function is increasing where the graph roughly goes from the lower left to the upper right (or stays horizontal), i.e., where the slope is positive or zero. The corresponding x-coordinates yield the interval on which f is increasing.

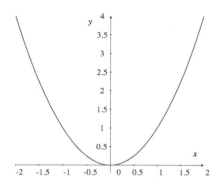

Figure 1.28: The graph of $y = x^2$.

In this example, we know, from Proposition 1.4.6, that $f'(x) = 2x$, and so Corollary 1.5.15 tells us that f is strictly decreasing on the interval $(-\infty, 0]$, and is strictly increasing on the interval $[0, \infty)$.

We can improve Corollary 1.5.15 somewhat.

Corollary 1.5.17. *Suppose that f is continuous on the interval I. Then,*

1. *if, for all but a finite number of x in I, $f'(x) > 0$, then f is strictly increasing on I;*

2. *if, for all but a finite number of x in I, $f'(x) < 0$, then f is strictly decreasing on I.*

Proof. We may apply the strictly monotone cases of Corollary 1.5.15 to each closed subinterval of the form $[x_1, x_2]$, where x_1 and x_2 are successive places where f' is not positive in the first case, or negative in the second case. The result follows. $\square$

Example 1.5.18. Note that f need not even be differentiable at the finite number of "bad" points allowed for in Corollary 1.5.17.

Consider the continuous function $y = x^{1/3}$. Then, y is not differentiable at $x = 0$ and, for $x \neq 0$, $y' = x^{-2/3}/3$, which is always positive. Thus, Corollary 1.5.17 tells us that $y = x^{1/3}$ is a strictly increasing function, which you can see from its graph in Figure 1.29.

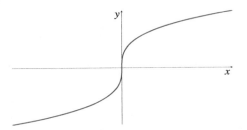

Figure 1.29: The graph of $f(x) = x^{1/3}$.

Corollaries 1.5.15 and 1.5.17 make it very important for us to know where the derivative of a function is positive and where it's negative. As we shall discuss at length in Section 3.2, the way in which one usually accomplishes this is by determining the zeroes of f', and then checking in-between the zeroes for whether f' is positive or negative. That is, we assume that, in-between successive zeroes of f', the sign of f' does not change, i.e., f' is always positive or always negative between successive zeroes. Why are we allowed to assume this?

If f' were continuous, then this result would follow from the Intermediate Value Theorem, Theorem 1.3.29; for, if a continuous function on an interval switches signs, then the function must hit zero in-between the two points where the function had opposite signs. However, even though functions must be continuous to have derivatives, **derivatives** themselves need not be continuous.

Admittedly, examples of differentiable functions which have discontinuous derivatives are hard to come by. One example is given by

$$f(x) \;=\; \begin{cases} x^2 \sin\left(\dfrac{1}{x}\right), & \text{if } x \neq 0; \\ 0, & \text{if } x = 0. \end{cases}$$

We leave it as an exercise for you to show that this function is differentiable everywhere, but that f' is not continuous at $x = 0$. If you do this exercise, what you will see is that $f'(0) = 0$ and $\lim_{x \to 0} f'(x)$ does not exist. As we show in Theorem 1.A.26 and its corollaries, this is the **only** way in which a differentiable function can fail to be continuous: the limit of the derivative must fail to exist at one or more points.

If a function f is differentiable, and f' is continuous, then f is said to be **continuously differentiable**. We shall not need the notion of continuously differentiable in this book for, even though derivatives need not be continuous, they must, in fact, satisfy the conclusion of the Intermediate Value Theorem; this is the content of the Intermediate Value Theorem for Derivatives, Theorem 1.A.25, which is in the Technical Matters section, Section 1.A. Here, we will state the conclusion of the Intermediate Value Theorem for Derivatives in the way in which we shall normally use it.

Theorem 1.5.19. *Suppose that f is differentiable on the open interval (a, b), and that, for all x in (a, b), $f'(x) \neq 0$.*

Then, either f' is positive at each point of (a, b), and so f is strictly increasing on (a, b), or f' is negative at each point of (a, b), and so f is strictly decreasing on (a, b).

We will conclude this section with a theorem that is useful for finding if/where functions attain global extrema. The idea of this theorem is easy: it basically says that if a continuous function is always increasing for $x < c$, and is always decreasing for $x > c$, then the function must be largest at $x = c$. Of course, there is the analogous statement for places where the function is smallest.

Theorem 1.5.20. (The First Derivative Test for Extrema) *Suppose that f is continuous on the interval I, that the interior of I is the open interval (a, b), and that c is a point in (a, b).*

If, for all x in (a, c), $f'(x)$ exists and is positive (resp., negative), and, for all x in (c, b), $f'(x)$ exists and is negative (resp., positive), then the restriction of the function f to the interval I attains a global maximum (resp., minimum) value at c; thus, (the unrestricted function) f attains a local maximum (resp., minimum) value at c.

Proof. This follows immediately from applying Corollary 1.5.15 to f on the interval consisting of those x in I such that $x \leq c$, and on the interval consisting of those x in I such that $x \geq c$. $\square$

Example 1.5.21. The First Derivative Test does **not** require that the derivative exists at the point where the sign of the derivative changes.

Recall the continuous function $g(x) = x^{2/3}$ from Example 1.4.19. The derivative of g at 0 does not exist. However, for $x \neq 0$, $g'(x) = 2x^{-1/3}/3$, which is negative when x is negative, and positive when x is positive. Therefore, the First Derivative Test is applicable, and it tells us that g attains a global minimum value at $x = 0$, as you can see in Figure 1.30.

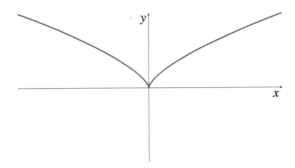

Figure 1.30: The graph of $g(x) = x^{2/3}$.

We shall return to local extrema and monotonic functions, and look at them in more depth in Section 3.2.

1.5.1 Exercises

Throughout these exercises, there are problems for which the calculation of the derivative, from the definition, is difficult. You may wish to revisit these exercises after you have read the first few sections of Chapter 2, and have some "easy" rules for calculating derivatives.

In each of Exercises 1 through 7, find the critical points and associated critical values, if any, of the given function.

1. $f(x) = \frac{1}{3}x^{1/3} - x$.

2. $g(x) = \dfrac{1}{1 + x^2}$.

3. $h(x) = x^3 + 4x^2 - x - 4$.

4. $s(x) = 2\sqrt{x} - 4x$.

5. $g(x) = (x - h)^2 + k$.

6. $j(x) = \sqrt{1 + x^2}$.

7. $p(x) = \sqrt{1 - x^2}$.

In each of Exercises **8** through **12**, you are given the derivative of a function f. Find the intervals (in the natural domain of f') on which f is increasing and those on which f is decreasing. You are **GIVEN** the derivative; you do not need to calculate it.

8. $f'(x) = \begin{cases} x + 5 & \text{if } x < 0; \\ -x^2 + 7x - 12 & \text{if } 0 < x < 10; \\ 0 & \text{if } x > 10. \end{cases}$

9. $f'(x) = \dfrac{x}{1 - x^2}$.

10. $f'(x) = \sqrt{x^2 - 1}$.

11. $f'(x) = \sqrt[3]{x^2 - 1}$.

12. $f'(x) = \left(x^3 + 3x^2 - 4x - 12\right)^2 \left(x^2 - 1\right)^3$.

In each of Exercises **13** through **16** , you are given a function f and an interval $[a, b]$. Find a c as is guaranteed to exist by the Mean Value Theorem, i.e., such that $a < c < b$ and the slope of the secant line between a and b is the same as the slope of the tangent line at c.

13. $f(x) = -x^2 + 9$, on the interval $[0, 3]$.

14. $f(x) = \sqrt{x + 1}$, on the interval $[0, 15]$.

15. $f(x) = x^3 + x^2 - 6x + 1$, on the interval $[-3, 2]$.

16. $f(x) = 3x + 7$, on the interval $[5, 9]$.

17. A man's normal systolic pressure S at age t years is modeled by the formula

$$S = 0.006t^2 - 0.02t + 120.$$

Pressure is measured in millimeters of mercury, or mm HG.

   a. What systolic pressure is predicted for a newborn?

   b. Use the Mean Value Theorem to show that the model predicts an age c in the interval $(20, 30)$ such that $S'(c) = 0.28$.

   c. Find the c, from part (b), explicitly, by calculating $S'(t)$.

In Exercises **18** through **21**, show that $f(x)$ is a solution to the initial value problem $df/dx = p(x)$, where $f(a) = b$.

18. $p(x) = 3x - 7$, $f(x) = 1.5x^2 - 7x + 4$, $f(3) = -3.5$.

19. $p(x) = \dfrac{2}{3}x^{(-1/3)} - x^{-2}$, $f(x) = x^{2/3} + x^{-1}$, $f(8) = \frac{33}{8}$.

20. $p(x) = x(x^2 - 1)^{-1/2} + 2$, $f(x) = \sqrt{x^2 - 1} + 2x + 7$, $f(-1) = 5$.

21. $p(x) = gx + v_0$, $f(x) = \frac{1}{2}gx^2 + v_0 x + h_0$, $f(0) = h_0$.

22. Find all critical points of the function $h(x) = x^4 - 4x^2$ on the specified interval.

    a. $(-3, 1)$.

    b. $[-3, 1)$.

    c. $[-1, 3]$.

    d. $[3, 6]$.

In each of Exercises 23 through 26, show that the function has exactly one zero in the interval. Accomplish this by using the Intermediate Value Theorem to show there is at least one zero, and Corollary 1.5.15 to show the root is unique.

23. $a(x) = 6x^3 + 9x - 10$, $I = (-\infty, \infty)$.

24. $b(x) = x^3 + 6x^2 - 3x - 18$, $I = [1, 2]$.

25. $c(x) = \sqrt{2x} - x^2$, $I = [1, 2]$.

26. $d(x) = \dfrac{1}{3x} - x$, $I = [0.25, 2]$.

27. Consider the following logic. $f(x)$ is continuous on $[a, b]$, differentiable on (a, b) and $f(a) = f(b)$. By Rolle's Theorem, there is exactly one value c in (a, b) such that $f'(c) = 0$. Is this statement true? If not, why not?

28. A toy company finds that the number, D, of action figures purchased (the *demand*), as a function of the price charged p (in dollars) is $D(p) = 10p^2 - 200p + 1000$, for $0 \le p \le 10$.

    a. Write the function $R(p)$ giving the net revenue – the total amount of money earned from sales of the action figures – as a function of the price charged.

    b. Find the intervals on which $R(p)$ is increasing/decreasing.

29. As we will derive (in a slightly more general context) in Example 4.2.11, the function describing the height of an object, above ground-level, in free-fall can be modeled by

$$p(t) = -\frac{1}{2}gt^2 + v_0 t + p_0,$$

where $g = 9.8$, in meters per second per second, v_0 is the initial velocity of the object, in meters per second, and p_0 is the initial height of the object, in meters. Note that, immediately after the object hits the ground, it is no longer in free-fall, and its height is no longer given by $p(t)$. ▶

    a. A stick is thrown upwards starting at a height of 10 meters, with an initial velocity of 5 m/s. Write the position as a function of time for the stick (for $t \geq 0$, and up until the stick hits the ground).

    b. When does the stick hit the ground (i.e., when does it attain its minimum height)?

    c. Find the intervals on which the height is increasing/decreasing.

    d. Given this information, when does the stick reach its maximum height?

30. The population of a certain ant colony in terms of time t (in years) is given by the following piecewise function $P(t) = \begin{cases} -t^3 + 10t^2 & \text{if } 0 < t \leq 7; \\ t^2 + 14t & \text{if } t > 7. \end{cases}$ ▶

    a. Show that this function is continuous for $t > 0$.

    b. Find the critical values of the function.

    c. Describe the intervals on which the function is increasing/decreasing.

31. Suppose that $f(x) = \sin^2(x)$ and $g(x) = -\cos^2(x)$. It's a fact that $f'(x) = g'(x)$, for all x (we'll prove this in later sections). What implication does this have for the relationship between $\sin^2(x)$ and $\cos^2(x)$?

32. Let
$$f(x) = \begin{cases} x^{1/3} & \text{if } x < 0; \\ x^2 & \text{if } x \geq 0; \end{cases}.$$

Take $I = (-\infty, \infty)$.

    a. What conclusion, if any, can be made about $f(x)$ using Corollary 1.5.15?

    b. What conclusion, if any, can be made about $f(x)$ using Corollary 1.5.17?

33. Show by example that there exists a function f that is continuous and strictly increasing on an interval, and that the set of x such that the condition $f'(x) > 0$ does *not* hold is infinite. This suggests that Corollary 1.5.17 may be generalized further.

34. Suppose that the position function of a particle at time t is given by

$$p(t) = \begin{cases} t & \text{if } 0 \leq t \leq 1; \\ at^2 + b & \text{if } t > 1; \end{cases}.$$

What are a and b if $t = 1$ is *not* a critical point?

35. Let $f_1(x) = |x|$, $f_2(x) = 0.5x^2 + 0.5$, and $f_3(x) = x^2$.

   a. If $h(x) = \min(f_1(x), f_2(x))$, what are the critical points of h?

   b. If $j(x) = \min(f_1(x), f_3(x))$, what are the critical points of j?

36. Suppose that the position function of a particle at time t is given by

$$p(t) = \begin{cases} (t-1)^2 & \text{if } 0 \le t \le 1; \\ at^n & \text{if } t > 1 \end{cases}.$$

   What must a be if $t = 1$ is non-critical?

37. Show by example that the Mean Value Theorem does not hold if the hypothesis "f is continuous on the closed interval $[a, b]$" is changed to "f is continuous on the open interval (a, b)."

38. "If f is differentiable on $(-\infty, \infty)$ and obtains a global extreme value at x, then $f'(x) = 0$." Is this statement true? Why or why not?

39. A typical liability insurance policy has a *deductible*, D, and a *limit*, L. If the total loss costs for a claim are x, then the policyholder must pay the first D dollars. The insurer will then pay the rest of the costs up to the limit L. Responsibility reverts back to the policyholder if the claim payment exceeds $D + L$. For example, suppose a policy has a deductible of \$500 and a limit of \$10,000. If a policyholder is in an accident with total loss costs of \$12,000, then the policyholder pays a \$500 deductible, the insurer pays \$10,000, and the policyholder pays the additional \$1500.

   a. Given a policy with deductible D and limit L, let $P(x)$ be the amount of a loss that a policyholder must pay. Write a formula for $P(x)$.

   b. Let $I(x)$ be the amount of a loss an insurer must pay. Write a formula for $I(x)$.

   c. Identify the critical points of $P(x)$. Assume a domain of $[0, \infty)$.

40. a. Suppose f and g are both continuous on $[a, b]$ and differentiable on (a, b). Show there exists a number c in (a, b) such that

$$[f(b) - f(a)]g'(c) = [g(b) - g(a)]f'(c).$$

   This statement is often called the *Generalized Mean Value Theorem*. Hint: Consider the function $h(x) = [f(b) - f(a)]g(x) - [g(b) - g(a)]f(x)$.

b. Show that the Mean Value Theorem is a consequence of the Generalized Mean Value Theorem by setting $g(x) = x$.

41. Suppose f is continuous on $[a, b]$, differentiable on (a, b), and that there is some positive number M such that $|f'(x)| \leq M$ f or all $x \in (a, b)$. Prove that for all x and y in (a, b),

$$|f(x) - f(y)| \leq M|x - y|.$$

This statement can be interpreted geometrically by saying the slope of the secant line between two points in (a, b) is uniformly bounded. In general, if a function f has the property that there exists a constant K such that $|f(x) - f(y)| \leq K|x - y|$ for all x and y in a given domain, we say f is *Lipschitz* continuous. In the special case that $f : [a, b] \to [a, b]$ and $K < 1$, f is called a *contraction mapping*.

42. Suppose that f and g are continuous on $[a, b]$, differentiable on (a, b), that $f(a) = g(a)$ and that $f'(x) < g'(x)$ for all x in (a, b). Prove that $g(b) > f(b)$. Hint: Consider the function $h = g - f$.

43. Prove that $\sqrt{1 + x} < 1 + \dfrac{x}{2}$ for $x > 0$. Hint: use the previous problem.

44. In this chapter, we used Rolle's Theorem to prove the Mean Value Theorem. Prove Rolle's Theorem using the Mean Value Theorem.

45. Suppose that the average temperature at a certain location is given by $T(t)$ degrees Celsius, where t is measured in years since 1990, and the domain of T is $[0, 12]$. Find the open intervals on which T is increasing, if $T'(t) = t^2 - 8t + 12$. ▶

46. Suppose that f is differentiable on an open interval I, and that, for all x in I, $f'(x) \geq 0$ (respectively, $f'(x) \leq 0$). Suppose further that there is no subinterval $[a, b]$ of I such that $a < b$ and f is constant on I. Prove that f is strictly increasing (respectively, decreasing) on I.

1.6   Higher-Order Derivatives

We have defined the instantaneous velocity and acceleration of an object: If $p = p(t)$ is the position of an object at time t, then the velocity $v = v(t)$ is equal to the derivative $p'(t)$, and the acceleration $a = a(t)$ is equal to the derivative $v'(t)$.

This means that $a(t)$ is the derivative, with respect to t, of the derivative, with respect to t, of the position $p(t)$. Such iterated derivatives occur often enough and are important enough that we adopt special terminology and notation.

Definition 1.6.1. *Suppose that we have a function $y = f(x)$. Then, the derivative of $f'(x)$, with respect to x, is called the* **second derivative of f, with respect to x.** *This is written as*
$$f''(x), \qquad (f(x))'', \qquad y'', \qquad \frac{d^2 y}{dx^2}, \qquad \frac{d^2 f}{dx^2}, \quad \text{or} \quad \frac{d^2}{dx^2}(f(x)).$$

The **third derivative of f, with respect to** x, *is defined to be the derivative of $f''(x)$, with respect to x.*

---

*More generally, for any positive integer n, the n-th derivative of f, **with respect to** x is obtained by iterating the derivative, with respect to x, n times, i.e., the n-th derivative of f, with respect to x, is the derivative, with respect to x, of the $(n-1)$-st derivative of f, with respect to x. This is written as*

$$f^{(n)}(x), \qquad (f(x))^{(n)}, \qquad y^{(n)}, \qquad \frac{d^n y}{dx^n}, \qquad \frac{d^n f}{dx^n}, \quad \text{or} \quad \frac{d^n}{dx^n}(f(x)).$$

The integer n is called the **order of the derivative.**

---

*If $n \geq 2$, $f^{(n)}(x)$ is referred to as a **higher-order derivative.***

Remark 1.6.2. It is sometimes convenient to refer to f itself as the 0-th derivative of f, i.e., $f^{(0)}(x)$ is simply $f(x)$.

It is a matter of personal taste at what order you stop using the prime notation. It is not uncommon to see $f'''(x)$ for the third derivative of f, with respect to x. Beyond the third derivative, it is usual to start superscripting by the order number in parentheses, as we wrote above, e.g., to write $f^{(4)}(x)$ for the fourth derivative.

The placement of the order indicators in $\dfrac{d^n y}{dx^n}$ may seem strange. They appear where they do because they are meant to indicate the application, n times, of the differentiation with respect to x operation, $\dfrac{d}{dx}$. For instance,

$$\frac{d}{dx}\left(\frac{d}{dx}(f(x))\right) = \left(\frac{d}{dx}\right)^2 (f(x)) = \frac{d^2}{dx^2}(f(x)) = \frac{d^2 f}{dx^2}.$$

Considering acceleration as the second derivative of position, with respect to time, is one of the most standard uses for higher-order derivatives. Actually, the third derivative of position, with respect to time, also has physical significance.

Recall Newton's 2nd Law of Motion from Example 1.4.13: if the mass m of an object is constant, then the net force, F, acting on the object is equal to ma, where a is the acceleration of the object. This means that the instantaneous rate of change of the net force on the object, with respect to time t, is given by

$$\frac{dF}{dt} = m\frac{da}{dt}.$$

In other words, the rate of change of the net force on the object is proportional to the rate of change of the acceleration. When you experience a "jerk", what you feel is a change in the net force acting on you; that is, dF/dt measures how much of a jerk you feel. Since dF/dt is proportional to da/dt, da/dt also measures how much of a jerk you feel.

This motivates the following definition.

Definition 1.6.3. *The derivative of acceleration, with respect to time, is called* **jerk**. *Thus, jerk is the third derivative of position, with respect to time.*

Remark 1.6.4. You should keep in mind the physical units on the higher-order derivatives. We know that the derivative $f'(x)$ of a function, f, with respect to a variable/quantity x, has units given by the units of f divided by the units of x. For instance, if $p = p(t)$ is the position, in meters, of an object at time t seconds, then the units of $p'(t)$ are m/s, meters per second.

What about the higher-order derivatives? The second derivatives $f''(x)$ or $p''(t)$ are the derivatives, with respect to x and t, respectively, of the derivatives. Hence, the units of the second derivative are the units of the derivative of the original function divided again by the units of x or t, respectively. This means that the units of the second derivative are: the units of the original function divided by the units of the variable, divided a second time by the units of the variable. For instance, the units of acceleration $a = p''(t)$ are (m/s)/s, meters per second per second, or m/s^2, meters per second squared.

By iterating this, we find that the units of $f^{(n)}(x)$ are the units of f divided by the units of x raised to the n-th power.

You may be wondering if there is a graphical way of "seeing" the higher-order derivatives or, at least, the second derivative. The answer is "yes", and we shall discuss this at length in Section 3.2, but we can have a preliminary discussion now.

The second derivative of f, with respect to x, is the instantaneous rate of change of the instantaneous rate of change of f, with respect to x. But, graphically, the instantaneous rate of change of f, with respect to x, is the slope of the tangent line to the graph of $y = f(x)$. Thus, the second derivative is the rate of change of the slope of the tangent lines to the graph of f, and it is relatively easy to see on a graph where this is positive or negative.

Example 1.6.5. Consider the graph in Figure 1.31, where we have cycled the colors blue, green, and red to help distinguish pieces of tangent lines (or you can click on the picture to view an animation).

What do you see in the graph? When $x = -2$, the slope of the tangent line is large. As you let x increase to -1, the slopes of the tangent lines decrease, i.e., the slopes get smaller. Therefore, the rate of change of the slope of the tangent lines is **negative** between $x = -2$ and $x = -1$. In other words, the second derivative is negative.

After $x = -1$ as x increase to $x = 0.5$, the slope of the tangent line becomes negative, and becomes even more negative as we closer to $x = 0.5$. Therefore, the second derivative is still negative until around $x = 0.5$.

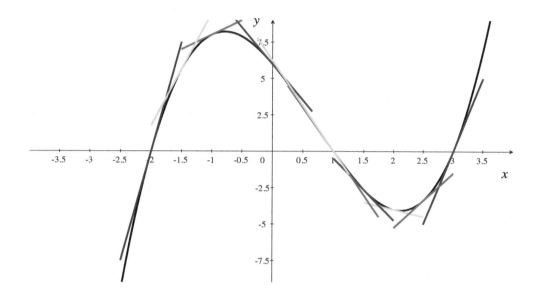

Figure 1.31: Tangent lines to a cubic graph.

Between $x = 0.5$ and $x = 2$, the slopes go from being pretty negative to being close to zero, i.e., the slopes are increasing, and so the rate of change of the slopes is positive. This means that the second derivative is positive between $x = 0.5$ and $x = 2$.

For $x > 2$, it appears that the slopes of the tangent lines continue to increase as x increases. Hence, the second derivative should be positive for these x values as well.

If you look at this example long enough, you should see that the second derivative, $f''(x)$, is negative where the graph of f is *concave down*, i.e., where the graph curves downward. What you're seeing when you observe that the graph is concave down is precisely that the slopes of the tangent lines are decreasing.

The graph is concave up, i.e., curves upward, where $f''(x)$ is positive; exactly what you see is the slopes of the tangent lines increasing.

In Section 2.1, we will discuss how to quickly calculate derivatives of polynomial functions. Then we shall return to this cubic graph in Example 2.1.11.

Let us record our observations above in a definition.

Definition 1.6.6. *We say that the graph of f is* **concave up** *on an open interval I if and only if $f''(x)$ exists and is positive for all x in I.*

The graph of f is **concave down** *on an open interval I if and only if $f''(x)$ exists and is negative for all x in I.*

If the concavity of the graph of f switches at $x = b$, that is, if b is in the domain of f, and the graph of f is concave up on some open interval (a, b) and concave down on some open interval (b, c), or vice-versa, then the point b (or the point $(b, f(b))$) is called an **inflection point of the graph of f.**

Note that, if the second derivative of f exists on an open interval around b, and b is an inflection point of the graph of f, then $f''(b)$ must be zero. However, it is **not** true that every place where $f'' = 0$ must be an inflection point.

An easy example is provided by $f(x) = x$, for which $f'(x) = 1$ and $f''(x) = 0$. The graph of $y = x$ is, of course, a straight line, and so is not concave up or down anywhere; hence, the concavity never switches signs, and there are no inflection points. To give a more interesting example, we have to look ahead and assume Theorem 2.1.1. Then it is easy for us to show that, if $g(x) = x^4$, then $g''(x) = 12x^2$. Thus, $g''(0) = 0$, but the graph of g is concave up for all $x < 0$ and for all $x > 0$, i.e., the concavity does not switch at $x = 0$, so $x = 0$ is not an inflection point of the graph of g.

Example 1.6.7. Suppose that the position of an object at time t hours is $p = p(t) = 5t^2 - 2t - 1$ miles.

The graph of this function appears in Figure 1.32. This graph is concave up everywhere (or, at least, in the portion shown), and so, from our discussion in the previous example, we expect to find that the second derivative is positive for all t. We shall calculate the second derivative, with respect to time, that is, we shall calculate the acceleration a and see that it is, in fact, always positive.

Using Proposition 1.4.14, Proposition 1.4.12, and Proposition 1.4.6, we quickly calculate the velocity v (mixing derivative notations, to continue helping you to adjust to using the various notations interchangeably):

$$v = \frac{dp}{dt} = (5t^2 - 2t - 1)' = (5t^2)' + (-2t)' + (-1)' =$$

$$5(t^2)' - 2(t)' + 0 = 5(2t) - 2 \cdot 1 = 10t - 2 \quad \text{mph.}$$

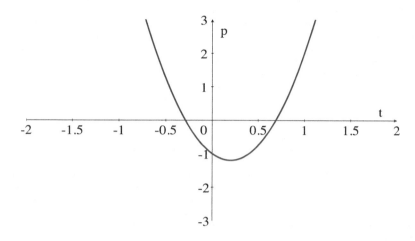

Figure 1.32: Position as a function of time.

Now, we find

$$a \;=\; \frac{dv}{dt} \;=\; \frac{d^2p}{dt^2} \;=\; (10t - 2)' \;=\; 10 \;\; \text{mph/hr}.$$

Therefore, the acceleration is constant and, as we expected from the graph, always positive.

The calculation of the jerk is easy:

$$j \;=\; \frac{da}{dt} \;=\; \frac{d^3p}{dt^3} = (10)' = 0 \;\; (\text{mph/hr})/\text{hr}.$$

Remark 1.6.8. While we do not yet have rules for differentiation that let us look at more complex examples of second derivatives in an easy manner, we should mention that second derivatives are talked about fairly often in the news.

When it is reported that "the rate of increase in unemployment is decreasing" or "the rate of increase in the average temperature of the Earth is increasing", these are references to the second derivative.

Think about it. If $U(t)$ is the number of unemployed people (belonging to some group of possible workers), then dU/dt is the (instantaneous) rate of change of U, with respect to t; when this rate is positive, it is frequently called the *rate of increase* in unemployment. To say that the rate of increase in unemployment is decreasing is to say that dU/dt is decreasing, i.e., the rate of change of dU/dt, with respect to time, is negative. Therefore, the phrase "the rate of increase in unemployment is decreasing" tells us implicitly that dU/dt is positive, and explicitly that d^2U/dt^2 is negative.

We conclude this section with a method for using second derivatives to detect local extrema and, in special cases, global extrema.

Theorem 1.6.9. (The Second Derivative Test for Extrema) *Suppose that $f'(c) = 0$. Then,*

1. if $f''(c) > 0$, then f attains a local minimum value at c;

2. if $f''(c) < 0$, then f attains a local maximum value at c.

Furthermore, suppose that f is continuous on an interval I, differentiable on the interior of I, and that c is the only critical point of f in the interior of I. Then,

a. if $f''(c) > 0$, then f attains a global minimum value at c;

b. if $f''(c) < 0$, then f attains a global maximum value at c.

Proof. We first show (1) and (2).

Assume $f'(c) = 0$ and $f''(c) > 0$. Note that the existence of $f''(c)$ implies that f' exists on some open interval around c, which also implies that f is continuous on an interval around c. We have

$$f''(c) \ = \ \lim_{h \to 0} \frac{f'(c+h) - f'(c)}{h} \ = \ \lim_{h \to 0} \frac{f'(c+h)}{h} \ > \ 0.$$

Looking at the two one-sided limits as $h \to 0^+$ and $h \to 0^-$, we see that there exists an open interval (c, b) on which f' must be positive, and an open interval (a, c) on which f' must be negative. Thus, we may apply the First Derivative Test, Theorem 1.5.20, to conclude that f attains a local minimum value at c.

The local maximum proof is completely analogous, and we leave it as an exercise.

We now show (a) and (b).

Now suppose that f is continuous on an interval I, differentiable on the interior of I, and that c is the only critical point of f in the interior of I. We wish to apply the First Derivative Test again.

Let (a, b) be the interior of I. Then, the sign of f' cannot change on the interval (a, c) or on the interval (c, b), for if the sign did change, then, by the Intermediate Value Theorem for Derivatives, Theorem 1.A.25, we would have to have $f'(x) = 0$ for some x in (a, b) other than $x = c$; a contradiction of our assumption. Now, either the conclusion or the proof of the local extrema cases of this theorem tells us that the sign of f' actually switches at c. Thus, we apply the First Derivative Test to finish the proof. $\qquad \square$

Remark 1.6.10. Note that the Second Derivative test says nothing about what happens if $f''(c) = 0$. There is a good reason for this; if $f'(c) = 0$ and $f''(c) = 0$, then f may attain a local maximum value, a local minimum value, or neither at c.

Assuming that enough derivatives of f exist to finally obtain one that is non-zero, the question is whether that first non-zero derivative $f^{(n)}(c) \neq 0$ occurs when n is even or n is odd.

If this n is odd, then f does not attain a local extreme value at c. If this n is even, then the situation is precisely like that of the Second Derivative Test: if $f^{(n)}(c) > 0$, then f attains a local minimum value at c; if $f^{(n)}(c) < 0$, then f attains a local maximum value at c. We leave the proof of this as an exercise.

A final comment on the Second Derivative Test: while the Second Derivative Test does not require the continuity of f'' on an open interval around c, this would, nonetheless, be the standard case. Assuming this to be true, it is easy to picture the conclusion of the Second Derivative Test, since we know that f'' being positive on an interval means that the graph of f is concave up on the interval, and f'' being negative on an interval means that the graph of f is concave down on the interval.

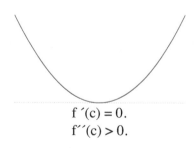

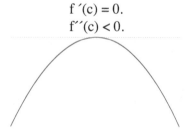

Figure 1.33: A typical local minimum.   Figure 1.34: A typical local maximum.

1.6.1   Exercises

For each of the functions in Exercises 1 through 5, find f', f''.

1. $f(x) = x^2 + x + 4$.

2. $f(x) = \dfrac{10}{x}$.

3. $f(x) = 5\sqrt{x} + x$.

4. $f(x) = 4x + \dfrac{1}{\sqrt{x}}$.

5. $f(x) = (x + 3)^2$.

Draw a rough sketch of the graphs of each of the functions in Exercises 6 through 8 as follows: first find the x- and y-intercepts (if they exist), next find and mark the critical points and intervals of increase and decrease, and finally find the intervals on which the graphs are concave up and down.

6. $f(x) = 9x - \sqrt{x}$, for $x \geq 0$.

7. $f(x) = \dfrac{1}{3}x^3 - x$.

8. $f(x) = 2x^2 + \dfrac{1}{x}$, for $x \neq 0$.

In the Exercises 9 through 11, you are given a position function $x(t)$, in feet, for a particle traveling along the x-axis, where t is measured in seconds. In each case, determine the open intervals of time when the particle is moving to the left (the negative direction) and those when the particle is moving to the right (the positive direction). Also determine the times during which the particle is gaining speed or losing speed. Recall that the instantaneous speed is the absolute value of the instantaneous velocity.

9. $x(t) = 5t^2 - 60t + 100$.

10. $x(t) = -t^3 + 12t^2 - 45t + 2$.

11. $x(t) = 2\sqrt[3]{t} - 96t^2$.

In Exercises 12 through 15, you are given a position function $x(t)$, in feet, of a particle, at time t seconds. Calculate the velocity and acceleration as functions of time.

12. $x(t) = 1/t$.

13. $x(t) = -\dfrac{1}{2}gt^2 + v_0 t + h_0$.

14. $x(t) = \sqrt[3]{t} + 3t^2 + 7$.

15. $x(t) = (t-1)^2 - (t+2)^3$

16. Consider the function
$$g(x) = \begin{cases} -x^2 & \text{if } x \le 0; \\ x^2 & \text{if } x > 0 \end{cases}.$$

   a. Show that g is continuous at $x = 0$.

   b. Show that $g'(0) = 0$.

   c. Does $g''(0)$ exist? If so, what is it? If not, why not?

17. Show that if $f(t) = \dfrac{1}{t} + t^2$, then $f''(t) \cdot t^2 = 2f(t)$.

18. Show that if $f(t) = 2\sqrt{t} + t$, then $f''(t)(t - f(t)) = 1/t$.

19. One of the solutions to the quadratic equation $y = ax^2 + bx + c$ is given by the quadratic formula
$$x = \frac{-b + \sqrt{b^2 - 4ac}}{2a}.$$

   Assume the discriminant $b^2 - 4ac$ is positive and calculate the second derivative of x with respect to c.

In each of Exercises 20 through 26, you are given the graph of a function. Approximate the intervals on which the function is increasing and decreasing, and approximate the intervals on which each function has a positive or negative second derivative.

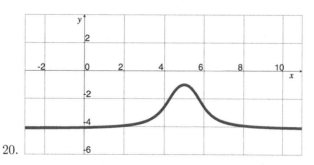

20.

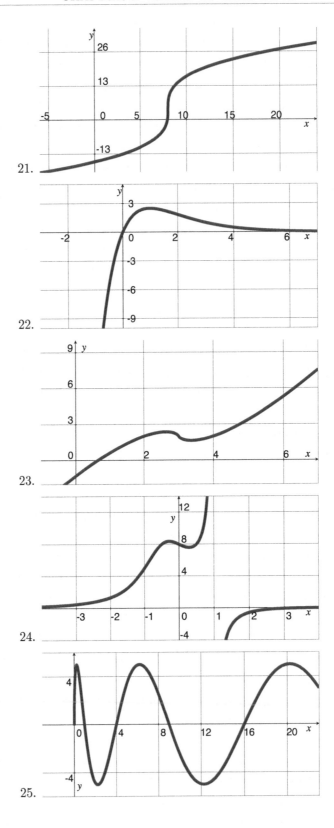

21.

22.

23.

24.

25.

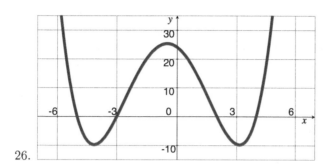

26.

In Exercises 27 through 30, calculate a general formula for $f^{(n)}(x)$.

27. $f(x) = 1/x$.

28. $f(x) = 1/x^2$.

29. $f(x) = x^k$, where k is a positive integer.

30. $f(x) = x^{-k}$, where k is a positive integer.

31. Let

$$g(x) = \begin{cases} -x^3 - 3x^2 - x + 2 & \text{if } -1 \leq x \leq 0; \\ x^3 - 3x^2 - x + 2 & \text{if } 0 \leq x \leq 1 \end{cases}.$$

   a. Show that $g(x)$ is continuous at $x = 0$.

   b. Show that $g'(x)$ is continuous at $x = 0$.

   c. Show that $g''(x)$ is continuous at $x = 0$.

   This problem is an example of a *cubic spline*. Cubic splines are used to develop smooth curves between finitely many data points. In this case, the data points are $(-1, 1)$, $(0, 2)$ and $(1, -1)$.

32. A rocket, carrying a satellite to orbit from its launch pad at sea level, is launched in such a way that its height, in meters above sea level, t seconds after the launch, is given by the function

$$y(t) = 4.9t^2 + 0.1667t^3.$$

   a. Calculate the velocity of the rocket 10 seconds after launch.

   b. What is the acceleration of the rocket at this time?

   c. At what time is the acceleration of the rocket twice that of the acceleration due to gravity (9.8 meters per second per second)?

   d. What is the jerk experienced by the rocket 10 seconds after launch?

In probability theory, the *cumulative distribution function*, $F(x)$, is the probability that a random variable is less than or equal to x. For example, the statement $F(20) = 0.3$ is the same as the statement that "the probability a random variable X is less than or equal to 20 is 0.3, or 30%". The *density function*, $f(x)$, is the derivative of the cumulative distribution function: $f(x) = \dfrac{d}{dx}F(x)$. Given the cumulative distribution functions in Exercises **33** through **35**, determine the local minima or maxima of the density function.

33. $F(x) = -2x^3 + 3x^2$. Domain: $[0, 1]$.

34. $F(x) = \dfrac{1}{10}x - 1$. Domain: $[10, 20]$.

35. $F(x) = \dfrac{15}{92}\left(x - \dfrac{2}{3}x^3 + \dfrac{1}{5}x^5\right) + \dfrac{1}{2}$. Domain: $[-2, 2]$.

36. Recall that an equation of the tangent line to the graph of $f(x)$ at $x = a$ is $P_1(x) = f(a) + f'(a)(x - a)$.

    a. Show that $P_1(a) = f(a)$.

    b. Show that $P_1'(a) = f'(a)$.

    c. Let $P_2(x) = f(a) + f'(a)(x - a) + \frac{1}{2}f''(a)(x - a)^2$. Show that $P_2(x)$ is a stronger approximation to $f(x)$ at the point $x = a$ in the sense that:

       c1. $P_2(a) = f(a)$.

       c2. $P_2'(a) = f'(a)$.

       c3. $P_2''(a) = f''(a)$.

In Exercises 37 through 39, calculate $P_2(x)$ at $x = a$, as defined in the previous problem.

37. $f(x) = x^3 + x^2 + x + 1$, $x = 0$.

38. $f(x) = 1/x$, $x = 1$.

39. $f(x) = \sqrt{x}$, $x = 1$.

40. Suppose f' is differentiable at x. Is f' continuous at x? Why or why not?

41. Let $f(x) = x^4 - 2(a + b)x^3 + 6abx^2 + a^2bx + 12$, where $a < b$. Find x values such that $f''(x) = 0$.

42. In each part of this exercise, give an example of a function f and a point c which satisfy the given criteria.

a. $f'(c) = 0$, $f''(c) = 0$ and f attains a local minimum at c.

b. $f'(c) = 0$, $f''(c) = 0$, and f attains a local maximum at c.

c. $f'(c) = 0$, $f''(c) = 0$, and f attains neither a local minimum nor a local maximum at c.

43. Let $g(x) = x^3 y + y^2 \sqrt{x}$ and $h(y) = x^3 y + y^2 \sqrt{x}$. This means that y is considered a constant in calculating $g(x)$ and x a constant for calculating h.

   a. Calculate $\dfrac{dg}{dx}$.

   b. Take the derivative with respect to y of your answer to part a.

   c. Calculate $\dfrac{dh}{dy}$.

   d. Take the derivative with respect to x of your answer to part c.

44. Suppose $f''(x) = (x - a_1)(x - a_2)(x - a_3)$ where $0 < a_1 < a_2 < a_3$. Are a_1, a_2 and a_3 all inflection points? Why or why not?

45. Let $U(t)$ be the number of unemployed people (among those who are eligible to work), in the USA, t months after August 1, 2008. Interpret the following statements in terms of the function $U(t)$ and its derivatives.

   a. The number of unemployed persons, as measured at the beginning of August, was 9.4 million.

   b. Unemployment was increasing by approximately $500,000$ people per month at the beginning of September.

   c. At the beginning of May of 2009, the number of unemployed persons was increasing, but the rate of unemployment was decreasing.

46. Suppose $f(x) = ax^4 + bx^3 + cx^2 + d$. Explain why f can have at most two inflection points.

47. Let $g(x) = x^3 + 3ax^2 + 3a^2 x + a^3$, where a is a positive constant. Does g have an inflection point?

48. Let $f(x) = x^4 + 4ax^3 + 6a^2 x^2 + 4a^3 x + a^4$. Does f have an inflection point?

49. Show that $h(x) = x^n$, where $n \geq 2$ is a positive integer, has a minimum at $x = 0$ if and only if x is even and an inflection point at $x = 0$ if and only if x is odd.

Appendix 1.A   Technical Matters

1.A.1   Properties of Real Numbers and Extended Real Numbers

We assume that you are familiar with the set $\mathbb{R}$ of real numbers; a real number may be zero, or a positive or negative number with a decimal expansion, which may require an infinite number of decimal places. A mathematically rigorous definition of the real numbers is beyond the scope of this text. We refer you to the treatment by Rudin in [2].

If $a, b \in \mathbb{R}$ and $a \leq b$, we define four types of *intervals from a to b*: the *closed interval from a to b*

$$[a, b] = \{x \in \mathbb{R} \mid a \leq x \leq b\};$$

the *open interval from a to b*

$$(a, b) = \{x \in \mathbb{R} \mid a < x < b\};$$

and the *half-open (half-closed) intervals from a to b*

$$(a, b] = \{x \in \mathbb{R} \mid a < x \leq b\} \quad \text{and} \quad [a, b) = \{x \in \mathbb{R} \mid a \leq x < b\}.$$

More generally, a subset I of $\mathbb{R}$ is an *interval* if and only if, for all $a, b \in I$, if $a \leq b$, then $[a, b] \subseteq I$. In other words, a subset of $\mathbb{R}$ is an interval if and only if, for every pair of numbers in I, all of the numbers in-between those two are also in I.

It is convenient to define the *extended real numbers* by considering the real numbers together with formal symbols $-\infty$ and ∞. Thus, we define the set of extended real numbers to be $\mathbb{R}^e = \mathbb{R} \cup \{-\infty, \infty\}$, where we declare that, for all $x \in \mathbb{R}^e$, $x \leq \infty$ and $-\infty \leq x$. In particular, since $-\infty < 0 < \infty$, we say that $-\infty$ is negative, and ∞ is positive; these symbols are referred to as *negative infinity* and *positive infinity*, respectively.

These infinity symbols give us a convenient way of writing the remaining types of interval of real numbers. Let $b \in \mathbb{R}$. We have the *open interval from b to ∞*

$$(b, \infty) = \{x \in \mathbb{R}^e \mid b < x < \infty\} = \{x \in \mathbb{R} \mid b < x < \infty\};$$

the *closed interval from b to* ∞, which is also half-open,

$$[b, \infty) = \{x \in \mathbb{R}^e \mid b \leq x < \infty\} = \{x \in \mathbb{R} \mid b \leq x < \infty\};$$

the *open interval from* $-\infty$ *to b*

$$(-\infty, b) = \{x \in \mathbb{R}^e \mid -\infty < x < b\} = \{x \in \mathbb{R} \mid -\infty < x < b\};$$

the *closed interval from* $-\infty$ *to b*, which is also half-open,

$$(-\infty, b] = \{x \in \mathbb{R}^e \mid -\infty < x \leq b\} = \{x \in \mathbb{R} \mid -\infty < x \leq b\};$$

and, finally, the entire real line

$$(-\infty, \infty) = \{x \in \mathbb{R}^e \mid -\infty < x < \infty\} = \{x \in \mathbb{R} \mid -\infty < x < \infty\} = \mathbb{R}.$$

A subset U of $\mathbb{R}$ is *an open subset of* $\mathbb{R}$ if and only if U is a union of open intervals, which is equivalent to: for all $x \in U$, there exists $\epsilon > 0$ such that the interval $(x - \epsilon, x + \epsilon)$ is contained in U. A subset E of $\mathbb{R}$ is *a closed subset of* $\mathbb{R}$ if and only if $\mathbb{R} \backslash E = \{x \in \mathbb{R} \mid x \notin E\}$ is an open subset of $\mathbb{R}$. It follows immediately that unions of open subsets of $\mathbb{R}$ are open subsets of $\mathbb{R}$, and intersections of a finite number of open subsets of $\mathbb{R}$ are open subsets of $\mathbb{R}$. This implies that intersections of closed subsets of $\mathbb{R}$ are closed subsets of $\mathbb{R}$, and unions of a finite number of closed subsets of $\mathbb{R}$ are closed subsets of $\mathbb{R}$.

Suppose that $X \subseteq Y \subseteq \mathbb{R}$. Then, X *is an open subset of* Y if and only if there exists an open subset U of $\mathbb{R}$ such that $X = Y \cap U$. We say that X *is a closed subset of* Y if and only if $Y \backslash X$ is an open subset of Y, which is equivalent to : there exists a closed subset E of $\mathbb{R}$ such that $X = Y \cap E$. Note that the open subsets of $\mathbb{R}$ which are also intervals are exactly what we call "open intervals" above, together with $(-\infty, \infty)$. The closed subsets of $\mathbb{R}$ which are also intervals are exactly what we call "closed intervals" above, together with $(-\infty, \infty)$. Yes - $\mathbb{R}$ is both an open and closed subset of itself, as is the empty set; these two sets are the only subsets of $\mathbb{R}$ which are both open and closed. Unions of open subsets of Y are open subsets of Y, and intersections of a finite number of open subsets of Y are open subsets of Y. As before, this implies that intersections of closed subsets of Y are closed subsets of Y, and unions of a finite number of closed subsets of Y are closed subsets of Y.

Suppose that $X \subseteq \mathbb{R}$. The set X is *bounded above* if and only if there exists $b \in \mathbb{R}$ such that $X \subseteq (-\infty, b]$; any such b is referred to as an *upper bound of X*. The set X is *bounded below* if and only if there exists $a \in \mathbb{R}$ such that $X \subseteq [a, \infty)$; any such a is referred to as an *lower bound of X*. The set X is *bounded* if and only if there exist $a, b \in \mathbb{R}$, where $a \leq b$, and $X \subseteq [a, b]$.

The set $Y \subseteq \mathbb{R}$ is *disconnected* if and only if there exist non-empty, open subsets X_1 and X_2 of Y such that $Y = X_1 \cup X_2$ and $X_1 \cap X_2 = \emptyset$; Y is *connected* if and only if it is not disconnected.

The set $Y \subseteq \mathbb{R}$ is *compact* if and only if, for all collections $\{U_j\}_{j \in J}$ of open subsets of Y such that $Y = \bigcup_{j \in J} U_j$ (an *open cover of Y*), there exists a **finite** subcollection $\{U_{j_1}, \ldots, U_{j_n}\}$ (a *finite subcover of Y*) such that $Y = \bigcup_{k=1}^{n} U_{j_k}$.

The following theorems express fundamental properties of the real numbers.

Theorem 1.A.1. (Least Upper Bound Property) *Suppose that $Y \subseteq \mathbb{R}$ is bounded above. Then, there exists a **least upper bound** for Y, i.e., there exists an upper bound b of Y such that, for all upper bounds c of Y, $b \leq c$.*

For example, the intervals $(-3, 5)$ and $(-3, 5]$ both have the same least upper bound, namely 5; note that the least upper bound of a set may or may not be an element of the set itself. We leave it to you to formulate the statement of the *Greatest Lower Bound Property*; that this property also holds for the real numbers can easily be proved by negating and using the Least Upper Bound Property.

Theorem 1.A.2. *The connected subsets of $\mathbb{R}$ are precisely the intervals.*

Theorem 1.A.3. (Heine-Borel) *The compact subsets of $\mathbb{R}$ are precisely the subsets of $\mathbb{R}$ which are both closed and bounded.*

1.A.2 Functions

A function consists of two sets (possibly the same), the *domain* and *codomain* of the function, and a rule for assigning to each element of the domain a unique element of the codomain. A function is also referred to as a *map*.

A function with domain A, codomain B, and a rule f is usually denoted by $f \colon A \to B$. For brevity, if the domain and codomain of the function have been clearly defined, one simply refers to the function by the name of its rule, that is, one would simply speak of the function f. We also say that f *maps A into B* or simply f *maps A to B*. If we have a function $f \colon A \to B$, and

$x \in A$, then the element of B that f assigns to x is denoted by $f(x)$, and one reads this as "f of x". We also refer to $f(x)$ as the *value of f at x*.

Most of the functions discussed in this book are *real functions*, functions $f \colon A \to B$ such that A and B are subsets of the set $\mathbb{R}$ of real numbers. One frequently specifies the variable names that one will use for elements of the domain and codomain by writing such things as "let $y = f(x)$ be the function from $\mathbb{R}$ to $\mathbb{R}$ given by $f(x) = x^2$". This specifies the domain and codomain of f as being the real numbers, gives the rule, and provides additional information that is not technically part of the function's definition; it tells the reader that x will be used to denote elements of the domain, and y will be used to denote elements of the codomain. In this setting, x is usually referred to as the *independent variable* and y is usually referred to as the *dependent variable*.

One also encounters phrases such as "the function $y = x^2$". This means that the reader is supposed to know what the domain and codomain are (frequently both $\mathbb{R}$) and that no function name is actually specified. It is standard in this setting to adopt the name of the codomain variable as the name of the function, and talk about the function $y = y(x)$. Of course, this means that when one writes "the function $y = f(x)$", the same function now has two possible names, f and y. We emphasize that the variable names are **not** part of the function definition; the functions $y = x^2$ and $z = t^2$ (as functions from $\mathbb{R}$ to $\mathbb{R}$) are the same.

In a further simplification (some would say "abuse") of notation and terminology, when dealing with real functions, one writes things such as "the function $y = \ln x + \sqrt{4 - x^2}$, and assumes that the reader knows that this is a real function and that the domain is "as big as possible". In other words, it is assumed that the reader is familiar enough with the functions involved in the definition of the y, and the domains of the functions involved, so that the domain can be taken to be the maximal set for which the given rule has meaning.

Suppose that we have a function $f \colon A \to B$. The *range* of f is the set $\{f(x) \mid x \in A\}$, i.e., the range of f is the set of elements of B that actually get "hit" by an element of A via the rule f. The range need not be the entire codomain. If the range of f equals its codomain, we say that the function f is *onto* or is a *surjection*, and we say that f *maps A onto B*.

The function f is *one-to-one* (or *1-to-1*), or an *injection*, if and only if, for all $x_1, x_2 \in A$, if $f(x_1) = f(x_2)$, then $x_1 = x_2$, i.e., if and only if every element in the range of f is hit, via f, by a **unique** element from the domain.

A function which is 1-to-1 and onto is called a *1-to-1 correspondence* or a *bijection*. The function f has an *inverse* if and only if f is a bijection; in this case, the *inverse of f* is the function $f^{-1} \colon B \to A$ given by the rule, for all $y \in B$, $f^{-1}(y)$ is the unique element $x \in A$ such that $f(x) = y$. Note that this means that f^{-1} is itself a bijection, and so possesses an inverse function, namely f, i.e., $(f^{-1})^{-1} = f$. Note that the notation for the inverse function can be easily confused with raising to the -1 power. When dealing with function names, the superscript of -1 is reserved for the inverse function. However, one may write $(f(x))^{-1}$ for the reciprocal of the **value** $f(x)$ (provided that such a value exists).

Given a set A, there is an important function associated with A, called the *identity function on A*. This is the function $\mathrm{id}_A \colon A \to A$ given by $\mathrm{id}_A(x) = x$, for all $x \in A$.

Given sets A, B, and C, and functions $f \colon A \to B$ and $g \colon B \to C$, one defines a new function, the *composition of g with f* or *g composed with f*, $g \circ f \colon A \to C$, given by, for all $x \in A$, $(g \circ f)(x) = g(f(x))$. One easily checks that, if $f \colon A \to B$ is a bijection, then $f^{-1} \circ f = \mathrm{id}_A$ and $f \circ f^{-1} = \mathrm{id}_B$. It is easy to verify, in fact, that $f \colon A \to B$ is a bijection if and only if there exists $g \colon B \to A$ such that $g \circ f = \mathrm{id}_A$ and $f \circ g = \mathrm{id}_B$, and when these equivalent conditions hold, g must be equal to f^{-1}.

If we have a function $f \colon A \to B$, and C is a subset of A, one defines a new function, the *restriction of f to C* or *f restricted to C*, $f_{|C} \colon C \to B$ given by, for all $x \in C$, $f_{|C}(x) = f(x)$, i.e., the restriction of f to C has the same rule and codomain as f, one simply restricts one's attention to what happens to elements of C. One may also restrict the codomain of $f \colon A \to B$ to any subset of B which contains the range of f; there is no standard notation for the restriction of the codomain of a function. For example, suppose that $f \colon \mathbb{R} \to \mathbb{R}$ is given by $f(x) = x^2$, for all $x \in \mathbb{R}$. Let $\mathbb{R}_{\geq 0} = \{ y \in \mathbb{R} \mid y \geq 0 \}$. Then, the function $g \colon \mathbb{R} \to \mathbb{R}_{\geq 0}$ given by $g(x) = x^2$, for all $x \in \mathbb{R}$, is technically not the same function as f, because g has a different codomain; many sources ignore the distinction between a function and that function with a restricted codomain.

Given $f \colon A \to B$ and subsets $X \subseteq A$ and $Y \subseteq B$, we define the *image of X by f*, $f(X)$, to be the set $\{ f(x) \mid x \in X \}$. We define the *inverse image of Y by f*, $f^{-1}(Y)$, to be the set $\{ a \in A \mid f(a) \in Y \}$; note that the inverse image of a set containing a single element could easily be a set containing more than one element, i.e., f need not be a bijection and so the **function** f^{-1} need not exist for us to discuss the inverse image of sets.

Both the image and inverse image via a map f take unions of sets to unions of sets, i.e., if I is some (possibly infinite) indexing set (a set whose elements are used solely for indexing purposes), and, for all $i \in I$, we have subsets $X_i \subseteq A$ and $Y_i \subseteq B$, then

$$f\left(\bigcup_{i \in I} X_i \right) = \bigcup_{i \in I} f(X_i) \qquad and \qquad f^{-1}\left(\bigcup_{i \in I} Y_i \right) = \bigcup_{i \in I} f^{-1}(Y_i).$$

The inverse image also respects intersections of sets, i.e.,

$$f^{-1}\left(\bigcap_{i \in I} Y_i \right) = \bigcap_{i \in I} f^{-1}(Y_i).$$

The image does **not**, in general, respect intersections.

If we have $f : A \to B$, $g : B \to C$, $X \subseteq A$, and $Y \subseteq C$, then $(g \circ f)(X) = g(f(X))$ and $(g \circ f)^{-1}(Y) = f^{-1}(g^{-1}(Y))$.

1.A.3 Proofs of Theorems on Limits and Continuity

Before beginning proofs on limits and continuity, we should remark that the *triangle inequality* gets used over and over again. This inequality simply says that, for all $a, b \in \mathbb{R}$, $|a+b| \leq |a| + |b|$.

Throughout this subsection, we shall prove theorems for real functions about two-sided limits, and about continuity at points which are in open intervals contained in the domain of the function under consideration. The proofs for one-sided limits, or for continuity when the domain is not open, are trivial variations on the proofs that we give. We leave many of them as exercises.

Finally, before we begin the proofs, note that the domain of f contains a deleted open interval around b if and only if the domain of f contains both a deleted right and left interval of b. Below, when taking limits as x approaches b, or approaches from the left or right, we assume that x is in a deleted open interval around b, or deleted right/left interval of b, as needed, and omit further references to the domain.

Theorem 1.A.4. (Theorem 1.3.5) *Suppose that* $\lim_{x \to b} f(x) = L_1$ *and* $\lim_{x \to b} f(x) = L_2$. *Then,* $L_1 = L_2$. *In other words,* $\lim_{x \to b} f(x)$, *if it exists, is unique.*

Proof. We shall show that, for all $\epsilon > 0$, $|L_1 - L_2| < \epsilon$; this shows that $L_1 = L_2$ since, otherwise, one would get a contradiction by taking $\epsilon = |L_1 - L_2|/2$.

Now, suppose $\epsilon > 0$. Then, $\epsilon/2 > 0$, and so, by definition of $\lim_{x \to b} f(x) = L_1$ and $\lim_{x \to b} f(x) = L_2$, there exist δ_1 and δ_2 such that $\delta_1 > 0$ and $\delta_2 > 0$, and such that, if $0 < |x - b| < \delta_1$ and $0 < |x - b| < \delta_2$, then $|f(x) - L_1| < \epsilon/2$ and $|f(x) - L_2| < \epsilon/2$. Select such δ_1 and δ_2, and let δ equal of the minimum value of δ_1 and δ_2. Now, select an x such that $0 < |x - b| < \delta$. Then,

$$|L_1 - L_2| = |(L_1 - f(x)) + (f(x) - L_2)| \leq |L_1 - f(x)| + |f(x) - L_2| <$$

$$\epsilon/2 + \epsilon/2 = \epsilon.$$

$\square$

Proposition 1.A.5. *The following are equivalent:*

1. $\lim_{x \to b} f(x) = L$;

2. $\lim_{x \to b} (f(x) - L) = 0$;

3. $\lim_{x \to b} |f(x) - L| = 0$.

Proof. All three statements are precisely that: for all $\epsilon > 0$, there exists $\delta > 0$ such that, if $0 < |x - b| < \delta$, then $|f(x) - L| < \epsilon$. $\qquad\square$

Theorem 1.A.6. (Theorem 1.3.6) *Suppose that f, g and h are defined on some deleted interval U around b, and that, for all $x \in U$, $f(x) \leq g(x) \leq h(x)$. Finally, suppose that $\lim_{x \to b} f(x) = L = \lim_{x \to b} h(x)$. Then, $\lim_{x \to b} g(x) = L$.*

Proof. Let $\epsilon > 0$. Let $\delta > 0$ be such that, for all x such that $0 < |x - b| < \delta$, we have that $x \in U$, $|f(x) - L| < \epsilon$, and $|h(x) - L| < \epsilon$. Then, if $0 < |x - b| < \delta$, we have

$$-\epsilon < f(x) - L \leq g(x) - L \leq h(x) - L < \epsilon,$$

i.e., $|g(x) - L| < \epsilon$. $\qquad\square$

Theorem 1.A.7. (Theorem 1.3.11) $\lim_{x \to b} f(x) = L$ *if and only if* $\lim_{x \to b^+} f(x) = L$ *and* $\lim_{x \to b^-} f(x) = L$.

Proof. If $\lim_{x \to b} f(x) = L$, it is immediate from the definitions that $\lim_{x \to b^+} f(x) = L$ and $\lim_{x \to b^-} f(x) = L$.

Suppose that $\lim_{x \to b^+} f(x) = L$ and $\lim_{x \to b^-} f(x) = L$. Let $\epsilon > 0$. Then, there exist $\delta_1 > 0$ and $\delta_2 > 0$ such that, if $0 < x - b < \delta_1$ or $0 < b - x < \delta_2$, then $|f(x) - L| < \epsilon$. Thus, if δ equals the minimum value of δ_1 and δ_2, then $0 < |x - b| < \delta$ implies that $|f(x) - L| < \epsilon$. $\qquad\square$

Proposition 1.A.8.

1. $\lim_{x \to 0^+} f(x) = L$ *if and only if* $\lim_{u \to 0} f(|u|) = L$;

2. $\lim_{x \to 0^-} f(x) = L$ *if and only if* $\lim_{u \to 0} f(-|u|) = L$.

Proof. We shall prove Item 1. The proof of Item 2 is completely analogous.

Suppose that $\lim_{x \to 0^+} f(x) = L$. Let $\epsilon > 0$. Let $\delta > 0$ be such that, if $0 < x < \delta$, then $|f(x) - L| < \epsilon$. It follows that, if $0 < |u| < \delta$, then $|f(|u|) - L| < \epsilon$. Thus, $\lim_{u \to 0} f(|u|) = L$.

Now suppose $\lim_{u \to 0} f(|u|) = L$. Let $\epsilon > 0$. Let $\delta > 0$ be such that, if $0 < |u| < \delta$, then $|f(|u|) - L| < \epsilon$. If $0 < x < \delta$, then $x = |x|$, and so $0 < |x| < \delta$; it follows that $|f(x) - L| = |f(|x|) - L| < \epsilon$. $\qquad\square$

Theorem 1.A.9. (Theorem 1.3.14) *Suppose that $c \in \mathbb{R}$ and $f(x) \equiv c$. Then, $\lim_{x \to b} f(x) = c$. In other words, $\lim_{x \to b} c = c$, and so constant functions are continuous.*

In addition, $\lim_{x \to b} x = b$, i.e., the identity function $f(x) = x$ is continuous.

Proof. Let $\epsilon > 0$. If $0 < |x - b| < \epsilon$, then $|c - c| < \epsilon$ and $|x - b| < \epsilon$. $\qquad\square$

Theorem 1.A.10. (Theorem 1.3.15) *Suppose that $\lim_{x \to b} f(x) = L_1$, $\lim_{x \to b} g(x) = L_2$, and that c is a real number. Then,*

1. *$\lim_{x \to b} cf(x) = cL_1$;*

2. *$\lim_{x \to b} [f(x) + g(x)] = L_1 + L_2$;*

3. *$\lim_{x \to b} [f(x) - g(x)] = L_1 - L_2$;*

4. *$\lim_{x \to b} [f(x) \cdot g(x)] = L_1 \cdot L_2$; and*

5. *if $L_2 \neq 0$, $\lim_{x \to b} [f(x)/g(x)] = L_1/L_2$.*

In addition, the theorem remains true if every limit, in the hypotheses and the formulas, is replaced by a one-sided limit, from the left or right.

Proof. We prove the two-sided limit statements. The proofs of the one-sided cases are straightforward modifications of these.

Item 1:

If $c = 0$, the result follows from Theorem 1.A.9. Assume now that $c \neq 0$. Suppose $\epsilon > 0$. Then, $\epsilon/|c| > 0$, and so there exists $\delta > 0$ such that, if $0 < |x - b| < \delta$, then $|f(x) - L_1| < \epsilon/|c|$, i.e., $|cf(x) - cL_1| < \epsilon$.

Item 2:

Suppose that $\epsilon > 0$. Then, $\epsilon/2 > 0$, and so there exist $\delta_1 > 0$ and $\delta_2 > 0$ such that, if $0 < |x - b| < \delta_1$, then $|f(x) - L_1| < \epsilon/2$, and $0 < |x - b| < \delta_2$, then $|g(x) - L_2| < \epsilon/2$. Let δ be the minimum of δ_1 and δ_2.

It follows that, if $0 < |x - b| < \delta$, then

$$|f(x) + g(x) - (L_1 + L_2)| \ \leq \ |f(x) - L_1| + |g(x) - L_2| \ < \ \epsilon/2 + \epsilon/2 \ = \ \epsilon.$$

Item 3:

This follows immediately from applying Items 1 and 2 to $f(x) + (-g(x))$.

Item 4:

First, suppose $L_1 = L_2 = 0$. Suppose $\epsilon > 0$. Then $\sqrt{\epsilon} > 0$. Let $\delta_1 > 0$ and $\delta_2 > 0$ be such that, if $0 < |x - b| < \delta_1$, then $|f(x)| < \sqrt{\epsilon}$, and $0 < |x - b| < \delta_2$, then $|g(x)| < \sqrt{\epsilon}$. Let δ be the minimum of δ_1 and δ_2. It follows that, if $0 < |x - b| < \delta$, then $|f(x)g(x)| < \epsilon$.

Now suppose that L_1 and L_2 are arbitrary. For all x,

$$f(x)g(x) - L_1 L_2 \ =$$

$$(f(x) - L_1)(g(x) - L_2) + L_1(g(x) - L_2) + L_2(f(x) - L_1).$$

Thus,

$$0 \leq |f(x)g(x) - L_1 L_2| \ \leq$$

$$|f(x) - L_1| \cdot |g(x) - L_2| + |L_1| \cdot |g(x) - L_2| + |L_2| \cdot |f(x) - L_1|.$$

By the Pinching Theorem and Proposition 1.A.5, we would be finished if we could show that the last summation above approaches 0 as $x \to b$. However, this is easy; Item 2 tells us that we can split up the sum (assuming the limits of the summands exist), the first summand approaches 0 by our special case where L_1 and L_2 were 0, while the second and third summands approach 0 by Item 1.

Item 5:

In light of Item 4, all that we need to show is that, if $L_2 \neq 0$, then $\lim_{x \to b}[1/g(x)] = 1/L_2$. We will assume that $L_2 > 0$; the proof when $L_2 < 0$ is essentially identical.

Suppose that $\epsilon > 0$. Then, $\epsilon L_2^2/(1 + \epsilon L_2) > 0$. Let $\delta > 0$ be such that, if $0 < |x - b| < \delta$, then $|g(x) - L_2| < \epsilon L_2^2/(1 + \epsilon L_2)$. We claim that, if $|g(x) - L_2| < \epsilon L_2^2/(1 + \epsilon L_2)$, then $|[1/g(x)] - [1/L_2]| < \epsilon$; this would finish the proof.

So, suppose that

$$|g(x) - L_2| < \epsilon L_2^2/(1 + \epsilon L_2),$$

that is,

$$\frac{-\epsilon L_2^2}{1 + \epsilon L_2} \;<\; g(x) - L_2 \;<\; \frac{\epsilon L_2^2}{1 + \epsilon L_2}.$$

Then,

$$\frac{L_2}{1 + \epsilon L_2} \;=\; L_2 + \frac{-\epsilon L_2^2}{1 + \epsilon L_2} \;<\; g(x) \;<\; L_2 + \frac{\epsilon L_2^2}{1 + \epsilon L_2} \;=\; \frac{L_2 + 2\epsilon L_2^2}{1 + \epsilon L_2}.$$

As all of the quantities above are positive, we may take reciprocals and reverse the inequalities to obtain

$$\frac{1 + \epsilon L_2}{L_2 + 2\epsilon L_2^2} \;<\; \frac{1}{g(x)} \;<\; \frac{1 + \epsilon L_2}{L_2},$$

and so

$$\frac{1 + \epsilon L_2}{L_2(1 + 2\epsilon L_2)} - \frac{1}{L_2} \;<\; \frac{1}{g(x)} - \frac{1}{L_2} \;<\; \frac{1 + \epsilon L_2}{L_2} - \frac{1}{L_2}.$$

This tells us that

$$\frac{-\epsilon}{1 + 2\epsilon L_2} \;<\; \frac{1}{g(x)} - \frac{1}{L_2} \;<\; \epsilon.$$

It remains for us to show that $-\epsilon \leq -\epsilon/(1 + 2\epsilon L_2)$ or, equivalently, that $1 \geq 1/(1 + 2\epsilon L_2)$, but this is trivial. $\qquad\square$

Theorem 1.A.11. (Theorem 1.3.23) *Suppose that the domain of f contains an open interval around b. Then, f is continuous at b if and only if $\lim_{x \to b} f(x) = f(b)$.*

If the domain of f is an interval of the form $[b, \infty)$, $[b, c]$, or $[b, c)$, where $c > b$, then f is continuous at b if and only if $\lim_{x \to b^+} f(x) = f(b)$.

If the domain of f is an interval of the form $(-\infty, b]$, $[a, b]$, or $(a, b]$, where $a < b$, then f is continuous at b if and only if $\lim_{x \to b^-} f(x) = f(b)$.

Proof. Suppose that the domain of f contains an open interval around b. Then the definitions of $\lim_{x \to b} f(x) = f(b)$ and of f being continuous at b exactly agree **except** for what happens when $x - b = 0$, i.e., when $x = b$. However, when $x = b$, $|f(x) - f(b)| = 0$, which is certainly less than any positive ϵ.

The one-sided statements follow in a similar manner. $\qquad\square$

Theorem 1.A.12. (Theorem 1.3.25) *Suppose that*

1. $\lim_{y \to M} f(y) = L$;

2. $\lim_{x \to b} g(x) = M$; *and*

3. *(a) f is continuous at M; or*

   (b) for all x in some deleted open interval around b, $g(x) \neq M$.

Then, $\lim_{x \to b} f(g(x)) = L$.

Proof. Suppose that $\epsilon > 0$. Then, there exists $\delta > 0$ such that, if $0 < |y - M| < \delta$, then y is in the domain of f and $|f(y) - L| < \epsilon$. As $\delta > 0$, there exists $\eta > 0$ such that, if $0 < |x - b| < \eta$, then x is in the domain of g and $|g(x) - M| < \delta$.

Therefore, if $0 < |x - b| < \eta$, and $g(x) \neq M$, then $0 < |g(x) - M| < \delta$, and so we may substitute $g(x)$ for our previous y to obtain that $|f(g(x)) - L| < \epsilon$. This proves the assertion under the hypothesis in Item 3b.

If f is continuous at M, then the second sentence of our proof could be replaced by: There exists $\delta > 0$ such that, if $|y - M| < \delta$, then y is in the domain of f and $|f(y) - L| < \epsilon$; that is, we can remove the condition that $0 < |y - M|$. The proof now proceeds as before. $\qquad \square$

Theorem 1.A.13. *A real function $f : A \to B$ is continuous (using Definition 1.3.20) if and only if, for every open subset V of B, $f^{-1}(V)$ is an open subset of A.*

Proof. Suppose that f is continuous, using the definition in Definition 1.3.20). Suppose that V is an open subset of B. Let $a \in f^{-1}(V)$. As V is open, there exists $\epsilon > 0$ such that $B \cap (f(a) - \epsilon, f(a) + \epsilon) \subseteq V$. As f is continuous at a, there exists $\delta > 0$ such that, if $x \in A$ and $|x - a| < \delta$, then $|f(x) - f(a)| < \epsilon$. In terms of intervals, this says: if $x \in A \cap (a - \delta, a + \delta)$, then $f(x) \in (f(a) - \epsilon, f(a) + \epsilon)$, i.e.,

$$A \cap (a - \delta, a + \delta) \subseteq f^{-1}\big(B \cap (f(a) - \epsilon, f(a) + \epsilon)\big) \subseteq f^{-1}(V).$$

Therefore, $f^{-1}(V)$ is an open subset of A.

Suppose now that, for every open subset V of B, $f^{-1}(V)$ is an open subset of A. Let $a \in A$, and let $\epsilon > 0$. Then, $B \cap (f(a) - \epsilon, f(a) + \epsilon)$ is an open subset of B. Hence,

$$f^{-1}\big(B \cap (f(a) - \epsilon, f(a) + \epsilon)\big)$$

is an open subset of A, which contains a. Therefore, there exists $\delta > 0$ such that

$$A \cap (a - \delta, a + \delta) \subseteq f^{-1}\big(B \cap (f(a) - \epsilon, f(a) + \epsilon)\big).$$

However, this says precisely that, if x is in A and $|x - a| < \delta$, then $|f(x) - f(a)| < \epsilon$. $\quad\square$

As an immediate corollary, we have:

Corollary 1.A.14. *Compositions of continuous functions are continuous.*

Theorem 1.A.15. *Suppose that f is continuous, and I is an interval contained in the domain of f. Then, $f(I)$ is an interval.*

Proof. By restricting the domain and/or codomain of f, we still obtain a continuous function. Thus, we may assume that I is the domain of f, and that f is onto. We will prove the theorem by contradiction.

Suppose that $f(I)$ is not an interval. Then, $f(I)$ is disconnected, i.e., there exist non-empty open subsets Y_1 and Y_2 of $f(I)$ such that $f(I) = Y_1 \cup Y_2$ and $Y_1 \cap Y_2 = \emptyset$. As f is continuous, $f^{-1}(Y_1)$ and $f^{-1}(Y_2)$ are open sets,

$$I = f^{-1}(f(I)) = f^{-1}(Y_1 \cup Y_2)) = f^{-1}(Y_1) \cup f^{-1}(Y_2),$$

and

$$\emptyset = f^{-1}(\emptyset) = f^{-1}(Y_1 \cap Y_2)) = f^{-1}(Y_1) \cap f^{-1}(Y_2).$$

In addition, neither $f^{-1}(Y_1)$ nor $f^{-1}(Y_2)$ is empty, as we are assuming f is onto, and neither Y_1 nor Y_2 is empty.

Therefore, I is disconnected; a contradiction of I being an interval. $\quad\square$

Corollary 1.A.16. (Intermediate Value Theorem) *Suppose that f is continuous, and that the domain, I, of f is an interval. Suppose that $a, b \in I$, $a < b$, and y is a real number such that $f(a) < y < f(b)$ or $f(b) < y < f(a)$. Then, there exists c such that $a < c < b$ (and, hence, c is in the interval I) such that $f(c) = y$.*

Proof. By Theorem 1.A.15, $f([a, b])$ is an interval, which certainly contains $f(a)$ and $f(b)$. If $f(a) < y < f(b)$ or $f(b) < y < f(a)$, then, by definition of an interval, $y \in f([a, b])$, i.e., there exists c such that $c \in [a, b]$ and $f(c) = y$. Now, c cannot be equal to a or b because of the strict inequalities. Thus, $c \in (a, b)$. $\qquad\square$

Theorem 1.A.17. *Suppose that f is continuous, and Y is a compact set contained in the domain of f. Then, $f(Y)$ is compact.*

Proof. We may once again assume that Y is the domain of f, and that f is onto.

Suppose that $f(Y) = \bigcup_{j \in J} U_j$, where each U_j is open in $f(Y)$. Then, $Y = \bigcup_{j \in J} f^{-1}(U_j)$, and each $f^{-1}(U_j)$ is open in Y. As Y is compact, there exists a finite subcollection $f^{-1}(U_{j_1})$, $\ldots$, $f^{-1}(U_{j_k})$ such that

$$Y = f^{-1}(U_{j_1}) \cup \cdots \cup f^{-1}(U_{j_k}).$$

Thus,

$$f(Y) = f\big(f^{-1}(U_{j_1})\big) \cup \cdots \cup f\big(f^{-1}(U_{j_k})\big) = U_{j_1} \cup \cdots \cup U_{j_k},$$

where the last equality uses that we are assuming that f is onto. $\qquad\square$

Corollary 1.A.18. **(Extreme Value Theorem)** *If the closed interval $[a, b]$ is contained in the domain of the a continuous function f, then $f([a, b])$ is a closed, bounded interval, i.e., an interval of the form $[m, M]$. Therefore, f attains a minimum value m and a maximum value M on the interval $[a, b]$.*

Proof. This follows immediately from Theorem 1.A.15 and the Heine-Borel Theorem, which tells us that the compact subsets of $\mathbb{R}$ are precisely the closed, bounded subsets of $\mathbb{R}$. $\qquad\square$

Theorem 1.A.19. *Suppose that $g : I \to J$ is a continuous, one-to-one, and onto real function, and I is an interval. Then, g is strictly monotonic, J is an interval of the same type as I (open, half-open, or closed), and $g^{-1} : J \to I$ is continuous.*

Proof. We shall prove the monotonicity (which has to be strict, since g is one-to-one) by contradiction. So, suppose to the contrary that we have a, b, c, and d in I such that $a < b$, $c < d$, $g(a) < g(b)$, and $g(c) > g(d)$.

As the domain I is an interval, for all $t \in [0, 1]$, $td + (1-t)b$ and $tc + (1-t)a$ are in I. Define the function $f : [0, 1] \to \mathbb{R}$ by

$$f(t) = g(td + (1-t)b) - g(tc + (1-t)a).$$

This is a difference of compositions of continuous functions and, hence, is continuous. Since $f(0) = g(b) - g(a) > 0$ and $f(1) = g(d) - g(c) < 0$, the Intermediate Value Theorem tells us that there exists $t_0 \in (0, 1)$ such that $f(t_0) = 0$. Thus,

$$g(t_0 d + (1 - t_0)b) = g(t_0 c + (1 - t_0)a).$$

As g is one-to-one, this implies that $t_0 d + (1 - t_0)b = t_0 c + (1 - t_0)a$, that is,

$$(1 - t_0)(b - a) = t_0(c - d).$$

However, t_0 and $1 - t_0$ are positive, $b - a$ is positive, while $c - d$ is negative; a contradiction. Therefore, g is strictly monotonic.

We will assume that g is monotonically increasing; the monotonically decreasing case is essentially the same.

It follows immediately that, for all intervals $[a, b]$ in I, $g([a, b]) = [g(a), g(b)]$, $g((a, b)) = (g(a), g(b))$, $g([a, b)) = [g(a), g(b))$, and $g((a, b]) = (g(a), g(b)]$. We conclude that J must be the same type of interval as I, and that g^{-1} of open subsets of J are open subsets of I, i.e., that g^{-1} is continuous. $\square$

Corollary 1.A.20. *Suppose that $q : I \to J$ is a continuous, one-to-one, and onto real function, and I is an open interval. Suppose that $b \in I$. Let m be a real function whose domain contains a deleted open interval A around $q(b)$. Then, there exists a deleted open interval B around b such that $q(B) \subseteq A$, and $\lim_{u \to q(b)} m(u) = L$ if and only if $\lim_{t \to b} m(q(t)) = L$.*

Proof. The statement about $q(B) \subseteq A$ is immediate, since q is continuous and one-to-one.

Assume that $\lim_{u \to q(b)} m(u) = L$. Then, Theorem 1.A.12, with $M = q(b)$ and using Item 3b, immediately implies that $\lim_{t \to b} m(q(t)) = L$.

Now, assume that $\lim_{t \to b} m(q(t)) = L$. By Theorem 1.A.19, q^{-1} is continuous. Suppose that $\epsilon > 0$. Let $\delta_1 > 0$ be such that, if $0 < |t - b| < \delta_1$, then $|m(q(t)) - L| < \epsilon$. Letting $u = q(t)$, we have that, if $0 < |q^{-1}(u) - b| < \delta_1$, then $|m(u) - L| < \epsilon$. As q^{-1} is continuous at $q(b)$, there exists $\delta > 0$ such that, if $|u - q(b)| < \delta$, then $|q^{-1}(u) - q^{-1}(q(b))| = |q^{-1}(u) - b| < \delta_1$. Note that, since q^{-1} is one-to-one, if $u \neq q(b)$, then $q^{-1}(u) \neq b$.

Therefore, we find that, if $0 < |u - q(b)| < \delta$, then $0 < |q^{-1}(u) - b| < \delta_1$, which implies that $|m(u) - L| < \epsilon$. Hence, $\lim_{u \to q(b)} m(u) = L$. $\square$

The theorems on limits involving the extended real numbers break up into many cases. The proofs in these cases are relatively easy, or are mild modifications of the proofs over the real numbers. We give two such proofs, and leave the remainder as exercises for the industrious reader.

Theorem 1.A.21. *Suppose that b is a real number and that $f(x) \neq 0$ for all x in some deleted open interval around b. Then, $\lim_{x \to b} f(x) = 0$ if and only if $\lim_{x \to b} \dfrac{1}{|f(x)|} = \infty$.*

Proof. Assume that $\lim_{x \to b} f(x) = 0$. Suppose that $N > 0$. Then, $1/N > 0$, and so there exists $\delta > 0$ such that, if $0 < |x - b| < \delta$, then $|f(x) - 0| < 1/N$. By our hypothesis on the deleted open interval around b, we may also choose δ small enough so that, if $0 < |x - b| < \delta$, then $f(x) \neq 0$. Therefore, if $0 < |x - b| < \delta$, we conclude that $1/|f(x)| > N$; this shows that $\lim_{x \to b} \dfrac{1}{|f(x)|} = \infty$.

Assume now that $\lim_{x \to b} \dfrac{1}{|f(x)|} = \infty$. Suppose that $\epsilon > 0$. Then, $1/\epsilon > 0$, and so there exists $\delta > 0$ such that, if $0 < |x - b| < \delta$, then $1/|f(x)| > 1/\epsilon$, i.e., $|f(x)| < \epsilon$. Therefore, $\lim_{x \to b} f(x) = 0$. $\qquad\square$

Theorem 1.A.22. *Suppose that b is a real number, that $\lim_{x \to b} f(x) = L$, where L is a positive real number, and that $\lim_{x \to b} g(x) = \infty$. Then, $\lim_{x \to b} f(x)g(x) = \infty$.*

Proof. Suppose that $N > 0$. Then, $2N/L > 0$, and certainly $L/2 > 0$. Thus, there exist δ_1.) and $\delta_2 > 0$ such that, if $0 < |x - b| < \delta_1$, then $|f(x) - L| < L/2$, and, if $0 < |x - b| < \delta_2$, then $g(x) > 2N/L$. Let δ be the minimum of δ_1 and δ_2.

Suppose that $0 < |x - b| < \delta$. Then, $-L/2 < f(x) - L < L/2$ and $g(x) > 2N/L$. It follows that $f(x) > L/2$, and $f(x)g(x) > (L/2)(2N/L) = N$. Therefore, $\lim_{x \to b} f(x)g(x) = \infty$. $\qquad\square$

1.A.4   Proofs of Theorems on Differentiability and Continuity

Lemma 1.A.23. *Suppose that f is differentiable at a, and let I be an open interval around a which is contained in the domain of f. Then, there exists a unique function $E(x)$, defined on I and continuous at a, such that, for all $x \in I$,*

$$f(x) = f(a) + f'(a)(x - a) + E(x)(x - a). \tag{1.6}$$

Moreover, $E(a) = 0$.

Proof. For $x \neq a$, we have

$$E(x) = \frac{f(x) - f(a)}{x - a} - f'(a).$$

Since we want $E(x)$ to be continuous at a, we need two things: for $E(a)$ to exist and for $\lim_{x \to a} E(x)$ to equal $E(a)$. As f is differentiable at a, $\lim_{x \to a} E(x) = 0$.

Thus, we need to define $E(a)$ to be equal to 0, and we are finished. $\qquad\square$

Theorem 1.A.24. *If f is differentiable at x, then f is continuous at x.*

Equivalently, if f is not continuous at x, then f is not differentiable at x.

Proof. This is immediate from the lemma, since the right-hand side of Formula 1.6 is continuous at a. $\qquad\square$

If f is differentiable, then f must be continuous; however, f' itself need not be continuous. On the other hand, f' always has the following intermediate value property:

Theorem 1.A.25. (Intermediate Value Theorem for Derivatives) *Suppose that f is differentiable, and that I is an interval contained in the domain of f. Then, $f'(I)$ is an interval.*

Proof. Suppose that a and b are in I, and $f'(a) < v < f'(b)$. We will assume that $a < b$; the case where $b < a$ is completely analogous.

We will show that there exists $c \in (a, b)$ (and, hence, in I) such that $f'(c) = v$.

Let $g(x) = f(x) - vx$. Then, $g'(a) < 0$ and $g'(b) > 0$. Since $\lim_{h \to 0} (g(a + h) - g(a))/h < 0$, there exists $h > 0$ such that $r := a + h < b$ and $g(r) < g(a)$. Since $\lim_{h \to 0} (g(b+h) - g(b))/h > 0$, there exists $h < 0$ such that $s = b + h > a$ and $g(s) < g(b)$. Therefore, the continuous function g does not attain its minimum value on $[a, b]$ at either a or b. Hence, by the Extreme Value Theorem, g attains its minimum value at some $c \in (a, b)$ and, by Theorem 1.5.4, $g'(c) = 0$, i.e., $f'(c) = v$. $\qquad\square$

Now we will show that, if f is differentiable, then the only way that f' can fail to be continuous is for the limit $\lim_{x \to a} f'(x)$ to not exist at one or more points a.

Theorem 1.A.26. *Suppose that f is differentiable on a deleted open left (respectively, right) interval of a, and that the limits, as x approaches a from the left (respectively, right), of $f'(x)$ and $\left[f(x) - f(a) \right]/(x - a)$ exist. Then, the two limits are equal.*

Proof. We will prove the result for the limits from the left; the proof from the right is completely analogous. Let I denote a deleted open left interval of a on which f is differentiable.

Let

$$L_1 = \lim_{x \to a^-} f'(x) \quad \text{and} \quad L_2 = \lim_{x \to a^-} \frac{f(x) - f(a)}{x - a}.$$

We will show that, for all $\epsilon > 0$, $|L_1 - L_2| < \epsilon$.

Suppose that $\epsilon > 0$. Then, $\epsilon/2 > 0$, and so, there exist $\delta_1 > 0$ and $\delta_2 > 0$ such that $(a - \delta_1, a)$ and $(a - \delta_2, a)$ are contained in I and

$$0 < a - x < \delta_1 \implies |f'(x) - L_1| < \frac{\epsilon}{2} \quad \text{and} \quad 0 < a - x < \delta_2 \implies \left| \frac{f(x) - f(a)}{x - a} - L_2 \right| < \frac{\epsilon}{2}.$$

Let $\delta > 0$ be the minimum of δ_1 and δ_2, so that $(a - \delta, a)$ is contained in I and

$$0 < a - x < \delta \implies |f'(x) - L_1| < \frac{\epsilon}{2} \quad \text{and} \quad \left| \frac{f(x) - f(a)}{x - a} - L_2 \right| < \frac{\epsilon}{2}. \tag{1.7}$$

Let x_0 be such that $0 < a - x_0 < \delta$, i.e., such that x_0 is in the interval $(a - \delta, a) \subseteq I$. Then, as f is differentiable on I, f is continuous on $[x_0, a)$. We claim, in fact, that f is continuous on the **closed** interval $[x_0, a]$, i.e., that $\lim_{x \to a^-} f(x) = f(a)$; we want this so that we can apply the Mean Value Theorem, Theorem 1.5.10, to f on $[x_0, a]$. We need to prove a one-sided version of Lemma 1.A.23.

Let $E(x)$ be the function defined on $(a - \delta, a]$ such that $E(a) = 0$ and, for all x in $(a - \delta, a)$,

$$E(x) = \frac{f(x) - f(a)}{x - a} - L_2.$$

Then, $E(x)$ is continuous on $(a - \delta, a]$ and, for all x in $(a - \delta, a]$,

$$f(x) = f(a) + (x - a)L_2 + (x - a)E(x).$$

It follows immediately that $f(x)$ is continuous on $(a - \delta, a]$.

Therefore, f is differentiable on (x_0, a) and continuous on $[x_0, a]$, and we may apply the

Mean Value Theorem to conclude that there exists c in the open interval (x_0, a) such that

$$f'(c) = \frac{f(x_0) - f(a)}{x_0 - a}. \tag{1.8}$$

Now, certainly $0 < a - c < \delta$ and $0 < a - x_0 < \delta$. Applying Formula 1.7, we conclude that

$$|f'(c) - L_1| < \frac{\epsilon}{2} \quad \text{and} \quad \left| \frac{f(x_0) - f(a)}{x_0 - a} - L_2 \right| < \frac{\epsilon}{2}.$$

By Formula 1.8, these are equivalent to

$$|f'(c) - L_1| < \frac{\epsilon}{2} \quad \text{and} \quad |f'(c) - L_2| < \frac{\epsilon}{2}.$$

Finally, the Triangle Inequality yields what we desired:

$$|L_1 - L_2| = \left| (L_1 - f'(c)) + (f'(c) - L_2) \right| \leq \left| L_1 - f'(c) \right| + \left| f'(c) - L_2 \right| < \frac{\epsilon}{2} + \frac{\epsilon}{2} = \epsilon$$

$\square$

The following two corollaries are immediate.

Corollary 1.A.27. *If f is differentiable on an open interval around a, and f' is not continuous at a, then $\lim_{x \to a} f'(x)$ does not exist.*

Corollary 1.A.28. *Suppose that f is differentiable on a deleted open interval around a. If the two one-sided limits $\lim_{x \to a^-} f'(x)$ and $\lim_{x \to a^+} f'(x)$ exist and are unequal, then f is not differentiable at a.*

Chapter 2

Basic Rules for Calculating Derivatives

In the last chapter, we developed the notion of the derivative, the instantaneous rate of change. We did not discuss many complicated examples, since the computation of the derivative required us to perform substantial algebraic manipulations.

In this chapter, we develop a few fairly simple rules for calculating derivatives, rules which make the calculation of the instantaneous rate of change far more manageable.

2.1   The Power Rule and Linearity

In this section, we shall see that, when you want to differentiate a sum or difference of functions, you can split up the sum or difference and differentiate each piece and, if you have a constant multiplied times a function f, the derivative is simply the constant times the derivative of f. The combination of these two properties is referred to as the *linearity of differentiation*.

We shall also look at our first case of the *Power Rule*, which tells us how to differentiate x^n, where n is a positive integer. By combining the Power Rule with linearity, we can quickly differentiate any polynomial function.

In Proposition 1.4.6, we showed that, if c is a constant, then $(c)' = 0$, $(x)' = 1$, and $(x^2)' = 2x$. These latter two formulas are special cases of the *Power Rule*:

Theorem 2.1.1. (Power Rule for Natural Exponents) *Suppose that n is a natural number (i.e., a positive integer). Then,*

$$(x^n)' = nx^{n-1},$$

with the understanding that, when $n = 1$, x^0 is the constant function 1 (i.e., we agree that, in this particular formula, even 0^0 is defined and equals 1).

Proof. This proof relies on understanding a couple of aspects of the *binomial expansion* of $(x + h)^n$. Before we discuss the general case, let's look at the cases where $n = 2$ and $n = 3$. We find

$$(x + h)^2 = x^2 + 2xh + h^2,$$

and

$$(x + h)^3 = (x + h)(x + h)^2 = (x + h)(x^2 + 2xh + h^2) = x^3 + 3x^2h + 3xh^2 + h^3.$$

What's important to us about this? In both of these cases, we see that $(x + h)^n$ is of the form $x^n + nx^{n-1}h + h^2p(x, h)$, where $p(x, h)$ is a polynomial in powers of h, with coefficients which involve x. When $n = 2$, the function $p(x, h)$ is simply the constant function 1; when $n = 3$, we see that $p(x, h)$ is equal to $3x + h$.

In general, it can be shown, by induction, that

$$(x + h)^n = x^n + nx^{n-1}h + h^2p(x, h),$$

where, again, $p(x, h)$ is a polynomial in powers of h with coefficients that involve x.

Thus,

$$(x^n)' = \lim_{h \to 0} \frac{(x + h)^n - x^n}{h} = \lim_{h \to 0} \frac{x^n + nx^{n-1}h + h^2p(x, h) - x^n}{h} =$$

$$\lim_{h \to 0} \left(nx^{n-1} + hp(x, h) \right) = nx^{n-1},$$

where we used that the limit of the polynomial (in h) $\lim_{h \to 0} p(x, h) = p(x, 0)$ exists.	$\square$

Example 2.1.2. Suppose that $f(t) = t^5$ and $g(w) = w^{97}$. Then, using the Power Rule, we quickly find that $f'(t) = 5t^4$, and $g'(w) = 97w^{96}$.

Remark 2.1.3. We usually extend the Power Rule to include the case where $n = 0$. That is, we think as follows:

$$(x^0)' = (1)' = 0 = 0x^{0-1}.$$

Of course, there is once again a problem when $x = 0$, but, nonetheless, we think of the fact that $(1)' = 0$ as a degenerate case of the Power Rule. This is helpful only in the sense that, when we say that a result follows from the Power Rule, we mean to include that we may also be using that $(c)' = c \cdot (1)' = 0$.

We shall revisit the Power Rule three more times: once for negative integer powers in Section 2.2, again for rational powers in Section 2.3, and a final time for arbitrary real powers (with domain restricted to the non-negative real numbers), again in Section 2.3.

In Proposition 1.4.12 and Proposition 1.4.14, we showed that, for a constant c, $(cf(x))' = cf'(x)$ and that $(f(x) + g(x))' = f'(x) + g'(x)$. The combination of these two results tells us that differentiation is what's known as a *linear operation*.

Theorem 2.1.4. (Linearity) *Suppose that a and b are constants, and that f and g are differentiable at x. Then, $af + bg$ is differentiable at x, and*

$$(af + bg)'(x) = af'(x) + bg'(x).$$

Linearity is commonly described by "when taking derivatives, you can split up sums and differences, and pull out constants".

Example 2.1.5. Let's calculate

$$\frac{d}{dt}\left(5t^3 - \sqrt{7}\,t^2\right).$$

By linearity, this derivative is equal to

$$5 \frac{d}{dt}\left(t^3\right) - \sqrt{7}\,\frac{d}{dt}\left(t^2\right).$$

Applying the Power Rule twice, we find that this equals

$$5 \cdot 3t^2 - \sqrt{7} \cdot 2t^1 \;=\; 15t^2 - 2\sqrt{7}\,t.$$

In fact, by combining the Power Rule for Natural Powers with linearity, and the fact that the derivatives of constant functions are zero, we can now differentiate any polynomial function $p(x) = a_n x^n + \cdots + a_1 x + a_0$, where $a_0, a_1, \ldots, a_n$ are constants.

Example 2.1.6. Consider the polynomial function

$$p(x) = 6x^8 - 3x^5 + \pi x^3 + 2x^2 - x + \sqrt{2}.$$

Linearity, applied several times, tells us that

$$p'(x) = 6(x^8)' - 3(x^5)' + \pi(x^3)' + 2(x^2)' - (x)' + (\sqrt{2})'.$$

Now, the Power Rule, together with the fact that the derivative of the constant $\sqrt{2}$ is 0, gives us

$$p'(x) = 6 \cdot 8x^7 - 3 \cdot 5x^4 + \pi \cdot 3x^2 + 2 \cdot 2x - 1 + 0 =$$
$$48x^7 - 15x^4 + 3\pi x^2 + 4x - 1.$$

Before we state the general result about differentiating polynomials as a corollary, it will be useful to introduce the *sigma notation* for summations.

> **Definition 2.1.7.** *Suppose that we have two integers m and n, where $m \leq n$, and we have a real function B, whose domain includes all of the integers i such that $m \leq i \leq n$. Then, we write $\sum_{i=m}^{n} B(i)$ for the summation, as i goes from m to n, of $B(i)$. This means exactly what it says:*
>
> $$\sum_{i=m}^{n} B(i) = B(m) + B(m+1) + \cdots + B(n-1) + B(n).$$

For example,

$$\sum_{i=-1}^{3} i^2 = (-1)^2 + 0^2 + 1^2 + 2^2 + 3^2 = 15,$$

or

$$p(x) = a_0 + a_1 x + \cdots + a_n x^n = \sum_{i=0}^{n} a_i x^i,$$

where we once again agree that x^0 in the summation is to be interpreted as equaling 1, even if $x = 0$.

Note that the indexing variable i is a *dummy variable*; if we replaced it with a j or k, or any other variable (which is not already present), the summation would not change.

From linearity and the Power Rule, we conclude:

> **Theorem 2.1.8.** *Suppose that $p(x)$ is the polynomial function*
>
> $$p(x) = a_0 + a_1 x + a_2 x^2 + \cdots a_{n-1} x^{n-1} + a_n x^n = \sum_{i=0}^{n} a_i x^i.$$
>
> *Then,*
>
> $$p'(x) = a_1 + 2a_2 x + \cdots + a_{n-1}(n-1)x^{n-2} + a_n n x^{n-1} = \sum_{i=1}^{n} a_i i x^{i-1}.$$

Example 2.1.9. In Example 1.1.11, we considered the radius of a spherical balloon as a function of its volume. In this example, consider the volume, V, in cubic inches, in terms of the radius R, in inches. We know that

$$V = \frac{4}{3}\pi R^3.$$

What is the instantaneous rate of change of V, with respect to R, when $R = 5$ in?

Solution:

We want to calculate dV/dR, when $R = 5$ in.

By linearity, we may "pull out" the constant $4\pi/3$:

$$\frac{dV}{dR} = \frac{4}{3}\pi \cdot \frac{d}{dR}(R^3).$$

By the Power Rule,

$$\frac{d}{dR}(R^3) = 3R^2.$$

Thus, we find

$$\frac{dV}{dR} = \frac{4}{3}\pi \cdot \frac{d}{dR}(R^3) = \frac{4}{3}\pi \cdot 3R^2 = 4\pi R^2 \ \text{in}^3/\text{in}.$$

Therefore,

$$\frac{dV}{dR}\bigg|_5 = 100\pi \ \text{in}^3/\text{in}.$$

Example 2.1.10. In Example 1.1.7, we considered the area A, in square inches, of a wide-screen television with a diagonal length of d inches. We found that $A = A(d) = 144d^2/337$.

It is now easy for us to calculate the instantaneous rate of change of A, with respect to d. We find

$$A'(d) = \left(\frac{144}{337}d^2\right)' = \left(\frac{144}{337}2d\right) = \frac{288\,d}{337} \ \text{in}^2/\text{in}.$$

Note that we did **not** write dA/dd. When d is a variable in the problem, Leibniz's notation is too confusing. Though, we could have first changed the name of the independent variable, and then used Leibniz's notation.

In Example 1.1.7, we also looked at the average rate of change, divided by the (initial) value of A, in order to get a *relative* or *fractional rate of change of A, with respect to d.* The fractional instantaneous rate of change of A, with respect to d, $A'(d)/A$ is a measure of the instantaneous rate of change in A as a fraction of A.

As we discussed in Example 1.1.7, the notion of the fractional rate of change is what you have in mind when you think "adding inches of diagonal length to a television increases the size of the tv screen by a bigger percentage for small screens than for large screens". This is easy for us to see now; $A(d)$ is equal to a constant c times d^2, and we find

$$\frac{A'(d)}{A(d)} = \frac{c \cdot 2d}{cd^2} = \frac{2}{d},$$

where the units are 1/in. It follows that, as d gets bigger, the fractional rate of change gets smaller.

Example 2.1.11. In Example 1.6.5, we looked at the concavity of a "cubic graph". We did not specify the cubic polynomial, as we did not want to differentiate a cubic before we had the Power Rule at our disposal.

The cubic polynomial whose graph appears in Example 1.6.5 is

$$y = f(x) = (x + 2)(x - 1)(x - 3) = x^3 - 2x^2 - 5x + 6.$$

We looked at the graph and estimated that the second derivative of $f(x)$ was negative for $x < 1/2$ and positive for $x > 1/2$. Let's see how well we estimated.

We calculate

$$y' = 3x^2 - 4x - 5, \qquad y'' = 6x - 4.$$

We see that, in fact, $y'' < 0$ if $x < 2/3$, and $y'' > 0$ if $x > 2/3$. So, our estimate that y'' switched signs at $x = 1/2$ was not so bad, but not so good.

Of course, the physical significance of such a calculation depends on what f and x represent. If x denotes the time in seconds, and f is the position of a moving object, in feet, then y' is the velocity of the object, in ft/s, and y'' is its acceleration, in (ft/s)/s. The fact that $y'' < 0$ for $x < 2/3$ would then mean that the object was decelerating at times before 2/3 of a second.

2.1.1 Exercises

Each of the functions in Exercises 1 through 14 is either a polynomial, simplifies to a polynomial function with a restricted domain, or is a piecewise polynomial function. For each function, determine its domain, and its first and second derivatives. You should also give the domains of these derivatives.

1. $f(x) = 9x^4 + 3x^2 + 2x + 1$.

2. $f(x) = 10x^{10} - 5x^8 + 2x^6$.

3. $f(x) = 2 + x$.

4. $f(x) = x(x - 2)(x + 2)$.

5. $f(x) = \dfrac{x^3 - 3x^2 - 9x + 27}{x^2 - 6x + 9}$.

6. $f(x) = 7x - 5x^5 + x^2 + cx^3$ (where c is a constant).

7. $f(x) = (x + a)^4$ (where a is a constant).

8. $f(x) = \dfrac{2x^2 - 6x + 10x^3}{x}$.

9. $f(x) = \dfrac{x^3 + 3x^2 - 4x - 12}{x^2 + x - 6}$.

10. $f(x) = (x^3 + 4)^2$.

11. $f(x) = \sqrt{x^2 + 4x + 4}$.

12. $f(x) = x^2 \cdot \left(x + \dfrac{1}{x}\right)^2$.

13. $f(x) = \left|x^2 + 3x - 10\right|$.

14.
$$f(x) = \begin{cases} x^2, & \text{if } x < 1; \\ 2x - 1, & \text{if } x \geq 1. \end{cases}$$

In Exercises 15-17, you are given the product of two functions. First, "multiply out" (distribute) the product to obtain a single polynomial, and then apply Theorem 2.1.8, to calculate the derivative. Next, calculate the derivative of each factor of the initial product, and verify that the derivative of the product (which you found first) is not the product of the derivatives of the factors.

15. $f(x) = (x^2 + 5x)(x + 2)$.

16. $f(x) = (x^3 + 2x + 3)(x^3 - 2x + 3)$.

17. $f(x) = x(x - 4)$.

18. The **degree** of a polynomial is the largest power of x that appears with a non-zero coefficient in the polynomial. Thus, if a_k denotes the coefficient of x^k in the polynomial, a degree n polynomial has $a_n \neq 0$, but $a_k = 0$ for all $k \geq n + 1$. What can you say about the n-th derivative of an n-th degree polynomial? What about the $(n+1)$-st derivative of an n-th degree polynomial?

In Exercises 19 - 21, use the Power Rule and linearity of the derivative to find a function that has the given derivative.

19. $f'(x) = 11x$.

20. $v'(t) = -40t^4$.

21. $m'(x) = 1 + 3x^2$.

22. Show that one function that has derivative x^n, where $n > 0$ is an integer, is $f(x) = \dfrac{x^{n+1}}{n+1}$.
    Is this the only function that has derivative x^n?

Recall that in probability theory, the density function of a distribution is the derivative of the cumulative distribution function. In Exercises 23 - 25 you are given the cumulative distribution function; calculate the density function.

23. $F(x) = (x+1)(x+2)$.

24. $F(x) = -\dfrac{1}{32}x^3 + \dfrac{3}{8}x$.

25. $F(x) = \dfrac{5}{9}x^6 - \dfrac{2}{3}x^5 + \dfrac{10}{9}x^3$.

In Exercises 26 - 29, use Definition 2.1.7 to write out each term of the series and calculate the sum.

26. $\displaystyle\sum_{i=4}^{8} 3i^2$.

27. $\displaystyle\sum_{i=1}^{5} \dfrac{(-1)^i}{i}$.

28. $\displaystyle\sum_{i=2}^{7} 6$.

29. $\displaystyle\sum_{j=-4}^{4} j^4$.

30. Show that summation is linear. Namely, show that if B and C are two real functions and k is any constant, then

    a.
    $$\sum_{i=m}^{n} [B(i) + C(i)] = \left[\sum_{i=m}^{n} B(i)\right] + \left[\sum_{i=m}^{n} C(i)\right]$$

b.

$$\sum_{i=m}^{n} [kB(i)] = k \left[\sum_{i=m}^{n} B(i) \right].$$

31. Let $F(n) = \sum_{i=1}^{n} i$. The function F produces the *triangular numbers*. Show that

$$F(n) = \frac{n(n+1)}{2}.$$

In Exercises 32 through 36, you are given the derivative of several functions that will be discussed in greater detail in later chapters. Use the given information and Theorem 2.1.4 to calculate the derivative of $f(x)$.

32. $(\sin x)' = \cos x$, $(\cos x)' = -\sin x$, $f(x) = 6 \sin x + 4x^6 - 9 \cos x$.

33. $(e^x)' = e^x$, $(e^{-x})' = -e^{-x}$, $f(x) = \dfrac{e^x - e^{-x}}{2}$.

34. $(x^n)' = nx^{n-1}$, where $n < 0$ is an integer, $f(x) = \dfrac{1}{x} + \dfrac{1}{2x^2} + \dfrac{1}{3x^3}$.

35. $(\tan^{-1} x)' = \dfrac{1}{1+x^2}$, $(\cot^{-1} x)' = -\dfrac{1}{1+x^2}$, $f(x) = 7 \tan^{-1} x - 11 \cot^{-1} x$.

36. Assume $a > 0, x > 0$. $(a^x)' = (a^x) \ln a$, $(\ln x)' = 1/x$, $f(x) = (\ln a)(\ln x) + (\ln a)(a^x)$.

37. Recall that the surface area of a closed right cylinder with radius r and height h is given by

$$A(r, h) = 2\pi rh + 2\pi r^2.$$

a. Treat r as a constant and calculate dA/dh.

b. Treat h as a constant and calculate dA/dr.

c. What is the rate of change of the area, with respect to the radius, when the radius is 4 inches and the height is 12 inches?

38. Does the Power Rule imply that the derivative of the function $f(x) = 2^x$ is $x \cdot 2^{x-1}$? Why or why not?

39. Use the Intermediate Value Theorem to argue that the function $h(x) = \dfrac{5}{4}x^3 - x^2 - 2x - 9$ has a critical point in the interval $(-1, 1)$.

40. Is it true that taking higher-order derivatives is a linear operation? Namely, is it true that if f and g are both n times differentiable functions that $(af+bg)^{(n)}(x) = af^{(n)}(x) + bg^{(n)}(x)$?

41. Use the linearity of the derivative and higher-order derivatives to prove that, if $y = f(x)$ and $y = g(x)$ are two sufficiently differentiable functions that both satisfy the differential equation $y'' + 5y' + y = 0$, then any function of the form $af + bg$ also satisfies the differential equation.

In Exercises 42 through 44, identify points where the second derivative is zero and determine whether or not those points are inflection points of the graph of the function.

42. $h(x) = \dfrac{1}{2}x^3 - \dfrac{9}{2}x^2 + 2x - 11$.

43. $k(t) = t^4 - 4t^3 + 6t^2 + 7t - 11$.

44. $j(t) = t^4 - 6t^2 + 7t - 11$.

45. Prove that any polynomial of the form $f(x) = ax^3 + bx^2 + cx + d$, where $a \neq 0$, has an inflection point and that the inflection point is independent of c and d.

46. A pile of sand sitting on a large floor forms the shape of a (right, circular) cone. The pile sits at a fixed height of 10 meters, and sand is added in such a way that only the radius grows. Find the rate of change of volume of the cone, as a function of the radius. (Recall that the volume of a cone is 1/3 of the area of the base times the height.) ▶

47. A piece of paper measuring 6 inches by 8 inches is used to make a box (with no top) in the following way: a square with sides of length x inches ($0 \leq x \leq 3$) is cut out from each corner and the flaps created in this way are folded up. (Note that we are allowing $x = 0$ and $x = 3$, in order to work on a closed interval, even though these extreme values lead to "boxes" of zero volume.)

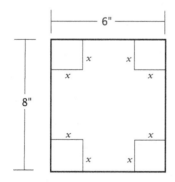

a. Write the volume $V(x)$ of the resulting box, as a function of x.

   b. Find the rate of change of volume, as a function of x.

   c. Find the critical points of $V(x)$, in order to determine the value of x which would maximize the volume (keep the restriction on the domain in mind).

   d. Now write the surface area of the box as a function of x.

   e. Determine the x value which would maximize the surface area.

48. Suppose the dimensions of a rectangular solid are changing in such a way that the length, width and height at time t are given by $l(t) = 1/t$, $w(t) = 1 + t^3$, and $h(t) = 2t^2 - t$, respectively. What is the rate of change of the volume, with respect to time?

49. A particle traveling on the x-axis is acted on by variable forces. The particle's position is given by $x(t) = 3t^4 - 20t^3 - 24t^2 + 240t + 1$, where t is in seconds and $x(t)$ is in meters from the origin, for $-3 \leq t \leq 10$.

   a. Find the velocity of the particle, as a function of the time t. Include units.

   b. Find the critical points for the function $x(t)$.

   c. Break the domain of x into regions on which x is increasing and on which x is decreasing.

   d. Use this information to determine the **total** distance traveled by the particle in the given time range.

   e. At what t value is the particle farthest to the right? To the left?

50. A pitcher atop a stand throws a ball from a height of 19.6 meters above sea level with vertical velocity 14.7 meters per second (where positive velocity indicates upward movement) and horizontal velocity 43 meters per second.

   a. Recall from Exercise 29 in Section 1.5 that, disregarding the horizontal velocity, the height, above ground, of an object in free-fall is $h(t) = -\frac{1}{2}gt^2 + v_0t + h_0$, where $g \approx 9.8$ m/s^2. Use this to determine the function giving the height of the ball as a function of time.

   b. At what time does the ball land on the ground?

   c. Assuming that there is no drag force on the ball (i.e., that its horizontal velocity is constant), how far does the ball travel horizontally before it hits the ground?

   d. Find the critical values for $h(t)$, to determine the intervals on which the height of the ball is increasing and decreasing. Find also the maximum height reached by the ball.

   e. How far has the ball traveled horizontally when it reaches its maximum height?

2.2 The Product and Quotient Rules

Suppose that we have two functions $f(x)$ and $g(x)$. We can form the more complicated functions $(f \cdot g)(x) = f(x)g(x)$ and $(f/g)(x) = f(x)/g(x)$, for those x values for which the product and quotient are defined (see Section 2.A for details).

We would like to have rules for calculating the derivatives the product $f(x)g(x)$ and quotient $f(x)/g(x)$. You might think that the rules would be something simple, like the derivative of $f(x)g(x)$ is $f'(x)g'(x)$, and the derivative of $f(x)/g(x)$ is $f'(x)/g'(x)$. These are absolutely **NOT** correct.

The actual rules are:

Theorem 2.2.1. *Suppose that f and g are differentiable at x. Then,*

1. **The Product Rule:** $f \cdot g$ *is differentiable at x, and*

$$(f \cdot g)'(x) = f(x)g'(x) + g(x)f'(x);$$

2. **The Quotient Rule:** *if $g(x) \neq 0$, then f/g is differentiable at x, and*

$$(f/g)'(x) = \frac{g(x)f'(x) - f(x)g'(x)}{(g(x))^2}.$$

Proof. From the definition of the derivative, we have

$$(f \cdot g)'(x) = \lim_{h \to 0} \frac{f(x+h)g(x+h) - f(x)g(x)}{h}.$$

We now add zero in a clever way in the numerator, and find

$$(f \cdot g)'(x) = \lim_{h \to 0} \frac{f(x+h)g(x+h) - f(x+h)g(x) + f(x+h)g(x) - f(x)g(x)}{h} =$$

$$\lim_{h \to 0} f(x+h) \cdot \frac{g(x+h) - g(x)}{h} \; + \; \lim_{h \to 0} g(x) \cdot \frac{f(x+h) - f(x)}{h} \; =$$

$$\lim_{h \to 0} f(x+h) \cdot \lim_{h \to 0} \frac{g(x+h) - g(x)}{h} \; + \; g(x) \cdot \lim_{h \to 0} \frac{f(x+h) - f(x)}{h} \; =$$

$$f(x)g'(x) + g(x)f'(x),$$

where we used that f is continuous at x, since f is differentiable at x (Theorem 1.4.17). This proves the Product Rule.

Now, suppose that $g(x) \neq 0$. As we are assuming that g is differentiable at x, it follows that g is continuous at x, and that the domain of g contains an open interval around x. From this, and the fact that $g(x) \neq 0$, we conclude, for all h in some open interval around 0, that $g(x+h) \neq 0$.

Now, we first find that

$$(1/g)'(x) = \lim_{h \to 0} \frac{1/g(x+h) - 1/g(x)}{h} = \lim_{h \to 0} \frac{g(x) - g(x+h)}{h \, g(x+h)g(x)}.$$

Once again, we use that, since g is differentiable at x, g is continuous at x, and find

$$(1/g)'(x) = \lim_{h \to 0} \frac{g(x) - g(x+h)}{h \, g(x+h)g(x)} = \frac{-1}{(g(x))^2} \cdot \lim_{h \to 0} \frac{g(x+h) - g(x)}{h} = -\frac{g'(x)}{(g(x))^2}.$$

Now, we combine the above with the Product Rule to obtain

$$\left(\frac{f}{g}\right)'(x) = \left(f \cdot \frac{1}{g}\right)'(x) = f(x) \cdot \left(\frac{1}{g}\right)'(x) + \left(\frac{1}{g}\right)(x) \cdot f'(x) =$$

$$-\frac{f(x)g'(x)}{(g(x))^2} + \frac{f'(x)}{g(x)} \; = \; \frac{g(x)f'(x) - f(x)g'(x)}{(g(x))^2}.$$

$\square$

Remark 2.2.2. Students sometimes ask "Why would you make up such complicated rules for derivatives of products and quotients?". The answer, of course, is that we don't **make up** the rules; we use the definition of the derivative, and see what rules we obtain.

It is not good to try to memorize the Product and Quotient Rules with the specific function names f and g; if you were to do this, it could easily cause problems when you have other function names to deal with. We suggest thinking of the Product Rule as saying "the derivative of the product is the first factor times the derivative of the second plus the second factor times the derivative of the first". Or even replace the two-syllable word "factor" with the one-syllable word "thing". Use whatever words make the rule easy for you to remember; it is worth sacrificing technical terminology, if you can use easier-to-remember words that convey the same meaning.

Similarly, the Quotient Rule should be remembered in some form like "the derivative of a quotient is the bottom times the derivative of the top minus the top times the derivative of the bottom, all over the bottom squared". Yes - using "numerator" and "denominator", instead of "bottom" and "top", would sound more sophisticated, but the "numerator" and "denominator" have too many syllables, and using the longer words really does seem to be an impediment for some people who try to memorize the Quotient Rule.

Example 2.2.3. Let $g(v) = (5v^2 - 2v)(v^7 + v - 3)$. Calculate $g'(v)$ two ways, first by expanding the product and second by using the Product Rule.

Solution:

First we find

$$g(v) \; = \; 5v^9 + 5v^3 - 15v^2 - 2v^8 - 2v^2 + 6v \; = \; 5v^9 - 2v^8 + 5v^3 - 17v^2 + 6v,$$

and so

$$g'(v) \; = \; 45v^8 - 16v^7 + 15v^2 - 34v + 6.$$

Now, we calculate $g'(v)$ using the Product Rule. We find

$$g'(v) \; = \; (5v^2 - 2v)(v^7 + v - 3)' + (v^7 + v - 3)(5v^2 - 2v)' \; =$$

$$(5v^2 - 2v)(7v^6 + 1) + (v^7 + v - 3)(10v - 2).$$

Of course, to compare this with our previous result, we need to expand the products and collect the powers of v. We find

$$g'(v) \ = \ (35v^8 + 5v^2 - 14v^7 - 2v) + (10v^8 - 2v^7 + 10v^2 - 2v - 30v + 6) \ =$$

$$45v^8 - 16v^7 + 15v^2 - 34v + 6,$$

which, of course, agrees with what we obtained without the Product Rule.

Example 2.2.4. Calculate
$$\left(\frac{w^3 + 4w}{w - 1} \right)'.$$

Solution:

Using the Quotient Rule, we find

$$\left(\frac{w^3 + 4w}{w - 1} \right)' \ = \ \frac{(w - 1)(w^3 + 4w)' - (w^3 + 4w)(w - 1)'}{(w - 1)^2} \ =$$

$$\frac{(w - 1)(3w^2 + 4) - (w^3 + 4w)(1)}{(w - 1)^2} \ = \ \frac{3w^3 + 4w - 3w^2 - 4 - w^3 - 4w}{(w - 1)^2} \ =$$

$$\frac{2w^3 - 3w^2 - 4}{(w - 1)^2}.$$

In Theorem 2.1.1, we gave the Power Rule for natural number exponents. We can now prove the Power Rule for the remaining integer powers.

Theorem 2.2.5. (Power Rule for Integer Powers) *Let n be an integer. Then, for all x in the domain of x^n,*
$$(x^n)' = nx^{n-1},$$
with the understanding that, in the cases where $n = 0$ or $n = 1$, x^0 is understood to be the constant function 1 (even when $x = 0$), and $0 \cdot x^{-1}$ is understood to be the constant function 0 (even when $x = 0$).

Proof. We need to prove this for $n \leq -1$, that is, when $n = -m$ and $m \geq 1$.

Using the Power Rule for Natural Powers and the Quotient Rule, we find

$$(x^{-m})' \;=\; \left(\frac{1}{x^m}\right)' \;=\; \frac{x^m \cdot (1)' - 1 \cdot (x^m)'}{x^{2m}} \;=\; \frac{0 - mx^{m-1}}{x^{2m}} \;=\; -mx^{-m-1}.$$

$\square$

Note that, if $n \leq -1$, then 0 is not in the domain of x^n. Thus, when $n \leq -1$, x^n is certainly not differentiable at $x = 0$ and so, of course, the Power Rule is not applicable at 0.

Example 2.2.6. Suppose that the position, in meters, of a particle moving along the x-axis is given by $x = x(t) = (5t^2 + t - 2)(t^2 - 7t + 3)$, where t is in seconds. Find the velocity of the particle, without "multiplying out" the product defining $x(t)$.

Solution:

The velocity, v, is the instantaneous rate of change of the position, with respect to time, i.e., $x'(t)$.

Using the Product Rule, we find

$$v = x'(t) \;=\; (5t^2 + t - 2)(t^2 - 7t + 3)' + (t^2 - 7t + 3)(5t^2 + t - 2)' \;=$$

$$(5t^2 + t - 2)(2t - 7) + (t^2 - 7t + 3)(10t + 1) \;\; \text{m/s}.$$

Example 2.2.7. Suppose that the velocity, in ft/s, of an object, moving in a straight line, is given by

$$v = v(t) = \frac{t^2 - 1}{t^2 + 1},$$

where t is in seconds. What is the acceleration of the object at time $t = 2$ seconds?

Solution:

The acceleration, a, is the instantaneous rate of change of the velocity, with respect to time, i.e., $v'(t)$.

Using the Quotient Rule, we find

$$a = \frac{dv}{dt} = \frac{(t^2 + 1)(t^2 - 1)' - (t^2 - 1)(t^2 + 1)'}{(t^2 + 1)^2} =$$

$$\frac{(t^2 + 1)2t - (t^2 - 1)2t}{(t^2 + 1)^2} = \frac{2t[(t^2 + 1) - (t^2 - 1)]}{(t^2 + 1)^2} = \frac{4t}{(t^2 + 1)^2} \quad (\text{ft/s})/\text{s}.$$

Example 2.2.8. The Law of Universal Gravitation says that the force of gravitation, F, in Newtons, exerted by an object of mass, M, in kg, on an object of mass m, in kg, at a distance (between the centers of mass) of r meters is given by

$$F = \frac{GMm}{r^2},$$

where G is the Universal Gravitational Constant, which is approximately equal to 6.67428×10^{-11} m^3/(kg $\cdot$ s^2).

Find the instantaneous rate of change of F, with respect to r.

Solution:

This is an easy exercise for us now. We could use the Quotient Rule, but since the numerator is constant, and the denominator is an integral power of r, it is easier simply to rewrite F as $GMmr^{-2}$ and use the Power Rule:

$$\frac{dF}{dr} = GMm(-2r^{-3}) = \frac{-2GMm}{r^3} \quad \text{N/m}.$$

Example 2.2.9. Consider the number, $P = P(t)$, of people in the population of an island, at time t years, where the population is constrained by available resources to a maximum size M (some fixed positive constant).

It is common to approximate the instantaneous rate of change of P, with respect to t, using the *logistic model*. This model assumes there exists a positive constant k such that

$$\frac{dP}{dt} = kP(M - P). \tag{2.1}$$

Such an equation is called a *differential equation*; it is an equality of functions, for all t in some interval, and a solution to such an equation is a function $P = P(t)$ for which the equation holds. We shall study such equations in depth, and explicitly solve the above differential equation, in Example 4.3.7 in Section 4.3.

Even now, we can examine a number of features of the logistic equation in Formula 2.1.

If $0 < P < M$, then, since $k > 0$, the rate of change of the population will be positive, i.e., the population will increase.

If P is equal to 0, then the rate of change of population is 0; this seems reasonable, since if there are no people, there will be no reproduction.

Also, if the population ever reaches the maximum M, then the rate of change of the population is also 0. This, too, should seem reasonable; if the population is increasing, and then hits its maximum, it must either remain at M or start to decrease. If the population stays at M, then the rate of change would be 0. If the population starts to decrease, then the derivative went from being positive to being negative, and so we would need for dP/dt to hit 0 in-between.

In fact, in Section 4.3, we shall see that, unless the population begins at $P = 0$ or at $P = M$, then Formula 2.1 implies that the population will never be 0 or M, but rather approaches the value M asymptotically.

As we mentioned above, if $0 < P < M$, then the population will increase. As P gets closer to M, Formula 2.1 tells us that dP/dt gets closer to 0. It would seem then that the second derivative d^2P/dt^2 must be negative as P gets close to M. Can we show this without solving explicitly for $P = P(t)$? Yes.

We can calculate d^2P/dt^2, in terms of P, by using Formula 2.1 and the Product Rule. We find

$$\frac{d^2P}{dt^2} = \frac{d}{dt}\big[kP(M-P)\big] = k\left[P \cdot \frac{d}{dt}(M-P) + (M-P) \cdot \frac{dP}{dt}\right] =$$

$$k\left[-P \cdot \frac{dP}{dt} + (M-P) \cdot \frac{dP}{dt}\right] = k \cdot \frac{dP}{dt} \cdot (M - 2P).$$

Using Formula 2.1 again, to replace dP/dt, we obtain:

$$\frac{d^2P}{dt^2} = k^2 P(M-P)(M-2P).$$

Therefore, we find that, if $0 < P < M$, then $d^2P/dt^2 > 0$ when $P < M/2$, and $d^2P/dt^2 < 0$ when $P > M/2$. In words, this says that the rate of increase of the population is itself increasing when $0 < P < M/2$, and then the rate of increase starts (and continues) decreasing once the population exceeds $M/2$.

Example 2.2.10. Suppose that we have a fixed number of atoms or molecules of an *ideal gas* in a container, where the container has a volume that can change with time, such as a balloon or the inside of a piston.

The Ideal Gas Law states that there is a relationship between the pressure P, in N/m^2, that the gas exerts on the container, the temperature T, in Kelvins (K), of the gas, and the volume V, in m^3, that the gas occupies; that relationship is

$$PV = kT, \tag{2.2}$$

where k is determined by the number of atoms or molecules of the ideal gas (together with the ideal gas constant). We shall assume a value of $k = 8$ N·m/K in this problem.

Now suppose that the pressure and volume of our ideal gas sample are changing with the time t, measured in seconds. Then, according to Formula 2.2, the temperature T is also a function of t, namely $T = PV/k$.

Suppose that at a certain time $t = t_0$ seconds, we know that the pressure is $100,000$ N/m^2, the volume is 0.02 m^3, the pressure is dropping at a rate of 100 (N/m^2)/s, and the volume is increasing at a rate of 0.005 m^3/s. At time t_0, is the temperature of the gas increasing or decreasing, and at what rate?

Solution:

Despite the fact that we don't know T explicitly as a function of t, we have enough information to answer this question. Omitting the units temporarily, we are told that, at time t_0, $P = 100,000$, $V = 0.02$, $dV/dt = 0.005$, and $dP/dt = -100$, where the minus sign is due to the fact that the pressure is "dropping".

Thus, using the Product Rule, we find that

$$\frac{dT}{dt} = \frac{1}{k}\left(P\frac{dV}{dt} + V\frac{dP}{dt}\right).$$

At $t = t_0$, we obtain

$$\frac{dT}{dt}\bigg|_{t_0} = \frac{1}{8}\left[(100,000)(0.005) + (0.02)(-100)\right] = 62.25 \text{ K/s}.$$

Hence, at time t_0, the temperature is increasing at a rate of 62.25 K/s.

Example 2.2.11. In Example 1.4.13, we looked at Newton's 2nd Law of Motion: the net force acting on an object equals the (instantaneous) rate of change, with respect to time, of the momentum of the object, when the mass is constant or when mass is added or subtracted with zero velocity. If F denotes the net force, m the mass, v the velocity, a the acceleration, and t the time, Newton's 2nd Law, combined with the Product Rule, tells us

$$F = \frac{d}{dt}(mv) = m\frac{dv}{dt} + v\frac{dm}{dt} = ma + v\frac{dm}{dt}.$$

If the mass is constant, which is the usual case, then dm/dt equals 0, and the formula above collapses to the familiar $F = ma$. However, if the mass is changing with time, then the dm/dt can be very important. See, for instance, Exercise 44 in this section.

2.2.1 Exercises

Find the first and second derivatives of the rational functions in Exercises 1 through 8.

1. $f(x) = \dfrac{x}{x-1}$.

2. $f(x) = \dfrac{x-1}{x}$.

3. $f(x) = \dfrac{2x^2 + 1.5x - 2}{3x^3 + 2x}$.

4. $f(x) = \dfrac{2x - \frac{1}{x}}{2x + \frac{1}{x}}$.

5. $f(x) = \dfrac{3x^2 + \pi x^4 + x}{x^{10} + 3}$.

6. $f(x) = \dfrac{x + a}{x + b}$.

7. $f(x) = \dfrac{4}{x^2 + 2}$.

8. $f(x) = \dfrac{(x - 4)^2}{(x - 3)^2}$ (either expand the numerator and denominator, or rewrite each of them as a product).

Find the critical points of each of the functions in Exercises 9 through 11, and describe the intervals on which the function is increasing and those on which it is decreasing.

9. $f(x) = (2x + 8)(-3x + 6)$.

10. $f(x) = \dfrac{x^2 - 2x - 1}{2x - 5}$

11. $f(x) = (x - 1)(x^3 + 3x^2 + 7x + 19)$.

In Exercises 12 - 15, determine if the statement is true or false. If it is true, what theorem(s) supports the conclusion? If it is false, present a counterexample.

12. If f and g are differentiable at x, then the function fg is differentiable at x.

13. If f and g are functions such that the function fg is differentiable at x, then at least one of f or g must be differentiable at x.

14. If f and g are differentiable at x, then the quotient f/g is differentiable at x.

15. If f and g are differentiable at x and $g(x) \neq 0$, then the quotient f/g is continuous at x.

16. What technique (or "trick") is used in the proof of the Product Rule?

17. Generalize the Product Rule to the product of 3 functions, i.e., if $f = f_1 f_2 f_3$, find f', in terms of f_1, f_2, f_3, and their individual derivatives.

    Can you see how this generalizes to a product of n functions?

18. Use your formula for the derivative of a product of n functions from the previous problem to determine the derivative of the following function: $f(x) = (2x + 5)^n$.

19. Let $k(x) = \dfrac{f(x)}{g(x)}$. Assume f and g are differentiable, and that f and g are everywhere non-zero. Show that $k' = k\left(\dfrac{f'}{f} - \dfrac{g'}{g}\right)$.

20. Suppose the cost of producing n cars is $\sqrt{8000 + 1000n}$ dollars.

    a. Write a formula for the cost per car of producing n cars.

    b. Use the Quotient Rule to determine the rate of change of the "cost per car" function.

In Exercises 21 through 23, give an equation for the tangent line to the graph of $y = f(x)$ at the given point. Leave your answers in point-slope form.

21. $f(x) = \dfrac{ax + b}{cx + d}$, $(0, b/d)$. Assume $d \neq 0$.

22. $f(x) = \dfrac{(x + 1)^3 (x + 2)^3}{x}$, $(-1, 0)$.

23. $f(x) = \dfrac{1}{1 + x^n}$, $(1, 1/2)$.

24. Let $g_n(x) = \dfrac{x^n - 1}{x - 1}$. What is $g_n'(1/2)$? What is the limit as $n \to \infty$ of $g_n'(1/2)$? Is there another way to see this?

25. Assume that f and g are differentiable. Derive expressions for

$$\frac{d^2}{dx^2}\left[f(x)g(x)\right] \quad \text{and} \quad \frac{d^3}{dx^3}\left[f(x)g(x)\right].$$

26. Assume that f and g are differentiable, and that g is non-zero. Derive an expression for

$$\frac{d^2}{dx^2}\left[\frac{f(x)}{g(x)}\right].$$

27. Recall that the magnitude F of the force of gravity between two masses M and m is given by $F = GMm/r^2$, where r is the distance between the two masses and G is the universal gravitational constant. Suppose that one object is increasing in mass and that the distance between the two objects is increasing. Let $M = 1000(1 + t^3)$ and $r(t) = t^2$. Assume m and G are constant and calculate $F'(t)$.

28. Assume f and g are differentiable, and that $g(x) > 0$, for all x. Suppose that, for all x,

$$g'f(g^2 + 1) + f'g(g^2 - 1) = 0.$$

Show that fg and f/g differ by a constant.

Calculate the derivatives in Exercises 29 through 33. Assume that f, g and h are all differentiable and positive valued functions.

29. $y = (x - a)^2 f(x)$.

30. $y = \dfrac{f(x) - g(x)}{f(x) + g(x)}$.

31. $y = \dfrac{f(x)}{x}$.

32. $y = \dfrac{f(x)}{x^n}$, n a positive integer.

33. $y = \dfrac{f(x)g(x)}{h(x)}$.

34. Let $f(x)$ be differentiable and let a be any point in the domain of f. Consider the AROC function: $AROC(x) = \dfrac{f(x) - f(a)}{x - a}$. Calculate $\dfrac{d}{dx} AROC(x)$ using the Quotient Rule.

In Exercises 35 through 38 you are given the velocity function of a particle at time t. Calculate the acceleration and jerk functions.

35. $v(t) = (t^2 + 3t + 1)(t^5 - 1)$.

36. $v(t) = \dfrac{t + 2}{\sqrt{t}}$.

37. $v(t) = \dfrac{t^4 - 1}{(t^2 - 1)(t + 3)}$.

38. $v(t) = \dfrac{t^7 - 1}{t - 1}$.

39. A root a of a polynomial $p(x)$ has *multiplicity* n if $(x - a)^n$ is the highest power of $(x - a)$ that's a factor of $p(x)$. For example, if $p(x) = (x - 7)^3(x + 4)^2$, then 7 is a root with multiplicity 3 and -4 is a root with multiplicity 2. Show that if $x = a$ is a root of multiplicity $n \geq 2$ of a polynomial $p(x)$, then a is also a root of $p'(x)$.

40. Suppose that f and g are differentiable functions and that, for all x, $f(x) \geq 0$, $g(x) > 0$, and $g'(x) \leq 0$. Suppose also that $h(x) = f(x) \cdot g(x)$ is an increasing function. Show that $f(x)$ must also be an increasing function.

41. Suppose that $f(x), g(x)$ are differentiable functions whose common domain is I, an open interval. Suppose further that neither function has a root in I (i.e., for all x in I, $f(x) \neq 0$, and $g(x) \neq 0$). Suppose further that, for all x in I, $\dfrac{f'(x)}{f(x)} = \dfrac{g'(x)}{g(x)}$.

    Show that $\dfrac{f(x)}{g(x)}$ is constant, and conclude that $f(x)$ is a multiple of $g(x)$.

42. Work is defined to be the product of force and displacement. Thus, if $p(t)$ is the position, in meters, of an object, which is traveling in a straight line, at time t seconds, and the object is acted on by a constant net force F, in Newtons, then we say that the work done by the force, at time t, is $W(t) = F \cdot (p(t) - p(0))$, in Joules. (Calculating work done by a variable force requires Integral Calculus.) ▶

    a. Write a formula describing the rate of change of work, as a function of time.

    b. If $p(t) = t^2 + 4t + 9$ meters, and $F = 16$ Newtons, find the rate of change of work, with respect to time, at $t = 20$ seconds.

43. Ohm's Law gives a relationship between current, resistance, and voltage in electrical circuits. One formulation of Ohm's Law is the relationship $I = V/R$, where I is current measured in amperes, V is potential difference (or voltage) across the circuit, measured in volts, and R is resistance, measured in ohms.

    a. Assuming that, in some circuit, both V and R are functions of time, find the rate of change of I, with respect to the time, t, in seconds, in terms of the rates of change of V and R, with respect to time.

    b. If $V = t^2 - 2t + 11$ and $R = t^2 - 2t + 3$, find the rate of change of current, with respect to time, at $t = 1$.

    c. At what time does the current reach its maximum?

    d. What are the voltage and resistance when the current is at its maximum?

44. A car is traveling along a road. At a particular time t_0 seconds, the velocity of the car is 10 m/s, and its acceleration is 2 m/s^2. At time t_0, the mass of the car is 2000 kg. However, due to a gas leak, the mass of the car is decreasing at a rate of $1/60$ kg/s. Assume (somewhat strangely) that the leaking gas exits the car with zero velocity (relative to whomever measures the car's velocity as being 10 m/s). What is the rate of change of the momentum of this car, with respect to time, at time t_0? What does this tell you about the net force acting on the car at time t_0?

45. Suppose that the population of a country is modeled using the logistic equation. This country can support a maximum population of 25 million people, and the proportionality constant is $k = 1.6$.

a. Write a formula for the rate of change of the population over time, $\dfrac{dP}{dt}$.

b. For what values of P is the population increasing?

c. For what values of P is the population decreasing?

d. Write a formula for $\dfrac{d^2P}{dt^2}$.

e. For what values of P is the rate of increase of the population increasing?

f. For what values of P is the rate of increase of the population decreasing?

46. Given a pair (h_1, h_2) of differentiable functions, define the derivative of the pair to be the pair of derivatives, i.e., define $(h_1, h_2)' = (h_1', h_2')$.

Now, consider two pairs of differentiable functions $(f_1(x), f_2(x))$ and $(g_1(x), g_2(x))$. Given such a pair, we define the *dot product* as follows:

$$(f_1(x), f_2(x)) \cdot (g_1(x), g_2(x)) = f_1(x)g_1(x) + f_2(x)g_2(x).$$

Prove that there is a Product Rule for the dot product, that is, prove that

$$[(f_1(x), f_2(x)) \cdot (g_1(x), g_2(x))]' =$$

$$(f_1(x), f_2(x)) \cdot (g_1(x), g_2(x))' + (f_1(x), f_2(x))' \cdot (g_1(x), g_2(x)).$$

Generalize the definition of the dot product to pairs of n functions (pairs of ordered n-tuples), and show that the Product Rule is still satisfied.

47. Verify that the Product Rule for dot products from the previous problem holds for the two triplets of functions $(x^2 + 1, 1/x, 5x + 7)$ and $(4x + 3, 1 + 1/x^2, 2x + 1)$.

48. Bob and Jane both know the Product Rule and, from Example 1.4.18, they both know that $|x|$ at not differentiable at 0. Bob and Jane are having an argument about the differentiability of $f(x) = x|x|$ at 0.

Bob says that $f(x)$ is not differentiable at 0, because the Product Rule tells us that

$$f'(x) = x\big(|x|\big)' + |x|(x)',$$

and $\big(|x|\big)'$ is not defined at 0.

Jane says that $f(x)$ is differentiable at 0, because, using the definition of the derivative, she finds:

$$f'(0) = \lim_{h \to 0} \frac{f(0+h) - f(0)}{h} = \lim_{h \to 0} \frac{h|h| - 0}{h} = \lim_{h \to 0} |h| = 0.$$

Who's right, and why is the wrong argument wrong?

2.3   The Chain Rule and Inverse Functions

Suppose that you have two differentiable functions $f(x)$ and $g(x)$, and you know their derivatives $f'(x)$ and $g'(x)$. Does this tell you the derivative of the composition $f(g(x))$? Yes, and the formula, or rule, that gives you the derivative of the composition is called the *Chain Rule*.

If $f(x)$ and $g(x)$ are inverse functions of each other, then $f(g(x)) = x$, and we can use the Chain Rule to find a formula for the derivative of g, provided that we know the derivative of f (or vice-versa). Thus, once we know the derivative of a invertible function, we can immediately obtain the derivative of its inverse. We use this here, and in later sections, to find derivatives of taking roots, logarithms, and inverse trigonometric functions.

Consider the function $m(x) = (x^3 - 4x + 7)^{100}$, where we have first applied the function $g(x) = x^3 - 4x + 7$ to x, and then taken the result and applied the function of raising to the 100-th power to that. Technically, we say that m is the composition of two functions, with the first, or "inside", function being $g(x) = x^3 - 4x + 7$, and the second, or "outside", function being $f(u) = u^{100}$. So, $m(x) = f(g(x))$ or, without the reference to x, we write $m = f \circ g$.

Regardless of exactly how you write it, our question is: how do you calculate $m'(x) = \left[(x^3 - 4x + 7)^{100}\right]'$?

Well...you could, in theory, multiply out the 100-th power, and then just use linearity and the Power Rule, but this would be **extremely** long and painful, and the odds on making a mistake would be high. Is there an easier way?

You might think "What's the problem? The Power Rule tells us that the derivative of raising to the 100-th power is raising to the 99-th power and multiplying by 100, so that $\left[(x^3 - 4x + 7)^{100}\right]'$ equals $100(x^3 - 4x + 7)^{99}$. Right?". **WRONG.** In fact, $100(x^3 - 4x + 7)^{99}$ is **part** of the answer, but there's another factor. Where does this extra factor come from? We'll see.

From the definition of the derivative, Definition 1.4.3, we have

$$\left[(x^3 - 4x + 7)^{100}\right]' = \lim_{\Delta x \to 0} \frac{\Delta\left[(x^3 - 4x + 7)^{100}\right]}{\Delta x}. \tag{2.3}$$

Now, it's true that the Power Rule is a result of the fact that

$$\lim_{\Delta x \to 0} \frac{\Delta x^{100}}{\Delta x} = 100x^{99},$$

but, in Formula 2.3, we don't have just x raised to the 100-th power, we have the quantity $u = g(x) = x^3 - 4x + 7$ raised to the 100-th power. To use the Power Rule, you don't need for the variable to be x, but you do need for the thing that you've got Δ of (or d of, as in dx) to match what you have raised to the 100-th power. Hmmmm...so what do we do?

Let's rewrite Formula 2.3, writing u in place of $x^3 - 4x + 7$; this will simplify the notation, and lead us to a very important algebraic "trick". Before we do this, it will be important to us in a minute that u is differentiable and, hence, continuous, which implies that, if Δx approaches 0, then the corresponding Δu also approaches 0.

Now, letting $u = g(x) = x^3 - 4x + 7$, Formula 2.3 becomes

$$\left[(x^3 - 4x + 7)^{100}\right]' = \lim_{\Delta x \to 0} \frac{\Delta u^{100}}{\Delta x} = \lim_{\Delta x \to 0} \left[\frac{\Delta u^{100}}{\Delta u} \cdot \frac{\Delta u}{\Delta x}\right].$$

This last equality results simply from the cancellation of the Δu's in the product, but is **very** nice, for, since $\Delta u \to 0$ as $\Delta x \to 0$, we have

$$\left[(x^3 - 4x + 7)^{100}\right]' = \lim_{\Delta x \to 0} \left[\frac{\Delta u^{100}}{\Delta u} \cdot \frac{\Delta u}{\Delta x}\right] = \lim_{\Delta u \to 0} \frac{\Delta u^{100}}{\Delta u} \cdot \lim_{\Delta x \to 0} \frac{\Delta u}{\Delta x},$$

and **now** the variable that we've got Δ of matches the variable that's raised to the 100-th power, and the Power Rule applies. In addition, the other limit is precisely the definition of the derivative du/dx. Thus, we obtain

$$\left[(x^3 - 4x + 7)^{100}\right]' = \frac{d}{dx}\left(u^{100}\right) = \frac{d}{du}\left(u^{100}\right) \cdot \frac{du}{dx} =$$

$$100u^{99}\frac{du}{dx} = 100(x^3 - 4x + 7)^{99}(3x^2 - 4).$$

Note that the second equality above looks like it results from the cancellation of the du's; this makes the result easy to remember. Of course, there is no separate quantity du that's really canceling, but the apparent cancellation is a result of the Δu's which actually cancelled in our derivation.

The method that we have just discussed works more generally, and goes by the name of the *Chain Rule*, which we state precisely in Theorem 2.3.1. The cancellation-like aspect of the Chain Rule becomes even more dramatic if we let $y = (x^3 - 4x + 7)^{100}$, for, then, our formula above can be written as

$$\frac{dy}{dx} = \frac{dy}{du} \cdot \frac{du}{dx},$$

where it **really** looks like the du's are canceling, but, keep in mind, that that is just a helpful memory device; there is no quantity du to cancel. Still, the fact that the Chain Rule in Leibniz's notation, the notation with the d's, looks like algebraic cancellation is one of the main reasons why Leibnitz's notation is frequently easier to use than the prime notation.

In the notation with the primes, the Chain Rule doesn't look like an algebraic operation; if $g(x) = x^3 - 4x + 7$ and $f(u) = u^{100}$, then, what we found above was that

$$[f(g(x))]' = f'(g(x)) \cdot g'(x).$$

As with the Product and Quotient Rules, it is not a good idea to memorize the Chain Rule, in this form, with the function names f and g in it. We suggest thinking of the Chain Rule, in this form, as "when you want to take the derivative of one function applied to another function, you take the derivative of the 'outside function', leaving the 'inside function' how it was, but then you have to multiply by the derivative of the 'inside function' ".

Our discussion above forms the heart of the proof of the Chain Rule; however, it omits a technical problem. The rigorous proof of the Chain Rule appears in Theorem 2.A.1 in Section 2.A.

Theorem 2.3.1. (The Chain Rule) *Suppose that $f : B \to C$ and $g : A \to B$ are real functions, and that x is a point in A such that g is differentiable at x, and f is differentiable at $g(x)$.*

Then, $f \circ g$ is differentiable at x, and

$$(f \circ g)'(x) = f'(g(x)) \cdot g'(x).$$

> If we let $u = g(x)$ and $y = f(u)$, then the Chain Rule can be written as
>
> $$\frac{dy}{dx} = \frac{dy}{du} \cdot \frac{du}{dx}.$$

Example 2.3.2. Find $\left(\left(2 - x^5\right)^9 + 4\left(2 - x^5\right)^7 - 13 \right)'$.

Solution:

We simply use the Chain Rule once or twice, depending on how you look at it.

$$\left(\left(2 - x^5\right)^9 + 4\left(2 - x^5\right)^7 - 13 \right)' =$$

$$9\left(2 - x^5\right)^8 \left(2 - x^5\right)' + 4 \cdot 7\left(2 - x^5\right)^6 \left(2 - x^5\right)' - 0 =$$

$$9\left(2 - x^5\right)^8 \left(-5x^4\right) + 28\left(2 - x^5\right)^6 \left(-5x^4\right) = -5x^4\left(2 - x^5\right)^6 \left[9\left(2 - x^5\right)^2 + 28\right],$$

where we used that the Chain Rule tells us that the derivative of $(2 - x^5)^n$ is the derivative of the outside function (raising to the power n), leaving the inside stuff, $2 - x^5$, exactly how it was, but then you have to multiply by the derivative of the inside stuff; thus, the derivative of $(2 - x^5)^n$ is $n(2 - x^5)^{n-1}(2 - x^5)'$.

We could also let $u = 2 - x^5$, so that our original function becomes $u^9 + 4u^7 - 13$, and use the Chain Rule, once, in Leibniz's notation:

$$\frac{d}{dx}\left(u^9 + 4u^7 - 13\right) = \frac{du}{dx} \cdot \frac{d}{du}\left(u^9 + 4u^7 - 13\right) = \left(-5x^4\right)\left(9u^8 + 4 \cdot 7u^6 - 0\right),$$

and then put back in that $u = 2 - x^5$, if we want the answer in terms of just x.

You could use the Chain Rule twice in Leibniz's notation, not that that is really different:

$$\frac{d}{dx}\left(u^9 + 4u^7 - 13\right) = 9u^8 \frac{du}{dx} + 4 \cdot 7u^6 \frac{du}{dx} - 0.$$

which, of course, yields what we found before, but looks more like what we discussed at the end of the previous example.

Example 2.3.3. You may wonder why the Chain Rule is called the Chain Rule, instead of the Composition Rule. It's because of how nicely the rule extends to a chain of compositions.

For instance, what if we compose a composition? What if we want to calculate $\big(m(f(g(x)))\big)'$? You apply the Chain Rule from the outside first, working your way inside. That is, you think of $m \circ f \circ g$ as $m \circ (f \circ g)$, then apply the Chain Rule, and then apply it again:

$$\big(m(f(g(x)))\big)' \;=\; m'(f(g(x))) \cdot (f(g(x))' \;=\; m'(f(g(x))) \cdot f'(g(x)) \cdot g'(x),$$

and the expression that expands on the right is a "chain" of derivatives.

For instance,

$$\left[\big((x^5 - 7x)^{100} + 4\big)^{13}\right]' \;=\; 13\big((x^5 - 7x)^{100} + 4\big)^{12} \cdot \big((x^5 - 7x)^{100} + 4\big)' \;=$$

$$13\big((x^5 - 7x)^{100} + 4\big)^{12}\big(100(x^5 - 7x)^{99}\big)(x^5 - 7x)' \;=$$

$$13\big((x^5 - 7x)^{100} + 4\big)^{12}\big(100(x^5 - 7x)^{99}\big)(5x^4 - 7).$$

Remark 2.3.4. We need to mention some issues with ways that functions are frequently written. The function $f(u) = u^{100}$ is the same as the function $f(x) = x^{100}$; the variable names are just placeholders. However, in physical problems, we frequently use a particular variable to denote a particular quantity, like time or distance, and we would never/rarely change the variable name used in the function. Using fixed variable names to denote the input and output of functions allows us to write fewer words, and just lets the variable names help keep track of what everything means. This use of specific variable names becomes especially important when composing functions.

For instance, we can write $y = u^{100}$ and $u = x^3 - 4x + 7$. Are there functions named y and u here? Yes and no. In this case, we would use u not only as a variable name, but also use it to denote a function name, and thus write $u = u(x) = x^3 - 4x + 7$. The "function" y is less well-defined; we could mean y as a function of u, given by $y = y(u) = u^{100}$, or we could mean y as a function of x, given by replacing the u in u^{100} by $x^3 - 4x + 7$, i.e., $y = y(x) = (x^3 - 4x + 7)^{100}$. Which one of these would someone mean if they referred to the function y? Either one; they would have to include the phrases "as a function of x" or "as a function of u" to be clear.

The good news is that, when you deal with functions that are specified using particular variable names, you usually treat things exactly as we discussed above, without really having to think about them. We made a big deal about this here because we wanted you to understand that, if you have something like $y = u^{100}$ and $u = x^3 - 4x + 7$, then y as a function of x actually means the composition of the functions $f(u) = u^{100}$ and $g(x) = x^3 - 4x + 7$. Furthermore, when we write dy/du and dy/dx, we mean that we are considering y as a function of u, and y as a function of x, respectively. In this example, that means that $dy/du = f'(u) = 100u^{99}$, while $dy/dx = (f \circ g)'(x)$, which is what we can calculate with the Chain Rule.

Perhaps the most important thing to be clear about is that, if you are dealing with $u = u(x)$, and are told to "find the derivative of u^{100}", you should realize that this instruction is ambiguous; it could mean the derivative with respect to u or x. As we wrote above, these two derivatives are different:

$$\frac{d}{du}\left(u^{100}\right) = 100u^{99},$$

while

$$\frac{d}{dx}\left(u^{100}\right) = \frac{d}{du}\left(u^{100}\right) \cdot \frac{du}{dx} = 100u^{99}\frac{du}{dx},$$

which will be different from the derivative with respect to u, unless $du/dx = 1$.

It would be confusing to write $(u^{100})'$ here. The prime notation $(u^{100})'$ should not be used in a situation in which you are also considering u as a function of another variable, since it doesn't specify what variable you are taking the derivative with respect to.

Example 2.3.5. The volume V, in cubic inches, of a spherical balloon is $4\pi R^3/3$, where R is the radius in inches. If someone is blowing air into the balloon at a constant rate of 64 in^3/s, how fast (i.e., at what rate) is the radius of the balloon increasing, when the radius is 3 in?

Solution:

In this problem, as in many physical problems, we know that the quantities involved are functions of the time t (here, in seconds), even though we do not explicitly know what the functions are. We may, nonetheless, think of the volume as a function of R or as a function of

t, and apply the Chain Rule to conclude that

$$\frac{dV}{dt} = \frac{dV}{dR} \cdot \frac{dR}{dt} = 4\pi R^2 \frac{dR}{dt}.$$

Thus, omitting the units temporarily, we find that, when $R = 3$,

$$64 = 4\pi(3)^2 \left.\frac{dR}{dt}\right|_{R=3}.$$

Therefore,

$$\left.\frac{dR}{dt}\right|_{R=3} = \frac{64}{36\pi} = \frac{16}{9\pi} \approx 0.5659 \text{ in/s}.$$

Note that we assumed above that R is a differentiable function of t, at least when $R = 3$ in. This is true, since the data given implies that V is differentiable function of t, and we can solve for R as a function of V. We really still need to know that cube roots of differentiable functions are differentiable; this will follow from the results later in this section.

Example 2.3.6. The amount of power, P, in watts, lost when a current of i amperes (amps) flows through a circuit of resistance R ohms is given by $P = i^2 R$.

Suppose that the resistance of the circuit is fixed at 5 ohms, and that the current is decreasing at a rate of 3 amps per second. Let t be the time measured in seconds. When the current is 7 amps, is the power loss, as a function of time, increasing or decreasing? At what rate?

Solution;

We are told that

$$\frac{di}{dt} = -3 \text{ amp/s},$$

where the minus sign is present since we are given a rate of **decrease**.

In particular, i is a differentiable function of t. Certainly, P is a differentiable function of i. Theorem 2.3.1 tells us that P is a differentiable function of t, and

$$\frac{dP}{dt} = \frac{dP}{di} \cdot \frac{di}{dt} = 2iR \cdot (-3) = -6iR = -30i \; watt/s$$

Thus, when $i = 7$ amps, we find that

$$\frac{dP}{dt}\bigg|_{i=7} = -210 \ \ \text{watt/s},$$

which means that, when the current is 7 amps, the rate at which power is lost is decreasing at a rate 210 watt/s.

Note the change in signs, due to the fact that we switched from dP/dt, which is the rate of increase, to referring to a rate of **decrease**.

$\overline{}$

Example 2.3.7. Consider now the function $f(x) = x^3$. The inverse function of f (see Definition 1.3.31) is $f^{-1}(x) = x^{1/3} = \sqrt[3]{x}$. Thus, for all x, $f(f^{-1}(x)) = x$. We want to use the Chain Rule to find the derivative of $x^{1/3}$.

Assuming that $f^{-1}(x)$ is differentiable, we may differentiate both sides of $f(f^{-1}(x)) = x$, using the Chain Rule on the left, to obtain

$$f'(f^{-1}(x)) \cdot (f^{-1})'(x) = 1,$$

and so, assuming that $f'(f^{-1}(x)) \neq 0$, we obtain a formula for the derivative of the inverse function:

$$(f^{-1})'(x) = \frac{1}{f'(f^{-1}(x))}.$$

Thus, in our specific example, where $f(x) = x^3$, we would obtain the derivative of the inverse function:

$$\left(x^{1/3}\right)' = \frac{1}{3(x^{1/3})^2} = \frac{1}{3}x^{-2/3}.$$

$\overline{}$

What's wrong with the calculation of the derivative of $x^{1/3}$ in the above example? Nothing, except that we assumed that, because f is differentiable, so is $f^{-1}(x)$, as long as $f'(f^{-1}(x)) \neq 0$. This is true, but doesn't follow from the Chain Rule; we need the *Inverse Function Theorem*:

Theorem 2.3.8. (Inverse Function Theorem) *Suppose that f is differentiable on an open interval I, and that, for all x in I, $f'(x) \neq 0$.*

Then, the range of the restriction of f to I is an open interval J, and the restricted function $f : I \to J$ is one-to-one and onto. The inverse $f^{-1} : J \to I$ is differentiable and, for all $b \in J$,

$$(f^{-1})'(b) = \frac{1}{f'(f^{-1}(b))};$$

this can also be written as

$$\frac{dx}{dy}\bigg|_{b} = \frac{1}{dy/dx}\bigg|_{f^{-1}(b)}.$$

For the proof, see Theorem 2.A.2 and Theorem 2.A.3

You cannot calculate higher-order derivatives simply by inverting, e.g., in general

$$\frac{d^2x}{dy^2} \neq \frac{1}{d^2y/dx^2}.$$

See Exercise 51 in this section for the correct relationship between $\dfrac{d^2x}{dy^2}$ and $\dfrac{d^2y}{dx^2}$.

Example 2.3.9. Let's return to the example of $y = f(x) = x^3$, and $y = f^{-1}(x) = x^{1/3}$, where we use the same independent and dependent variable names, x and y, respectively, for both functions, so that we may graph the two functions on a common set of coordinate axes. The graphs appear in Figure 2.1.

Note that a point (a, b) is on the graph of f if and only if (b, a) is on the graph of f^{-1}. Thus, the graph of f^{-1} is obtained from the graph of f by swapping all of the $x-$ and y-coordinates. The effect of this is that the graph of f^{-1} is the "reflection about the line $y = x$" of the graph of f.

Theorem 2.3.8 tells us that the slope of the graph of f^{-1}, where $x = b$, is the reciprocal of the slope of the graph of f, where $x = f^{-1}(b)$. In our current example, this implies that the slope of the graph of $y = x^{1/3}$ where $x = 1/8$ is the reciprocal of the slope of the graph of $y = x^3$, where $x = \sqrt[3]{1/8} = 1/2$, which is $3(1/2)^2 = 3/4$.

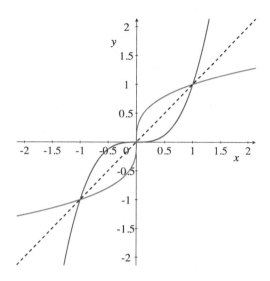

Figure 2.1: The graphs of $y = x^3$ and $y = x^{1/3}$.

Note that the horizontal tangent line to $y = f(x)$ at the origin gets flipped about $y = x$ to become a vertical tangent line. The slope of a vertical line is undefined; which is a reflection of the fact that $f^{-1}(x) = x^{1/3}$ is not differentiable at 0, though we still say the tangent line exists.

As in the example above, we can use Theorem 2.3.8, plus the Power Rule for Integer Exponents, to differentiate all of the root functions $x^{1/n}$, where $n \neq 0$ is an integer. By using the Chain Rule, and composing extracting roots and raising to powers, we obtain the Power Rule for x^r, where r can be any rational number. (You may wish to look again at our discussion of the function x^r in Remark 1.3.39.)

Corollary 2.3.10. (Power Rule for Rational Exponents) *Suppose that r is a rational number, and that x is in the domain of x^r. Then, if $x \neq 0$, the function x^r is differentiable at x, and*

$$(x^r)' = rx^{r-1}.$$

If $r > 1$, then we may eliminate the restriction that $x \neq 0$.

Example 2.3.11. Suppose that $z = \dfrac{1}{\sqrt{5w^2 - 7}}$. Calculate $\dfrac{dz}{dw}$.

Solution:

Rewrite z as $(5w^2 - 7)^{-1/2}$. Using the Power Rule and the Chain Rule, we find

$$\frac{dz}{dw} = \frac{d}{dw}\left[(5w^2 - 7)^{-1/2}\right] = -\frac{1}{2}(5w^2 - 7)^{(-1/2)-1} \cdot \frac{d}{dw}(5w^2 - 7) =$$

$$-\frac{1}{2}(5w^2 - 7)^{-3/2}(10w) = \frac{-5w}{(5w^2 - 7)^{3/2}}.$$

Example 2.3.12. Recall Example 1.4.15, in which we had a straight rod of length 1 meter, with a circular cross section of constant radius 0.04 meters. An x-axis was marked off along the rod, with the origin at one of the rod and the other end at $x = 1$ meter.

We assumed that the mass of the rod between 0 and any x value less than or equal to 1 was given by

$$m(x) = 5x^3 + x \ \text{ kg},$$

and assumed that the density, the mass per volume, of the rod was constant at any fixed x value, i.e., we assumed that the density at each point in a cross section perpendicular to the rod was constant.

We wished to determine the density $\rho(x)$ of the rod at each point in the perpendicular cross section of the rod at coordinate x, for $0 < x < 1$.

When we first looked at this problem, we had not established any rules for differentiation. The problem is much easier for us now.

Density is mass per volume. Thus, we would like to calculate the instantaneous rate of change of the mass, with respect to volume, i.e., we want to calculate dm/dV, where V denotes the volume.

How is the volume, V, a variable quantity? The volume, V, is the volume of the rod between 0 and x meters, which is $V = V(x) = \pi(0.04)^2 x$ cubic meters.

Now, to find $\rho(x)$, we could solve for x in terms of V, and substitute that in for x in our formula for $m(x)$, in order to find m as a function of V. We could then calculate dm/dV, and replace any V's in our result by $\pi(0.04)^2 x$, so that we would obtain the density as a function of x.

However, the Chain Rule gives us a better way to proceed.

First, we use the Power Rule and linearity to quickly calculate

$$\frac{dm}{dx} = 5 \cdot 3x^2 + 1 = 15x^2 + 1 \ \ \text{kg/m},$$

and find that this agrees with our calculation of $m'(x)$ in Example 1.4.15.

We also find easily that

$$\frac{dV}{dx} = \pi(0.04)^2 \ \ \text{m}^3/\text{m}.$$

By the Chain Rule (and knowing that we **could** write m and a function of V), we have

$$\frac{dm}{dx} = \frac{dm}{dV} \cdot \frac{dV}{dx}, \ \ \text{i.e.,}$$

$$\rho(x) = \frac{dm}{dV} = \frac{dm/dx}{dV/dx} = \frac{15x^2 + 1}{\pi(0.04)^2} \ \ \text{kg/m}^3,$$

which, of course, is what we found in Example 1.4.15.

As our final example in this section, we would like to look ahead to an important use of the Chain Rule that we will consider in Section 2.10.

Example 2.3.13. We wish to amplify some of our comments from Remark 2.3.4 by considering a differentiable function $y = f(x)$ and asking: what is the derivative of y^5?

Hopefully, you remember Remark 2.3.4 and immediately said or thought "Ha. Do you mean the derivative with respect to y or with respect to x?". These are different (in general):

$$\frac{d}{dy}\left(y^5\right) = 5y^4 \ \ \ \ \text{and} \ \ \ \ \frac{d}{dx}\left(y^5\right) = \frac{d}{dy}\left(y^5\right) \cdot \frac{dy}{dx} = 5y^4 \frac{dy}{dx}.$$

where, of course, we used the Chain Rule to find the derivative of y^5 with respect to x (it looks like the dy's cancel).

It's interesting that this sort of use of the Chain Rule enables us, at times, to find a formula for the derivative dy/dx even when you are not explicitly given a function in the form $y = f(x)$.

Suppose that you are told (or can prove) that there is a function $y = f(x)$ which satisfies the equation

$$xy + 7 = x^5 + y^5, \tag{2.4}$$

and are asked to calculate the derivative dy/dx.

Then, despite the fact that we cannot solve this equation algebraically to write $y = f(x)$, we can, nonetheless, calculate dy/dx in terms of x **and** y.

How does this work? We just keep in mind that y is some function of x, and differentiate both sides of Formula 2.4 with respect to x; derivatives of y will yield dy/dx's, and then we solve algebraically for dy/dx.

In this example, we proceed as follows:

Differentiate each side, with respect to x:

$$\frac{d}{dx}(xy + 7) = \frac{d}{dx}(x^5 + y^5).$$

Apply the Product Rule on the left and the Chain Rule on the right to obtain

$$x\frac{dy}{dx} + y\frac{dx}{dx} = 5x^4 + 5y^4\frac{dy}{dx}.$$

Use that $dx/dx = 1$, and put the terms that are multiplied by dy/dx on one side of the equation and the other terms on the other side:

$$x\frac{dy}{dx} - 5y^4\frac{dy}{dx} = 5x^4 - y.$$

Factor out the dy/dx and divide to get the answer:

$$\frac{dy}{dx} = \frac{5x^4 - y}{x - 5y^4}.$$

The price that we pay for not starting with y as an explicit function of x is that we end up with a function of two variables, x and y, in our formula for dy/dx.

In a problem such as this, Formula 2.4 is said to define the function y *implicitly*, and the process that we have just demonstrated is called *implicit differentiation*. We will look at implicitly defined functions and implicit differentiation in depth in Section 2.10.

2.3.1 Exercises

Calculate the derivatives of the functions in Exercises 1 through 19.

1. $f(x) = (2x + 1)^3$.

2. $g(t) = (5t^2 + 3t - 4)^{10}$.

3. $h(x) = (\frac{1}{3}x + 12)^7$.

4. $p(x) = \sqrt{x^2 + 2x + 3}$.

5. $f(s) = s^5 + 2s + \dfrac{1}{5}$.

6. $r(x) = (x^2 + 1)^5(2x + 3)$.

7. $g(p) = (3p^2 + 2p)^9 + (4p^3 + 72)^{11}$.

8. $h(s) = 15(5s^2 + 3s + 1)^{\frac{2}{3}}$.

9. $f(r) = \dfrac{(4r^7 + 6r^5)^{100}}{50}$.

10. $h(x) = (13x^2 + 29x + 70)^{\frac{-3}{5}} + 54$.

11. $t(x) = \dfrac{30}{(5x + 1)^2}$.

12. $k(x) = \sqrt[7]{x^2 + 3x} + 12$.

13. $f(t) = \dfrac{(3t + 2)^4 - 7}{t^2 + t + 2}$.

14. $h(x) = \dfrac{4}{x^2} - \dfrac{8}{\sqrt{x}}$.

15. $m(x) = \left[(x + 5)^3 + (2x + 5)^5\right]^4$.

16. $r(x) = \left(x^2 - \dfrac{1}{x^2}\right)^7$.

17. $g(x) = ((x^2 + 1)^5 + x^6 + x^3)^4$.

18. $j(x) = \sqrt{x^2 + \sqrt{x^2 + 1}}$.

19. $h(x) = \sqrt{x + \sqrt{x + \sqrt{x}}}$.

20. Suppose that f and g are differentiable functions and that $h(x) = f(g(x))$. Suppose further that $g(b) = b$ and $g(a) = a$, where $a < b$. Prove that there exists $c \in (a, b)$ such that $f(b) - f(a) = f'(g(c))g'(c)(b - a)$.

21. Consider the functions $f_n(x) = x^n$, for $n = 0, 1, 2, 3, 4$. For which values of n does f_n possess a differentiable inverse function for all x in the range of f_n?

In Exercises 22 through 24, find points, if any, where the second derivative of the function is zero.

22. $f(x) = (ax + b)^{10}$, $a \neq 0$.

23. $g(x) = \sqrt{x^2 + 5x + 6}$.

24. $h(x) = (x^2 + 6)^2$.

25. The total capacitance, C_{total}, of two capacitors, with capacitance C_1 and C_2, in series, is given by
$$C_{total} = \frac{1}{\frac{1}{C_1} + \frac{1}{C_2}}.$$

Suppose that in such a circuit the capacitance of one capacitor is fixed at 50 farads ($C_1 = 50$) and we vary the capacitance of the second capacitor. Show that C_{total} has an inverse by finding an explicit formula for C_2 in terms of C.

26. Given the setup in the previous problem, suppose that $C_1 = 3C_2 + \dfrac{1}{C_2}$. What is $\dfrac{dC_{total}}{dC_2}$?

27. Which of the following answer choices correctly completes the statement: "The graph of f^{-1} is the graph of f reflected over ____."

    a. The x-axis.

    b. The y-axis.

    c. The line $y = x$.

    d. The origin.

28. Suppose that $f : I \to J$ is a continuous, one-to-one and onto function, that f is differentiable at $x = a$, $f(a) = b$, and $f'(a) \neq 0$. Is it true that that f^{-1} is continuous at b? Why or why not?

29. Suppose that a differentiable function f has the property that $f'(x) = f(x)$ for all x. Let g be another differentiable function.

    a. What is $[f(g(x))]'$?

    b. What is $[g(f(x))]'$?

30. Let $P(x) = (x + a)^k(x + b)^n$ where $a \neq b$ and n and k are both integers greater than or equal to 2. When does $P'(x) = 0$?

In Exercises 31 through 35, you are given a function $f(x)$, its domain I, and a point b in the range of f. Use the Inverse Function Theorem, as given in Corollary 2.3.8, to calculate $(f^{-1})'$ at b.

31. $f(x) = x^3$, $I = (-\infty, \infty)$, $b = 27$.

32. $f(x) = x^5 + x^3 + 1$, $I = (-\infty, \infty)$, $b = 3$.

33. $f(x) = x^2 + x^4$, $I = (0, \infty)$, $b = 20$.

34. $f(x) = \sqrt{1 - x^2} + x^2$, $I = (0, 3/4)$, $b = \dfrac{\sqrt{3}}{2} + \dfrac{1}{4}$.

35. $f(x) = \dfrac{1}{3 + x + x^2}$, $I = (-1/2, 1/2)$, $b = 1/3$.

In Exercises 36 and 37, find the critical points of $f(x)$, $g(x)$ and $h(x) = f(g(x))$.

36. $f(x) = x^2 - 1$, $g(x) = 2x^3 - 3x^2$.

37. $f(x) = \dfrac{x^2}{2} - 3x + 1$, $g(x) = -x^2 + 6x$.

38. Show that if $p(x)$ is a polynomial and $f(x) = x^2$, then the critical points of $h(x) = f(p(x))$ are either zeros of $p(x)$ or $p'(x)$.

39. Suppose that, as in Example 2.3.6, the amount of power, P, lost when a current of i amperes flows through a circuit of resistance R ohms is $P = i^2 R$. Suppose that $R = 8$ ohms and that the current is increasing at a rate of 5 amps per second. What is the rate of power loss, in watts per second, when the current is 11 amps?

The kinetic energy, in joules, of a particle in motion with mass m and velocity v is given by the equation $E = \frac{mv^2}{2}$. In Exercises 40 through 42, you are given the mass in kg, and velocity function, in meters per second, of the particle. Calculate dE/dt at the given time.

40. $m = 8$, $v(t) = 4$, $t = 9$.

41. $m = 7$, $v(t) = 3t + 2$, $t = 4$.

42. $m = 6$, $v(t) = \sqrt{t^2 + \sqrt{t}}$, $t = 9$.

43. Now suppose that the mass also varies with time. What is dE/dt at $t = 2$ if $m(t) = t^2 + 3t + 1$ kg and $v(t) = t^5 + t^3 + 6$ m/s?

44. An ant is marching along a hill in such a way that the ant's altitude, A, is a function of its distance r, in centimeters, from the peak, and its distance from the peak is a function of time, t, in seconds.

    a. Find the rate of change of altitude, with respect to distance from the peak, if $dr/dt = -0.5$ cm/s, and the ant's altitude is increasing at a rate of 2 cm/s.

    b. Find the rate of change of altitude, with respect to time, if the rate of change of r, with respect to time, is 0.9 cm/s, and the rate of change of altitude, with respect to distance from the peak, from the peak is -0.7 cm/s.

45. Recall that the surface area of a closed, right circular cylinder, of radius r and height h, is given by $A = 2\pi r^2 + 2\pi rh$, and that the volume of the cylinder of the same dimensions is $V = \pi r^2 h$.

    a. Suppose the cylinder is growing in such a way that both its height and its radius are functions of time. Find an expression for the rate of change of the surface area, with respect to time, in terms of the height, the radius, and their rates of change.

    b. How quickly is the volume growing, if the height of the cylinder is 4 in, and is growing at 5 in/hr, while the radius is 6 in, and is growing at a rate of 1 in/hr?

    c. Suppose now that $r(h) = \sqrt{h}$, so that radius is a function of the height. Find an expression giving the rates of change of surface area and volume, with respect to height.

    d. With the same conditions as part (c), find the rate of change of the ratio A/V, with respect to height. Is A/V always increasing as a function of h ($h > 0$)? Always decreasing? Sometimes increasing and sometimes decreasing?

46. Einstein's Special Theory of Relativity predicts that the mass of an object, measured by some observer, grows with the speed at which the object is moving (again, relative to the observer). This can be written in the form:

$$m(v) = \frac{m_0}{\sqrt{1 - \dfrac{v^2}{c^2}}},$$

where $c \approx 3.0 \times 10^8$ meters per second is the (constant) speed of light and m_0 is the *rest mass* of the object (i.e., the mass measured by the observer if the object were not moving relative to them).

    a. Find an expression for the rate of change of the mass, as a function of velocity.

b. Find the rate of change of the mass, with respect to velocity, if the velocity is: 100 m/s, then 1000 m/s.

c. If the velocity is a function of time, find an expression giving the rate of change of mass, with respect to time.

d. Recall that the net force (according to our observer) on an object is the rate of change of the (observed) momentum, mv, with respect to time. Assuming that our object's mass is changing solely due to the relativistic effect of it changing velocity, how much force does our observer calculate is acting on the mass, in terms of c, m_0, v, and the acceleration, a, of the object?

e. According to our observer, how much force would it require to accelerate an object of rest mass 10 kg which is traveling at nine-tenths the speed of light (i.e., $v = 9c/10$) by a rate of 1 m/s^2? How much force would be required to accelerate the same object by 1 m/s^2, if its velocity were 0?

47. A general formula for a circle of radius r centered at the origin is $x^2 + y^2 = r^2$. Clearly the y-coordinate of a point on the circle is not a function of the x-coordinate (the graph fails the vertical line test), but the upper semi-circle can be written as $y = \sqrt{r^2 - x^2}$. In this exercise, we fix the radius at $r = 10$.

a. Find an equation of the tangent line to our upper semi-circle at the point $(6, 8)$.

b. Two curves (or lines) are said to be *orthogonal* at a point P, where the curves intersect, provided that their tangent lines exist at P, and their tangent lines at P are perpendicular. Recall that the slopes of perpendicular lines are negative reciprocals of each other.

   Find an equation of the line orthogonal to our upper semi-circle at the point $(6, 8)$

c. Find an equation of the parabola passing through the point $(0, 1)$ and the point $(6, 8)$, such that the parabola is orthogonal to our upper semi-circle at the point $(6, 8)$.

48. The following functions are written in terms of another (differentiable) function $f(x)$. Write the derivative of each in terms of x, f, and f'.

a. $g(x) = \dfrac{f\left(\frac{1}{x}\right)}{x}$.

b. $h(x) = \sqrt{f(x)^2 + 2}$.

c. $p(x) = x \cdot f\left(\sqrt[3]{2x + 1}\right)$.

d. $q(x) = \dfrac{[f(x)]^2 - 1}{f(x) + 1}$.

49. Suppose that the position of a particle, in feet, traveling on the x-axis is given by $x(t) = t\sqrt{1 - t^3}$, where t is the time, in seconds, and $0 \leq t \leq 1$.

    a. How fast is the particle traveling at time $t = 1/8$ seconds?

    b. At what point in time is the particle at rest (i.e., have 0 velocity)?

    c. Does the particle change directions at the time that you found above?

    d. Where is the particle when it is at rest?

50. A function f is called *even* if, for all x in the domain of f, $-x$ is also in the domain of f, and $f(-x) = f(x)$. A function f is called *odd* if, for all x in the domain of f, $-x$ is also in the domain of f, and $f(-x) = -f(x)$. For example, x^2 is an even function, and x^3 is odd.

Suppose that f is differentiable. Show, by using the Chain Rule, that:

- if $f(x)$ is even, then $f'(x)$ is odd.

- if $f(x)$ is odd, then $f'(x)$ is even.

51. Suppose that $y = f(x)$ is differentiable, and has an inverse function $x = f^{-1}(y)$, which is also differentiable.

    a. Suppose that d^2y/dx^2 and d^2x/dy^2 exist, and that $dy/dx \neq 0$. Show that

$$\frac{d^2x}{dy^2} = -\left(\frac{dy}{dx}\right)^{-3}\frac{d^2y}{dx^2}.$$

    b. Verify the formula above in the special case of $y = x^3$ (where we take the domain and codomain/range to be all real numbers other than 0).

Use the second derivative formula in the previous problem to calculate $\dfrac{d^2x}{dy^2}$ at the given y value in Exercises 52 and 53.

52. $y = x^5 + 2x^3 + 5x + 7$, $y = 7$.

53. $y = (5x + 3)^3$, $y = 27$.

54. The *curvature* of a function $f(x)$ at a point is given by the equation

$$\kappa(x) = \frac{f''(x)}{(1 + [f'(x)]^2)^{3/2}}.$$

Assume $\kappa(x)$ is differentiable and calculate its derivative.

2.4 The Exponential Function

In this section, we will define the *exponential function*, exp, one of the most important functions in all of mathematics, and we will derive many of its properties.

What we will show is that there exists a number e, which is approximately 2.718, such that the exponential function $\exp(x) = e^x$ is its own derivative, i.e., $(e^x)' = e^x$. However, it is not so easy to show that such a number e exists. In fact, we will define the function exp first, then define the number e to be $\exp(1)$, and then show that $\exp(x) = e^x$. Of course, this just pushes back the question to: how do we define exp?

There are many approaches to defining exp. The approach that we shall take is that exp is a limit of special polynomial functions. This limit is similar to the real limits that we have looked at, but is slightly different; we will need limits of *sequences*. We will define this concept, but put most of the details in Section 2.A.

Sequences of Real Numbers and Functions

Definition 2.4.1. *Suppose that m is an integer, i.e., is in the set $\mathbb{Z}$. Denote by $\mathbb{Z}_{\geq m}$ the set of integers which are greater than or equal to m.*

*A function $b : \mathbb{Z}_{\geq m} \to \mathbb{R}$ is called a **sequence (of real numbers)**. In place of $b(n)$, it is standard to write b_n.*

*We say that **the sequence b_n converges to (a real number)** L, and write $\lim_{n \to \infty} b_n = L$ if and only if, for all $\epsilon > 0$, there exists an integer $N \geq m$ such that, for all integers $n \geq N$, $|b_n - L| < \epsilon$.*

*If a sequence does not converge to some L, then we say that the sequence **diverges**.*

Example 2.4.2. The technical definition above may make sequences look horrifyingly complicated. They are not.

For instance, consider the sequence

$$b_n = 7 - \frac{1}{n-2},$$

for all integers $n \geq 3$. Then, $b_3 = 7 - 1/1 = 6$, $b_4 = 7 - 1/2 = 13/2$, $b_5 = 7 - 1/3 = 20/3$, etc. Since $1/(n-2)$ approaches 0 as n approaches infinity, it is easy to see that the sequence b_n converges to 7.

Remark 2.4.3. Recall Definition 1.3.45, and note that, if b is a general real function, the definition of $\lim_{x \to \infty} b(x) = L$ is precisely the same as the definition for the sequence b_n to converge to L, **except** that, for a sequence, N and n are restricted to being integers.

In particular, if you start with a real function f, whose domain contains $\mathbb{Z}_{\geq m}$, and $\lim_{x \to \infty} f(x) = L$, then the sequence obtained from f by restricting its domain to $\mathbb{Z}_{\geq m}$ must converge to L.

We are now going to define a special *sequence of real functions*; that is, for each n in some $\mathbb{Z}_{\geq m}$, we are going to specify a real function f_n. First, however, we need to make another definition.

Definition 2.4.4. *For n in $\mathbb{Z}_{\geq 0}$, define $n!$ (read: n **factorial**) as follows:*
If $n \geq 1$, $n!$ is equal to the product of all of the integers between n and 1, including n and 1. In the special case where $n = 0$, we define $0!$ to equal 1.

Remark 2.4.5. For example, $5! = 5 \cdot 4 \cdot 3 \cdot 2 \cdot 1 = 120$. Note that $5! = 5 \cdot (4!)$. This is a general property of factorials, that is, for all integers $n \geq 1$, $n! = n \cdot [(n-1)!]$; note that this equality is true when $n = 1$ precisely because we made the definition that $0! = 1$.

Recall the sigma notation for summations from Definition 2.1.7. For each n in $\mathbb{Z}_{\geq 0}$, define a degree n polynomial function $P_n : \mathbb{R} \to \mathbb{R}$ by

$$P_n(x) = \sum_{k=0}^{n} \frac{x^k}{k!} = \frac{x^0}{0!} + \frac{x^1}{1!} + \frac{x^2}{2!} + \frac{x^3}{3!} + \cdots + \frac{x^n}{n!},$$

where, as before, by x^0, we mean the constant function 1, even when $x = 0$ (we must make this special definition, since 0^0 is undefined).

Thus,

$$P_0(x) = \sum_{k=0}^{0} \frac{x^k}{k!} = \frac{x^0}{0!} = \frac{1}{1} = 1,$$

$$P_1(x) = \sum_{k=0}^{1} \frac{x^k}{k!} = \frac{x^0}{0!} + \frac{x^1}{1!} = 1 + x,$$

$$P_2(x) = \sum_{k=0}^{2} \frac{x^k}{k!} = \frac{x^0}{0!} + \frac{x^1}{1!} + \frac{x^2}{2!} = 1 + x + \frac{x^2}{2},$$

and so on.

For each fixed x, we then have a sequence of real numbers, $P_n(x)$, and it can be shown (we will not do so here) that this sequence converges to a limit which, of course, depends on x; denote this limit by $L(x)$.

Thus, $L(x)$ is a real function, and we give this function a new name:

Definition 2.4.6. *The **exponential function**, exp, is defined by*

$$\exp(x) = \lim_{n \to \infty} P_n(x) = \lim_{n \to \infty} \sum_{k=0}^{n} \frac{x^k}{k!}.$$

This definition implies that, to approximate $\exp(x)$ to any desired accuracy, we can calculate $P_n(x)$ for sufficiently large n.

Knowing how large to pick n, given a desired amount of accuracy, is beyond our discussion at this point. What we have actually defined here is an *infinite series*, an important topic in Calculus that is not covered in this textbook. If you wish to read about this topic, we recommend [3], [2], or Worldwide Integral Calculus, [1].

The convergence of $P_n(x)$ to $\exp(x)$ has a special property: restricted to any bounded subset B of $\mathbb{R}$ (see Section 1.A), the sequence P_n *converges uniformly* to exp. This means that, for any given desired accuracy, we can estimate $\exp(x)$ by $P_n(x)$, to within the desired accuracy, by using sufficiently large n, and **you can use the same large n for every x in B.**

Now, we wish to look at two special properties of the sequence of polynomial functions P_n.

First, for all n, $P_n(0) = 1$. Therefore, $\exp(0) = 1$.

Secondly, related to differentiation, we have

$$P'_0(x) \;=\; (1)' \;=\; 0;$$

$$P'_1(x) \;=\; (1+x)' \;=\; 1;$$

$$P'_2(x) \;=\; \left(1 + x + \frac{x^2}{2}\right)' \;=\; 1 + x;$$

and, generally, for $n \geq 1$, we use linearity, the Power Rule, and the fact that $k/(k!) = 1/[(k-1)!]$, for $k \geq 1$, to conclude

$$P'_n(x) \;=\; \left(\sum_{k=0}^{n} \frac{x^k}{k!}\right)' \;=\; \sum_{k=1}^{n} \frac{kx^{k-1}}{k!} \;=\; \sum_{k=1}^{n} \frac{x^{k-1}}{(k-1)!} \;=\;$$

$$\frac{x^0}{0!} + \frac{x^1}{1!} + \frac{x^2}{2!} + \frac{x^3}{3!} + \cdots + \frac{x^{n-1}}{(n-1)!} \;=\; P_{n-1}(x).$$

This means that, for $n \geq 1$, the sequence of derivatives P'_n is the same as the sequence P_n (with a shift in indices). Therefore, not only does the sequence of functions P_n converge uniformly to exp on any bounded subset of $\mathbb{R}$, the sequence of derivatives P'_n also converges uniformly to exp. From Theorem 7.17 of [2], we conclude

---

Theorem 2.4.7. *The function* $\exp : \mathbb{R} \to \mathbb{R}$ *is differentiable,* $\exp(0) = 1$, *and*

$$\exp'(x) = \exp(x).$$

Thus, **the exponential function is a function with a remarkable property: it is its own derivative.**

We can iterate this to obtain:

Corollary 2.4.8. *All of the higher derivatives* $\exp^{(n)}(x)$ *exist and are equal to* $\exp(x)$.

Of course, we can combine our formula for the derivative of exp with our other derivative formulas.

Example 2.4.9. Let $y = \exp(5x)$. Then, using that $\exp' = \exp$ and the Chain Rule, we find

$$\frac{dy}{dx} = [\exp'(5x)](5x)' = [\exp(5x)] \cdot 5 = 5\exp(5x).$$

You can see that, each time you take a derivative, another factor of 5 appears; hence,

$$\frac{d^n y}{dx^n} = 5^n \exp(5x).$$

Example 2.4.10. Let $p(w) = 5w^3 + w\exp(w^2 - 7w)$. Then,

$$p'(w) = 15w^2 + w\big(\exp(w^2 - 7w)\big)' + [\exp(w^2 - 7w)] \cdot (w)' =$$

$$15w^2 + w[\exp(w^2 - 7w)] \cdot (w^2 - 7w)' + \exp(w^2 - 7w) =$$

$$15w^2 + w[\exp(w^2 - 7w)](2w - 7) + \exp(w^2 - 7w) =$$

$$15w^2 + (2w^2 - 7w + 1)\exp(w^2 - 7w).$$

Example 2.4.11. Calculate

$$\frac{d}{dx}\left(\frac{\exp(x)}{1 + \exp(x)}\right).$$

Solution:

$$\frac{d}{dx}\left(\frac{\exp(x)}{1 + \exp(x)}\right) = \frac{(1 + \exp(x))\exp'(x) - \exp(x)(1 + \exp(x))'}{(1 + \exp(x))^2} =$$

$$\frac{(1 + \exp(x))\exp(x) - \exp(x)\exp(x)}{(1 + \exp(x))^2} = \frac{\exp(x)}{(1 + \exp(x))^2}.$$

We can use the facts that $\exp' = \exp$ and $\exp(0) = 1$ to derive many properties of the exponential function.

Proposition 2.4.12. *For all x, $\exp(x)\exp(-x) = 1$. Hence, $\exp(x)$ is never zero, and must, in fact, always be positive, that is, for all x, $\exp(x) > 0$.*

Proof. We first show that $\exp(x)\exp(-x)$ is a constant function; we accomplish this by showing that its derivative is zero, and then apply Theorem 1.5.11. Using the Product Rule and the Chain Rule, we calculate

$$\big(\exp(x)\exp(-x)\big)' = \exp(x) \cdot \big(\exp(-x)\big)' + \exp(-x)\exp'(x) =$$

$$\exp(x)\exp'(-x)(-1) + \exp(-x)\exp(x) = \exp(x)\exp(-x)(-1) + \exp(-x)\exp(x) = 0.$$

Thus, there exists a constant C such that, for all x, $\exp(x)\exp(-x) = C$. When $x = 0$, we have $C = \exp(0)\exp(0) = 1$, and so $\exp(x)\exp(-x) = 1$.

Since $\exp(x)$ is never zero, and is continuous (since it's differentiable), the Intermediate Value Theorem, Theorem 1.3.29, implies that $\exp$ is either always positive or is always negative. As $\exp(0) = 1 > 0$, $\exp(x)$ must always be positive. $\qquad\square$

Corollary 2.4.13. *The exponential function is strictly increasing, and its graph is always concave up.*

Proof. We have $\exp(x) = \exp(x) > 0$ and $\exp''(x) = \exp(x) > 0$. $\qquad\square$

We can prove that the properties in Theorem 2.4.7 completely characterize the exponential function.

Theorem 2.4.14. *Suppose that f is defined on an open interval I and that, there exists a constant k such that, for all x in I, $f'(x) = kf(x)$. Then, there exists a constant C such that, for all x in I, $f(x) = C\exp(kx)$.*

Therefore, there is exactly one differentiable function from $\mathbb{R}$ to $\mathbb{R}$ whose value at 0 is equal to 1, and which is its own derivative. That function is the exponential function, $\exp$.

Proof. Suppose that $f : I \to \mathbb{R}$ is differentiable, and $f'(x) = kf(x)$. By Proposition 2.4.12, $\exp(kx)$ is never zero. Thus, $g(x) = f(x)/\exp(kx)$ is a differentiable function. We wish to show that $g'(x) = 0$, so that $g(x)$ is equal to some constant C by Theorem 1.5.11. This would show that $f(x) = C\exp(kx)$.

Using the Quotient Rule and the Chain Rule, we calculate

$$\left(\frac{f(x)}{\exp(kx)} \right)' = \frac{\exp(kx)f'(x) - f(x)(\exp(kx))'}{\left(\exp(kx) \right)^2} =$$

$$\frac{\exp(kx)kf(x) - f(x)\exp(kx) \cdot k}{\left(\exp(kx) \right)^2} = 0.$$

This proves the first statement.

If $f : \mathbb{R} \to \mathbb{R}$ is such that $f'(x) = f(x)$ and $f(0) = 1$, then, by the above, with $k = 1$, $f(x) = C\exp(x)$, and $1 = f(0) = C\exp(0) = C$. Therefore, $C = 1$, and $f(x) = \exp(x)$. □

Example 2.4.15. A number of physical quantities change, approximately, at a rate, with respect to time, that is proportional to the quantity present at that time.

For instance, some populations change at a rate that is roughly proportional to the current population; this seems reasonable, for if the population is 2, 3, or m times bigger, then there are 2, 3, or m times as many people who can reproduce or die and, hence, the rate of change of the population is also 2, 3, or m times as big.

As another example, radioactive material decays at a rate that is approximately proportional to the amount (measured in atoms, or by mass, or by weight) present. This, too, seems reasonable, since, when there are 2, 3, or m times as many radioactive atoms presents, there are 2, 3, or m times as many that will decay at that time.

In both of these cases, and in any problem where the rate of change dA/dt of a quantity $A(t)$ is proportional to the quantity present, there exists a constant k such that $dA/dt = kA$, and Theorem 2.4.14 tells us that

$$A(t) = C\exp(kt),$$

for some constant C. Thus, a quantity whose rate of change is proportional to the quantity present is said to **grow exponentially** or **decay exponentially**, depending on whether A is increasing or decreasing, respectively, with respect to time.

In addition, we can give physical meaning to the constant C (assuming that $t = 0$ is in the domain of A). At $t = 0$, we find $A(0) = C \exp(k \cdot 0) = C \cdot 1 = C$, so that C is the quantity present at $t = 0$, i.e., the initial amount of the quantity. This is frequently written as A_0. Thus, we have

$$A(t) \ = \ A_0 \exp(kt),$$

Assuming that A is positive (which it would be it A represents the amount of some quantity that's present), then a positive proportionality constant k corresponds to the case where A is increasing, since then $dA/dt > 0$. A negative k value (for positive A) indicates that A is decreasing.

In the radioactive decay example, k is certainly negative. In the population example, k could be positive or negative, depending on which is bigger, the birth rate or the death rate.

One final word on the proportionality constant involved in exponential growth and decay: it is standard to make proportionality constants positive, provided that this can easily be accomplished by inserting explicit minus signs if necessary. For example, in an exponential decay problem, we usually write $dA/dt = -kA$, where k is **positive**.

We now give two important algebraic properties of the exponential function.

Theorem 2.4.16. *Let a be a real number, and let r be a rational number. Then, for all x,*

$$\exp(x + a) \ = \ \exp(x) \exp(a)$$

and

$$\exp(rx) \ = \ (\exp(x))^r.$$

Proof. In a similar fashion to the proof of Theorem 2.4.14, we consider the quotient of the left-hand side of each equality over the right-hand side. We then take derivatives, using the Quotient Rule, the Chain Rule, and linearity, and show that the derivatives of the quotients are both zero. Hence, the quotients are constants, and by plugging in $x = 0$, we find that those constants are 1. We leave the actual calculations to you in Exercise 26 in this section.   □

Remark 2.4.17. By letting $x = 1$ in $\exp(rx) = (\exp(x))^r$, we find that, for all rational numbers r, $\exp(r) = (\exp(1))^r$. Since exp is differentiable, exp is continuous. It follows that $\exp(x) = (\exp(1))^x$, for all real x, **provided** that you know what it means to raise a positive base to arbitrary real powers, and that doing so is a continuous function of the exponent. These things are not so easy to define and prove. For instance, what does $3^{\sqrt{2}}$ mean? Since $\sqrt{2}$ is not a ratio of integers, we cannot calculate $3^{\sqrt{2}}$ by taking a root and raising to an integer power.

The slick way around these issues is to use $\exp(x)$ itself, together with its inverse function $\ln(x)$ (see Section 2.5), to **define** what raising a positive base to a real power means. We take this approach for general positive bases in Section 2.6, but for now, we make this definition only when the base is $\exp(1)$.

In the special case of base $\exp(1)$, we make the following definition:

Definition 2.4.18. *We define the constant e to be* $\exp(1)$, *and define e^x to equal* $\exp(x)$. *Thus, for all real x,*
$$e^x = \exp(x) = \lim_{n \to \infty} \sum_{k=0}^{n} \frac{x^k}{k!}.$$

We want to reiterate that, if x is a rational number, then $e^x = \exp(x)$, by Theorem 2.4.16. For irrational numbers x, Definition 2.4.18 **defines** what e^x means for us.

Remark 2.4.19. It is extremely important to note the difference between a power function, like x^3, and the exponential function e^x. A power function has a variable base and a fixed exponent. The exponential function has a fixed base and a variable exponent. You differentiate power functions using the Power Rule; thus, $(x^3)' = 3x^2$. The Power Rule does **not** apply to e^x; the correct derivative is $(e^x)' = e^x$. It is a **horrible**, but common, mistake to try to apply the Power Rule in calculating $(e^x)'$, and arrive at xe^{x-1}. This is **completely wrong**.

The constant e is not equal to a finite or repeating decimal; e is irrational. In fact, e is much "worse" (or "better", depending on your point of view) than just being irrational; e is transcendental, i.e., not a root of a polynomial with integer coefficients. This is different from something like $\sqrt{2}$ which is irrational, but not transcendental, since $\sqrt{2}$ is a root of $x^2 - 2$.

Still, we would like to know **something** about the value of the constant e, one of the most important constants in all of mathematics. If you check on a calculator or computer, you will find that e is approximately 2.71828182846. How can we see, by hand, that $2 < e \leq 3$, or obtain an even better approximation? We discuss this below.

Remark 2.4.20. By definition,

$$e = \exp(1) = \lim_{n \to \infty} P_n(1) = \lim_{n \to \infty} \sum_{k=0}^{n} \frac{1}{k!} = \lim_{n \to \infty} \left(1 + 1 + \frac{1}{2!} + \frac{1}{3!} + \cdots + \frac{1}{n!} \right).$$

The sequence $P_n(1)$ strictly increases as n increases, and approaches e. It follows that e is what's known as the *least upper bound* (see Theorem 1.A.1) of the sequence $P_n(1)$. This, together with the fact that $P_n(1)$ strictly increases, implies that for all n, $P_n(1) < e$, and, if B is a real number and N is an integer such that, for all $n \geq N$, $P_n(1) \leq B$, then $e \leq B$.

To see how this is used, let's suppose that we wish to show that $2 < e \leq 3$.

To show that $2 < e$, it suffices to show, for some n_0, that $2 \leq P_{n_0}(1)$, for then we have $2 \leq P_{n_0}(1) < e$. We may use $n_0 = 1$, since $P_1(1) = 2$, to conclude that $2 \leq P_1(1) < e$.

To show that $e \leq 3$ is harder. We need to first discuss *geometric finite sums*. Fix a real number $r \neq 1$, and, for all integers $n \geq 0$, define the sum

$$\sigma_n(r) = 1 + r + r^2 + \cdots + r^{n-1} + r^n = \sum_{k=0}^{n} r^k.$$

Then,

$$r\sigma_n(r) = r + r^2 + \cdots + r^n + r^{n+1}.$$

and, except for the missing first term, and the extra final term, this is the same sum as that for $\sigma_n(r)$. Thus, we find that

$$r\sigma_n(r) = \sigma_n(r) - 1 + r^{n+1}.$$

Solving for $\sigma_n(r)$, we obtain

$$\sigma_n(r) = \frac{r^{n+1} - 1}{r - 1}. \tag{2.5}$$

Why did we derive this formula for finite geometric sums?

First, notice that $2 \geq 2$, $3 \cdot 2 \geq 2 \cdot 2$, $4 \cdot 3 \cdot 2 \geq 2 \cdot 2 \cdot 2$, and, in general, $k! \geq 2^{k-1}$, for all integers $k \geq 2$ (and, in fact, the inequality also holds for $k = 0$ and 1). Therefore,

$$P_n(1) = 1 + 1 + \frac{1}{2!} + \frac{1}{3!} + \cdots + \frac{1}{n!} \leq 1 + 1 + \frac{1}{2} + \frac{1}{2^2} + \cdots + \frac{1}{2^{n-1}} =$$

$$1 + \sigma_{n-1}\left(\frac{1}{2}\right) = 1 + \frac{\left(\frac{1}{2}\right)^n - 1}{\frac{1}{2} - 1} = 1 + 2\left(1 - \frac{1}{2^n}\right) \leq 3.$$

As we discussed above, since $P_n(1) \leq 3$, for all n, it follows that $e \leq 3$.

How can you obtain better bounds on e than $2 < e \leq 3$? You need to go "out farther" in the summations on each side. What do we mean by this? For instance, $P_4(1) < e$, so

$$2.70833\overline{3} = 1 + 1 + \frac{1}{2!} + \frac{1}{3!} + \frac{1}{4!} = P_4(1) < e.$$

To get an upper-bound on e, we use

$$1 + 1 + \frac{1}{2!} + \frac{1}{3!} + \cdots + \frac{1}{n!} =$$

$$1 + 1 + \frac{1}{2} + \frac{1}{2 \cdot 3}\left(1 + \frac{1}{4} + \frac{1}{4 \cdot 5} + \frac{1}{4 \cdot 5 \cdot 6} + \cdots + \frac{1}{4 \cdots n}\right) \leq$$

$$1 + 1 + \frac{1}{2} + \frac{1}{6} \cdot \sigma_{n-3}\left(\frac{1}{4}\right),$$

for $n \geq 3$ (where $1/(4 \cdots n)$ is set equal to 1 when $n = 3$). Therefore, for $n \geq 3$

$$P_n(1) \leq 1 + 1 + \frac{1}{2} + \frac{1}{6} \cdot \left(\frac{\left(\frac{1}{4}\right)^{n-2} - 1}{\frac{1}{4} - 1}\right) = 2.5 + \frac{2}{9} \cdot \left(1 - \left(\frac{1}{4}\right)^{n-2}\right) \leq 2.722\overline{2}.$$

Thus, what we have shown is that $2.70833\overline{3} < e \leq 2.722\overline{2}$.

As we mentioned above, and as your calculator will tell you,

$$e \approx 2.71828182846, \tag{2.6}$$

which agrees with our inequalities above.

Using the types of arguments that we used above, it can be shown that, for all integers $n \geq 0$,

$$0 \; < \; e - P_n(1) \; \leq \; \frac{n+2}{(n+1)! \cdot (n+1)}.$$

By taking $n = 14$, we see that the error in approximating e by $P_{14}(1)$ is at most $16/(15!)(15)$, which is less than 10^{-12}. Therefore, $P_{14}(1)$ will give us the approximation found in Formula 2.6 (using the number of shown decimal places).

Let us summarize what we know about the exponential function. The variable n in the theorem below denotes one which takes on only non-negative integer values.

Theorem 2.4.21. *The exponential function* $\exp(x) = e^x$ *has the following properties:*

1. *The domain of* e^x *is the set of all real numbers;*

2. *for all* x,
$$e^x \; = \; \lim_{n \to \infty} \left(1 + x + \frac{x^2}{2!} + \frac{x^3}{3!} + \cdots + \frac{x^n}{n!} \right);$$

3. $e^0 = 1$;

4. $(e^x)' = e^x$, *and so, all of the derivatives of* e^x *equal* e^x;

---

5. *for all* x, $e^{-x} = 1/e^x$;

6. *for all* x, $e^x > 0$,

7. e^x *is strictly increasing, and its graph is always concave up;*

8. *for all* x *and* a, $e^{x+a} = e^x \cdot e^a$;

9. *for all* x *and for all rational numbers* r, $e^{rx} = (e^x)^r$;

10. $e \approx 2.71828182846$;

11. $\lim_{x \to -\infty} e^x = 0$, and $\lim_{x \to \infty} e^x = \infty$;

12. *the range of e^x is all positive real numbers;*

13.

$$e^x = \lim_{n \to \infty} \left(1 + \frac{x}{n}\right)^n.$$

Proof. We have already discussed or proved all of these properties, except the last three.

Because $e > 1$, the limits of the sequences e^n and e^{-n}, as n approaches ∞, are ∞ and 0, respectively. Since e^x is an increasing function, it follows that $\lim_{x \to \infty} e^x = \infty$ and $\lim_{x \to -\infty} e^x = 0$. This proves Item 11.

As the domain of e^x is $(-\infty, \infty)$, e^x is continuous (since it's differentiable), and e^x is strictly increasing, the range of e^x is an open interval by Theorem 1.A.19. By Items 6 and 10, that interval must be $(0, \infty)$. This proves Item 12.

We shall defer the proof of Item 13 until Theorem 2.5.10. $\qquad \square$

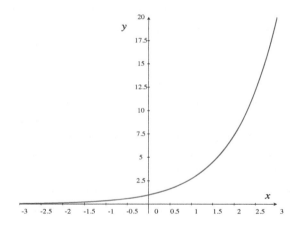

Figure 2.2: The graph of $y = e^x$.

From the properties above, we can sketch a reasonable graph of $y = e^x$. The function e^x is always positive, is always increasing, and its graph is always concave up. The function is continuous, since it's differentiable, and its domain is an interval; hence, the graph is in one connected piece. As $x \to -\infty$, the graph approaches the x-axis as an asymptote, and e^x goes to ∞ as $x \to \infty$. We also know that $e^0 = 1$, and $e^1 \approx 2.7$. Therefore, the graph of $y = e^x$ looks like what we see in Figure 2.2.

Before we leave this section, we would like to look at a few more examples.

Example 2.4.22. We usually think of the Chain Rule as saying "if we have one function composed with another, then we find the derivative of the composition by differentiating the 'outside' function, while leaving the 'inside' function the way it was, but then we have to multiply by the derivative of the 'inside' function". When the outside function is $\exp(x)$, written as e^x, this can get a little confusing, since the terms "inside" and "outside" don't make much sense here.

Consider $m(x) = e^{x^2+1}$. We would like to find $m'(x)$. The Chain Rule looks more applicable if we write $m(x) = \exp(x^2 + 1)$. Then, we quickly find

$$m'(x) \;=\; [\exp'(x^2+1)] \cdot (x^2+1)' \;=\; [\exp(x^2+1)] \cdot 2x \;=\; e^{x^2+1} \cdot 2x.$$

However, you should not **have** to rewrite things in terms of exp. You simply need to get used to the fact that, in a function of the form $e^{g(x)}$, the "outside" function, the function which is applied second, is raising e to a power, i.e., the exponential function, while the "inside" function, the function which is applied first, is the function in the exponent. Thus, the Chain Rule tells us that

$$\left(e^{g(x)}\right)' \;=\; e^{g(x)} \cdot g'(x).$$

Example 2.4.23. Let

$$f(x) \;=\; \frac{e^{2x} - 1}{e^x + 5}.$$

Find $\lim_{x\to\infty} f(x)$, $\lim_{x\to-\infty} f(x)$, $f'(x)$, and determine where f is an increasing function.

Solution:

$$\lim_{x\to\infty} f(x) \;=\; \lim_{x\to\infty} \frac{e^{2x}-1}{e^x+5} \;=\; \lim_{x\to\infty} \frac{e^x - 1/e^x}{1 + 5/e^x}.$$

Since $\lim_{x\to\infty} e^x = \infty$, it follows that, as $x \to \infty$, $1/e^x \to 0$ and $5/e^x \to 0$. Thus,

$$\lim_{x\to\infty} f(x) = \lim_{x\to\infty} \frac{e^x - 1/e^x}{1 + 5/e^x} = \frac{\infty - 0}{1 + 0} = \infty.$$

As $x \to -\infty$, $2x \to -\infty$, and so $e^x \to 0$ and $e^{2x} \to 0$. Thus,

$$\lim_{x\to-\infty} f(x) = \lim_{x\to-\infty} \frac{e^{2x} - 1}{e^x + 5} = \frac{0 - 1}{0 + 5} = -\frac{1}{5}.$$

We calculate the derivative:

$$f'(x) = \frac{(e^x + 5)(e^{2x} - 1)' - (e^{2x} - 1)(e^x + 5)'}{(e^x + 5)^2} =$$

$$\frac{(e^x + 5)(e^{2x} \cdot 2) - (e^{2x} - 1)e^x}{(e^x + 5)^2} = \frac{2e^{3x} + 10e^{2x} - e^{3x} + e^x}{(e^x + 5)^2} =$$

$$\frac{e^{3x} + 10e^{2x} + e^x}{(e^x + 5)^2}.$$

As e^x is always positive, we see that, for all x, $f'(x) > 0$, and so f is strictly increasing on the entire real line $(-\infty, \infty)$.

Example 2.4.24. Recall Example 2.2.9, in which we considered the number, $P = P(t)$, of people in the population of an island, at time t years, where the population is constrained by available resources to a maximum size M (some fixed positive constant).

We assumed that the rate of change of P was given by the logistic model differential equation:

$$\frac{dP}{dt} = kP(M - P), \tag{2.7}$$

where k is a positive constant.

Show that

$$P = \frac{MP_0}{(M - P_0)e^{-Mkt} + P_0} \tag{2.8}$$

is a solution to Formula 2.7 and that, if the population is given by Formula 2.8, then P_0 must be the population at time $t = 0$, i.e., the initial population.

Solution:

First, we find $P(0) = MP_0/[(M - P_0) \cdot 1 + P_0] = P_0$. Thus, P_0 is the population at time $t = 0$.

To show that the given function P is a solution to Formula 2.7, we must calculate dP/dt, calculate $kP(M - P)$, and show that the two are equal.

First, let's look at $M - P$. We have

$$M - P = M - \frac{MP_0}{(M - P_0)e^{-Mkt} + P_0} =$$

$$\frac{M(M - P_0)e^{-Mkt} + MP_0}{(M - P_0)e^{-Mkt} + P_0} - \frac{MP_0}{(M - P_0)e^{-Mkt} + P_0} =$$

$$\frac{M(M - P_0)e^{-Mkt}}{(M - P_0)e^{-Mkt} + P_0}.$$

Therefore, we find, if P is given by Formula 2.8, then

$$kP(M - P) = \frac{kP_0M^2(M - P_0)e^{-Mkt}}{\left((M - P_0)e^{-Mkt} + P_0\right)^2}. \tag{2.9}$$

Now, we need to look at dP/dt. If P is given by Formula 2.8, then

$$\frac{dP}{dt} = \left(\frac{MP_0}{(M - P_0)e^{-Mkt} + P_0}\right)' = MP_0 \cdot \frac{d}{dt}\left([(M - P_0)e^{-Mkt} + P_0]^{-1}\right) =$$

$$MP_0(-1)[(M - P_0)e^{-Mkt} + P_0]^{-2} \cdot \frac{d}{dt}[(M - P_0)e^{-Mkt} + P_0] =$$

$$MP_0(-1)[(M - P_0)e^{-Mkt} + P_0]^{-2} \cdot (M - P_0) \cdot \frac{d}{dt}\left(e^{-Mkt}\right) =$$

$$MP_0(-1)[(M - P_0)e^{-Mkt} + P_0]^{-2} \cdot (M - P_0) \cdot e^{-Mkt} \cdot \frac{d}{dt}(-Mkt) =$$

$$MP_0(-1)[(M - P_0)e^{-Mkt} + P_0]^{-2} \cdot (M - P_0) \cdot e^{-Mkt} \cdot (-Mk).$$

Now, some relatively easy rearrangement/simplification of the final expression shows that it agrees with what we obtained for $kP(M-P)$ in Formula 2.9.

Of course, the real question is: how did we come up with the function $P(t)$ in the first place? We will show how to derive that the given $P(t)$ is the general solution to the logistic equation in Example 4.3.7 in Section 4.3.

2.4.1 Exercises

Differentiate the functions in Exercises 1 through 16.

1. $f(x) = 5e^x + 3$.

2. $g(t) = e^{2t}$.

3. $h(x) = 3e^{x^2+1}$.

4. $p(x) = xe^x$.

5. $q(t) = (12t + 75)e^t$.

6. $r(x) = \dfrac{1}{3}e^{21x}$.

7. $k(x) = 7e^{(x^2)} + 15$.

8. $f(s) = 2e^s + \dfrac{1}{s} + 8$.

9. $f(x) = \dfrac{1}{12e^x}$.

10. $g(x) = (e^{3x} + 19)^{\frac{2}{3}}$.

11. $f(r) = 6r^5 + 15e^r - 25\sqrt[5]{r^7}$.

12. $f(x) = \dfrac{100}{1 + 15e^{1.1x}}$.

13. $g(t) = 18.4e^{-0.3t} + 9.7$.

14. $h(x) = \dfrac{e^x + e^{-x}}{2}$.

15. $q(t) = e^t + \dfrac{1.5}{e^{2t}} + e^{\frac{1}{3}}$.

16. $r(s) = e^{-\frac{1}{s^2}}$.

17. Sketch the graph of the function e^{kx} for the following values of k:

    a. $k = 0.2$.

    b. $k = 1$.

    c. $k = 2$.

Write out b_1, b_2, b_3 and b_4 for the sequences in problems 18 through 20.

18. $b_n = 2^{-n}$.

19. $b_n = n!$.

20. $b_n = (1 + 1/n)^n$.

21. The famous Fibonacci sequence is defined recursively (in terms of previous elements of the sequence) as follows $a_1 = 1$, $a_2 = 1$ and, for $n \geq 3$, $a_n = a_{n-1} + a_{n-2}$. For example, $a_3 = a_2 + a_1 = 2$. What are a_4, a_5 and a_6?

22. Calculate $P_n(x)$ for the following values of n and x.

    a. $n = 3, x = 2$.

    b. $n = 4, x = 0$.

    c. $n = 3, x = -1$.

23.   a. Let $r = 0.8$ and calculate $\sigma_n(r)$ for $n = 5, 10, 50$.

    b. What is $1/(1 - r)$ when $r = 0.8$?

    c. Let $r = 1.4$ and calculate $\sigma_n(r)$ for $n = 5, 10, 50$.

    d. What is $1/(1 - r)$ when $r = 1.4$?

    e. What general observation can you make based on the relationships between parts (a) and (b) and between parts (c) and (d)?

24. Sketch the graphs of e^x and e^{-x}. Find their slopes at $x = 2$ and $x = -2$. Decide which function is increasing and which is decreasing over the interval $[-2, 2]$. What can you say about their behaviors over larger intervals?

25. Each of the following functions is defined in terms of a differentiable function $f(x)$. Find the derivative of each in terms of $f'(x)$ and $f(x)$.

    a. $g(x) = e^{f(x)}$.

    b. $g(x) = e^{f(e^x)}$.

    c. $g(x) = f\left(e^{\left(\frac{1}{x}\right)}\right)$.

    d. $g(x) = f(x) \cdot e^x$.

26. Prove Theorem 2.4.16 by showing that, if a and r are constants, and r is rational, then

$$\frac{\exp(x+a)}{\exp(x)\exp(a)}$$

and

$$\frac{\exp(rx)}{(\exp(x))^r}$$

are constant functions of x, and then substitute $x = 0$ to show that the constant is 1 in both cases.

27. Determine the domains of the following functions:

    a. $f(x) = \sqrt{e^{-x}}$

    b. $h(t) = e^{\sqrt{x}}$

    c. $y(x) = \dfrac{1}{e^x - e}$

28. Let $f(x) = e^x + x - 2$.

    a. Calculate $f(0)$ and $f(1)$.

    b. Show that $f'(x) > 0$ for all x.

    c. Conclude that the equation $e^x + x = 2$ has one and only one solution, and that the solution is in the interval $(0, 1)$.

29. The following function appears in connection with the *Poisson distribution*: $\phi(t) = \exp[\lambda(\exp(t) - 1)]$.

    a. Calculate $\phi'(0)$.

    b. Calculate $\phi''(0)$.

30. Prove that the set of critical points of $f(x) = e^{g(x)}$ is equal to the set of critical points of g. Assume that g is differentiable.

31. Prove that, if g is a twice-differentiable function whose graph is concave up everywhere, then the graph of $f(x) = e^{g(x)}$ is also concave up everywhere. ▶

32. Prove that, if g is twice-differentiable and a is an inflection point of $f(x) = e^{g(x)}$, then $g''(a) = -[g'(a)^2]$.

33. Calculate an upper bound for $e - P_n(1)$ for $n = 3, 4, 5$.

34. If $f(x) = e^{kx}$, what is $f^{(n)}(x)$?

35. Assume that all you know about exp is that its domain is $(-\infty, \infty)$, that it's differentiable at 0, and that, for all a and b, $\exp(a + b) = \exp(a)\exp(b)$. Show that exp is differentiable on $(-\infty, \infty)$ and
$$\exp'(x) = \exp(x) \cdot \exp'(0).$$

36. Let $f(x) = e^x$ and suppose g is a differentiable function defined on $(0, \infty)$ with the property that $f(g(x)) = x$.

    a. Use the Inverse Function Theorem, Corollary 2.3.8, to prove that $g'(x) = e^{-g(x)}$.

    b. Calculate $g''(x)$ in terms of $g(x)$ (as in part (a)).

    c. Use parts (a), (b) and properties of the exponential function to draw a conclusion about the concavity of the graph of $g(x)$.

37. Suppose that the population of a bacteria at time t is modeled by the equation $P(t) = 50,000e^{4t}$. What is $P'(t)/P(t)$?

38. An object is referred to as a *blackbody* if it absorbs all electromagnetic radiation. While a blackbody emits no electromagnetic radiation, it will emit thermal radiation or *blackbody radiation*. Planck's Law expresses the energy density of blackbody radiation in terms of the wavelength λ, the temperature T and the constants π, h, c and k via the equation

$$f(\lambda) = \frac{8\pi hc\lambda^{-5}}{\exp\left(\frac{hc}{\lambda kT}\right) - 1}.$$

    a. What is $\dfrac{df}{d\lambda}$?

    b. The function f can also be interpreted as a function of temperature when λ is held constant. What is $\dfrac{df}{dT}$?

39. The famous bell curve is the graph of the density of a normal distribution and is given by the equation
$$f(x) = \frac{1}{\sigma\sqrt{2\pi}}e^{-\frac{(x-\mu)^2}{2\sigma^2}},$$

where μ is the mean of the distribution, and σ is the standard deviation.

    a. Calculate $f'(x)$ and show that f has a critical point at $x = \mu$.

    b. Calculate $f''(x)$.

    c. Does the graph of f have any inflection points? If so, where?

40. As we mentioned in Example 2.4.15, populations frequently grow or die at a rate that is (roughly) proportional to the current population, that is, populations are frequently assumed to grow or "decay" exponentially.

    Suppose that a certain bacterial population $P = P(t)$ has an (instantaneous) death rate of 5% (of the current population) and an (instantaneous) spawning (birth) rate of 15% (of the current population) per minute.

    a. The above information leads to a differential equation of the form

    $$\frac{dP}{dt} = kP,$$

    where k is equal to what?

    b. Apply Theorem 2.4.14 to find the general form of $P(t)$.

    c. Find the particular expression for $P(t)$, if at time $t = 0$, there are 1000 bacteria living.

41. Show that the function $f(x) = e^{-2x} + e^{3x}$ satisfies the differential equation $f''(x) - f'(x) - 6f(x) = 0$.

42. The function $\cosh(x) = \dfrac{e^x + e^{-x}}{2}$ is called the *hyperbolic cosine*.

    a. Find the critical point(s), if there are any, of $\cosh(x)$.

    b. Find the inflection point(s), if there are any, of the graph of $\cosh(x)$.

    c. Show that $\cosh(x)$ satisfies the equation $f''(x) = f(x)$.

    d. Show that $\cosh(x)$ is an even function.

43. The function $\sinh(x) = \dfrac{e^x - e^{-x}}{2}$ is called the *hyperbolic sine*.

    a. Find the critical point(s), if there are any, of $\sinh(x)$.

    b. Find the inflection point(s), if there are any, of $\sinh(x)$.

    c. Show that $\sinh(x)$ satisfies the equation $f''(x) = f(x)$.

    d. Show that $\sinh(x)$ is an odd function.

44. The function $\tanh(x) = \dfrac{\sinh(x)}{\cosh(x)}$ is called the *hyperbolic tangent*. Calculate the derivative of $\tanh(x)$.

45. Show that $\cosh^2(x) - \sinh^2(x) = 1$.

46. Identify any inflection points of the $f(x) = e^{-1/x}$.

47. Use a calculator and plug in values of x close to 0 in order to determine/guess the value of $\lim\limits_{x \to 0} \dfrac{e^x - 1 - x}{x^2}$.

48. Calculate $\lim\limits_{x \to 0} \dfrac{e^x - 1}{x^4 + 5x^3 + 6x}$.

49. Suppose that the resources of an island can support no more than 300 million inhabitants and that the population of the island in 2009 is 150 million.

     a. Using the logistic model, where $k = 0.0002$ (with units of 1/person-yr), what is the predicted population in 2014?

     b. When does the model predict the population will reach 200 million?

50. If $f(x) = e^x$, explain why it is incorrect to say $f'(x) = xe^{x-1}$.

51. The *demand* D for a product is a function of price p; $D = D(p)$ is the number of sales of the product when the price charged per unit is p.

     a. Write the expression for revenue (net income from sales of the product) as a function of price and demand.

     b. The exponential function is one which is often used to model demand. Thus, the formula for demand is often of the form $D(p) = Ce^{-kp}$ (where k is positive, since demand should drop as price increases, and C is positive to reflect that when the product is free, at least some will be taken). Find the instantaneous rate of change of the revenue, as a function of price, given this demand function.

     c. Under this assumption on demand, find the price, in terms of k, which would maximize revenue.

52. *Newton's Law of Cooling* can be stated in the following way: if $T = T(t)$ is the temperature of an object, which is placed in surroundings which are kept at a constant (ambient) temperature, T_{amb}, then the rate of change of the temperature of the object, with respect to time, is proportional to the difference between the ambient temperature and the temperature of the object.

Thus, we obtain the differential equation

$$\frac{dT}{dt} = k\left(T_{\text{amb}} - T\right),$$

where $k > 0$ is some constant inherent to the environment and the object.

     a. Let $v = dT/dt$, take the derivative of both sides of differential equation, with respect to t, and write your answer in terms of v and dv/dt.

    b. According to Theorem 2.4.14, what form must the function $v = v(t)$ have?

    c. Now, use that $v = dT/dt$, together with the fact that the original differential equation must be satisfied, to obtain the general form for $T = T(t)$.

    d. Can you explain why we obtain the same form for the differential equation in Newton's Law of Cooling, regardless of what units are used for t, and regardless of what temperature scale is being used for T?

53. Recall that two curves are said to be *orthogonal* at a point P, where they intersect, provided that their tangent lines exist at P, and their tangent lines at P are perpendicular.

    Find a family of parabolas $y = ax^2 + bx + c$ that are orthogonal to the graph of the exponential function $f(x) = e^{kx}$, at the point $(0, 1)$; that is, determine what relations must be satisfied between a, b, c, and k.

54. Suppose that n is an integer. Show that $\lim\limits_{x \to \infty} \dfrac{x^n}{e^x} = 0$.

55. Use the ideas in the proof of Theorem 2.4.14, to prove the following:

    Theorem: Suppose that f and g are differentiable on the open interval I, and that, for all x in I,
    $$\frac{df}{dx} = g'(x)f(x).$$
    Then, on the interval I, $f(x) = Ce^{g(x)}$, where C is a constant.

56. Let $f_1(x) = e^x$, $f_2(x) = e^{f_1(x)}$ and $f_n(x) = e^{f_{n-1}(x)}$ for $n = 2, 3, 4, \ldots$. Prove that $f'_n(x) = f_1(x) \cdot f_2(x) \cdot \ldots \cdot f_n(x)$.

2.5   The Natural Logarithm

In the previous section, we defined the exponential function, $\exp(x) = e^x$. The domain of exp is the entire real line $(-\infty, \infty)$ and its range is the set of positive real numbers $(0, \infty)$. As exp is strictly increasing, it is one-to-one. Therefore, $\exp : (-\infty, \infty) \to (0, \infty)$ has an inverse function. As $\exp(x) = e^x$, it is hardly surprising that we refer to the inverse function of exp as a type of logarithm; it is the base e logarithm, the function that "undoes" raising e to an exponent.

Definition 2.5.1. *The* **natural logarithm***,* $\ln : (0, \infty) \to (-\infty, \infty)$ *is the inverse function of* $\exp : (-\infty, \infty) \to (0, \infty)$.

The function $\ln$ *is also called the* **base** e **logarithm** *and is sometimes denoted by* $\log_e$.

The following properties are immediate from the properties of $\exp(x) = e^x$, and the fact that the natural logarithm function is the inverse of the exponential function.

Theorem 2.5.2. *The natural logarithm has the following properties:*

1. *The domain of* $\ln$ *is* $(0, \infty)$ *and its range is* $(-\infty, \infty)$;

2. *for all* x, *and for all* $y > 0$,
$$\ln(e^x) = x \qquad \text{and} \qquad e^{\ln y} = y;$$

3. $\ln(1) = 0$ *and* $\ln(e) = 1$; *and*

4. $\lim_{x \to \infty} \ln x = \infty$, *and* $\lim_{x \to 0+} \ln x = -\infty$.

Theorem 2.5.3. *For all* $x > 0$, $\ln'(x) = 1/x$, *and so* $\ln''(x) = -1/x^2$.

Thus, $\ln x$ *is strictly increasing on its domain, and its graph is always concave down.*

Proof. By Theorem 2.3.8, for all $x > 0$,

$$\ln'(x) \;=\; \frac{1}{\exp'(\ln x)} \;=\; \frac{1}{\exp(\ln x)} \;=\; \frac{1}{x}.$$

The second derivative formula follows from the Power Rule, applied to x^{-1}. $\square$

We could sketch the graph of $y = \ln x$ from the properties above or, since ln is the inverse function of exp, we can just "reflect the graph" of $y = e^x$ across the line $y = x$.

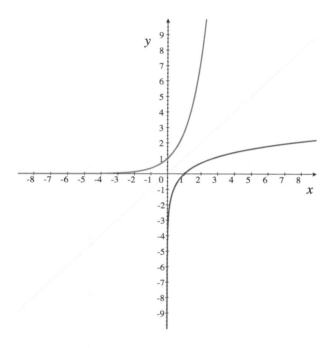

Figure 2.3: The graphs of $y = e^x$ and $y = \ln x$.

Let's look at a few examples of calculating derivatives involving $\ln x$.

Example 2.5.4. Calculate $(-x + x \ln x)'$.

Solution:

We use linearity and the Product Rule to find:

$$(-x + x \ln x)' \;=\; -1 + x \cdot \frac{1}{x} + (\ln x) \cdot 1 \;=\; -1 + 1 + \ln x \;=\; \ln x.$$

Thus, recalling our definition of *anti-derivative* in Example 1.5.13, we see that $-x + x \ln x$ is an anti-derivative of $\ln x$.

Example 2.5.5. Calculate

$$\left[\frac{\ln(e^r + r)}{5 + \ln r}\right]'.$$

Solution:

$$\left[\frac{\ln(e^r + r)}{5 + \ln r}\right]' = \frac{(5 + \ln r)[\ln(e^r + r)]' - [\ln(e^r + r)](5 + \ln r)'}{(5 + \ln r)^2} =$$

$$\frac{(5 + \ln r)\left[\dfrac{1}{e^r + r} \cdot (e^r + r)'\right] - [\ln(e^r + r)] \cdot \dfrac{1}{r}}{(5 + \ln r)^2} =$$

$$\frac{(5 + \ln r)\left(\dfrac{e^r + 1}{e^r + r}\right) - \dfrac{\ln(e^r + r)}{r}}{(5 + \ln r)^2}.$$

This final expression can be manipulated further algebraically, but nothing is going to make the answer look terribly nice.

Example 2.5.6. Consider the function $g(x) = \ln(-x)$, with its "natural domain", i.e., the domain of g is the set of $x < 0$, so that $-x > 0$ and we may apply the natural logarithm.

Applying the Chain Rule, we find

$$g'(x) = \frac{1}{-x} \cdot (-x)' = \frac{1}{-x} \cdot -1 = \frac{1}{x}.$$

It is tempting to look at this result and say "Ah ha! The functions $\ln x$ and $\ln(-x)$ have the same derivative, $1/x$, and so Corollary 1.5.12 tells us that there exists a constant C such that $\ln(-x) = \ln x + C$." This is **not** correct. There is no open interval on which both $\ln x$ and $\ln(-x)$ are defined; Corollary 1.5.12 does not apply.

We can use the absolute value function $|x|$ to write one formula for both $(\ln x)'$ and $\big(\ln(-x)\big)'$. By definition, for all x, if $x \geq 0$ then $|x| = x$, and, if $x \leq 0$, then $|x| = -x$.

Therefore, what we have shown above is that, for all $x \neq 0$, we have

$$\big(\ln(|x|)\big)' \;=\; \frac{1}{x}. \tag{2.10}$$

Example 2.5.7. Suppose that an object of constant mass m is moving in a straight line through a fluid, and the only force acting on the object is the force of resistance, F_{resist}, from the fluid.

Suppose that b and v_0 are positive constants, x_0 is a constant, and that, for all $t > -1/(bv_0)$, the position x at time t is given by

$$x \;=\; \frac{1}{b}\,\ln(bv_0 t + 1) + x_0.$$

Show that x_0 is the position of the object at $t = 0$, v_0 is the velocity of the object at $t = 0$, and that F_{resist} is always proportional to the square of the velocity of the object, but acts (pushes) in the direction opposite that of the motion of the object (as a force of resistance should).

Solution:

Newton's 2nd Law of Motion, for a constant mass, tells us that

$$F_{\text{resist}} \;=\; ma,$$

where $a = a(t)$ is the acceleration of the object. Therefore, we need to calculate the velocity and acceleration of the object, in addition to verifying that x_0 and v_0 are the position and velocity, respectively, of the object at $t = 0$.

We find that

$$x(0) \;=\; \frac{1}{b}\,\ln(1) + x_0 \;=\; 0 + x_0 \;=\; x_0;$$

thus, x_0 is the position of the object at $t = 0$.

The velocity $v = v(t)$ of the object is the rate of change of the position, with respect to time.

We calculate

$$v \;=\; \frac{dx}{dt} \;=\; \frac{1}{b} \cdot \frac{1}{bv_0 t + 1} \cdot (bv_0 t + 1)' + 0 \;=\; \frac{1}{b} \cdot \frac{1}{bv_0 t + 1} \cdot (bv_0) \;=\; \frac{v_0}{bv_0 t + 1}.$$

When $t = 0$, we find that $v(0) = v_0/(0 + 1) = v_0$, and so v_0 is, indeed, the velocity of the object at $t = 0$.

Note that, as v_0 is positive, v is always positive (for $t > -1/(bv_0)$). Thus, the object is always moving in the positive direction, i.e., the x-coordinate of the object is always increasing.

We calculate the acceleration of the object by finding the rate of change, with respect to time, of the velocity:

$$a \;=\; \frac{dv}{dt} \;=\; \left(\frac{v_0}{bv_0 t + 1} \right)' \;=\; v_0 \big[(bv_0 t + 1)^{-1} \big]' \;=$$

$$v_0 \cdot -1(bv_0 t + 1)^{-2}(bv_0 t + 1)' \;=\; -v_0(bv_0 t + 1)^{-2}(bv_0) \;=\; \frac{-bv_0^2}{(bv_0 t + 1)^2}.$$

Comparing this with what we found for v, we see that

$$a \;=\; -bv^2.$$

Multiplying by the mass m, we find

$$F_{\text{resist}} \;=\; ma \;=\; -mbv^2.$$

As $-mb$ is a constant, we conclude that F_{resist} is proportional to the square of the velocity. In addition, we see that the force of resistance is negative, i.e., acts in the direction opposite that of the velocity.

We shall return to this example in Example 4.4.4, where we will start with the fact that F_{resist} is proportional to the square of the velocity, and determine that the position function must be of the form we began with here.

In addition to the properties of $\ln x$ given in Theorem 2.5.2 and Theorem 2.5.3, natural logarithm has all of the algebraic properties that you expect from a logarithm. Note that we continue to refer to exponents that are rational numbers, as we have yet to define any base, other than e, raised to irrational powers.

Theorem 2.5.8. *The natural logarithm has the following algebraic properties:*

1. *if $x > 0$, $\ln(1/x) = -\ln(x)$*

2. *for all $x > 0$ and $a > 0$, $\ln(ax) = \ln a + \ln x$;*

3. *for all $x > 0$ and $a > 0$, $\ln(a/x) = \ln a - \ln x$; and*

4. *for all $x > 0$ and all rational numbers r, $\ln(x^r) = r \ln x$.*

Proof. These properties follow from the corresponding properties for e^x.

Suppose that $x > 0$. Let $y = \ln x$, so that $e^y = x$. Then, $1/x = e^{-y}$, and we find

$$\ln(1/x) \;=\; \ln(e^{-y}) \;=\; -y \;=\; -\ln x.$$

Suppose that $x, a > 0$. Let $y = \ln x$, and let $b = \ln a$, so that $e^y = x$, $e^b = a$, and $ax = e^b \cdot e^y = e^{b+y}$. Then,

$$\ln(ax) \;=\; \ln(e^{b+y}) \;=\; b+y \;=\; \ln a + \ln x.$$

The formula for $\ln(a/x)$ follows by combining Items 1 and 2.

Finally, suppose that $x > 0$ and r is a rational number. Let $y = \ln x$, so that $e^y = x$. Then, $e^{ry} = (e^y)^r = x^r$, and so

$$\ln(x^r) \;=\; \ln(e^{ry}) \;=\; ry \;=\; r \ln x.$$

$\square$

We can use the algebraic properties of $\ln x$, combined with its derivative formula, to quickly differentiate complicated functions which involve many products, quotients, and exponents. What we are about to demonstrate usually goes by the name *logarithmic differentiation*.

Example 2.5.9. Suppose that

$$y = \frac{(x^2+1)^6 \cdot e^{5x^3-2x}}{\sqrt{x^4+7}}.$$

Then, using the algebraic properties of ln, we obtain

$$\ln y = 6\ln(x^2+1) + 5x^3 - 2x - \frac{1}{2}\ln(x^4+7). \qquad (2.11)$$

Now, y is a function of x, so that $\ln y$ is a composition of functions, and we have

$$\frac{d}{dx}(\ln y) = \frac{d}{dy}(\ln y) \cdot \frac{dy}{dx} = \frac{1}{y}\frac{dy}{dx}.$$

Therefore, if we differentiate both sides of Formula 2.11, with respect to x, we obtain

$$\frac{1}{y}\frac{dy}{dx} = 6 \cdot \frac{1}{x^2+1} \cdot 2x + 15x^2 - 2 - \frac{1}{2} \cdot \frac{1}{x^4+7} \cdot 4x^3.$$

Finally, multiplying both sides by y, we are finished:

$$\frac{dy}{dx} = y \cdot \left(\frac{12x}{x^2+1} + 15x^2 - 2 - \frac{2x^3}{x^4+7} \right) =$$

$$\frac{(x^2+1)^6 \cdot e^{5x^3-2x}}{\sqrt{x^4+7}} \cdot \left(\frac{12x}{x^2+1} + 15x^2 - 2 - \frac{2x^3}{x^4+7} \right).$$

We can use our results about the derivative of $\ln x$ and its algebraic properties to give another well-known and useful characterization of the exponential function. We stated this characterization, without proof, as Item 13 in Theorem 2.4.21.

Theorem 2.5.10. *For all x,*

$$e^x = \lim_{n \to \infty} \left(1 + \frac{x}{n}\right)^n,$$

where the limit is over natural numbers n.

Proof. If $x = 0$, then the equality certainly holds. So, we will assume that $x \neq 0$ throughout the remainder of the proof.

We know that $\ln'(1) = 1/1 = 1$. Therefore, since $\ln(1) = 0$,

$$\lim_{h \to 0} \frac{\ln(1+h)}{h} = \lim_{h \to 0} \frac{\ln(1+h) - \ln(1)}{h} = \ln'(1) = 1.$$

Let $\hat{h} = h/x$, so that $h = x\hat{h}$. Then, from the above, we have

$$1 = \lim_{h \to 0} \frac{\ln(1+h)}{h} = \lim_{\hat{h} \to 0} \frac{\ln(1+x\hat{h})}{x\hat{h}} = \frac{1}{x} \cdot \lim_{\hat{h} \to 0} \frac{\ln(1+x\hat{h})}{\hat{h}}.$$

Therefore,

$$\lim_{\hat{h} \to 0} \frac{\ln(1+x\hat{h})}{\hat{h}} = x.$$

As this limit exists and equals x, as $\hat{h}$ approaches 0, through an interval around 0, the sequence obtained by replacing $\hat{h}$ with $1/n$ must converge to x also, i.e.,

$$\lim_{n \to \infty} \frac{\ln\left(1 + \frac{x}{n}\right)}{1/n} = x.$$

Thus,

$$\lim_{n \to \infty} \frac{\ln\left(1 + \frac{x}{n}\right)}{1/n} = x.$$

Hence,

$$\lim_{n \to \infty} \ln\left[\left(1 + \frac{x}{n}\right)^n\right] = \lim_{n \to \infty} \frac{\ln\left(1 + \frac{x}{n}\right)}{1/n} = x.$$

As exp is continuous, we obtain

$$\lim_{n \to \infty} \left(1 + \frac{x}{n}\right)^n = \lim_{n \to \infty} \exp\left(\ln\left[\left(1 + \frac{x}{n}\right)^n\right]\right) =$$

$$\exp\left(\lim_{n\to\infty} \ln\left[\left(1 + \frac{x}{n}\right)^n\right]\right) = \exp(x).$$

$\square$

Example 2.5.11. Theorem 2.5.10 actually comes up in an important way in calculations of interest; for instance, the interest on money in a savings account.

Banks typically specify a yearly interest rate as a percentage. In this example, we'll assume an annual 2% interest rate.

If this were a *simple* interest rate, then, if you had $100 in the bank at the beginning of the year (and didn't deposit more or withdraw any), at the end of the year, you would have $102 in the bank, i.e., $1 + 0.02$ times what you started with.

However, banks typically *compound* interest quarterly, monthly, or over some other period. Here, we will assume monthly. What does it mean to compound the annual 2% interest rate monthly? It means that each month, you receive $(2/12)\%$ interest. You may wonder what the difference is, since it sounds like, after 12 months, you'd still just have $(12 \cdot 2/12)\% = 2\%$. The difference is that, when the interest is compounded, the interest from the prior months is already placed into your account, so that in following months, you receive interest not just on your original deposit, but also on the interest that you've already accrued up to that point.

How much money would you end up with after one year, if you started with $100, and you received an annual 2% interest rate, compounded monthly? At the end of each month, you would have $(1 + 0.02/12)$ times however much you had at the beginning of the month. Thus, at the end of the first month, you'd have

$$\$100 \cdot \left(1 + \frac{0.02}{12}\right),$$

after 2 months,

$$\$100 \cdot \left(1 + \frac{0.02}{12}\right) \cdot \left(1 + \frac{0.02}{12}\right) = \$100 \cdot \left(1 + \frac{0.02}{12}\right)^2,$$

and, after 12 months,

$$\$100 \cdot \left(1 + \frac{0.02}{12}\right)^{12} \approx \$102.02.$$

Okay. So, compounding the interest monthly only gave you slightly less than 2 cents more over a year than a simple 2% interest rate, but you might think that if the interest were compounded daily, or hourly, or every second, then you'd end up with a lot of money. This is **not** the case.

Look at the formula that we obtained for the amount of money that we would have with a 2% interest rate, compounded over 12 periods. The exact same reasoning would tell us that, if we compound the 2% interest over n periods, then, after a year, the amount of money in the bank would be

$$\$100 \cdot \left(1 + \frac{0.02}{n}\right)^n .$$

Even if the interest were compounded arbitrarily often, Theorem 2.5.10 tells us that we would never end up with more money than

$$\lim_{n \to \infty} \$100 \cdot \left(1 + \frac{0.02}{n}\right)^n \; = \; \$100 e^{0.02} \; \approx \; \$102.020134,$$

where we have shown a large number of decimal places to emphasize that compounding infinitely often still doesn't get you even an extra 3 cents over what simple interest gets you.

From our reasoning above, it is easy to derive a general formula for the amount of money $A = A(t)$ that you would have in the bank after t years, assuming you start with an amount A_0, make no withdrawals, that the annual interest rate is r, and that the interest is compounded n times each year. We find

$$A(t) \; = \; A_0 \left[\left(1 + \frac{r}{n}\right)^n\right]^t \; = \; A_0 \left(1 + \frac{r}{n}\right)^{nt},$$

where, strictly speaking, this formula is valid only when nt is a non-negative integer, since the bank is not continuously depositing interest, but rather deposits interest at n specified times each year.

In such financial calculations, when n is large, it is fairly standard to use the approximation

$$\left(1 + \frac{r}{n}\right)^n \; \approx \; e^r,$$

which yields the approximation

$$A(t) \; \approx \; A_0 e^{rt}.$$

How good is this approximation? Let's start with $A_0 = \$1000$, assume an annual interest rate of $2\% = 0.02$, and look at the amount money you would have after 10 years, if the interest were compounded quarterly (4 times a year) and, again, if the interest were compounded monthly. We'll compare both of these with the exponential approximation.

If the interest were compounded quarterly, you would end up with

$$1000\left(1+\frac{0.02}{4}\right)^{40} \approx 1220.79$$

dollars after 10 years.

If the interest were compounded monthly, you would end up with

$$1000\left(1+\frac{0.02}{12}\right)^{120} \approx 1221.20$$

dollars after 10 years.

Our exponential approximation gives

$$1000e^{0.2} \approx 1221.40$$

dollars after 10 years. That's a pretty good approximation to either actual amount. Not surprisingly, the approximation is better for the larger number of compounding periods.

2.5.1 Exercises

Differentiate the functions in Exercises 1 through 14. It may be useful to use the properties of the natural logarithm to simplify some of the functions before calculating the derivatives.

1. $f(x) = 1.5x^2 + 3\ln x + 2.8$.

2. $g(x) = x\ln x + e^{2x}$.

3. $h(x) = 11\ln(x^2 + x + 1) + e^9$.

4. $r(x) = \dfrac{x-1}{\ln x}$.

5. $p(x) = 6\ln(e^{x^2+1})$.

6. $g(x) = \dfrac{1 + 2\ln x}{3x^2 + 2x + 5}$.

7. $f(t) = \ln\left(t + \dfrac{1}{t} + 2\right)$.

8. $p(s) = 1.8e^{\ln(12s+3)}$.

9. $f(x) = \dfrac{1}{x^2 \ln x}$.

10. $g(t) = \ln(10 + 1.5e^{-3t})$.

11. $h(x) = \dfrac{\ln x}{x} + \dfrac{1}{x} + \dfrac{1}{\ln(x)}$.

12. $p(x) = \sqrt[3]{\ln x} + 1.2$.

13. $r(x) = \ln(\sqrt{x^2 - 1})$.

14. $k(x) = (\ln x)(\ln x)$.

Use logarithmic differentiation to find $\dfrac{dy}{dx}$ **for each of the functions in Exercises 15 through 21.**

15. $y = \dfrac{75}{12e^{-1.8x}}$.

16. $y = \dfrac{1}{3}\sqrt{x}\,e^{x^2+1}$.

17. $y = \dfrac{e^{3x^2+2x+1}\sqrt{x+1}}{2x^2 + 5}$.

18. $y = \sqrt{(x^2 - 4)(x^2 + 7)}$.

19. $y = \sqrt[3]{(x - a)(x - b)(x - c)}$, where a, b and c are three distinct constants.

20. $y = \dfrac{\sqrt{x^2 - 11}}{x^6\sqrt{x^2 + 5}}$.

21. $g(x) = x^x$.

22. What are the domain and range of the natural logarithm?

23. Suppose the position of a particle, in meters, at time $t > 0$ seconds, is given by $p(t) = \ln t$. What is the jerk?

24. Let $a_n = \left(1 + \frac{1}{n}\right)^n$.

a. Calculate a_1, a_2 and a_3.

b. What is $\lim_{n \to \infty} a_n$?

In Exercises 25 through 27, locate points where the second derivative is zero and determine whether or not they are inflection points.

25. $g(t) = \frac{1}{2}t^{-1} - \ln t$.

26. $h(t) = \frac{1}{2}t^{-1} + \ln t$.

27. $j(t) = \frac{1}{2}t^2 - (5t + 6)\ln t - 5t$.

28. Reprove the Power Rule for $y = x^r$, where $x > 0$ and r is rational, by using logarithmic differentiation.

29. Consider the function $f(x) = \dfrac{(x-2)^{10}}{(x-10)^2}$, defined on the interval $I = (10, \infty)$.

    a. Use logarithmic differentiation to determine all critical points of $f(x)$.

    b. Determine, by using the First Derivative Test, Theorem 1.5.20, whether each critical point yields a maximum or a minimum value of f.

30. This problem presents a second method of proving Theorem 2.5.8. Suppose that a and b are positive. For x in $(0, \infty)$, define $f(x) = \ln(ax) - \ln(x)$. Show that $f'(x) = 0$ for all x in $(0, \infty)$, and conclude that f is a constant function. Choose a convenient value for x to determine the constant value of f (in terms of a and/or b), and then conclude that $\ln(ab) = \ln(a) + \ln(b)$.

In Exercises 31 through 33, assume that an object is moving in a straight line through a fluid according to the setup in Example 2.5.7.

31. If the velocity of the object moving through the fluid at time $t = 3$ seconds is 6 m/s and v_0 is 9 m/s, then what is b?

32. What are the units of b?

33. If $b = 3$ and $v_0 = 5$, find expressions for the $v(t)$ and $a(t)$.

34. Suppose you deposit \$50,000 in a savings account with a 6% interest rate. Calculate the amount of money you'll have in the account after one year if compounding occurs:

    a. Annually.

    b. Monthly.

    c. Weekly.

    d. Daily (assume 365 days in a year).

    e. Continuously.

35. Prove that there is one and only solution to the equation $\ln x + x = 0$.

36. Suppose you deposit \$60,000 in a savings account with an 8% interest rate. The amount $A(n)$ in the account after one year where the compounding frequency is n is given by $A(n) = 60000(1 + 0.08/n)^n$. Calculate the AROC of $A(n)$ over the given intervals.

    a. $[1, 12]$.

    b. $[12, 52]$.

    c. $[52, 365]$.

In Exercises 37 through 40, find an equation of the tangent line to the graph of the function at $x = a$. Leave your answer slope-intercept form.

37. $g(x) = \ln(x + 1)$, $a = 0$.

38. $h(x) = \ln\left(\dfrac{9 + x^2}{9 - x^2}\right)$, $a = 0$.

39. $w(x) = x^9 e^x$, $a = 1$.

40. $f(x) = \ln(\ln x)$, $a = e$.

41. Use logarithmic differentiation to calculate the derivative of $y = k^{(k^x)}$ where $k > 0$.

42. Show that $y = 5 + 7e^{-t/3}$ satisfies the differential equation $\dfrac{d}{dt}[\ln(y - 5)] = -\dfrac{1}{3}$.

43. The *Gompertz equation* is often used to predict population growth, and is given by

$$\frac{dy}{dt} = cy \ln\left(\frac{y}{a}\right),$$

where $a > 0$ and $c < 0$ are constants. Verify that, for all constants b, $y = ae^{be^{ct}}$ is a solution to the Gompertz equation.

44. Use the fact that $(\ln x)' = 1/x$ and the Inverse Function Theorem, Corollary 2.3.8, to prove that $(e^x)' = e^x$.

45. Let $f_1(x) = \ln x$ and, for integers $n \geq 2$, let $f_n(x) = \ln[f_{n-1}(x)]$. Prove that

$$f_n'(x) = \frac{f_1'(x)}{f_1(x) \cdot f_2(x) \cdot \ldots \cdot f_{n-1}(x)}.$$

46. Use the definition of the derivative to prove $\lim\limits_{x \to 0} \dfrac{\ln(1+x)}{x} = 1$. Hint: if $f(x) = \ln x$, what is $f'(1)$?

47. Find a formula for the n-th derivative of $\ln x$.

48. Suppose that the spread of a disease in a scarcely populated area is modeled by the function $D(t) = t \ln t + t + 1$ for $t > 0$, where $D(t)$ is the number of people (in hundreds) infected by the disease, and t is measured in years. It is an easy application of l'Hôpital's Rule, Theorem 3.5.1, that $\lim_{t \to 0^+} D(t) = 1$; thus, for all practical purposes, we may assume that, at time $t = 0$, 100 people have the disease.

   a. Use a calculator to approximate how many people are infected at $t = 10$ years.

   b. Show that the number of people with the disease decreases for a while after $t = 0$, then attains a global minimum, and, after that, is always increasing. Find the unique critical point for the function $D(t)$.

   c. How many cases of the disease are there at the time given by the critical point?

   d. We know that (essentially) 100 people have the disease at $t = 0$. There is a unique $t > 0$ at which the number of people with the disease is once again 100. Find this t value.

49. Calculate the average rate of change of $f(x) = \ln x$ on the interval $[e^n, e^m]$ (where $n < m$) in terms of n and m, and call it $\chi_{n,m}$. Then, for each such interval, find a c value guaranteed by the Mean Value Theorem, i.e., a value c such that $f'(c) = \chi_{n,m}$ and $e^n < c < e^m$. For each fixed $\chi_{n,m}$, is this c value unique?

50. Consider the family of functions

$$f(x) = a \ln x + bx^2 + cx.$$

   a. Calculate $f'(x)$ and $f''(x)$ in terms of a, b, and c.

   b. Find the values of a, b, and c such that $f(1) = f'(1) = f''(1) = 1$.

2.6 General Exponential and Logarithmic Functions

Suppose that $b > 0$. If r is a rational number, a quotient of integers, then we certainly know what b^r means. If we put things in terms of base e, we get

$$b^r = (e^{\ln b})^r = e^{r \ln b}.$$

Therefore, if we want to define b^x, for arbitrary real x, and we want b^x to be a continuous function of x, there's only one way to do it.

Definition 2.6.1. *Suppose that $b > 0$ and x is a real number. Then, we define the **base b exponential function**, b^x, by*

$$b^x = \exp(x \ln b) = e^{x \ln b}.$$

The domain of b^x is $(-\infty, \infty)$ and, if $b \neq 1$, its range is $(0, \infty)$.

*A function of the form $f(x) = c\, b^x$ is called a **general exponential function**.*

Note that this definition, when $b = e$, agrees with our earlier definition of e^x, since $e^{x \ln e} = e^{x \cdot 1} = e^x$.

We can now demonstrate familiar algebraic properties.

Theorem 2.6.2. *Suppose that $a > 0$, $b > 0$, and x and y are real numbers. Then,*

1. $b^0 = 1$;
2. $b^{x+y} = b^x \cdot b^y$;
3. $b^{-x} = 1/b^x$;
4. $(b^x)^y = b^{xy}$;
5. $\ln(b^x) = x \ln b$;
6. $a^x \cdot b^x = (ab)^x$.

Proof. Of course, these properties all follow from the definition of raising b to a power, and the corresponding properties for raising e to powers. We will demonstrate Items 4, 5, and 6, and leave the other items as exercises.

We find

$$(b^x)^y \ = \ (e^{x \ln b})^y \ = \ e^{y \ln \left(e^{x \ln b}\right)} \ = \ e^{yx \ln b} \ = \ b^{xy},$$

$$\ln(b^x) \ = \ \ln(e^{x \ln b}) \ = \ x \ln b,$$

and

$$a^x \cdot b^x \ = \ e^{x \ln a} \cdot e^{x \ln b} \ = \ e^{x \ln a + x \ln b} \ = \ e^{x(\ln a + \ln b)} \ = \ e^{x \ln(ab)} \ = \ (ab)^x.$$

$\square$

The derivative of b^x is easy to calculate.

Theorem 2.6.3. *Suppose that $b > 0$. Then,*

$$(b^x)' \ = \ b^x \ln b.$$

Proof. This follows from the Chain Rule:

$$(b^x)' \ = \ (e^{x \ln b})' \ = \ e^{x \ln b}(x \ln b)' \ = \ e^{x \ln b} \ln b \ = \ b^x \ln b.$$

$\square$

Example 2.6.4. Calculate

$$\left(\ln(3^t + 1)\right)'.$$

Solution:

$$\left(\ln(3^t + 1)\right)' \ = \ \frac{1}{3^t + 1}\,(3^t + 1)' \ = \ \frac{3^t \ln 3}{3^t + 1}.$$

Example 2.6.5. Let $f(x) = x\,2^x$. Calculate df/dx, and determine where f is increasing and where it is decreasing. Show that f attains a global minimum value, and determine that value.

Solution:

$$\frac{df}{dx} \;=\; \frac{d}{dx}\,(x\,2^x) \;=\; x\,\frac{d}{dx}(2^x) \;+\; 2^x\,\frac{d}{dx}(x) \;=\; x\,2^x \ln 2 + 2^x \;=\; 2^x(x \ln 2 + 1).$$

As 2^x is always positive, df/dx will be positive or negative precisely when $x \ln 2 + 1$ is positive or negative, respectively. Note that, since $2 > 1$ and $\ln x$ is a strictly increasing function, $\ln 2 > \ln 1 = 0$.

Therefore, f is decreasing where $x \ln 2 + 1 < 0$, i.e., on the interval $x < -1/\ln 2$, and f is increasing on the interval $x > -1/\ln 2$. The First Derivative Test, Theorem 1.5.20, tells us that f attains a global minimum value at the critical point $x = -1/\ln 2$, and that minimum value is

$$f\!\left(-\frac{1}{\ln 2}\right) \;=\; -\frac{1}{\ln 2}\cdot 2^{-1/\ln 2} \;=\; -\frac{1}{\ln 2}\cdot (e^{\ln 2})^{-1/\ln 2} \;=\; -\frac{1}{e \ln 2}.$$

As an immediate corollary to Theorem 2.6.3, we have:

Corollary 2.6.6. *Suppose that $b > 0$. Then,*

1. *if $b = 1$, $b^x = 1^x = 1$;*

2. *if $b > 1$, b^x is strictly increasing, and its graph is always concave up; and*

3. *if $b < 1$, b^x is strictly decreasing, and its graph is always concave up.*

Proof. This is immediate from the theorem, combined with the fact that $\ln b > 0$ (resp., $\ln b < 0$) if $b > 1$ (resp., $b < 1$), and

$$(b^x)'' \;=\; (b^x \ln b)' \;=\; b^x (\ln b)^2.$$

$\square$

Remark 2.6.7. In fact, since $b^{-x} = (b^{-1})^x = (1/b)^x$, you can obtain the graphs of exponential functions with base less than 1 by reflecting, across the y-axis, the graph of the exponential function with the reciprocal base, which will be greater than 1.

For example, here we graph both $y = 2^x$ and $y = (1/2)^x = 2^{-x}$.

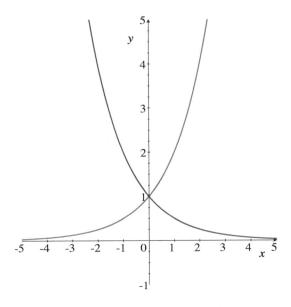

Figure 2.4: The graphs of $y = 2^x$ and $y = (1/2)^x$.

You may wonder why we ever use any constant base other than e in an exponential function; after all, we can convert any other base to base e easily. The reason is that some problems just present themselves in forms that lead naturally to other bases.

Example 2.6.8. Recall Example 2.4.15, in which we discussed radioactive decay.

Let $A = A(t)$ denote the mass of a radioactive substance at time t. Assuming that radioactive material decays at a rate that is proportional to the amount present, we have $dA/dt = -kA$, where $k > 0$ is a constant. As we saw in Example 2.4.15, Theorem 2.4.14 tells us that

$$A(t) \;=\; A_0 e^{-kt} \;=\; A_0 (e^{-k})^t, \qquad\qquad (2.12)$$

where $A_0 = A(0)$ is the initial mass of the radioactive material.

Thus, we know that $A(t)$ is an exponential function, but e^{-k} could be any real number b (actually, any real number less than 1, since $k > 0$); so we could just as easily write $A(t) = A_0 b^t$. Is there some base b that would be better to use than e? **Yes**, because of the way that radioactive decay rates are specified.

The decay rate of radioactive material is usually specified by giving the material's *half-life*. Assuming that $A_0 \neq 0$, the half-life, H, is the amount of time that it takes for half of the material to decay, i.e., H is the time such that $A(H) = A_0/2$. Specifying H determines the proportionality constant k in Formula 2.12, for you solve

$$A_0/2 \; = \; A(H) \; = \; A_0 e^{-kH},$$

and find

$$k \; = \; \frac{\ln(1/2)}{-H} \; = \; \frac{\ln 2}{H}.$$

Note that the A_0 disappears from the calculation; the half-life does not depend on the initial quantity of radioactive material.

We obtain that

$$A(t) \; = \; A_0 e^{-t\ln 2/H} \; = \; A_0 \big(e^{\ln 2}\big)^{-t/H} \; A_0 \cdot 2^{-t/H} \; = \; A_0 \left(\frac{1}{2}\right)^{t/H}.$$

However, this last formula is the obvious formula that we should have arrived at with no work, and no reference to base e; it simply says that every time another H amount of time passes, the amount of material present is multiplied by another $1/2$, which, of course, is exactly what half-life means.

The point is that it was somewhat silly to use the exponential function and natural logarithm in the first place; if an exponential decay rate is specified by being given the half-life of the material, the nicest base for the problem is either base $1/2$ or base 2.

Example 2.6.9. Radioactive materials are not the only types of substances that decay exponentially, and whose decay rates are typically specified by giving half-lives. When some drugs are given to humans, or other animals, the rate of "decay" of the amount of the drug, by mass, in the body, is exponential, with decay rates that are usually specified in terms of half-lives.

For example, digitalis, a medication for various heart conditions, decays exponentially after being administered, with a half-life of (approximately) 36 hours. If a patient is given 250 μg (micrograms) of digitalis, how long will it be before only 25 μg of digitalis remain in the patient?

Solution:

From our discussion in Example 2.6.8, we immediately have that $A(t)$, the number of micrograms of digitalis remaining in the patient, after time t hours, is

$$A(t) = 250 \cdot 2^{-t/36}.$$

We want to set $A(t)$ equal to 25, and solve for t. Dividing both sides of the equation by 250, we obtain

$$\frac{1}{10} = 2^{-t/36}.$$

Even though this is a base 2 problem, it is easier to use the natural logarithm on most calculators. So, we take the natural logarithm of each side, and obtain

$$\ln(1/10) = \ln\left(2^{-t/36}\right) = -\frac{t}{36}\ln 2.$$

Using that $\ln(1/10) = -\ln(10)$, we find

$$t = \frac{36\ln 10}{\ln 2} \approx 119.59 \text{ hours},$$

or roughly 5 days.

Corollary 2.6.6 tells us that if $b > 0$ and $b \neq 1$, then the function b^x is strictly monotonic and, in particular, is one-to-one. Thus, the inverse function exists.

Definition 2.6.10. *Suppose that $b > 0$ and $b \neq 1$. Then, the* **base b logarithm***,* $\log_b :$ $(0, \infty) \to (-\infty, \infty)$ *is defined to be the inverse function of b^x.*

Thus, for all $x > 0$, $b^{\log_b x} = x$, and, for all y, $\log_b(b^y) = y$. In words, $\log_b x$ is the exponent that you have to raise b to in order to get x.

We shall, in fact, have little occasion to want to use logarithms with any base other than e; essentially, the only logarithm that we shall use throughout this book is $\ln x$. Why? Because the result below tells us that $\log_b$ is a constant times $\ln x$, and so we can easily change algebra and Calculus questions about base b logarithms into $\ln x$ questions.

Theorem 2.6.11. *Suppose that $b > 0$ and $b \neq 1$. Then,*

$$\log_b x = \frac{\ln x}{\ln b},$$

and so,

$$(\log_b x)' = \frac{1}{x \ln b}.$$

Proof. The first equality is easy. We have

$$b^{(\ln x / \ln b)} = e^{(\ln x / \ln b)(\ln b)} = e^{\ln x} = x,$$

which shows that $\ln x / \ln b = \log_b x$.

As $\ln b$ is a constant, the derivative formula follows. $\square$

Example 2.6.12. Calculate $\bigl(\log_{10}(x^2 + 1)\bigr)'$.

Solution:

$$\bigl(\log_{10}(x^2 + 1)\bigr)' = \frac{1}{(x^2 + 1)\ln 10} \cdot (x^2 + 1)' = \frac{2x}{(x^2 + 1)\ln 10}.$$

At long last, we can prove the general Power Rule for arbitrary real exponents.

Theorem 2.6.13. (General Power Rule) *Consider the function x^p, where p is constant, and $x > 0$.*
Then,

$$(x^p)' = px^{p-1}.$$

Proof. This is easy now.

$$(x^p)' = (e^{p \ln x})' = e^{p \ln x}(p \ln x)' = x^p \cdot \frac{p}{x} = px^{p-1}.$$

$\square$

Example 2.6.14. Calculate $(t^\pi + \pi^t)'$.

Solution:

$$\left(t^\pi + \pi^t\right)' \;=\; \pi\, t^{\pi-1} + \pi^t \ln \pi.$$

You may wonder why the Power Rule and the derivative of general exponential functions look so different; the formulas $(x^p)' = px^{p-1}$ and $(b^x)' = b^x \ln b$ seem completely unrelated. It is true that power functions, with a variable base and fixed exponent, are very different from general exponential functions, with a fixed base and a variable exponent, but this answer isn't really satisfying. Shouldn't there be **one** general formula that gives us both cases?

In fact, there is, and we give it below. However, it is complicated enough that it is not worth memorizing; you should simply re-derive it if you need it.

Theorem 2.6.15. (**Power-Exponential Rule**) *Suppose that $f(x)$ and $g(x)$ are differentiable functions, defined on an open interval I, and that, for all x in I, $g(x) > 0$.*
   Then, $(g(x))^{f(x)}$ is differentiable on I, and

$$\left((g(x))^{f(x)}\right)' \;=\; f(x)(g(x))^{f(x)-1}g'(x) + (g(x))^{f(x)}f'(x)\ln(g(x)).$$

Proof. This is actually surprisingly easy.

$$\left((g(x))^{f(x)}\right)' \;=\; \left(e^{f(x)\ln(g(x))}\right)' \;=\; e^{f(x)\ln(g(x))}\left(f(x)\ln(g(x))\right)' \;=$$

$$(g(x))^{f(x)}\left[f(x)\cdot\frac{1}{g(x)}\cdot g'(x) + f'(x)\ln(g(x))\right] \;=$$

$$f(x)(g(x))^{f(x)-1}g'(x) + (g(x))^{f(x)}f'(x)\ln(g(x)).$$

$\square$

You get the Power Rule from the Power-Exponential Rule by letting $g(x) = x$, for $x > 0$, and letting $f(x) = p$ be a constant. You get the general exponential derivative formula from the Power-Exponential Rule by letting $g(x) = b$ be a positive constant, and letting $f(x) = x$.

2.6.1   Exercises

Differentiate the functions in Exercises 1 through 20.

1. $f(x) = 5(2^x)$.

2. $g(t) = 2.7 + 5^{13t}$.

3. $h(x) = \ln(3^{2x+1}) + 15$.

4. $p(x) = 12e^{3x} + 4^{3x}$.

5. $r(t) = 5^t \ln t$.

6. $g(x) = \dfrac{1 + x + x^2}{14^x + 3.2}$.

7. $f(x) = (\log_3 x)^2$.

8. $h(x) = \dfrac{1}{(x+1)^3} + 7^{\frac{x}{3}}$.

9. $r(x) = 17 + 6^{x^2+1}$.

10. $g(x) = (\ln x)^{x^2}$.

11. $f(t) = \dfrac{1.1 + 7^t}{\ln t}$.

12. $f(x) = \log_7 x + \ln x + 3^x$.

13. $h(x) = (x^2 + 5x + 3)^{(7x-2)}$.

14. $g(t) = \dfrac{9^t \sqrt{t+1}}{16}$.

15. $f(t) = 93e^{-3.5t} + 71(13^{-4t})$.

16. $p(x) = 12^{x \ln x}$.

17. $g(x) = \log_b(\log_b x)$, $b > 0$, $b \neq 1$.

18. $f(x) = \log_3 \frac{\sqrt{x^2+1}}{\sqrt[3]{2+x}}$.

19. $h(x) = \log_2 \left(x + \sqrt{1+x^2}\right)$.

For the functions in Exercises 21 and 22, find the critical points and the intervals on which the functions are increasing or decreasing.

20. $f(x) = x^2 2^{-x} + 1$.

21. $g(x) = (x^2)^x$. (Hint: Use logarithmic differentiation, or that $e^{\ln x} = x$).

22. Prove Item 1 of Theorem 2.6.2, that $b^0 = 1$, where b is a real number,

23. Prove Item 2 of Theorem 2.6.2, that $b^{x+y} = b^x b^y$, where $b > 0$ and x and y are real numbers.

24. Prove Item 3 of Theorem 2.6.2, that $b^{-x} = 1/b^x$, where $b > 0$ and x is a real number.

25. Show that if the derivative of $f(x)^{f(x)}$ is $f'(x)f(x)^{f(x)}(1 + \ln f(x))$ by setting $g(x) = f(x)$ in Theorem 2.6.15.

Use the result of the previous problem to calculate the derivatives of the functions in Exercises 27 through 29. In each exercise, specify the domain where the formula is valid.

26. $h(x) = x^x$.

27. $g(x) = (\ln x)^{\ln x}$.

28. $p(x) = (x^2 - 1)^{x^2-1}$.

29. Describe the relationship between the graphs of a^x and $\log_a x$ in terms of a reflection.

30. Describe the relationship between the graphs of a^x and $\left(\frac{1}{a}\right)^x$ in terms of a reflection.

31. Let $f(x) = \log_{\ln x} x$. What is $f'(x)$?

32. What is the natural domain of the function $f(x) = \log_{\sqrt{4-x^2}} 7$?

33. Let $f_i(x) = \log_{a_i} x$ for $i = 1, 2, ...n$, where $a_i > 0$. Let $f(x) = f_1(x) \cdot f_2(x) \cdot ... \cdot f_n(x)$. Calculate $f'(x)$.

34. Show that $\lim\limits_{h \to 0} \frac{\log_b(1+h)}{h} = \frac{1}{\ln b}$. You may use Theorem 2.6.11.

35. What is an equation for the tangent line to the graph of $y = a^x$ at the point $(0, 1)$?

(▶)

36. What is an equation for the tangent line to the graph of $y = \log_b x$ at the point $(1, 0)$?

37. A farmer opens a savings account with a $1000 initial investment in a bank that is offering an annual interest rate of 3%, compounded quarterly. Assume that no money is removed from the account. Recall our discussion of such a situation in Example 2.5.11.

   a. Write a formula for $A(t)$, the actual amount of money in the bank at time t years after the principal investment.

   b. Determine constants c and b such that the general exponential function $E(t) = cb^t$ gives a reasonable approximation of $A(t)$.

   c. Assume that your formula from part (a) is valid for all $t \geq 0$ (not just when $4t$ is an integer). What is the instantaneous rate of change of $A(t)$, with respect to time, one month after opening the account?

   d. What is the instantaneous rate of change of $E(t)$, with respect to time, one month after opening the account?

   e. Use your approximation $E(t)$ to estimate the amount of money in the account after 25 years. Compare this with the actual result using $A(t)$.

38. Consider the function $f(x) = (1.2)^x$.

   a. Calculate the average rate of change of f, with respect to x, on the interval $[0, b]$.

   b. Find the value of c in the above interval guaranteed by the Mean Value Theorem, i.e., the c such that $f'(c)$ coincides with the average rate of change from part (a).

39. Remark 2.6.7 illustrates the graphs of the functions $f(x) = b^x$ and $g(x) = b^{-x}$ ($b > 0$).

   a. At what point do the graphs of $f(x)$ and $g(x)$ intersect?

   b. Show that there are exactly two values of b such that $f(x)$ and $g(x)$ are orthogonal (i.e., have perpendicular tangent lines) at the intersection point.

40. Explain why the following statement is false: If $b > 0$, then $f(x) = b^x$ is strictly monotonic.

41. Prove that if $f(x) = b^x$, then $f^{(n)}(x) = f(x) \cdot (\ln b)^n$.

42. Justify the following statement: If $f(x) = b^x$ where $b > 0$ and $b \neq 1$ then the graph of f on the interval $[a, b]$ lies below the secant line between $(a, f(a))$ and $(b, f(b))$.

In Exercises 43 through 45, show that $y = f(x)$ is a solution to the given differential equation.

43. $y' = (\ln 3)(y - 5)$, $f(x) = 5 - 4(3)^x$.

44. $y' = \frac{1}{x}\left(\frac{1}{x\ln 7} - y\right)$, $f(x) = \frac{\log_7 x}{x}$.

45. $y' = \frac{y}{x}\left(\frac{1}{(\log_{10} x)\ln 10} + 10\right)$.

Physical *sound pressure* is measured in Newtons per square meter, or pascals. The perceived loudness of a given sound pressure p is measured by comparing its magnitude to the perceived loudness of a reference sound pressure p_0, where p_0 is the threshold of human hearing, according to the formula

$$L = 10\log\left(\frac{p^2}{p_0^2}\right),$$

where L is measured in decibels, or dB. Here we adopt the convention that when no explicit base is given for a logarithm, the base is assumed to be 10. Use this formula to answer Exercises 46 through 48.

46. Prove that the formula for perceived loudness can be rewritten as $L = 20\log\left(\frac{p}{p_0}\right)$.

47. Justify the following rule of thumb: a doubling of the physical sound pressure leads to an increase of 6.02 dB in perceived loudness.

48. Calculate dL/dp.

49. Let $A(t)$ denote the mass, in kilograms, of radioactive material at a time t years after some initial measurement.

    a. Assuming that there are initially 100 kilograms of radioactive material and, after 10 years, there remain only 50 grams, write a function $A(t)$, in base e, for the amount of material remaining after t years.

    b. What is the half-life of this material?

    c. Express $A(t)$ using the base $1/2$.

50. Recall from Example 2.2.9 and Example 2.4.24 that a population constrained by some maximum size M can be modeled by a logistic differential equation

$$\frac{dP}{dt} = kP(M - P).$$

This maximum M is called the *carrying capacity*. A solution to this logistic equation is given by

$$P(t) = \frac{MP_0}{(M - P_0)e^{-Mkt} + P_0},$$

where t denotes the units of time after the initial population P_0 is measured.

a. If a bacterial colony in a petri dish has a carrying capacity, M, of 1000 bacteria, with an initial population P_0 of 100 bacteria and, after 5 days, there are 400 bacteria, find k.

b. Write $P'(t)$ and $P''(t)$ explicitly, i.e., without reference to $P(t)$.

c. Use the differential equation to find the value of P at which P has an inflection point, then use the formula for $P(t)$ to find the corresponding t value.

51. Suppose that n is a positive integer. The number of digits $D = D(n)$ in the base 10 representation of n is the unique positive integer D such that

$$(10)^{D-1} \leq n < (10)^D.$$

For instance, $(10)^2 = 100 \leq 374 < 1000 = (10)^3$, and 374 has 3 digits. Applying the strictly increasing function $\log_{10}$ to these inequalities, we obtain

$$D - 1 \leq \log_{10} n < D,$$

or, equivalently,

$$\log_{10} n < D \leq 1 + \log_{10} n.$$

Thus, for a positive integer n, $1 + \log_{10} n$ is an approximation of $D(n)$, the number of digits in the base 10 representation of n.

For real numbers $x > 0$, we can try to approximate the above base 10 digits function for integers; define

$$E(x) = 1 + \log_{10} x = 1 + \frac{\ln x}{\ln 10}.$$

This is a continuous approximation of a discontinuous function which yields the number of digits only when x is a positive integral power of 10.

a. Find the derivative of $E(x)$, with respect to x.

b. Determine the rate of change of $E(x)$, with respect to x, at $x = 1000$.

c. Find $E(100)$ and $E(10)$.

d. According to the Mean Value Theorem, there exists a c such that $10 < c < 100$ with the property that

$$E'(c) = \frac{E(100) - E(10)}{100 - 10}.$$

Find such a c value.

52. There is a mathematical fable regarding the invention of the game of chess, played on an 8 by 8 board. According to the story, the king asked the inventor of the game to name his reward. The inventor's request was simple; it was that one grain of rice be placed on the first square of the chessboard, two on the second, four on the third, etc. Each of the 64 squares has twice as many grains as the previous square.

a. Show that the total number of grains of rice on all squares is $2^{64} - 1$.

b. Use the previous problem to determine how many digits this number has.

2.7 Trigonometric Functions: Sine and Cosine

We assume that you have some familiarity with trigonometric (trig) functions and, in particular, with sine and cosine. We need trig functions not merely in the context of triangles, but also as fundamental *periodic functions*, defined in the context of circles. The fact that trig functions are basic periodic functions is what makes them so important in many physics and engineering applications.

Let's first define periodic functions; these are functions whose values repeat if we change the argument by a fixed value.

Definition 2.7.1. *Suppose that $p > 0$. Let f be a real function. Suppose that, for all x in the domain of f, $x+p$ and $x-p$ are also in the domain of f, and $f(x+p) = f(x) = f(x-p)$. Then, we say that f is* **periodic** *or p-**periodic**. If there is a smallest such p, that smallest p is called the* **period** *of f.*

Remark 2.7.2. Some sources would reserve the term *p-periodic* for a periodic function whose period is p. With our definition, a 5-periodic function, with period 5, would also be 10-periodic (still with period 5). Note also that, by our definition, a constant function is p-periodic for **every** positive p; though, a constant function has no period, since there is no such smallest p.

We now need to review/summarize all of the properties of trig functions that we will need. There are aspects of trigonometry that will not be particularly relevant for us, as far as Calculus is concerned; for instance, we will not use the Law of Sines or the Law of Cosines. Until we get to limits and Calculus involving trig functions, we will not prove any results; we refer you to any Pre-Calculus textbook which contains trigonometry.

We assume that you are familiar with using degrees to measure angles, but we shall need *radians*. It is **extremely** important that Calculus with trig functions is always performed with radians, not degrees.

Consider the unit circle, the circle of radius 1, centered at the origin, together with a circle of arbitrary radius r centered at the origin. See Figure 2.5.

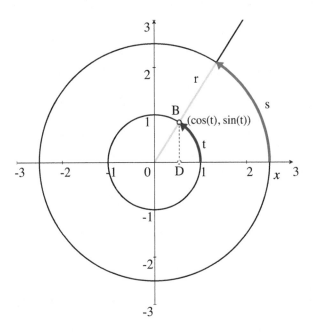

Figure 2.5: Sine and cosine as coordinates on the unit circle.

You think of $(1, 0)$ as a "starting point" (or as an "origin") on the unit circle. The positive direction on the unit circle is the counter-clockwise direction; the negative direction is the clockwise direction. Since the circle of radius 1 has circumference 2π, we see that if you start at $(1, 0)$ and move clockwise $2\pi/4 = \pi/2$ units, then you move a quarter of the way counter-clockwise around the circle and, thus, end up at $(0, 1)$. If you were to move $-\pi/2$ units along the unit circle, this would mean that you start at $(1, 0)$ and move one quarter of the way **clockwise** around the unit circle and, thus, you would end up at $(0, -1)$. Of course, if you were to travel 2π or -2π units around the unit circle, then you would arrive back at the starting point $(1, 0)$.

Essentially by using similar triangles (curved triangles, or similar sectors or arcs), we see in Figure 2.5 that the length t of the arc on the unit circle can be related the corresponding length s on the circle of radius r by $t/1 = s/r$. As this quantity, the length of an arc of the circle, divided by the radius of the circle, does not depend on the radius, it can be used as a measure of the angle that the (green) ray $\overline{OB}$ in the figure makes with the positive x-axis; we say that the **angle is (or, measures)** $t/1 = s/r$ *radians*, which is abbreviated *rad*. Note that an angle of t radians should be thought of as $t/1$, the length of the arc divided by the length of the radius, so that the radian measure of an angle does **not** have length units.

It is easy to convert between radians and degrees. There are 2π radians in a circle (since the circumference of the unit circle is 2π), which must equal $360°$. Therefore,

$$2\pi \text{ rad} = 360°.$$

Unless we explicitly state otherwise, all angles in this book are measured in radians. We shall frequently write something about "the angle t"; this means t radians, even if we do not explicitly mention radians.

Now, consider the point (x, y) on the unit circle that is t radians along the unit circle (from the starting point $(1, 0)$, with positive t meaning counter-clockwise movement, and negative t meaning clockwise movement of $|t|$ units). Then, the x- and y-coordinates are functions of t.

Definition 2.7.3. *The* **cosine,** $\cos(t)$, **of the angle of** t **radians** *is the corresponding x-coordinate on the unit circle, and the* **sine,** $\sin(t)$, **of the angle of** t **radians is the corresponding y-coordinate on unit circle.**

Note that, for angles t such that $0 < t < \pi/2$ (that is, angles between $0°$ and $90°$), this definition of sine and cosine agree with the usual ones for right triangles. That is, if you form the right triangle $\triangle ODB$ in Figure 2.5, by dropping a perpendicular line segment from $B = (\cos(t), \sin(t))$ to the x-axis, then sine is the length of the "opposite side" (the y-coordinate on the unit circle), divided by the length, 1, of the hypotenuse, while cosine is the length of the "adjacent side" (the x-coordinate on the unit circle), divided by the length, 1, of the hypotenuse.

We now list a number of properties of sine and cosine; some are immediate from the discussion and definitions above, some are not. We should mention that it is standard to write $\cos^n(t)$ and $\sin^n(t)$ in place of $(\cos(t))^n$ and $(\sin(t))^n$, respectively, for all $n \neq -1$. The superscript of -1 is reserved for the inverse trig functions. It is also common to not enclose the independent variable in parentheses, if no confusion is likely.

1. The Fundamental Trigonometric Identity: for all t, $\sin^2 t + \cos^2 t = 1$;

2. sine and cosine are periodic, with period 2π. It follows that, for all integers k, we have equalities of functions $\sin(t + 2\pi k) = \sin t$ and $\cos(t + 2\pi k) = \cos t$.

3. sine is an odd function, and cosine is an even function, i.e., for all t, $\sin(-t) = -\sin t$ and $\cos(-t) = \cos t$.

4. The Angle Addition Formulas: for all α and β,

$$\sin(\alpha + \beta) = \sin\alpha\cos\beta + \sin\beta\cos\alpha$$

and

$$\cos(\alpha + \beta) \; = \; \cos\alpha\cos\beta - \sin\alpha\sin\beta.$$

5. From letting $\beta = \alpha$ in the Angle Addition Formulas, we obtain the Double Angle Formulas: for all α,

$$\sin(2\alpha) \; = \; 2\sin\alpha\cos\alpha$$

and

$$\cos(2\alpha) \; = \; \cos^2\alpha - \sin^2\alpha \; = \; 2\cos^2\alpha - 1 \; = \; 1 - 2\sin^2\alpha.$$

6. For all α,

$$\sin\alpha \; = \; \cos(\pi/2 - \alpha) \quad \text{and} \quad \cos\alpha \; = \; \sin(\pi/2 - \alpha)$$

From our discussion above, and our knowledge of right isosceles triangles and $30°$-$60°$-$90°$ triangles, and using that 2π rad $= 360°$, we have the following table for special angles (in radians) in the 1st quadrant:

t	0	$\pi/6$	$\pi/4$	$\pi/3$	$\pi/2$
$\sin t$	0	$1/2$	$\sqrt{2}/2$	$\sqrt{3}/2$	1
$\cos t$	1	$\sqrt{3}/2$	$\sqrt{2}/2$	$1/2$	0

The values of sine and cosine at other special angles, and in other quadrants, can be obtained from the fundamental ones in the table above, combined with the Angle Addition Formulas.

We want to look at the graphs of sine and cosine. We will, as usual, use x for the independent variable, and y for the dependent variable. Do not confuse this x and this y with the x- and y-coordinates on the unit circle; x will now represent the angle (in radians, of course) and y will be the value of $\sin x$ or $\cos x$. The graphs appear in Figure 2.6 and Figure 2.7.

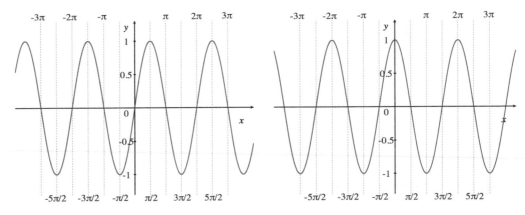

Figure 2.6: The graph of $y = \sin x$.   Figure 2.7: The graph of $y = \cos x$.

We need to investigate the continuity and differentiability of sine and cosine. It is convenient to once again denote the angle under consideration by t, which is also the (signed) arc length on the unit circle, measured from the starting point $(1, 0)$. For t between 0 and $\pi/2$, non-inclusive, there are three inequalities that we obtain from considering Figure 2.8.

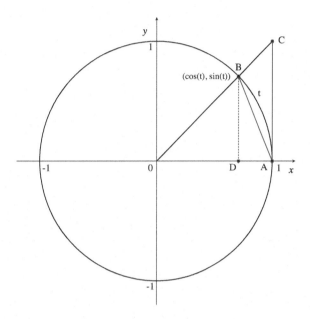

Figure 2.8: Comparing t, $\sin t$, and $\cos t$.

First, note that the area of the sector $\triangleleft OAB$ is equal to the same fraction of the area of the entire circle as the fraction that t is of the entire perimeter, i.e., we have

$$\text{area of } \triangleleft OAB \;=\; \frac{t}{2\pi}(\pi \cdot 1^2) \;=\; t/2.$$

Now, the area of the triangle $\triangle OAB$ is one half the length of its base times the height, and is less than the area of the sector $\triangleleft OAB$; thus, we have

$$0 \;<\; \frac{1}{2}(1)(\sin t) \;<\; \frac{t}{2}.$$

However, we also see that the sector $\triangleleft OAB$ has smaller area than the triangle $\triangle OAC$. As the triangle $\triangle OAC$ is similar to the triangle $\triangle ODB$, we have that

$$\frac{\sin t}{\cos t} = \frac{\text{length of } \overline{DB}}{\text{length of } \overline{OD}} = \frac{\text{length of } \overline{AC}}{\text{length of } \overline{OA}} = \text{length of } \overline{AC}.$$

Hence, the area of the triangle $\triangle OAC$ is $(1/2)(1)(\sin t)/(\cos t)$.

Combining all of the above, we find, for $0 < t < \pi/2$,

$$0 < \frac{\sin t}{2} < \frac{t}{2} < \frac{1}{2} \cdot \frac{\sin t}{\cos t},$$

or, multiplying by 2,

$$0 < \sin t < t < \frac{\sin t}{\cos t}. \tag{2.13}$$

As

$$0 \leq \sin t \leq t,$$

we may use a one-sided version of the Pinching Theorem, Theorem 1.3.6, to find that

$$\lim_{t \to 0^+} \sin t = 0.$$

An analogous calculation for $-\pi/2 < t < 0$ lets us conclude that

$$\lim_{t \to 0^-} \sin t = 0.$$

Combining the two one-sided limit equalities above, we obtain:

Lemma 2.7.4.
$$\lim_{t \to 0} \sin t = 0,$$
and
$$\lim_{t \to 0} \cos t = 1.$$

Proof. The first limit follows from our discussion. The Fundamental Trigonometric Identity tells us that

$$\cos t = \pm\sqrt{1 - \sin^2 t}.$$

If $-\pi/2 \le t \le \pi/2$, then $\cos t > 0$, and so

$$\cos t \;=\; \sqrt{1 - \sin^2 t}.$$

Hence, the second limit follows from the first.   □

Now, look again at Formula 2.13. For $0 < t < \pi/2$, we see that, since $\sin t < t$,

$$\frac{\sin t}{t} \;<\; 1.$$

On the other hand, since $t < \sin t / \cos t$, we have

$$\cos t \;<\; \frac{\sin t}{t}.$$

Therefore,

$$\cos t \;\le\; \frac{\sin t}{t} \;\le\; 1.$$

Applying the Pinching Theorem (as $t \to 0^+$) and the second limit in Lemma 2.7.4, and looking at a similar argument for $t < 0$, we find that

---

Lemma 2.7.5.
$$\lim_{t \to 0} \frac{\sin t}{t} = 1 \qquad \text{and} \qquad \lim_{t \to 0} \frac{1 - \cos t}{t} = 0.$$

Proof. The first limit follows from our discussion. The second limit follows from the first, and Lemma 2.7.4, since

$$\lim_{t \to 0} \frac{1 - \cos t}{t} = \lim_{t \to 0} \left[\frac{1 - \cos t}{t} \cdot \frac{1 + \cos t}{1 + \cos t} \right] = \lim_{t \to 0} \frac{1 - \cos^2 t}{t(1 + \cos t)} =$$

$$\lim_{t \to 0} \frac{\sin^2 t}{t(1 + \cos t)} = \lim_{t \to 0} \frac{\sin t}{t} \cdot \lim_{t \to 0} \frac{\sin t}{1 + \cos t} = 1 \cdot (0/2) = 0.$$

                                                                                    □

We can now find the derivatives of sine and cosine.

Theorem 2.7.6. *For all x, the functions, $\sin x$ and $\cos x$ are differentiable and, hence, continuous. The derivatives are*

$$\sin' x = \cos x \quad \text{and} \quad \cos' x = -\sin x.$$

Proof. We have

$$\sin' x = \lim_{h \to 0} \frac{\sin(x+h) - \sin x}{h} = \lim_{h \to 0} \frac{\sin x \cos h + \cos x \sin h - \sin x}{h} =$$

$$\lim_{h \to 0} \left[-\sin x \left(\frac{1 - \cos h}{h} \right) + \cos x \left(\frac{\sin h}{h} \right) \right],$$

which, by Lemma 2.7.5, is equal to $-(\sin x)(0) + (\cos x)(1) = \cos x$.

For the derivative of $\cos x$, we find

$$\cos' x = \lim_{h \to 0} \frac{\cos(x+h) - \cos x}{h} = \lim_{h \to 0} \frac{\cos x \cos h - \sin x \sin h - \cos x}{h} =$$

$$\lim_{h \to 0} \left[-\cos x \left(\frac{1 - \cos h}{h} \right) - \sin x \left(\frac{\sin h}{h} \right) \right],$$

which, by Lemma 2.7.5, is equal to $-(\cos x)(0) - (\sin x)(1) = -\sin x$. $\square$

As an immediate corollary, we have

Corollary 2.7.7. *The second derivatives of sine and cosine are*

$$\sin'' x = -\sin x \quad \text{and} \quad \cos'' x = -\cos x.$$

Thus, $y = \sin x$ and $y = \cos x$ are solutions to the differential equation

$$y'' + y = 0.$$

You should practice combining these new differentiation rules with our previously established rules.

Example 2.7.8. Let $z = r\sin(5r)$ and $w = e^{\cos u}$. Find dz/dr and dw/du.

Solution:

Using the Product and Chain Rules, we find

$$\frac{dz}{dr} \;=\; r\frac{d}{dr}\big(\sin(5r)\big) + \big(\sin(5r)\big)\frac{d}{dr}(r) \;=\; r\cos(5r)\frac{d}{dr}(5r) + \big(\sin(5r)\big)\cdot 1 \;=$$

$$5r\cos(5r) + \sin(5r).$$

Using the Chain Rule, we calculate

$$\frac{dw}{du} \;=\; e^{\cos u}\frac{d}{du}(\cos u) \;=\; -e^{\cos u}\sin u.$$

Example 2.7.9. Calculate $\big(\ln(\cos x)\big)'$ and $\left(5\cos(3x) + \dfrac{\sin x}{x}\right)'$.

Solution:

$$\big(\ln(\cos x)\big)' \;=\; \frac{1}{\cos x}\,(\cos x)' \;=\; -\frac{\sin x}{\cos x}.$$

$$\left(5\cos(3x) + \frac{\sin x}{x}\right)' \;=\; 5\big(-\sin(3x)\big)3 \;+\; \frac{x(\sin x)' - \sin x \cdot 1}{x^2} \;=$$

$$-15\sin(3x) \;+\; \frac{x\cos x - \sin x}{x^2}$$

While sine and cosine are defined in terms of the unit circle, they frequently arise in the context of bodies that are in no way moving in a circle. Sine and cosine are the fundamental periodic functions that arise in many contexts, such as when an object's position is oscillating, as in the example below.

Example 2.7.10. The left end of a horizontal spring is attached to a wall, and a block is attached to the right end of the spring. Suppose that the center of the block is at $x = 0$ meters, when the spring is at its natural length. This is called the *equilibrium position* for the block.

Suppose that the block is set in motion, and that the floor exerts no friction force on the block. Hence, the only force acting on the block is due to the spring, which pushes the block to the right when the block is to the left of the equilibrium position, so that the spring is compressed, and which pulls the block to the left when the block is to the right of the equilibrium position, so that the spring is stretched. See Figures 2.9 and 2.10.

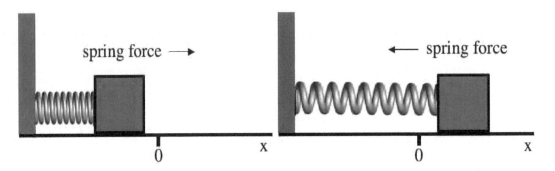

Figure 2.9: A compressed spring.   Figure 2.10: A stretched spring.

Suppose that at time t seconds, the position of the block, in meters, is

$$x = 2\cos(3t).$$

Find the position, velocity, and acceleration of the (center of the) block at $t = 0$ and $t = \pi/2$ seconds. What happens to the position of the block after a long period of time?

Solution:

The position of the block at $t = 0$ seconds is $x(0) = 2\cos(0) = 2$ meters to the right of the equilibrium position (where we have, as is conventional, taken to the right as the positive direction). At $t = \pi/2$ seconds, $x = 2\cos(3\pi/2) = 2 \cdot 0 = 0$ meters, i.e., the block is at the equilibrium position.

We used here that an angle of $3\pi/2$ on the unit circle would put you at the point $3/4$ of the way around the circle from the starting point of $(1, 0)$; that is the point $(0, -1)$. Since cosine is the x-coordinate on the unit circle, $\cos(3\pi/2) = 0$.

Now, let's calculate the velocity, v, and the acceleration, a. By the Chain Rule and linearity, we find

$$v = \frac{dx}{dt} = 2 \cdot \left(-\sin(3t) \right) \cdot (3t)' = -6\sin(3t) \ \text{m/s},$$

and

$$a = \frac{dv}{dt} = -6\big(\cos(3t) \big) \cdot 3 = -18\cos(3t) \ \text{(m/s)/s}.$$

At $t = 0$ seconds, the velocity of the block is $v(0) = -6\sin(0) = 0$ m/s and the acceleration is $a = -18\cos(0) = -18$ (m/s)/s. Thus, at $t = 0$ seconds, the block is 2 meters to the right of the equilibrium position, with 0 velocity, but accelerating to the left (because of the minus sign) at 18 (m/s)/s.

At $t = \pi/2 \approx 1.57$ seconds, we find $v = -6\sin(3\pi/2) = (-6)(-1) = 6$ m/s and $a = -18\cos(3\pi/2)0$ (m/s)/s. Thus, at $t = \pi/2$ seconds, the block is at the equilibrium position, moving to the right at 6 m/s, and not accelerating.

Intuitively, it is clear that the block must have "turned around" between $t = 0$ and $t = \pi/2$ seconds, for at $t = 0$ seconds, the mass was to the right of equilibrium, stopped, but accelerating to the left, and then, at $t = \pi/2$ seconds, the mass is at equilibrium position, but is moving to the right.

Actually, as we discussed in Section 1.5, it is easy for us to find exactly when the block reaches its maximum displacements to the left and right of the equilibrium position; these will be when x attains local maxima and minima values, which is when $v = x' = 0$. This occurs at each time t such that $\sin(3t) = 0$. Now, looking at the unit circle, we see that $\sin\theta = 0$ if and only if $\theta = k\pi$, where k is an integer, and so $\sin(3t) = 0$ if and only if $t = k\pi/3$ for some integer k. When $k = 0$, $t = 0$ seconds, and the block is at its extreme right-hand position. When $k = 1$, $t = \pi/3$ seconds, and the block is at its extreme left-hand position of $x = 2\cos(\pi) = -2$ meters.

If we think of the motion as starting at $t = 0$ seconds, then the position, velocity, and acceleration of the block will repeat every time the argument of cos and sin, namely $3t$, is an integer multiple of 2π. The first time this happens, after $t = 0$, is when $3t = 2\pi$, i.e., when $t = 2\pi/3$ seconds. In other words, $\cos(3t)$ and $\sin(3t)$ are periodic, with period $2\pi/3$. The period of these functions, the amount of time that it takes for the block to make one complete oscillation, is (not surprisingly) called the *period* of the oscillation(s); it is usually denoted by T. It follows that, for any constant b, the block goes through b oscillations in bT seconds and, thus, the block goes through $1/T$ oscillations in 1 second. An oscillation (or *cycle*) per second is a unit of measurement called Hertz, Hz, and the number of cycles per second of an oscillation is called the *frequency* of the oscillation, which is $1/T$ Hz.

What happens to the position of the block after a long period of time? As time goes on, $\cos(3t)$ continues to oscillate between -1 and 1. This is called *simple harmonic motion*. In theory (but not in practice), the position of the block continues to oscillate forever between -2

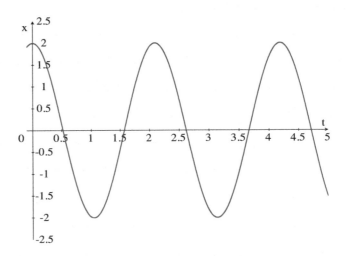

Figure 2.11: The motion of a block on a spring.

meters and 2 meters, i.e., between 2 meters to the left of the equilibrium position and 2 meters to the right of the equilibrium position. We say that the *amplitude* of the oscillation is 2 meters.

Example 2.7.11. Assume that we have the same situation as Example 2.7.10, except that we now assume that one more force acts on the block: the floor exerts a friction force on the block proportional to the velocity of the block.

Suppose that, in this situation, at time t seconds, the position of the block, in meters, is

$$x = 2e^{-t}\cos(3t).$$

Find the position, velocity, and acceleration of the (center of the) block at $t = 0$ and $t = \pi/2$ seconds. What happens to the position of the block after a long period of time?

Solution:

Let's first calculate the velocity, v, and acceleration, a, of the block. Using the Product and Chain Rules, we find

$$v = \frac{dx}{dt} = 2\big(e^{-t}(\cos(3t))' + (\cos(3t))(e^{-t})'\big) =$$

$$2\big(e^{-t}(-\sin(3t))(3) + (\cos(3t))e^{-t}(-1)\big) \;=\; -2e^{-t}\big(3\sin(3t) + \cos(3t)\big) \;\;\text{m/s},$$

and, writing out fewer steps for calculating the acceleration, we find

$$a \;=\; \frac{dv}{dt} \;=\; -2\big[e^{-t}\big(9\cos(3t) - 3\sin(3t)\big) + \big(3\sin(3t) + \cos(3t)\big)(-e^{-t})\big] \;=$$

$$-2e^{-t}\big(8\cos(3t) - 6\sin(3t)\big) \;\;\text{(m/s)/s}.$$

When $t = 0$ seconds, we obtain

$$x \;=\; 2 \cdot 1 \cdot 1 \;=\; 2 \;\;\text{m}, \qquad v \;=\; -2 \cdot 1 \cdot (0 + 1) \;=\; -2 \;\;\text{m/s},$$

and

$$a \;=\; -2 \cdot 1 \cdot (8 - 0) \;=\; -16 \;\;\text{(m/s)/s}.$$

When $t = \pi/2$ seconds, we find

$$x \;=\; 2 \cdot e^{-\pi/2} \cdot \cos(3\pi/2) \;=\; 0 \;\;\text{m},$$

$$v \;=\; -2 \cdot e^{-\pi/2} \cdot \big(3\sin(3\pi/2) + \cos(3\pi/2)\big) \;=\; 6e^{-\pi/2} \;\approx\; 1.24728 \;\;\text{m/s},$$

and

$$a \;=\; -2 \cdot e^{-\pi/2} \cdot \big(8\cos(3\pi/2) - 6\sin(3\pi/2)\big) \;=\; -12e^{-\pi/2} \;\approx\; -2.49456 \;\;\text{(m/s)/s}.$$

What happens to the position of the block after a long period of time? Since $|\cos(3t)| \le 1$, it follows that $|x| = 2e^{-t}|\cos(3t)| \le 2e^{-t}$, and, as $t \to \infty$, $e^{-t} \to 0$, i.e., $\lim_{t \to \infty} e^{-t} = 0$. Therefore, as t gets arbitrarily big, x gets arbitrarily close to 0 meters, i.e., the position of the block approaches the equilibrium position, as you would expect. You can see this easily in Figure 2.12.

In Example 2.7.10 and Example 2.7.11, we described physical situations and then just handed you supposedly reasonable position functions. In fact, these types of position functions can be

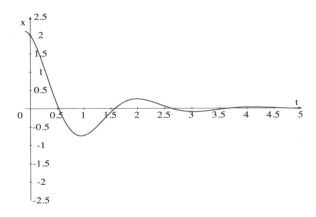

Figure 2.12: The motion of a block on a spring, with friction.

derived from more basic physical principles and assumptions, but involve solving differential equations. While we shall not derive the position function that we gave in Example 2.7.11, we **will** show you how you obtain the type of position function that we gave in Example 2.7.10.

We first need two lemmas.

Lemma 2.7.12. *Suppose that, for all x in some open interval I, $g''(x) = -g(x)$. (In particular, we are assuming that g, g', and g'' are defined on I.)*
Then, there exists a constant C such that, for all x in I,

$$(g(x))^2 + (g'(x))^2 \ = \ C.$$

Proof. As you probably guessed, you prove this by showing that the derivative of the expression on the left is zero. Using the Chain Rule, and that $g''(x) = -g(x)$, we find

$$\left[(g(x))^2 + (g'(x))^2\right]' \ = \ 2g(x)g'(x) + 2g'(x)g''(x) \ = \ 2g(x)g'(x) - 2g'(x)g(x) \ = \ 0.$$

$\square$

Lemma 2.7.13. *Suppose that $y = f_1(x)$ and $y = f_2(x)$ are both solutions to the differential equation $y'' + y = 0$, on an open interval I.*

Then, for all constants a and b, $y = af_1(x) + bf_2(x)$ is a solution to $y'' + y = 0$, on the open interval I.

Proof. This is easy. We have

$$[af_1(x) + bf_2(x)]'' + \big(af_1(x) + bf_2(x)\big) \ = \ af_1''(x) + bf_2'' + af_1(x) + bf_2(x) \ =$$

$$a\big(f_1''(x) + f_1(x)\big) + b\big(f_2''(x) + f_2(x)\big) \ = \ a \cdot 0 + b \cdot 0 \ = \ 0.$$

$\square$

Now we can prove:

Theorem 2.7.14. *Suppose that, for all x in some open interval I around the origin, $f''(x) = -f$, and that $f(0) = 0$, and $f'(0) = 1$.*

Then, on the interval I, $f(x) = \sin x$.

Proof. Let $g(x) = f(x) - \sin x$. By Lemma 2.7.13, $g''(x) = -g(x)$. Therefore, by Lemma 2.7.12, there exists a constant C such that, for all x in I,

$$(g(x))^2 + (g'(x))^2 = C.$$

As $g(0) = 0$ and $g'(0) = f'(0) - \cos(0) = 0$, the constant C must be 0. However, if the sum of two non-negative quantities is zero, both quantities must be zero. Thus, for all x in I, $g(x) = 0$ (and also $g'(x) = 0$). Thus, $f(x) = \sin x$ on I. $\square$

Corollary 2.7.15. *Suppose that, for all x in some open interval I around the origin, $f''(x) = -f$. Let $a = f(0)$ and let $b = f'(0)$.*

Then, on the interval I,

$$f(x) = a\cos x + b\sin x.$$

Proof. First, suppose that $a = b = 0$. Then, since Lemma 2.7.12 implies that $(f(x))^2 + (f'(x))^2$ is a constant, by plugging in 0 for x, we see that the constant must be zero. It follows that $f(x)$ is 0, for all x in I. Therefore, in this trivial case, $f(x) = a\cos x + b\sin x = 0$.

Now, suppose that not both a and b are 0. Then $a^2 + b^2 \neq 0$. Define, for all x in I,

$$g(x) = \frac{b}{a^2 + b^2}\, f(x) - \frac{a}{a^2 + b^2}\, f'(x).$$

We leave it as an exercise for you to show that $g''(x) = -g(x)$, $g(0) = 0$, and $g'(0) = 1$.

Hence, it follows from Theorem 2.7.14 that

$$\sin x = g(x) = \frac{b}{a^2 + b^2}\, f(x) - \frac{a}{a^2 + b^2}\, f'(x).$$

Differentiating, and using that $f''(x) = -f(x)$, we find

$$\cos x = \frac{b}{a^2 + b^2}\, f'(x) - \frac{a}{a^2 + b^2}\, f''(x) = \frac{b}{a^2 + b^2}\, f'(x) + \frac{a}{a^2 + b^2}\, f(x).$$

Finally, multiplying the above equalities for $\sin x$ and $\cos x$ by b and a, and adding, we obtain

$$a\cos x + b\sin x =$$

$$\frac{ab}{a^2 + b^2}\, f'(x) + \frac{a^2}{a^2 + b^2}\, f(x) + \frac{b^2}{a^2 + b^2}\, f(x) - \frac{ab}{a^2 + b^2}\, f'(x) = f(x).$$

$\square$

Corollary 2.7.16. *Let ω be a non-zero constant. Suppose that, for all x in some open interval I around the origin, $f''(x) = -\omega^2 f$. Let $a = f(0)$ and let $b = f'(0)$.*
Then, on the interval I,

$$f(x) = a\cos(\omega x) + \frac{b}{\omega}\sin(\omega x).$$

Proof. For all x in I, define $g(x) = f(x/\omega)$. We leave it as an exercise for you to show that $g''(x) = -g(x)$, and that $g(0) = a$ and $g'(0) = b/\omega$. Apply Corollary 2.7.15 to g, and then use that $f(x) = g(\omega x)$. $\qquad\qquad\qquad\qquad\qquad\qquad\qquad\qquad\qquad\qquad$ $\square$

Example 2.7.17. Finally, we are in a position to discuss the physics behind the position function that you were given in Example 2.7.10.

Springs, if they are not compressed or stretched too far, are typically assumed to obey *Hooke's Law*: the force that the spring exerts is proportional to the displacement from the equilibrium position, and acts in the direction opposite the displacement.

Let $F = F(t)$ denote the force exerted by the spring on the block at time t, let m denote the mass of the block, and let $x = x(t)$ denote the position of the center of the block at time t, as measured from its equilibrium position. Note that the displacement of the center of the block from its equilibrium position and the displacement of the right end of the spring from its equilibrium position are the same; thus, we may use the block's displacement and equilibrium position in applying Hooke's Law. Hence, what Hooke's Law tells us is that there exists a positive constant k such that

$$F = -kx.$$

The constant k depends on the spring in question, and is called the *spring constant*.

By Newton's 2nd Law of Motion, we have

$$mx'' = -kx, \text{ or, equivalently, } x'' = -\left(\sqrt{k/m}\right)^2 x.$$

By Corollary 2.7.16, all of the solutions to this differential equation are of the form

$$x = C_1 \cos(\omega t) + C_2 \sin(\omega t),$$

where C_1, C_2, and ω are constants, namely, $\omega = \sqrt{k/m}$, $C_1 = x(0)$, and $C_2 = x'(0)/\omega$.

In Example 2.7.10, we gave the position of the mass as $x = 2\cos(3t)$, where x was in meters and t in seconds. This corresponds to an initial position of 2 meters, and initial velocity of 0 m/s, and the value of $\sqrt{k/m}$ is 3 rad/s.

Let's look at another example, this time involving electrical circuits.

Example 2.7.18. In a *simple RLC circuit*, there is, in series, a resistor, with constant resistance R ohms, an inductor, with constant inductance L henrys, and a capacitor, with constant capacitance C farads. The resistor, inductor, and capacitor are referred to as *circuit elements*. See Figure 2.13. We assume that $R \geq 0$, $L \geq 0$, and $C > 0$.

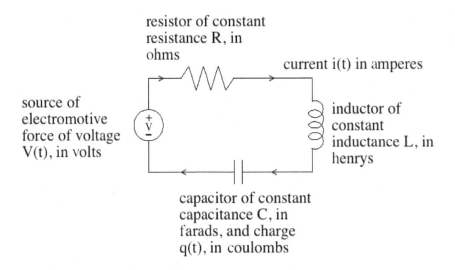

Figure 2.13: A simple RLC circuit.

There is usually a source of electromotive force (EMF), such as a battery, in the circuit, which provides a voltage of $V = V(t)$ volts, where V may vary with time t, in seconds. There is a charge across the capacitor of $q = q(t)$ coulombs, and current of $i = i(t)$ amperes (amps) runs through the circuit.

The current i is equal to the rate of change of the charge across the capacitor, with respect to time, i.e., $i = dq/dt$.

There are voltage drops associated with the current passing through each circuit element. The voltage drops are Ri, $L\,di/dt$, and q/C for the resistor, inductor, and capacitor, respectively. Now, one of *Kirchhoff's Laws* states that the sum of the voltage drops across the circuit elements is equal to the applied voltage, i.e., the voltage from the EMF source.

Therefore, we have

$$L\frac{di}{dt} + Ri + \frac{1}{C}q = V(t). \tag{2.14}$$

If we differentiate both sides of this equation, with respect to t, and use that $i = dq/dt$, we obtain

$$L\frac{d^2i}{dt^2} + R\frac{di}{dt} + \frac{1}{C}i = V'(t). \tag{2.15}$$

Now, suppose that $R = 0$ ohms, i.e., there is no resistor in the circuit, and that the voltage $V(t)$ is constant. We assumed at the beginning that $C > 0$; assume now that $L > 0$. In this case, Formula 2.15 reduces to:

$$L\frac{d^2i}{dt^2} + \frac{1}{C}i = 0, \quad \text{or, equivalently,} \quad \frac{d^2i}{dt^2} = -\frac{1}{LC}i. \tag{2.16}$$

By Corollary 2.7.16, all of the solutions to this differential equation are of the form

$$i = C_1\cos(\omega t) + C_2\sin(\omega t),$$

where $\omega = 1/\sqrt{LC}$, $C_1 = i(0)$, and $C_2 = i'(0)/\omega$.

This is interesting to think about. In Example 2.7.17, we saw that an object attached to a spring, with no friction from the floor, will oscillate back and forth in simple harmonic motion. This is easy to imagine. It is a little more unusual to think about the current in a circuit with an inductor, and capacitor, and a battery (of constant voltage), in series, with no resistance (analogous to no friction). What we have just seen is that the current in this situation will also undergo simple harmonic oscillations, i.e., the current would be *alternating current*.

We would like to end this section by discussing sequences of polynomials which converge uniformly (on any bounded set) to the sine and cosine functions. This is analogous to how we defined the exponential function at the beginning of Section 2.4. As with the exponential function, these polynomials, which converge to sine and cosine, allow us to approximate sine and cosine very well, at least, when x is close to zero.

Consider the sequence of even polynomial functions E_n:

$$E_0(x) = 1;$$

$$E_1(x) = 1 - \frac{x^2}{2!};$$

$$E_2(x) = 1 - \frac{x^2}{2!} + \frac{x^4}{4!};$$

$$E_3(x) = 1 - \frac{x^2}{2!} + \frac{x^4}{4!} - \frac{x^6}{6!};$$

and, in general,

$$E_n(x) = \sum_{k=0}^{n} (-1)^k \frac{x^{2k}}{(2k)!}.$$

Consider also the sequence of odd polynomial functions O_n:

$$O_0(x) = x;$$

$$O_1(x) = x - \frac{x^3}{3!};$$

$$O_2(x) = x - \frac{x^3}{3!} + \frac{x^5}{5!};$$

$$O_3(x) = x - \frac{x^3}{3!} + \frac{x^5}{5!} - \frac{x^7}{7!};$$

and, in general,

$$O_n(x) = \sum_{k=0}^{n} (-1)^k \frac{x^{2k+1}}{(2k+1)!}.$$

It can be shown that, for all real x, the sequences $E_n(x)$ and $O_n(x)$ converge; thus, we obtain functions $E(x) = \lim_{n \to \infty} E_n(x)$ and $O(x) = \lim_{n \to \infty} O_n(x)$. It can also be shown that, on any bounded set, the sequence of functions E_n converges uniformly to E and the sequence O_n converges uniformly to O.

Now, we leave it to you to verify that, for all positive integers n,

$$E_n'(x) = -O_{n-1}(x), \quad E_n''(x) = -E_{n-1}(x), \quad O_n'(x) = E_n(x), \quad \text{and } O_n''(x) = -O_{n-1}(x).$$

It follows that, on any bounded set, E_n'' converges uniformly to $-E$, and O_n'' converges uniformly to $-O$.

From these facts, it can be concluded that

$$E'(x) = -O(x), \quad O'(x) = E(x), \quad E''(x) = -E(x), \quad \text{and } O''(x) = -O(x).$$

In addition, we see easily that $E(0) = 1$, $E'(0) = 0$, $O(0) = 0$, and $O'(0) = 1$. Therefore, Corollary 2.7.15 tells us that $E(x) = \cos x$ and $O(x) = \sin x$.

The point of this is that

$$\cos x = \lim_{n \to \infty} \left(1 - \frac{x^2}{2!} + \frac{x^4}{4!} - \frac{x^6}{6!} + \cdots + (-1)^n \frac{x^{2n}}{(2n)!} \right)$$

and

$$\sin x = \lim_{n \to \infty} \left(x - \frac{x^3}{3!} + \frac{x^5}{5!} - \frac{x^7}{7!} + \cdots + (-1)^n \frac{x^{2n+1}}{(2n+1)!} \right).$$

Note that, when $|x|$ is close to 0, e.g., $|x| \leq 0.1$, the larger powers of x will be significantly smaller, in absolute value, than the smaller powers of x and, even when added, will not contribute much to the sums. To convince yourself of this, when $|x| \leq 0.1$, think about the decimal places being affected by the terms of the summations.

Thus, knowing that $\cos x$ and $\sin x$ are limits of polynomial functions lets us approximate them well. For instance, we know that $\cos x$ and $\sin x$ are continuous everywhere, in particular, at $x = 0$. So, $\lim_{x \to 0} \cos x = \cos 0 = 1$ and $\lim_{x \to 0} \sin x = \sin 0 = 0$, which means that when x is "close to" 0, $\cos x$ is "close to" 1 and $\sin x$ is "close to" 0. What if we want better approximations? We could use the first non-constant polynomials in the sequences of functions above: when x is "close to" 0, $\cos x \approx E_1(x) = 1 - x^2/2$ and $\sin x \approx O_0(x) = x$.

Are these really better approximations? You can check on a calculator; make sure that it's set to use radians! When $x = 0.001$, your calculator should tell you

$$\sin(0.001) \text{ "} = \text{" } 0.0009999998$$

and

$$\cos(0.001) \text{ "} = \text{" } 0.9999995.$$

We have placed these equal signs in quotes, since, really, what your calculator gives you are approximations, which may be only as good as the number of decimal places displayed.

How close are these calculator answers to what we get from the approximations $\sin x \approx x$ and $\cos x \approx 1 - x^2/2$? Certainly, what your calculator yields for $\sin(0.001)$ is very, very close to the value of x itself, 0.001. Furthermore, when $x = 0.001$,

$$1 - \frac{x^2}{2} = 1 - \frac{(0.001)^2}{2} = 0.9999995,$$

which is exactly what our calculator answer from above was. (We are **not**, however, claiming that the calculator answer is the exact value of $\cos(0.001)$.)

As a final comment, we should remark that the approximation $\sin x \approx x$, when $|x|$ is close to 0, is used very often in physics and engineering applications.

2.7.1   Exercises

Differentiate the functions in Exercises 1 through 19.

1. $f(t) = 5 \sin(2t)$.

2. $g(\theta) = 4.3 + 0.6 \cos(1.2\theta)$.

3. $h(x) = 3 \sin^2 x + \dfrac{2}{5} \sin x + e^{0.9}$.

4. $p(x) = e^x \sin(\sqrt{3x})$.

5. $q(\theta) = \ln(e^{2\pi\theta})(2 \sin^2 \theta + \cos 2\theta)$.

6. $f(t) = \dfrac{\cos\left(\frac{\pi}{2} - t\right)}{2 \sin^5 t}$.

7. $g(x) = \sin\left(\dfrac{\pi}{6} + x\right) - \cos\dfrac{\pi}{6}$.

8. $h(t) = \dfrac{\sin(2t)}{14 \sin t \cos^2 t}$.

9. $f(x) = x^2 \sin x + \sin\dfrac{2\pi}{3}$.

10. $g(t) = e^{-t \sin(t^2+1)}$.

11. $h(x) = \dfrac{\cos(\sqrt{x})}{x}$.

12. $p(x) = \dfrac{1}{x} \sin\left(\dfrac{1}{x}\right)$.

13. $f(\theta) = \dfrac{\sin \theta}{\cos \theta}$.

14. $g(x) = \ln\left(\dfrac{1}{\sin x}\right)$.

15. $q(x) = \dfrac{1 - \sin x}{\cos x}$.

16. $f(x) = (x + \sin(3x))^4$.

17. $g(x) = 2x\cos(2x) + \sin(2x)$.

18. $r(x) = \dfrac{2 + 3\sin x}{3 + 2\sin x}$.

19. $p(\theta) = 4\sin\left(\dfrac{\theta}{4}\right) - \cos(3\theta)$.

In Exercises 20 and 21, graph the given functions on the specified intervals and answer the following questions: (a) On which subintervals is the function increasing? Decreasing? (b) What are the critical points of the function?

20. $f(x) = \sin(2x)$ over $[-\pi, \pi]$

21. $g(x) = \cos(-x)$ over $[0, 2\pi]$

In Exercises 22 through 24, calculate $f^{(n)}(x)$.

22. $f(x) = a\cos(bx)$.

23. $f(x) = a\sin(bx)$.

24. $f(x) = a\sin(x)\cos(x)$.

25. What is the 33rd derivative of $g(x) = \sin(x) + \cos(x)$?

Find the limits in Exercises 26 through 30.

26. $\displaystyle\lim_{t \to 0} \frac{e^{3t} - 1}{\sin(3t)}$.

27. $\displaystyle\lim_{t \to 0} \frac{\sin(5t)}{\sin(10t)}$.

28. $\displaystyle\lim_{t \to 0} \frac{\cos(t)\sin^2(t)}{t}$.

29. $\displaystyle\lim_{t \to 0} \frac{1 - \sin^2(t) - \cos(t)}{t}$.

30. $\displaystyle\lim_{t \to 0} \frac{\sin(ax)}{(bx)}$, where $a \neq 0$ and $b \neq 0$.

31. Find a continuous function f with a half-open interval domain $[a, b)$ and a closed interval range $[c, d]$. Your answer should be in terms of a, b, c, and d.

   An interesting fact is that a continuous function in the reverse direction does not exist. There is no continuous function g with domain $[c, d]$ and with a range of the form $[a, b)$, $(a, b]$ or (a, b).

In Exercises 32 through 34, write an equation for the solution of the current $i(t)$ based on the information given for a simple RLC circuit, in which $R = 0$.

32. $L = 12$, $C = 8.5$, $i(0) = 5$.

33. $\omega = 0.5$, $i(0) = 10$, $i'(0) = 6$.

34. $C_1 = 3$, $C_2 = 8$, $i'(0) = 4$

35. Let $i(t)$ be the current in a simple RLC circuit, in which $R = 0$, and suppose that

$$\frac{d^2i}{dt^2} = -48\cos(4t) - 80\sin(4t).$$

What is $i(t)$?

36. Find the maximum value of the function $h(x) = \cos(2x) - 2\cos(x)$ on the interval $[0, 2\pi]$.

37. Let $f(x) = \sin^4 x$.

    a. At what points is the second derivative zero?

    b. Which of these points, if any, correspond to inflection points on the graph?

38. Let $h(t) = t + \sin t$. Show that h has critical points, but that h attains neither relative minimum nor relative maximum values at the critical points.

39. Let $f_1(x) = \sin x$, $f_2(x) = \sin(\sin x)$, $f_3(x) = \sin(\sin(\sin x))$ and, in general, $f_n(x) = \sin\big(f_{n-1}(x)\big)$, for all $n \geq 2$. Prove that $f_n'(\pi/2) = 0$, for all $n \geq 1$.

40. Suppose that functions f and g have periods h and k, respectively, where h and k are both positive integers and f and g are both defined for all real numbers. Prove that fg is hk-periodic.

41.   a. Calculate $O_0(\pi/6), O_1(\pi/6)$ and $O_2(\pi/6)$.

      b. What value does $O_n(\pi/6)$ approach as n increases?

42. Write out $E_4(x)$ and calculate $E_4'(x)$ and $E_4''(x)$.

Calculate dy/dx in Exercises 43 through 45 using implicit differentiation, as discussed in Example 2.3.13.

43. $y^2 = \cos(xy)$.

44. $\sin(x + y) = e^y$.

45. $\ln(x + y) = \sin(xy)$.

46. Suppose (c, d) is an interval on which $\cos(x)$ is positive. Suppose further that $f(x)$ is a differentiable function with range (c, d) and the property that $\sin(f(x)) = x$ for all x in the domain of f. Calculate $f'(x)$ explicitly using the Inverse Function Theorem, Corollary 2.3.8.

47. *Interference* refers to the physical effects of superimposing two waves.

  a. Prove the identity $\sin B + \sin C = 2\sin(B/2 + C/2)\cos(B/2 - C/2)$.

  b. Let $y_1 = A\sin(x)$ and $y_2 = A\sin(x + \theta)$, where θ is some constant. Let $y = y_1 + y_2$. Show that $y = M\sin(x + \theta/2)$, where M is a function of A and θ.

  c. Calculate $\dfrac{dM}{d\theta}$.

  d. The waves given by y_1 and y_2 are said to interfere *constructively* when $|M|$ achieves its maximum as a function of θ, and *destructively* when $|M|$ achieves its minimum as a function of θ. What values of θ lead to constructive and destructive interference?

48. Let $y_1 = A\sin(x - t)$ and $y_2 = A\sin(x + t)$. These equations are used to model the positions of two waves moving in opposite directions along a string, where x denotes a location on the string and t is a time parameter.

  a. If the two waves are traveling simultaneously, the resultant wave is $y = y_1 + y_2$. Show that $y = 2A\sin x \cos t$. The function y is an example of a *standing* wave.

  b. The *nodes* of a standing wave are the values of x such that $y = 0$ for all values of t. Physically, nodes are locations on the string where energy cannot be transported left or right. What are the nodes in this case?

  c. Calculate $\dfrac{dy}{dx}$.

49. Let $y_1(t) = e^{-t/2}\cos\left(\dfrac{\sqrt{3}t}{2}\right)$ and $y_2(t) = e^{-t/2}\sin\left(\dfrac{\sqrt{3}t}{2}\right)$. Show that y_1 and y_2 are both solutions of the differential equation $y'' + y' + y = 0$.

50. The *Wronskian* of two differentiable functions $y_1(t)$ and $y_2(t)$ is defined as

$$W(y_1, y_2)(t) = y_1(t)y_2'(t) - y_2(t)y_1'(t)$$

Show that if $y_1(t) = e^{at}\cos(bt)$ and $y_2(t) = e^{at}\sin(bt)$, where a and b are two real numbers and $a \neq 0$, then the Wronskian of y_1 and y_2 is non-zero for all t.

51. Assume a mass m is attached to a spring which obeys Hooke's law. Recall that if there is no damping, Newton's 2nd Law of Motion implies $mx'' + kx = 0$ and $\omega = \sqrt{k/m}$. Now assume that an external force is periodically applied to the mass and the magnitude of force applied at time t is given by $F(t) = A \cos \omega t$. Newton's 2nd Law now implies the slightly more complicated equation

$$mx'' + kx = A \cos \omega t.$$

   a. Show that $x(t) = C_1 \cos \omega t + C_2 \sin \omega t + \dfrac{A}{2m\omega} t \sin \omega t$ is a solution to the differential equation above.

   b. What happens to $x(t)$ as $t \to \infty$?

   This type of behavior is called *resonance* and is a result of the fact that the forcing frequency and the frequency of the system are both ω.

52. Data collected from a midwest city shows that an appropriate model for the mean daily temperature t months from the beginning of the year is

$$T(t) = -49 \cos \left(\frac{\pi t}{6} - \frac{\pi}{6} \right) + 42,$$

   measured in degrees Fahrenheit.

   a. Find the mean temperature at the end of May.

   b. Approximately which day is the coldest in the city? Which is the hottest?

   c. During which month is the mean temperature increasing the most rapidly? During which month is it decreasing the most rapidly?

53. The voltage supplied by an electrical outlet in France is given by

$$V(t) = 220 \sin(100\pi t),$$

   (measured in volts). Here, the sign of the voltage function gives the direction of the flow of the current, and t is the time, in seconds, from some initial measurement.

   a. What is the maximum voltage supplied by the outlet? What is the first time after the initial measurement that the maximum voltage is achieved?

b. What is the first time after the initial measurement at which the voltage provided by the outlet is increasing the most rapidly?

54. Suppose that

$$D(t) = 10\sin(0.16\pi t) + 20$$

is a model for the encroachment, measured in feet from some fixed line, of the sea into a bay, t hours after midnight on January 1st, 2001.

   a. What is the smallest value of $D(t)$ and how long after midnight on January 1, 2001 does it occur?

   b. What is the first time at which $D(t)$ is increasing as rapidly as possible? Decreasing as rapidly as possible?

55. Reprove the identity that, for all x, $\sin^2 x + \cos^2 x = 1$, by using the following two steps:

   (a) Differentiate $\sin^2 x$ and $\cos^2 x$ separately to conclude that the derivative of $\sin^2 x + \cos^2 x$ is zero. Conclude that $\sin^2 x + \cos^2 x$ is a constant.

   (b) Substitute a convenient value for x to conclude that the constant from part (a) must be 1.

56. Suppose that r and b are constants. Show that $f(x) = e^{-rx}\sin(x+b)$ is a solution to the differential equation

$$y'' + 2ry' + (r^2 + 1)y = 0.$$

57. Express the derivative of the function $f(x) = \sin x \cos x$, in terms of only $\cos(2x)$.

58. A disk of radius 2 meters is spinning. A Cartesian coordinate system is set up so that the center of the disk is located at the origin. Suppose that the x-and y-coordinates of a dot on the edge of the disk, at time t seconds after the wheel is put in motion, are given by the equations:

$$x(t) = 2\sin\left(\frac{\pi}{3}t\right) \qquad \text{and} \qquad y(t) = 2\cos\left(\frac{\pi}{3}t\right).$$

   a. Where is the dot at times $t = 0$ and $t = 3$?

   b. What are the dot's horizontal and vertical velocities at these times?

   c. What are the dot's horizontal and vertical accelerations at these times?

   d. Judging from the velocity functions at $t = 0$, was the wheel spun clockwise or counterclockwise?

59. A particle travels along the x-axis with position function

$$x(t) = \frac{1}{3}\sin^3 t - \sin t,$$

for $0 \leq t \leq 2\pi$.

    a. Find all critical points of $x(t)$.

    b. At what t value(s) does the particle change directions?

    c. At what t value(s) is the particle the farthest to the right (i.e., possess the largest x value)?

    d. At what t value(s) is the particle the farthest to the left?

    e. At what t values does the graph of $x(t)$ have inflection points?

60. An object of mass $m = 0.2$ kilograms is glued to the end of a spring which is sitting horizontally on a table. We assume that the spring obeys Hooke's Law. When lying on the table, the object is motionless (i.e., at equilibrium) 1 meter from the other (fixed) end of the spring. When the fixed end of the spring is held in the air, the force of gravity $F = mg$ on the object acts downward, causing the object's vertical equilibrium position to be 1.1 meters from the fixed end. Assume the value of g to be 9.8 m/s^2.

    a. Calculate the spring constant k for this spring.

    b. The spring and object are placed back on the table (so that the object's equilibrium position is again 1 meter from the fixed end), the object is pulled to 1.5 meters from the fixed end, and released. Disregarding the effects of friction, find the function $x(t)$ for the distance of the object from the fixed end t seconds after the object is released.

    c. At what velocities is the object traveling at the times when it passes through the equilibrium point?

61. An object of mass m kilograms is attached to the free end of a spring, which has a spring constant of k Newtons per meter (this means, in particular, that we are assuming that the spring obeys Hooke's Law). We set up the x-axis, in meters, along the spring, so that the equilibrium position is at $x = 0$. Here, by "equilibrium position", we mean the equilibrium position **if there are no forces acting on the object other than the spring force**.

However, suppose that, in addition to the force exerted by the spring on the object, there is also a constant force of F_0 Newtons acting on the object. This would be the case, for instance, if we had a vertical spring, for then gravity would also act on the object.

In this case, Newton's 2nd Law of Motion tells us that

$$mx'' = -kx + F_0,$$

where the primes denote derivatives with respect to t. Thus, we have

$$x'' = -\frac{k}{m}x + \frac{F_0}{m}. \tag{2.17}$$

The presence of the F_0/m term keeps us from being able to immediately apply Corollary 2.7.16 to solve this differential equation. However, a clever substitution will let us solve for x.

Let $y = x + b$, where b is a constant. Hence, $x = y - b$. Substitute this into Formula 2.17 to obtain a differential equation involving y, instead of x. Now determine the value of b that will let you use Corollary 2.7.16 to solve for y, and then find $x = x(t)$. Use x_0 and v_0 for the initial position and velocity, respectively, of the object.

2.8   The Other Trigonometric Functions

In this section, we will look at the remaining four trigonometric functions: tangent, cotangent, secant, and cosecant. For angles t between 0 and $\pi/2$ radians, i.e., $0 < t < 90°$, the definitions of these other four trigonometric functions agree with the definitions that are used in the context of right-triangles, but we make the definitions using sine and cosine, and so obtain new periodic functions.

As we will be dividing by $\cos x$ and $\sin x$, we need to know where these functions are equal to zero. Looking at the unit circle, we see that $\cos x = 0$ if and only if $x = \pi/2 + k\pi$, where k is an integer. Similarly, $\sin x = 0$ if and only if $x = k\pi$ for some integer k.

Therefore, letting k below denote arbitrary integers, we define:

the **tangent of** x, $\tan x$, by

$$\tan x = \frac{\sin x}{\cos x}, \quad \text{if } x \neq \pi/2 + k\pi,$$

the **cotangent of** x, $\cot x$, by

$$\cot x = \frac{\cos x}{\sin x}, \quad \text{if } x \neq k\pi,$$

the **secant of** x, $\sec x$, by

$$\sec x = \frac{1}{\cos x}, \quad \text{if } x \neq \pi/2 + k\pi,$$

and the **cosecant of** x, $\csc x$, by

$$\csc x = \frac{1}{\sin x}, \quad \text{if } x \neq k\pi.$$

From the Fundamental Trigonometric Identity, it is easy to derive two other identities, which hold for all x in the domains:

$$1 + \tan^2 x = \sec^2 x, \quad \text{and} \quad 1 + \cot^2 x = \csc^2 x.$$

You may wonder about the use of the prefix "co-" in the terms *cosine*, *cotangent*, and *cosecant*. Two angles α and β are called *complementary* if and only if $\alpha + \beta = \pi/2$ (in degrees, this means that the angles add up to 90°). As we discussed when reviewing the properties of sine and cosine, if α and β are complementary, then $\sin\alpha = \cos\beta$. It follows that the value of any "co-" trig function on an angle β is the same as the value of the corresponding "non-co-" trig function on the complementary angle α.

As $\sin(x \pm \pi) = -\sin x$ and $\cos(x \pm \pi) = -\cos x$, we see that $\tan x$ and $\cot x$ are periodic, with period π. Using the even- and odd-ness of $\cos x$ and $\sin x$, respectively, it is easy to see that $\tan x$ and $\cot x$ are odd functions. The graphs of tangent and cotangent are in Figure 2.14 and Figure 2.15.

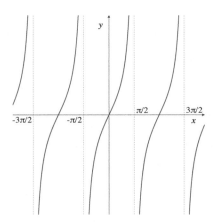

Figure 2.14: The graph of $y = \tan x$.

Figure 2.15: The graph of $y = \cot x$.

Clearly, $\sec x$ and $\csc x$ are periodic, with period 2π, and are even and odd functions, respectively. Also, as $|\sin x| \leq 1$ and $|\cos x| \leq 1$, it follows that $|\sec x| \geq 1$ and $|\csc x| \geq 1$. The graphs of secant and cosecant are in Figure 2.16 and Figure 2.17.

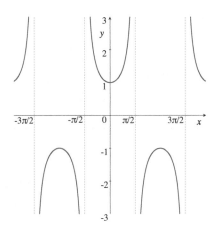

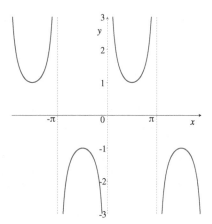

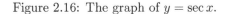

Figure 2.16: The graph of $y = \sec x$.

Figure 2.17: The graph of $y = \csc x$.

Since we know the derivatives of $\sin x$ and $\cos x$, and we know the Quotient Rule, the Chain Rule, and the Fundamental Trigonometric Identity, it easy to calculate the derivatives of $\tan x$, $\cot x$, $\sec x$, and $\csc x$.

Theorem 2.8.1. *The functions $\tan x$, $\cot x$, $\sec x$, and $\csc x$ are differentiable, and their derivatives are given by:*

$$\tan' x \;=\; \frac{1}{\cos^2 x} \;=\; \sec^2 x;$$

$$\cot' x \;=\; -\frac{1}{\sin^2 x} \;=\; -\csc^2 x;$$

$$\sec' x \;=\; \frac{\sin x}{\cos^2 x} \;=\; \tan x \sec x;$$

and

$$\csc' x \;=\; -\frac{\cos x}{\sin^2 x} \;=\; -\cot x \csc x.$$

Proof. We will derive the formulas for the derivatives of tangent and secant, and leave the other two as exercises.

We calculate, using the Quotient Rule,

$$\tan' x \;=\; \left(\frac{\sin x}{\cos x}\right)' \;=\; \frac{\cos x \cos x - (\sin x)(-\sin x)}{\cos^2 x} \;=\;$$

$$\frac{\cos^2 x + \sin^2 x}{\cos^2 x} \;=\; \frac{1}{\cos^2 x} \;=\; \sec^2 x.$$

Next, we calculate, using the Chain Rule,

$$\sec' x \;=\; \left[(\cos x)^{-1}\right]' \;=\; -1 \cdot (\cos x)^{-2}(-\sin x) \;=\;$$

$$\frac{\sin x}{\cos^2 x} \;=\; \tan x \sec x.$$

$\square$

Remark 2.8.2. Note that the derivative of tangent is always positive. Thus, as you can see in Figure 2.14, the function $\tan x$ is strictly increasing on each interval contained in its domain. In particular, when restricted to the fundamental interval $(-\pi/2, \pi/2)$, $\tan x$ is strictly increasing and, hence, one-to-one on that interval. Of course, $\tan x$ itself is far from one-to-one, since it is π-periodic.

Similarly, the derivative of cotangent is always negative. Thus, as you can see in Figure 2.15, the function $\cot x$ is strictly decreasing on each interval contained in its domain. In particular, when restricted to the fundamental interval $(0, \pi)$, $\cot x$ is strictly decreasing and, hence, one-to-one on that interval. However, $\cot x$ itself is not one-to-one, since it is π-periodic.

Example 2.8.3. Calculate $(x \tan x)'$, $\big(7 \csc(3x)\big)'$, and $\left(\sec x + \dfrac{\cot x}{x}\right)'$.

Solution:

$$(x \tan x)' \;=\; x(\tan x)' + (\tan x)(x)' \;=\; x \sec^2 x + \tan x.$$

$$\big(7 \csc(3x)\big)' \;=\; 7\big(-\cot(3x) \csc(3x)\big)(3x)' \;=\; -21 \cot(3x) \csc(3x).$$

$$\left(\sec x + \frac{\cot x}{x}\right)' \;=\; \tan x \sec x \;+\; \frac{x(\cot x)' - (\cot x)(x)'}{x^2} \;=$$

$$\tan x \sec x \;+\; \frac{-x \csc^2 x - \cot x}{x^2}.$$

Example 2.8.4. Let $w = e^{-\theta} \sec \theta$. Find the critical points of w.

Solution:

Recalling Definition 1.5.6, what we need to do is find the values of θ, in the domain of w, at which $dw/d\theta$ is zero or undefined. As $w = e^{-\theta} \sec \theta = e^{-\theta}/\cos \theta$, w is differentiable at all θ at which $\cos \theta \neq 0$, i.e., at all θ in the domain of w. Thus, we need to find those θ at which $dw/d\theta = 0$.

We calculate

$$\frac{dw}{d\theta} \;=\; e^{-\theta}\big(\sec \theta\big)' + \big(\sec \theta\big)(e^{-\theta})' \;=\; e^{-\theta} \sec \theta \tan \theta - e^{-\theta} \sec \theta \;=$$

$$e^{-\theta} \sec \theta (\tan \theta - 1).$$

As $e^{-\theta} \sec \theta$ is never zero, the critical points of w occur where $\tan \theta = 1$. Where does $\tan \theta = 1$? As you probably know, this happens in a right isosceles triangle, i.e., when the angle is $45°$. In radians, this means that $\theta = \pi/4$. As we pointed out in Remark 2.8.2, tangent is one-to-one on the open interval $(-\pi/2, \pi/2)$. Therefore, $\theta = \pi/4$ is the only angle between $-\pi/2$ and $\pi/2$ whose tangent is equal to 1. However, tangent is π-periodic, and so, the complete collection of critical points is given by $\theta = \pi/4 + k\pi$, where k is any integer.

It is extremely difficult to see these critical points by looking at the graph and looking for horizontal tangent lines; it is essentially impossible to select a good scale and graphing window that let you see many of these critical points at once.

Example 2.8.5. A lighthouse is 0.25 miles from a beach; let's call the point on the beach, closest to the lighthouse, P. Viewed from above, the lighthouse turns counter-clockwise, and completes one full revolution every 8 seconds.

A boy, playing on the beach, decides that he will run along the beach and try to keep up with where the beam appears to be on the beach. See Figure 2.18.

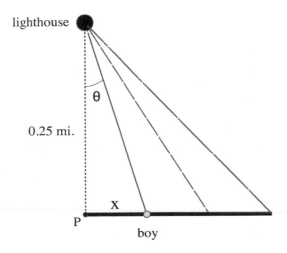

Figure 2.18: A lighthouse, and a boy on the beach. Top view.

How fast must the boy run to keep up with the light, when the beam from the lighthouse is hitting the beach at a point 0.5 miles from P?

Solution:

Let t denote the time, in seconds, x denote the distance, in miles, from P to the boy, and θ the angle, in radians, that the beam of light makes with the line segment from the lighthouse to P, measured counter-clockwise. See Figure 2.18.

As there are 2π radians in a full revolution, we are told that

$$\frac{d\theta}{dt} = \frac{2\pi}{8} = \frac{\pi}{4} \text{ rad/s}.$$

We also see that

$$\tan\theta = (\text{opposite length})/(\text{adjacent length}) = \frac{x}{0.25}.$$

Thus, $x = 0.25\tan\theta$.

Now, θ and x are functions of time, and we may differentiate both sides of this last equation, with respect to t, and find

$$\frac{dx}{dt} = 0.25\frac{d}{dt}(\tan\theta) = 0.25\frac{d}{d\theta}(\tan\theta)\cdot\frac{d\theta}{dt} = 0.25(\sec^2\theta)(\pi/4) \text{ mi/s}.$$

We want to calculate $\dfrac{dx}{dt}\Big|_{x=0.5mi}$. Hence, we need to know the value of $\sec^2\theta$ when $x = 0.5$ miles. When $x = 0.5$ miles, $\tan\theta = 0.5/0.25 = 2$, and $\tan^2\theta + 1 = \sec^2\theta$. Thus, $\sec^2\theta = 5$ when $x = 0.5$ miles, and we find

$$\frac{dx}{dt}\Big|_{x=0.5mi} = (0.25)(5)(\pi/4) \approx 0.9817477 \text{ mi/s}.$$

That is, the boy would need to run at a speed of about 1 **mile per second** to keep up with the beam of light, sweeping along the beach. Good luck!

2.8.1   Exercises

Differentiate the functions in Exercises 1 through 16.

1. $y(\theta) = \cot\theta$.

2. $j(\omega) = \csc\omega$.

3. $h(u) = 7\cot(3u - 9)$.

4. $f(x) = 1 + \tan^2 x$.

5. $f(\theta) = e^{-2\theta}\csc\theta$.

6. $g(x) = \cot(x^3 - 2x^2)$.

7. $g(x) = \tan(\sqrt{1 + x})$.

8. $f(t) = 4\cos^3 t + 3\sin^2 t - \tan t$.

9. $p(x) = 7x^4\csc x - 2x^3\sin x$.

10. $h(t) = \tan(\cos t)$.

11. $f(x) = \sqrt{\sec x}$.

12. $q(\theta) = \dfrac{\cos\theta}{\csc\theta}$.

13. $p(x) = (\tan(2x))^{\frac{3}{2}}$.

14. $r(t) = (\sin^2 t)\tan t$.

15. $r(x) = 3^{-x}\tan(2x^3)$.

16. $p(\theta) = \sec^3(2\theta)$.

17. Consider the function $g(\theta) = |\tan\theta|$ on the interval $\left(-\dfrac{\pi}{2}, \dfrac{\pi}{2}\right)$.

    (a) On which subintervals is the function decreasing? Increasing?

    (b) Find $\dfrac{dg}{d\theta}$ (where it is defined). Identify critical points (if any).

Calculate the second derivatives of the functions in Exercises 18 through 21.

18. $y = \tan x$.

19. $g(t) = \sec t$.

20. $f(z) = \csc z$.

21. $p = \cot u$.

22. Let $f(t) = \sec^2 t - \tan^2 t$.

    a. Calculate $f'(t)$.

    b. Conclude that $f(t)$ is a constant function on each interval in the domain of $f(t)$, namely on each interval $I_k = (-\pi/2 + \pi k, \pi/2 + \pi k)$, for all integers k. Let C_k denote the constant value of f on I_k.

    c. Show, in fact, that all of the C_k's are equal by evaluating $f(t)$ at one convenient value of t in each interval I_k. Conclude that $f(t)$ is a constant, for all t in its domain.

23. Show that $y = \tan x$ satisfies the differential equation $y' = 1 + y^2$. Hint: use the identity established in the previous problem.

In Exercises 24 through 26, you are given a function f and a point $x = a$. Determine if the function has a differentiable inverse at the point $b = f(a)$. Use Theorem 2.3.8 or Corollary 2.3.8.

24. $f(x) = \tan x$, $x = 0$.

25. $f(x) = \sec x$, $x = 0$.

26. $f(x) = \cot x$, $x = 0$.

27. Prove that $\displaystyle\lim_{x \to 0} \frac{\tan ax}{x} = a$.

28. Calculate $\displaystyle\lim_{x \to 0} \frac{\tan x}{ax}$, $a \neq 0$.

29. Calculate $\displaystyle\lim_{x \to 0} \frac{\sin x}{x + \tan x}$.

You are given the position function of a particle in Exercises 30 through 32. Calculate the velocity and acceleration functions.

30. $p(t) = \dfrac{\sqrt{t}}{1 - \tan t}$.

31. $p(t) = \sin(\csc(t))$.

32. $p(t) = \tan t \sin t$.

33. What is the period of each function?

      a. $f(x) = \sin x$.   *b.* $g(x) = \cos x$.   *c.* $h(x) = \tan x$.

      d. $j(x) = \cot x$.   *e.* $k(x) = \csc x$.   *f.* $m(x) = \sec x$.

Recall that the hyperbolic sine, cosine and tangent are defined as

$$\sinh x = \frac{e^x - e^{-x}}{2}, \qquad \cosh x = \frac{e^x + e^{-x}}{2}, \qquad \tanh x = \frac{\sinh x}{\cosh x}.$$

We now add the hyperbolic secant, cosecant and cotangent to the list. Their definitions are analogous to the ordinary trig functions.

$$\operatorname{csch} x = \frac{1}{\sinh x}, \qquad \operatorname{sech} x = \frac{1}{\cosh x}, \qquad \operatorname{coth} x = \frac{1}{\tanh x}.$$

Use these definitions to derive the formulas in Exercises 34 through 36.

34. $\dfrac{d}{dx}(\operatorname{sech} x) = -\operatorname{sech} x \tanh x$.

35. $\dfrac{d}{dx}(\operatorname{csch} x) = -\operatorname{coth} x \operatorname{csch} x$.

36. $\dfrac{d}{dx}(\operatorname{coth} x) = -\operatorname{csch}^2 x$.

37. Show that $y = \operatorname{csch} x$ has a global maximum at $x = 0$.

Locate inflection points of the functions in Exercises 38 through 41. Review the graphs in this chapter to see if your answers make sense.

38. $f(x) = \tan x$.

39. $g(x) = \cot x$.

40. $h(x) = \sec x$.

41. $j(x) = \csc x$.

In Exercises 42 through 44, show that the given function is a solution to the given differential equation.

42. $y = \tan x$, $y'' = 2y'y$.

43. $y = \tan x$, $y''' = 2y'(y^2 + 1)$. Hint: use the previous problem.

44. $y = \sec x$, $\dfrac{d^2y}{dx^2} = y^3 + (\tan x)y'$.

45. Recall that a function is even if $f(x) = f(-x)$ and odd if $f(-x) = -f(x)$. Say whether each function is odd, even, or neither.

    a. $a(x) = \sin x$.   *b.* $b(x) = \cos x$.   *c.* $c(x) = \tan x$.

    d. $d(x) = \cot x$.   *e.* $e(x) = \sec x$.   *f.* $f(x) = \csc x$.

46. Consider the function $f(u) = \csc\left(2u + \frac{\pi}{2}\right)$ and the interval $[-3\pi, 3\pi]$.

    a. Graph the function for u values in the given interval. What u values in the given interval are **not** in the domain of f?

    b. By looking at the graph, identify the critical points of f which lie in $[-3\pi, 3\pi]$.

    c. Verify the critical points identified above using algebraic/trigonometric methods.

47. Consider the functions $f(x) = \cos x$ and $g(x) = \cot x$.

    a. Find the points of intersection of the graphs of $f(x)$ and $g(x)$.

    b. At which of their points of intersection do the graphs of $f(x)$ and $g(x)$ intersect tangentially (that is, share a common tangent line)?

    c. At which of their points of intersection do the graphs of $f(x)$ and $g(x)$ intersect orthogonally (i.e., have tangent lines which are perpendicular)?

    d. Graph both curves to convince yourself that your answer to part (c) is accurate.

48. In geometry, you learn that the tangent line to a circle at a point p is perpendicular to the radial line drawn from the center of the circle to p. Use the following hints to prove this using Calculus:

    a. Set up a Cartesian coordinate system. Draw a circle of radius r centered at the origin, and a radial line at some angle θ in the interval $\left(0, \dfrac{\pi}{2}\right)$ to the positive x-axis.

    b. Find the slope of this radial line, as well as the coordinates of the intersection point p (in the first quadrant) in terms of θ and r.

c. Use the standard Cartesian equation defining a circle to find the slope of the tangent line at the point p. Is the tangent line perpendicular to the radial line?

49. Consider the following diagram of a unit circle (a circle whose radius is 1), centered at the origin, a radial line at an angle of θ to the positive x-axis, where θ is in the interval $(0, \frac{\pi}{2})$, and a tangent line to the circle at the point where the radial line and circle intersect. Let s and t be the lengths indicated in the diagram.

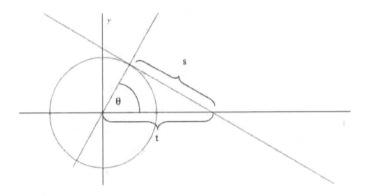

a. Describe s and t as functions of the angle θ.

b. What is the instantaneous rate of change of s with respect to θ, when $\theta = \frac{\pi}{4}$?

c. What is the instantaneous rate of change of t with respect to θ, when $\theta = \frac{\pi}{3}$?

2.9   Inverse Trigonometric Functions

The inverse trigonometric functions are very important. However, if you've never seen the inverse trig functions before, you may be a bit confused; part of the importance of the trigonometric functions stems from the fact that they are fundamental **periodic** functions, which, in particular, means that the trig functions are very explicitly **not** one-to-one, and so have no inverses.

We deal with this in a manner similar to what we do for the function x^n when n is an even integer; we restrict the domain of the original function to some fundamental subset of the domain on which the function **is**, in fact, one-to-one. We then invert this restricted function. Of course, we would like to use restricted domains on which the trig functions assume all of their possible values.

Consider the graphs of $y = \sin x$ and $y = \cos x$:

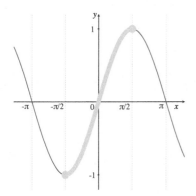

Figure 2.19: The graph of $y = \sin x$.   Figure 2.20: The graph of $y = \cos x$.

We have indicated in green a portion of each graph where $\sin x$ and $\cos x$ take on all of their possible values, that is, all of the values y such that $-1 \le y \le 1$. Moreover, we see that the green portions of the graph satisfy a horizontal line test; there are no horizontal lines that would intersect the green part of either graph more than once. This means that if we restrict x to the corresponding subsets of the domains, then we obtain one-to-one functions, which possess inverses. If you are worried that perhaps the graphs do not really show what's going on, you may look at $(\sin x)' = \cos x$ and $(\cos x)' = -\sin x$ on the intervals $[-\pi/2, \pi/2]$ and $[0, \pi]$, respectively, and apply Corollary 1.5.15 to conclude that sin and cos are strictly monotonic on the given intervals and, hence, one-to-one.

We will denote the trig functions with "nicely restricted" domains (and codomains) by their usual names, but we will capitalize the first letter. These restricted trig functions have inverses. The first letters in the names of theses inverses are **not** capitalized, since there are no other inverse trigonometric functions to confuse them with.

> **Definition 2.9.1.** *We define* $\text{Sin} : [-\pi/2, \pi/2] \to [-1, 1]$ *and* $\text{Cos} : [0, \pi] \to [-1, 1]$ *to be the restrictions of* $\sin$ *and* $\cos$, *respectively, to the given domains and codomains.*
>
> *We define* **inverse sine** *and* **inverse cosine** *to be the inverse functions of* Sin *and* Cos, *respectively. We denote these inverse functions by* $\sin^{-1}$ *and* $\cos^{-1}$, *respectively.*

Remark 2.9.2. Thus, the domains of $\sin^{-1}$ and $\cos^{-1}$ are both the interval $[-1, 1]$. The range of $\sin^{-1}$ is the interval $[-\pi/2, \pi/2]$, and the range of $\cos^{-1}$ is the interval $[0, \pi]$.

In words, $\sin^{-1} v$ is the angle between $-\pi/2$ and $\pi/2$, inclusive, whose sine equals v, and $\cos^{-1} v$ is the angle between 0 and π, inclusive, whose cosine equals v.

You must be very careful here. This means that, for all v in the interval $[-1, 1]$, $\sin(\sin^{-1} v) = v$ and $\cos(\cos^{-1} v) = v$, but, if x is not in the interval $[-\pi/2, \pi/2]$, then $\sin^{-1}(\sin x) \neq x$ and, if x is not in the interval $[0, \pi]$, then $\cos^{-1}(\cos x) \neq x$.

The function $\sin^{-1}$ is sometimes denoted arcsin, and $\cos^{-1}$ is sometimes denoted arccos. We shall not use this notation.

Note that, while $\sin^k x$ and $\cos^k x$ mean $(\sin x)^k$ and $(\cos x)^k$, for all integers $k \neq -1$, things are different when $k = -1$; $\sin^{-1} x$ and $\cos^{-1} x$ do **not** indicate raising to the -1 power. They denote the inverse functions of Sin and Cos, respectively. If you wish to raise $\sin x$ or $\cos x$ to the -1 power, you simply have to write $(\sin x)^{-1}$ and $(\cos x)^{-1}$.

> **Theorem 2.9.3.** *The functions* $\sin^{-1}$ *and* $\cos^{-1}$ *are continuous, and are differentiable on the interval* $(-1, 1)$. *The derivatives are given by*
>
> $$(\sin^{-1} x)' \;=\; \frac{1}{\sqrt{1 - x^2}}, \qquad \text{and} \qquad (\cos^{-1} x)' \;=\; \frac{-1}{\sqrt{1 - x^2}}.$$

Proof. That $\sin^{-1}$ and $\cos^{-1}$ are continuous is immediate from Theorem 1.3.35.

That $\sin^{-1}$ and $\cos^{-1}$ are differentiable on the interval $(-1, 1)$ follows at once from Theorem 2.3.8, which also tells us how to calculate the derivatives.

For x in the interval $(-1, 1)$, we find

$$(\sin^{-1} x)' \;=\; \frac{1}{\sin'(\sin^{-1} x)} \;=\; \frac{1}{\cos(\sin^{-1} x)},$$

and

$$(\cos^{-1} x)' \;=\; \frac{1}{\cos'(\cos^{-1} x)} \;=\; \frac{1}{-\sin(\cos^{-1} x)}.$$

We need to show that $\cos(\sin^{-1} x) = \sqrt{1 - x^2}$ and $\sin(\cos^{-1} x) = \sqrt{1 - x^2}$, and we'll be finished. We will show one of these, and leave the other as an exercise for you.

Consider $\cos(\sin^{-1} x)$. Let $\theta = \sin^{-1} x$. Thus, by the definition of $\sin^{-1}$, $\sin \theta = x$ and $-\pi/2 \leq \theta \leq \pi/2$. Now, $\cos(\sin^{-1} x) = \cos \theta$, which by the Fundamental Trigonometric Identity, is equal to $\pm\sqrt{1 - \sin^2 \theta} = \pm\sqrt{1 - x^2}$. However, for $-\pi/2 \leq \theta \leq \pi/2$, $\cos \theta \geq 0$ and, thus, $\cos \theta = +\sqrt{1 - \sin^2 \theta}$. Hence,

$$\cos(\sin^{-1} x) \;=\; \cos \theta \;=\; \sqrt{1 - x^2}.$$

$\square$

Remark 2.9.4. Note that the derivatives of $\sin^{-1}$ and $\cos^{-1}$ are totally algebraic in nature, and do not involve trigonometric or inverse trigonometric functions. This should be contrasted with the derivatives of the trigonometric functions, which yielded other trigonometric functions.

Example 2.9.5. Calculate $\left(x^2 \cos^{-1} x\right)'$ and $\left(\sin^{-1}\left(\sqrt{1 - x^2}\right)\right)'$.

Solution:

$$\left(x^2 \cos^{-1} x\right)' \;=\; x^2 \cdot \frac{-1}{\sqrt{1 - x^2}} \;+\; (\cos^{-1} x)(2x) \;=\; \frac{-x^2}{\sqrt{1 - x^2}} \;+\; 2x \cos^{-1} x.$$

$$\left(\sin^{-1}\left(\sqrt{1-x^2}\right)\right)' = \frac{1}{\sqrt{1-\left(\sqrt{1-x^2}\right)^2}}\cdot\left(\sqrt{1-x^2}\right)' =$$

$$\frac{1}{\sqrt{x^2}}\cdot\left((1-x^2)^{1/2}\right)' = \frac{1}{|x|}\cdot\frac{1}{2}(1-x^2)^{-1/2}(-2x) = -\frac{x}{|x|}\cdot\frac{1}{\sqrt{1-x^2}}.$$

Note that, when $x < 0$, $|x| = -x$ and this derivative is simply $1/\sqrt{1-x^2}$; when $x > 0$, this derivative is $-1/\sqrt{1-x^2}$.

Now let's look at tangent and cotangent. Their graphs are:

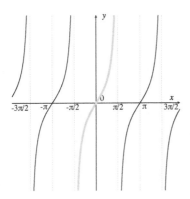

Figure 2.21: The graph of $y = \tan x$.   Figure 2.22: The graph of $y = \cot x$.

We have once again indicated, in green, portions of the graph that correspond to fundamental subsets of the domains, over which the functions are one-to-one.

Definition 2.9.6. *We define* Tan $: (-\pi/2, \pi/2) \to (-\infty, \infty)$ *and* Cot $: (0, \pi) \to (-\infty, \infty)$ *to be the restrictions of* tan *and* cot, *respectively, to the given domains and codomains.*

  We define **inverse tangent** *and* **inverse cotangent** *to be the inverse functions of* Tan *and* Cot, *respectively. We denote these inverse functions by* $\tan^{-1}$ *and* $\cot^{-1}$, *respectively.*

Theorem 2.9.7. *The functions* $\tan^{-1}$ *and* $\cot^{-1}$ *are differentiable on the interval* $(-\infty, \infty)$. *The derivatives are given by*

$$(\tan^{-1} x)' \;=\; \frac{1}{1+x^2}, \qquad \text{and} \qquad (\cot^{-1} x)' \;=\; \frac{-1}{1+x^2}.$$

Proof. That $\tan^{-1}$ and $\cot^{-1}$ are differentiable on the interval $(-1, 1)$ follows at once from Theorem 2.3.8, which also tells us how to calculate the derivatives.

We find

$$(\tan^{-1} x)' \;=\; \frac{1}{\tan'(\tan^{-1} x)} \;=\; \frac{1}{\sec^2(\tan^{-1} x)},$$

and

$$(\cot^{-1} x)' \;=\; \frac{1}{\cot'(\cot^{-1} x)} \;=\; \frac{1}{-\csc^2(\cot^{-1} x)}.$$

We need to show that $\sec^2(\tan^{-1} x) = 1 + x^2$ and $\csc^2(\cot^{-1} x) = 1 + x^2$, and we'll be finished. We will show one of these, and leave the other as an exercise for you.

Consider $\sec^2(\tan^{-1} x)$. Let $\theta = \tan^{-1} x$. Thus, by the definition of $\tan^{-1}$, $\tan \theta = x$ (and $-\pi/2 < \theta < \pi/2$). Since $\tan^2 \theta + 1 = \sec^2 \theta$, we find that

$$\sec^2(\tan^{-1} x) \;=\; \sec^2 \theta \;=\; 1 + \tan^2 \theta \;=\; 1 + x^2.$$

$\square$

Finally, we need to look at secant and cosecant.

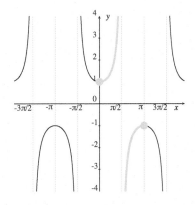

Figure 2.23: The graph of $y = \sec x$.

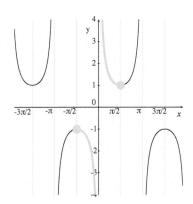

Figure 2.24: The graph of $y = \csc x$.

We have indicated, in green, portions of the graph that correspond to fundamental subsets of the domains, over which the functions are one-to-one.

Definition 2.9.8. *We define* $\text{Sec} : [0, \pi/2) \cup (\pi/2, \pi] \rightarrow (-\infty, -1] \cup [1, \infty)$ *and* $\text{Csc} :$ $[-\pi/2, 0) \cup (0, \pi/2] \rightarrow (-\infty, -1] \cup [1, \infty)$ *to be the restrictions of* $\sec$ *and* $\csc$*, respectively, to the given domains and codomains.*

We define **inverse secant** *and* **inverse cosecant** *to be the inverse functions of* Sec *and* Csc*, respectively. We denote these inverse functions by* $\sec^{-1}$ *and* $\csc^{-1}$*, respectively.*

Theorem 2.9.9. *The functions* $\sec^{-1}$ *and* $\csc^{-1}$ *are continuous, and are differentiable on the interior,* $(-\infty, -1) \cup (1, \infty)$*, of their domains. The derivatives are given by*

$$(\sec^{-1} x)' \;=\; \frac{1}{|x|\sqrt{x^2 - 1}}, \qquad \text{and} \qquad (\csc^{-1} x)' \;=\; \frac{-1}{|x|\sqrt{x^2 - 1}}.$$

Proof. Theorem 1.3.35 implies that $\sec^{-1}$ and $\csc^{-1}$ are continuous when restricted to either of $(-\infty, -1]$ or $[1, \infty)$; thus, they are continuous on the union.

That $\sec^{-1}$ and $\csc^{-1}$ are differentiable on $(-\infty, -1) \cup (1, \infty)$ follows at once from Theorem 2.3.8, which also tells us how to calculate the derivatives.

We find

$$(\sec^{-1} x)' \;=\; \frac{1}{\sec'(\sec^{-1} x)} \;=\; \frac{1}{\sec(\sec^{-1} x) \cdot \tan(\sec^{-1} x)} \;=\; \frac{1}{x \tan(\sec^{-1} x)}$$

and

$$(\csc^{-1} x)' \;=\; \frac{1}{\csc'(\csc^{-1} x)} \;=\; \frac{1}{-\csc(\csc^{-1} x) \cdot \cot(\csc^{-1} x)} \;=\; \frac{-1}{x \cot(\csc^{-1} x)}.$$

We need to show that, if $x > 1$, then $\tan(\sec^{-1} x) = \sqrt{x^2 - 1}$ and, if $x < -1$, then $\tan(\sec^{-1} x) = -\sqrt{x^2 - 1}$. We need to show something analogous for $\cot(\csc^{-1} x)$. We shall deal with $\tan(\sec^{-1} x)$, and leave $\cot(\csc^{-1} x)$ as an exercise.

Consider $\tan(\sec^{-1} x)$. Let $\theta = \sec^{-1} x$. Thus, by the definition of $\sec^{-1}$, $\sec\theta = x$ and $0 \leq \theta \leq \pi$ and $\theta \neq 0$. As $\tan^2\theta + 1 = \sec^2\theta$, we find that

$$\tan(\sec^{-1} x) \;=\; \tan\theta \;=\; \pm\sqrt{\sec^2\theta - 1} \;=\; \pm\sqrt{x^2 - 1}.$$

We need to show that $\tan\theta > 0$ when $x > 1$, and $\tan\theta < 0$ when $x < -1$. However, this is easy, since, when $x > 1$, $\theta = \sec^{-1} x$ is between 0 and $\pi/2$ (noninclusive), and, when $x < -1$, $\theta = \sec^{-1} x$ is between $\pi/2$ and π (noninclusive). Now look at the sign of tangent in these regions. $\qquad\square$

Different sources select different fundamental domains for secant and cosecant; this changes the ranges of the inverse functions, and the formulas for their derivatives. For this reason, many books (and people) make an effort to avoid using inverse secant and cosecant.

Example 2.9.10. Let $z = \tan^{-1}(e^v)$ and $w = r\sec^{-1} r$. Find dz/dv and dw/dr.

Solution:

Using the Chain Rule, we find

$$\frac{dz}{dv} \;=\; \frac{1}{(e^v)^2 + 1} \cdot \frac{d}{dv}(e^v) \;=\; \frac{e^v}{e^{2v} + 1}.$$

Using the Product Rule, we find

$$\frac{dw}{dr} \;=\; r \cdot (\sec^{-1} r)' + (\sec^{-1} r) \cdot (r)' \;=\; \frac{r}{|r|\sqrt{r^2 - 1}} + \sec^{-1} r.$$

Example 2.9.11. This example is similar to Example 2.8.5, but the given data and what is asked for are essentially reversed from the earlier example.

A tower, with a spotlight, is 150 feet from a prison wall; let's call the point on the wall, closest to the tower, P. Viewed from above, with the tower on the top and the wall on the bottom, a prisoner is running from left to right at 22 feet per second (which is 15 mph). See Figure 2.25

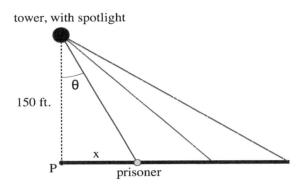

Figure 2.25: A spotlight, and a prisoner along a wall. Top view.

If the spotlight is turning in such a way that it keeps pointing at the prisoner, at what rate is the spotlight turning when the prisoner is 50 feet to the right of the point P?

Solution:

Let x denote the distance, in feet, from the prisoner to the point P. We are told that $dx/dt = 22$ ft/s, where, of course, t is the time, in seconds. We see that $\tan\theta = x/150$ and, as we clearly have $-\pi/2 < \theta < \pi/2$, it follows that $\theta = \tan^{-1}(x/150)$.

Now, x and θ are both functions of time, and we may differentiate both sides of this last equation, with respect to t, using the Chain Rule:

$$\frac{d\theta}{dt} = \left[\frac{d}{dx}\left(\tan^{-1}(x/150)\right)\right] \cdot \frac{dx}{dt} \quad \text{rad/s},$$

which equals

$$\frac{1}{(x/150)^2 + 1} \cdot \frac{1}{150} \cdot 22 \quad \text{rad/s}.$$

We are asked for $d\theta/dt$, when $x = 50$ ft. We find

$$\frac{11}{75} \cdot \frac{1}{(1/3)^2 + 1} = \frac{11}{75} \cdot \frac{9}{10} = 0.132 \quad \text{rad/s}.$$

Given the definitions of the inverse trigonometric functions, and the identities that, for all x, $\sin(\pi/2 - x) = \cos x$ and $\cos(\pi/2 - x) = \sin x$, it is easy to show:

Theorem 2.9.12. *For all x in the domains of the functions involved:*

$$\sin^{-1} x + \cos^{-1} x = \pi/2;$$

$$\tan^{-1} x + \cot^{-1} x = \pi/2;$$

and

$$\sec^{-1} x + \csc^{-1} x = \pi/2.$$

Remark 2.9.13. The theorem above explains why the derivatives of the co-inverse trig functions are just the negations of the derivatives of the corresponding non-co-inverse trig functions.

Alternatively, you can start with the derivative formulas, and derive the identities. For instance, since $(\sin^{-1} x)' = (-\cos^{-1} x)'$, Corollary 1.5.12 tells us that there is a constant C such that, for all x in the open interval $(-1, 1)$,

$$\sin^{-1} x = -\cos^{-1} x + C.$$

If we plug in $x = 0$, we obtain that $0 = -\pi/2 + C$, and so $C = \pi/2$. To obtain the identity when $x = \pm 1$, we need to use the continuity of $\sin^{-1}$ and $\cos^{-1}$.

2.9.1 Exercises

Differentiate the functions in Exercises 1 through 20.

1. $f(x) = 2\sin^{-1} x + x^2 + 1$.

2. $p(z) = 17\csc^{-1}\left(\dfrac{z}{2}\right)$.

3. $h(u) = 3\ln(\sin^{-1} u)$.

4. $q(t) = \dfrac{\tan^{-1} t}{1 + t^2}$.

5. $g(x) = (\cos^{-1} x)\sqrt{1 - x^2}$.

6. $p(\theta) = \sin(\cos^{-1} \theta)$.

7. $r(t) = \tan^{-1}(\sqrt{t})$.

8. $f(\theta) = \dfrac{e^{-2\theta}}{\sin^{-1} \theta}$.

9. $q(u) = \tan^{-1}(\cot u) - \cot^{-1}(\tan u)$.

10. $g(x) = \sin^{-1}(\cos 3x)$.

11. $h(\theta) = 7^{\theta} \csc^{-1} \theta$.

12. $f(x) = \cos^{-1}(\sqrt{1 - x^2})$.

13. $r(t) = (\sec^{-1} t)^3 + 12 \sec^{-1} t - \tan^{-1} \pi$.

14. $g(u) = \tan^{-1}(3u^2 - 1)$.

15. $h(x) = \sin^{-1}\left(\dfrac{1}{x}\right) + \cos^{-1}\left(\dfrac{1}{x}\right)$.

16. $p(x) = \dfrac{1 + \tan^{-1} x}{1 - 2\tan^{-1} x}$.

17. $q(u) = \cot^{-1}(e^u - 1)$.

18. $c(t) = e^{\tan^{-1} t}$.

19. $g(z) = \dfrac{\tan^{-1}(az)}{1 + a^2 z^2}$, where $a > 0$ is a constant.

20. $h(t) = \tan^{-1}(\cos(1/t))$.

Find all inflection points of the graphs of the functions in Exercises 21 through 23.

21. $f(x) = \sin^{-1} x$.

22. $g(x) = \cos^{-1} x$.

23. $h(x) = \tan^{-1} x$.

Prove the composition formulas in Exercises 24 through 26.

24. $\sin(\tan^{-1} x) = \dfrac{x}{\sqrt{1 + x^2}}$.

25. $\cos(\tan^{-1} x) = \dfrac{1}{\sqrt{1 + x^2}}$.

26. $\tan(\sin^{-1} x) = \dfrac{x}{\sqrt{1 - x^2}}$.

27. Say whether the equation is true or false. If it is false, why?

    a. $\sin(\sin^{-1} 1.5) = 1.5$.

    b. $\sin^{-1}(\sin 1.5) = 1.5$.

    c. $\cos(\cos^{-1} 1.5) = 1.5$.

    d. $\cos^{-1}(\cos 1.5) = 1.5$.

28. What is the domain of the function $h(x) = \sqrt{\sin^{-1} x - 1}$?

Recall the definitions of the *hyperbolic trigonometric* functions.

$$\sinh x = \frac{e^x - e^{-x}}{2}, \qquad \cosh x = \frac{e^x + e^{-x}}{2}, \qquad \tanh x = \frac{\sinh x}{\cosh x}.$$

The functions $\sinh x$ and $\tanh x$ are both one-to-one. The range of $\sinh x$ is the entire real line $(-\infty, \infty)$. The range of $\tanh x$ is the interval $(-1, 1)$. The function $\cosh x$ is not one-to-one; we must restrict its domain to $x \geq 0$ before we can produce an inverse function. The range of $\cosh x$ is the interval $[1, \infty)$. Use these formulas and information to explicitly verify the formulas in Exercises 29 through 31 below for the inverse hyperbolic trigonometric functions.

29. $\sinh^{-1} x = \ln\left(x + \sqrt{x^2 + 1}\right)$.

30. $\cosh^{-1} x = \ln\left(x + \sqrt{x^2 - 1}\right)$.

31. $\tanh^{-1} x = \dfrac{1}{2} \ln\left(\dfrac{1 + x}{1 - x}\right)$.

32. Explain why $\sin x$ does not have an inverse on the interval $\left[\dfrac{\pi}{4}, \dfrac{3\pi}{4}\right]$.

33. Show that $\sinh^{-1} x$ and $\tan^{-1} x$ have the same tangent line at $x = 0$.

34. Calculate $\displaystyle\lim_{t \to 0} \dfrac{\cos^{-1} t - \frac{\pi}{2}}{t}$.

35. For $-\pi/2 < x < \pi/2$, let $f(x) = \tan x$ and $g(x) = 1/f(x)$. Show that

$$\left(\frac{d}{dx}f^{-1}(x)\right)^2 = \left(\frac{d}{dx}g^{-1}(x)\right)^2.$$

36. Prove that

$$\tan^{-1} x + \tan^{-1}\left(\frac{1}{x}\right) = \begin{cases} \pi/2 & x > 0; \\ -\pi/2 & x < 0. \end{cases}$$

37. Consider the function defined below:

$$h(x) = \begin{cases} \sin^{-1}(x) & \text{if } -1 \leq x \leq 1; \\ \sin^{-1}(x-2) + \pi & \text{if } 1 < x \leq 2. \end{cases}$$

Show that h is continuous, one-to-one, and that h has a continuous inverse, but not a differentiable inverse.

38. Use the identity

$$\tan^{-1} a - \tan^{-1} b = \tan^{-1}\left(\frac{a-b}{1-ab}\right)$$

to argue that

$$\lim_{h \to 0} \frac{1}{h} \tan^{-1}\left(\frac{h}{1 - x^2 - hx}\right) = \frac{1}{1+x^2}.$$

In Exercises 39 and 40, prove inverse trigonometric addition formulas. Do so by showing that for a fixed y value, the left- and right-hand sides of the equation have the same derivative with respect to x, and so, the two sides differ by a constant (on any interval, which leads to restrictions on the allowed (x, y) pairs). Then, plug in a convenient x value to determine that the constant is zero.

39. $\tan^{-1} x + \tan^{-1} y = \tan^{-1}\left(\frac{x+y}{1-xy}\right)$, provided that $-\pi/2 < \tan^{-1} x + \tan^{-1} y < \pi/2$.

40. $\sin^{-1} x + \sin^{-1} y = \sin^{-1}\left(x\sqrt{1-y^2} + y\sqrt{1-x^2}\right)$, provided that $-\pi/2 \leq \sin^{-1} x + \sin^{-1} y \leq \pi/2$.

In Exercises 41 and 42 prove the identities by showing the derivatives of the two sides are equal so the left and right-hand sides must differ by a constant. Evaluate the two sides of the identities at a convenient point to conclude that the constant is zero.

41. $\tan^{-1} x = 2\tan^{-1} \dfrac{x}{1 + \sqrt{1+x^2}}.$

42. $\sin^{-1} x = 2 \tan^{-1} \dfrac{x}{1 + \sqrt{1 - x^2}}.$

43. Find the maximum and minimum values of $f(x) = \sin^{-1} x$ on its domain.

44. A bird is hovering at a height of 100 feet above the ground, and drops a stick. The bird flies away horizontally at a rate of 1.5 feet per second. As it continues to fly horizontally, the bird looks back longingly at the stick. Let x denote the horizontal position of the bird, measured from where the stick is dropped, and let y denote the height above the ground of the stick. Let θ denote the positive acute angle from the horizontal path of the bird to the line segment from the bird to the stick.

    a. Draw a diagram of this situation, which includes sketches of the ground, the path of the bird, the distance between the ground and the path of the bird, the path of the stick, and the initial point at which the bird drops the stick at time $t = 0$ seconds. Include in your diagram the relevant data at some typical time $t > 0$ seconds: the positions of the stick and bird, the line segment connecting the bird and the stick, and the quantities x, y, and θ.

    b. Recall that in free-fall, an object's height can be modeled by

    $$y(t) = -\frac{1}{2} g t^2 + v_0 t + y_0,$$

    where g is the acceleration due to gravity, and y_0 and v_0 are the initial height and velocity, respectively, of the object. Assume that $g = 32$ ft/s^2. Determine the height function of the stick.

    c. Give a function for the angle θ in terms of x and y.

    d. Regarding y and x as functions of time, find the rate of change of θ, with respect to time, when $t = 2$ seconds.

45. Let $f_1(x) = \tan^{-1} x$ and, for $n = 2, 3, 4, 5, ...$, let $f_n(x) = \tan^{-1}(f_{n-1}(x))$.

    a. Show that $f_n(x) \le f_{n-1}(x)$ for $x \ge 0$, and $f_n(x) \ge f_{n-1}(x)$ for $x \le 0$.

    b. For all positive integers n, show that an equation of the tangent line at $x = 0$ of the graph of $f_n(x)$ is given by the equation $y = x$.

46. The diagram below shows two circles of radii r_1 and r_2, whose centers are separated by a distance $s > r_1 + r_2$. In this case, it can be shown that there are exactly two pairs of points p_1, q_1 (on the first circle) and p_2, q_2 (on the second circle) so that the tangent line to the first circle at p_1 (respectively, p_2) coincides with the tangent line to the second circle at q_1 (respectively, q_2), and such that this mutual tangent line crosses the line segment

connecting the centers of the two circles. The diagram shows one such mutual tangent line.

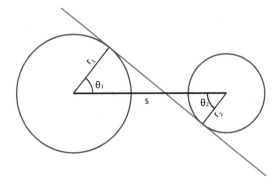

a. The tangent line is perpendicular to the radial line, in each circle, that intersects it. Use this (and high school geometry) to conclude that $\theta_1 = \theta_2$.

b. Find an equation relating θ (shorthand for θ_1 and θ_2, since they are equal) and s, the distance between the centers of the circles. Solve this relation for $\theta = \theta(s)$.

c. What is $\lim_{s \to \infty} \theta(s)$?

d. Find a formula for $d\theta/ds$.

47. A child (represented by the star) is sliding down a slide in the shape of the parabola $y = x^2$, as shown in the diagram below.

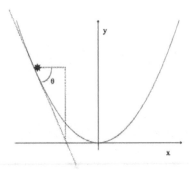

a. Find an expression for the angle of decline θ, as a function of x.

b. What is the rate of change of θ, with respect to x, when $x = -1$?

48. Solve parts (a) and (b) of the previous problem when $y = x^k$ where k is an even positive integer. ▶

49. Let $f(x)$ be a differentiable function whose domain is $(-\infty, \infty)$. Define $\theta(x)$ to be the angle in the open interval $(-\pi/2, \pi/2)$ that the tangent line to the graph of the function f at the point $(x, f(x))$ makes with the right-hand portion of a horizontal line through $(x, f(x))$, as shown in the diagram.

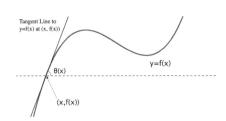
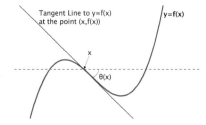

    a. Find an explicit expression for $\theta(x)$ in terms of $f(x)$ and/or $f'(x)$. (Hint: First, do this for x equal to a fixed value a, and then replace a with x.)

    b. At what values of x does the function $\theta(x)$ have critical points (again, relative to information about $f(x)$ and $f'(x)$)?

50. Prove the second statement of Theorem 2.9.12, that $\tan^{-1}(x) + \cot^{-1}(x) = \pi/2$, by showing the derivative of the left-hand side is zero and evaluating the left-hand side for a convenient value. ▶

51.   a. Let $f(x) = \tan^{-1} x$. Calculate $f''(x)$ and $f^{(3)}(x)$.

    b. Use this information to calculate the third degree Taylor polynomial (see Definition 3.A.2) for f centered at $a = 0$:

$$P_3(x; a) = f(a) + f'(a)(x - a) + \frac{f''(a)(x - a)^2}{2} + \frac{f^{(3)}(a)(x - a)^3}{3!}.$$

    c. Use the fact that $f(1) = \pi/4$ to arrive at the approximation $\pi \approx 4(1 - 1/3)$.

2.10   Implicitly Defined Functions

In this section, we will consider functions $y = y(x)$ that are not defined explicitly, but are instead defined by stating that some equation involving x and y is satisfied; such functions are said to be *implicitly defined*. We wish to calculate derivatives of implicitly defined functions, **without** solving explicitly for $y = y(x)$ first. This process, which we discussed briefly in Example 2.3.13, is called *implicit differentiation*.

Let's look at a few examples.

Example 2.10.1. Suppose that you are given that

$$x^2 - y = 7.$$

How do you find dy/dx?

Hopefully, this seems like a silly question. You solve for y, and find $y = x^2 - 7$. Then, you differentiate and obtain $dy/dx = 2x$.

However, we did not **have** to solve for y first. We could have thought "x^2 is a function of x, y is a function of x (which we haven't solved for), and so $x^2 - y$ is a function of x." Then, the equality $x^2 - y = 7$ would tell us that the function of x given by $y - x^2$ is equal to the constant function 7. What's the point of this? If the two functions of x are equal, we can differentiate each side of the equality, with respect to x, and obtain an equality of the derivatives.

That is, we find

$$\frac{d}{dx}(x^2 - y) = \frac{d}{dx}(7),$$

and, hence,

$$2x - \frac{dy}{dx} = 0,$$

i.e., $dy/dx = 2x$. Of course, this is exactly what we obtained by first solving for y in terms of x.

Why bother solving for dy/dx without first explicitly solving for y as a function of x? In our current example, it makes very little difference. But, in more complicated examples, like those below, there is a HUGE difference; it may be difficult or impossible to solve for y explicitly as a function of x and, yet, you can still calculate dy/dx.

Example 2.10.2. Suppose that you are told that $x^2 + y^2 = 1$. How do you find dy/dx?

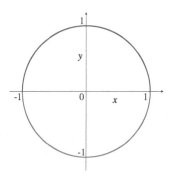

Figure 2.26: The graph of $x^2 + y^2 = 1$.

If you solve for y in terms of x, you obtain

$$y = \pm\sqrt{1 - x^2}.$$

This is not a **function**. Functions are required to give back a single value at every point in the domain; $y = \pm\sqrt{1 - x^2}$ gives two different y values for each x in the interval $(-1, 1)$. On the graph, this is reflected by fact that the *vertical line test* fails; that is, the graph of $x^2 + y^2 = 1$ is not the graph of a function because there are vertical lines which intersect the graph more than once.

However, in this example, we can at least "split" y into two separate functions. If we have $x^2 + y^2 = 1$ **and require that** $y \geq 0$, then we obtain that $y = \sqrt{1 - x^2}$. The graph of this function is the blue upper-half of the graph in Figure 2.26.

If we have $x^2 + y^2 = 1$ **and require that** $y \leq 0$, then we obtain that $y = -\sqrt{1 - x^2}$. The graph of this function is the red lower-half of the graph in Figure 2.26.

In order to distinguish which y we are discussing, we will let $u(x) = \sqrt{1 - x^2}$ and $l(x) = -\sqrt{1 - x^2}$. As we now have **functions**, we can ask about the differentiability of $u(x)$ and $l(x)$, and ask for formulas for the derivatives.

We find
$$u'(x) = \left[(1 - x^2)^{1/2}\right]' = \frac{1}{2}(1 - x^2)^{-1/2}(-2x) = \frac{-x}{\sqrt{1 - x^2}}.$$

Note that the denominator on the right above is precisely $u(x)$ itself. Therefore, if we have $y = u(x) = \sqrt{1 - x^2}$, we find that $y = u(x)$ is differentiable wherever $y = \sqrt{1 - x^2} \neq 0$ and, at all such points

$$dy/dx = -x/y.$$

Now let's look at $y = l(x) = -\sqrt{1-x^2}$. Then, $l(x)$ is simply $-u(x)$, and our work above tells us that $y = l(x)$ is differentiable wherever $\sqrt{1-x^2} \neq 0$ and, at all such points $l'(x) = -u'(x) = x/\sqrt{1-x^2}$. We can rewrite this; if $y = -\sqrt{1-x^2}$, then, what we just found was that

$$dy/dx = x/(-y) = -x/y.$$

This is exactly the formula we found for $y = u(x) = \sqrt{1-x^2}$!

What we have just seen is that the equation $x^2 + y^2 = 1$ can be used to define y as a function of x, provided that we specify that $y \geq 0$ or $y \leq 0$, and, in either case, if $y \neq 0$, you find

$$\frac{dy}{dx} = -\frac{x}{y}.$$

Could we have found this one common formula without splitting things up into the cases where $y \geq 0$ and $y \leq 0$? Yes.

Assume that we know, near a given point (a, b) on the unit circle, that we **can** solve for y as a differentiable function of x; that is, we have $y = f(x)$ for **some** function $f(x)$ (whose domain includes an open interval around a, and whose range includes an open interval around b). As we saw above, the only points on the unit circle where we can not find such a $y = f(x)$ is where $y = 0$, i.e., at the two points $(-1, 0)$ and $(1, 0)$.

If we let y be such a function $f(x)$, that means precisely that we have an identity, an equality of functions of x, given by

$$x^2 + y^2 = x^2 + (f(x))^2 = 1.$$

This means that we can differentiate each side of the equation, with respect to x, and get a new, valid, equation. As we differentiate, we simply keep in mind that y is some function of x, and use the Chain Rule:

$$\frac{d}{dx}\left[x^2 + y^2\right] = \frac{d}{dx}(1);$$

$$2x + \frac{d}{dx}(y^2) = 0;$$

$$2x + \frac{d}{dy}(y^2) \cdot \frac{dy}{dx} = 0; \text{ and}$$

$$2x + 2y \frac{dy}{dx} = 0.$$

Solving now for dy/dx, we find

$$\frac{dy}{dx} = -\frac{x}{y},$$

as we found (twice) before. Cool, huh?

Note that, to actually get numbers out of this formula, you would need to be given the values of both x **and** y, satisfying $x^2 + y^2 = 1$, or, at least, given x and told whether y is positive or negative.

We would like to use the method from the last example to find dy/dx when we are given more-complicated equations involving x and y, for instance, something like $y - x = e^{xy}$, where we cannot algebraically solve to find y as a function of x. We can do this, and we will in the next example, but there is the question of how we even know that there is one, and only one, function $y = f(x)$ which satisfies $y - x = e^{xy}$, so that it makes sense to talk about dy/dx.

Thus, there are two issues at hand right now. First, we need to define what it means for an equation involving x and y to *implicitly define y as a function of x*; given an equation $F(x, y) = G(x, y)$, what we want it to mean is that there's one, and only one, function $y = f(x)$ that makes the equation true for all x, or at least all x close to some x value that we care about. Second, we need a theorem that tells us conditions that guarantee that such implicitly defined functions exist. For "easy" functions F and G, you would be able to solve algebraically for an explicit function $y = f(x)$, but, for more-complicated F and G, we would like to know that a unique solution $y = f(x)$ exists, even if we can't find a formula for it.

We define *implicitly defined functions* below. Then we give another example, the example $y - x = e^{xy}$, mentioned above. After that example, we give the basic theorem that tells us when $F(x, y) = G(x, y)$ defines a function implicitly, the *Implicit Function Theorem*, Theorem 2.10.5.

Before we give this definition, for a point (a, b) in the xy-plane, we need something that plays a role analogous to that of an open interval around a real number; thus, we use the term an *open ball around* (a, b) to mean the points inside of any circle of positive radius, centered at (a, b), not including the points on the circle itself (just as an open interval does not include the endpoints).

Definition 2.10.3. *Let (a, b) be an ordered pair of real numbers (i.e., (a, b) is a point in the xy-plane). Suppose that $F(x, y)$ and $G(x, y)$ are functions from a subset of the xy-plane into the real numbers, and that F and G are defined on an open ball $\mathcal{U}$ around the point (a, b). Suppose also that $F(a, b) = G(a, b)$.*

We say that the equation $F(x, y) = G(x, y)$ **implicitly defines** y **as a function of** x, **near** (a, b), *if and only if*

1. *there exists an open interval I around a, an open interval J around b, and a function $f : I \rightarrow J$ such that $f(a) = b$ and, for all x in I, $(x, f(x))$ is in $\mathcal{U}$, and*

$$F(x, f(x)) = G(x, f(x));$$

2. *such a function, f, as given in Item 1, is essentially unique; that is, any two such functions must be the same when restricted to some open intervals around a and b.*

When these conditions are satisfied, we also say that $y = f(x)$ is **implicitly defined by** $F(x, y) = G(x, y)$, **near** (a, b).

For instance, what we saw in Example 2.10.2 was that $x^2 + y^2 = 1$ implicitly defines y as a function of x, near any point (a, b) such that $a^2 + b^2 = 1$ and $b \neq 0$. If $b > 0$, then the implicitly defined function can be solved for explicitly; we obtain $y = \sqrt{1 - x^2}$. If $b < 0$, then the implicitly defined function can again be solved for explicitly; we obtain $y = -\sqrt{1 - x^2}$.

Example 2.10.4. Let's consider a more complicated equation, where we have no hope of solving algebraically for y in terms of x.

Consider

$$y - x = e^{xy}. \tag{2.18}$$

As in the last part of the previous example, we will assume that y is some differentiable function of x, one that we simply can't find algebraically. What's really true is that we assume that we are only considering the problem near points where Formula 2.18 actually does implicitly define y as a differentiable function of x; we shall discuss this further below.

Thus, we assume that y is some function of x such that, for all x in some open interval, Formula 2.18 holds for all x. This means that Formula 2.18 is an equality of functions of x, and we may once again differentiate both sides, **with respect to** x. You just have to apply all of our differentiation rules, keeping in mind that y is **some** function of x (that we don't know explicitly).

$$\frac{d}{dx}(y - x) = \frac{d}{dx}(e^{xy});$$

$$\frac{dy}{dx} - 1 = e^{xy} \cdot \frac{d}{dx}(xy);$$

$$\frac{dy}{dx} - 1 = e^{xy}\left(x\frac{dy}{dx} + y\frac{dx}{dx}\right) = e^{xy}\left(x\frac{dy}{dx} + y\right).$$

We now expand the right-hand side of this equation, then put all of the terms with factors of dy/dx on one side of the equation, and the terms without a factor of dy/dx on the other side of the equation.

$$\frac{dy}{dx} - 1 = xe^{xy}\frac{dy}{dx} + ye^{xy};$$

$$\frac{dy}{dx} - xe^{xy}\frac{dy}{dx} = 1 + ye^{xy}.$$

Now, factor out the dy/dx on the left, and divide by the other (messy) factor.

$$\left(1 - xe^{xy}\right)\frac{dy}{dx} = 1 + ye^{xy},$$

and so,

$$\frac{dy}{dx} = \frac{1 + ye^{xy}}{1 - xe^{xy}}. \tag{2.19}$$

Given that we have no idea how to solve, explicitly, for y in the equation $y - x = e^{xy}$, how do you use the formula for dy/dx that we found in Formula 2.19? You need to be given the x **and** y values for a point (x, y) which satisfies $y - x = e^{xy}$. A typical question might be: what is dy/dx at $(x, y) = (-1, 0)$?

To answer this, it is a good idea to first verify that $(-1, 0)$ actually satisfies $y - x = e^{xy}$. We find that, yes, $0 - (-1)$ does, in fact, equal $e^{(-1)(0)}$, since both are equal to 1. Then, from Formula 2.19, we find

$$\frac{dy}{dx}\bigg|_{(x,y)=(-1,0)} = \left(\frac{1 + ye^{xy}}{1 - xe^{xy}}\right)\bigg|_{(x,y)=(-1,0)} = \frac{1 + 0 \cdot e^0}{1 - (-1)e^0} = \frac{1}{2}.$$

In the whole discussion above, we assumed that $y - x = e^{xy}$ implicitly defined y as a differentiable function of x. In fact, Theorem 2.10.5 below will tell us that this assumption is correct, near any point (a, b) for which our final answer for dy/dx is defined (assuming we started with "reasonable" functions of x and y on each side of the initial equation).

Everything that we did in all of the examples above just uses the Chain Rule and algebra, **except** where we assumed that our equations implicitly defined y as a function of x. As we said earlier, the primary result that gives conditions that guarantee the existence of such implicitly defined functions is called the *Implicit Function Theorem*.

We give below an informal statement of the Implicit Function Theorem, which we state, here, only for elementary functions. We give a rigorous statement of the general theorem in Theorem 2.A.4 in Section 2.A. However, both our informal statement of the Implicit Function Theorem and our formal statement use notions and technical terms from *multi-variable Calculus*; notions which we do not define here or even in Section 2.A. We recommend [2], or any rigorous textbook on multi-variable Calculus.

We first need a couple of quick definitions.

Suppose, as in the examples above, we have a function $F(x, y)$, a function of two variables. An *interior point of the domain* of such a function is a point such that there is an open ball around the point which is also contained in the domain. As with functions of one variable, an *elementary function of two variables* is a function which is a constant function, a power function, a polynomial function, an exponential function, a logarithmic function, a trigonometric function, or inverse trigonometric function, or any finite combination of such functions using addition, subtraction, multiplication, division, or composition, where all of the functions may involve both variables. Just like functions of a single variable, an elementary function of two variables is continuous at all points in its domain.

Theorem 2.10.5. (Implicit Function Theorem, informal version) *Let (a, b) be an ordered pair of real numbers (i.e., (a, b) is a point in the xy-plane). Suppose that $F(x, y)$ and $G(x, y)$ are elementary functions, defined on an open ball $\mathcal{U}$ around (a, b), such that $F(a, b) = G(a, b)$. Consider the equation*

$$F(x, y) = G(x, y).$$

Then, you can apply the following process to determine if the equation above implicitly defines y as a differentiable function $f(x)$, near (a,b), and, if so, how to find dy/dx at (a,b), i.e., how to find $f'(a)$ without ever solving explicitly for $y = f(x)$:

1. *Assume that y is a differentiable function of x, and differentiate both sides of $F(x,y) = G(x,y)$ with respect to x; the Chain Rule will produce dy/dx's every time that you differentiate an expression containing y's.*

2. *Write the resulting equation above in the form $A(x,y)\dfrac{dy}{dx} = B(x,y)$, so that $dy/dx = B(x,y)/A(x,y)$.*

3. *Verify that $A(x,y)$ and $B(x,y)$ are defined on some open ball around (a,b). Verify also that $A(a,b) \neq 0$.*

In this case, $F(x,y) = G(x,y)$ implicitly defines y as a function $f(x)$, near (a,b). Moreover, there exist open intervals I and J as in Definition 2.10.3 such that the function $f : I \to J$ is differentiable, and, for all $x \in I$, $A(x, f(x)) \neq 0$, and, for $y = f(x)$,

$$\frac{dy}{dx} = f'(x) = \frac{B(x, f(x))}{A(x, f(x))} = \frac{B(x,y)}{A(x,y)}.$$

In particular,

$$\frac{dy}{dx}\bigg|_{(x,y)=(a,b)} = f'(a) = \frac{B(a,b)}{A(a,b)}.$$

The only part of the above theorem which requires proof is that, if $A(a,b) \neq 0$, then $F(x,y) = G(x,y)$ implicitly defines y as a function $f(x)$, near (a,b); the remainder of the result follows from the Chain Rule. See Remark 2.A.5 for a discussion of why the formal Implicit Function Theorem, Theorem 2.A.4, gives us what we need.

Let's look at one more example, one in which we find the first **and second** derivatives implicitly.

Example 2.10.6. Suppose that y is implicitly defined by

$$y^5 + xy = 1.$$

Find $\dfrac{dy}{dx}$ and $\dfrac{d^2y}{dx^2}$.

Solution:

We differentiate, always with respect to x, and always assuming that y is some function of x for which the given equation holds.

We find

$$5y^4 \frac{dy}{dx} + x \frac{dy}{dx} + y \cdot 1 = 0.$$

Therefore,

$$\frac{dy}{dx} = -\frac{y}{5y^4 + x}.$$

Now, we differentiate again, still with respect to x, using the Quotient Rule:

$$\frac{d^2y}{dx^2} = -\frac{(5y^4 + x)\frac{dy}{dx} - y\left(20y^3 \frac{dy}{dx} + 1\right)}{(5y^4 + x)^2}.$$

Replacing the two occurrences of dy/dx on the right-hand side with what we calculated above, we obtain

$$\frac{d^2y}{dx^2} = -\frac{-y - y\left(20y^3\left(-\frac{y}{5y^4+x}\right) + 1\right)}{(5y^4 + x)^2} = \frac{y + y\left(20y^3\left(-\frac{y}{5y^4+x}\right) + 1\right)}{(5y^4 + x)^2}.$$

There are a number of different ways to write this result in a more attractive form. We leave it as an exercise for you to verify that the above formula "simplifies" to

$$\frac{d^2y}{dx^2} = \frac{2xy - 10y^5}{(5y^4 + x)^3}.$$

In this problem, we proceeded in a straightforward way with implicit differentiation. However, as we can easily solve the original equation for x, in terms of y, we have another, easier, way of computing dy/dx. We calculate dx/dy and take the reciprocal. We will take this approach now, to see how it compares with implicit differentiation.

Solving the original equation for x, we find

$$x = \frac{1 - y^5}{y} = y^{-1} - y^4,$$

and so

$$\frac{dx}{dy} = -y^{-2} - 4y^3 = -\left(\frac{1}{y^2} + 4y^3\right) = -\frac{1 + 4y^5}{y^2}.$$

Thus,

$$\frac{dy}{dx} = -\frac{y^2}{1 + 4y^5}.$$

We leave it as an exercise for you to show that this agrees with what we found above by implicit differentiation, at points which satisfy $y^5 + xy = 1$.

As we remarked after Theorem 2.3.8, you cannot calculate d^2y/dx^2 simply by inverting d^2x/dy^2. However, as we indicated in Exercise 51 in Section 2.3 (with the roles of x and y reversed), we have

$$\frac{d^2y}{dx^2} = -\left(\frac{dx}{dy}\right)^{-3}\frac{d^2x}{dy^2} = -\left(-y^{-2} - 4y^3\right)^{-3}\left(2y^{-3} - 12y^2\right).$$

Once again, we leave it as an exercise for you to show that this agrees with what we found above by implicit differentiation, at points which satisfy $y^5 + xy = 1$.

2.10.1 Exercises

In each of Exercises 1 through 17, calculate $\dfrac{dy}{dx}$, in terms of x and y. (For those interested in seeing these curves, programs such as Grapher, MatLab, Maple, or Mathematica will plot implicitly defined functions.)

1. $x^2 + y^2 = 25$.

2. $4x^2 + 9y^2 = 1$.

3. $y^2 + 4x = 5$.

4. $16x^2 - 4y^2 = 1$.

5. $3x^2 + 2y^2 + 5xy - 8xy + y = 12$.

6. $x = \sin(3x + y)$.

7. $y = e^{x+y}$.

8. $x = \tan^{-1} y$.

9. $y^2 + \cos(xy) = 0$.

10. $x = \cot^{-1}(x - y)$.

11. $(x - y)^2 = x^2 y^3$.

12. $x^2 = \ln(xy)$.

13. $e^{xy} = e^x + e^y$.

14. $x^{\frac{2}{3}} + y^{\frac{2}{3}} - xy = 1$.

15. $x = \sqrt{y^2 - x^2}$.

16. $\sin^2 x + \sin^2 y = \sin(x + y)$.

17. $y = \sec(x^2) + \tan(y^2)$.

If we have a function $F(x, y)$, there is a notion of F being differentiable, and there are the *partial derivatives* $\partial F/\partial x$ and $\partial F/\partial y$. The partial derivative $\partial F/\partial x$ means the derivative of F, as a function of x, assuming y is constant. Naturally, $\partial F/\partial y$ means the derivative of F, as a function of y, assuming x is constant.

Given an equation in the form $F(x, y) = 0$, where F is differentiable and y is a differentiable function of x, a quick way to calculate $y'(x)$ is to use the formula $\dfrac{dy}{dx} = -\dfrac{\partial F/\partial x}{\partial F/\partial y}$. Use this method to calculate $y'(x)$ in Exercises 18 through 22, and compare your work and answers with what you obtain by applying Theorem 2.10.5. In fact, these two methods are essentially the same; see Theorem 2.A.4 and Remark 2.A.5 for more information.

18. $x^4 + x^3 y + y^4 = 9$.

19. $x - \sin(xy) = y$.

20. $x \cos(xy) = 0$.

21. $e^{x+y} + y - x = 0$.

22. $xy + \ln(xy) = 1$.

23. Consider the implicitly defined function $\sqrt{xy} = \dfrac{x+y}{2}$. Note that this equality defines the set of points (x, y) such that the geometric mean of x and y is equal to the arithmetic mean. ▶

    a. Calculate $\dfrac{dy}{dx}$ using implicit differentiation.

    b. Square both sides of the original equation and solve explicitly for y as a function of x.

    c. Compare your answers to part (a) and part (b). This example shows the danger in blindly applying implicit differentiation; sometimes explicit methods are better. It also shows that the geometric and arithmetic means of two positive numbers are equal only if $x = y$.

24. Calculate $\dfrac{dy}{dx}$ if $\dfrac{2}{\frac{1}{x} + \frac{1}{y}} = \dfrac{x+y}{2}$. Hint: see the previous problem. The left hand side of this equation is the *harmonic mean* of two numbers. Beware of textbooks that use the terms 'harmonic mean' and 'geometric mean' interchangeably.

25. Consider the hyperbola defined by

$$\frac{x^2}{a^2} - \frac{y^2}{b^2} = 1.$$

    In this exercise, you will show that no tangent line to any point on this hyperbola passes through the origin. ▶

    a. Use implicit differentiation to find $\dfrac{dy}{dx}$ for the y in the first equation, in terms of x, y, a and b.

    b. Find an equation for the tangent line to any point (x_0, y_0) on the hyperbola.

    c. Follow these steps to show by contradiction that no tangent line passes through the origin:

        i. Plug $(0, 0)$ into the equation you found in part (b) to find y_0 in terms of x_0, a, and b.

        ii. Now plug x_0 and your expression for y_0 into the equation which defines the hyperbola to show that the equation is **not** satisfied, i.e., to show that this point does not, in fact, lie on the hyperbola.

26. Consider the simple implicit equation $y^2 = x^2$.

   a. Decide whether or not implicit differentiation can be applied to calculate dy/dx at the following points: $(0,0)$, $(1,1)$, $(-1,1)$, and $(0,1)$. Explain.

   b. Describe the set of solutions to this equation.

In each of Exercises 27 through 31, show that any solution $y = y(x)$ to the given implicit equation is a solution to the given differential equation.

27. $3y^2 - 2x^3 = 1$, $y' = \dfrac{x^2}{y}$.

28. $\dfrac{1}{2}\tan(2y) = \dfrac{x}{2} + \dfrac{1}{12}\sin(6x)$, $y' = \cos^2(3x) \cdot \cos^2(2y)$.

29. $\dfrac{1}{y} + 2x + \dfrac{x^2}{2} = 1$, $y' = 2y^2 + xy^2$.

30. $2y^3 = 3\left(\sin^{-1}(x)\right)^2$, $y'(x)\sqrt{1-x^2}y^2 = \sin^{-1}(x)$.

31. $2e^x(x-1) + y^2 = C$, $y'(x) = -\dfrac{xe^x}{y}$.

32. The *folium of Descartes* is a curve defined by the equation $x^3 + y^3 = 6xy$. Verify that $(3,3)$ is a point on the folium of Descartes, and find the slope of the tangent line to the curve at that point.

33. Carefully explain why or why not implicit differentiation to find dy/dx can be used in the following cases.

    a. $x^2 + y^2 = 1$,   b. $x^2 + y^2 = 0$,   c. $x^2 + y^2 = -1$.

34. Suppose $y^3 + yx^2 + xy^2 - 3x^3 + y - x = 0$. Show that this equation implicitly defines y as a differentiable function of x, **for all real** x. That is, for all a, show that there is a b such that $(x,y) = (a,b)$ satisfies the given equation, and such that the given equation implicitly defines y as a differentiable function $f(x)$ near (a,b).

35. An *astroid* is defined by the equation $x^{2/3} + y^{2/3} = 4$.

   a. Use implicit differentiation to determine points where $y'(x)$ is defined.

   b. Solve for y explicitly and determine the points where $y'(x)$ is undefined.

In Exercises 36 through 38, assume that the position, $p(t)$, of a particle at time t is defined implicitly. Determine the velocity function of the particle at the given time in terms of t and p.

36. $p^2 t + t^2 p = 6t$.

37. $\cos(t)\sin(p) = 1/2$.

38. $\sqrt{t + p + 3} + \sqrt{t + p} = 6$.

39. Show that an equation of the tangent line to the curve $ax^2 + by^2 + cxy + dx = 0$ at the point (p, q) is

$$2apx + 2bqy + cqx + cpy - 2cpq - 2bq^2 - 2ap^2 - pd = 0$$

40. Consider the equation
$$x^2 + 2xy - y^2 = 50.$$

   a. Find $\dfrac{dy}{dx}$ and $\dfrac{d^2y}{dx^2}$, using implicit differentiation.

   b. Find all points of horizontal and vertical tangency to the graph of the given equation, that is, find all points where $dy/dx = 0$ or dy/dx is infinite (i.e., $dx/dy = 0$).

   c. Find all inflection points.

41. Consider the graph defined by the equation

$$5y^2 + (2x - 6)y + (2x + y) = 0.$$

   a. Use implicit differentiation to find the derivative $\dfrac{dy}{dx}$ in terms of x and y.

   b. Does the graph have any horizontal tangent lines? (Be careful here; the question is: is there a point (x, y) such that $dy/dx = 0$ **and** such that (x, y) satisfies the given equation.)

   c. Solve for x in terms of y. Then, use a calculator or computer to graph $x = x(y)$. Does what you see in the graph agree with your answer in part (b)?

   d. Find the x-coordinate of the point on the graph with y-coordinate 1, and use your answer from part (a) to calculate the slope of the tangent line at this point.

42. Consider the ellipse
$$\frac{x^2}{25} + \frac{y^2}{100} = 1. \qquad \blacktriangleright$$

   a. Calculate $\dfrac{dy}{dx}$ and $\dfrac{d^2y}{dx^2}$, in terms of x and y for this ellipse.

   b. Verify that the point $(3, 8)$ is on this curve, and find an equation for the tangent line to the ellipse at this point. Find the x-intercept of this tangent line.

    c. For $-5 < x < 5$, $x \neq 0$, find an expression for the function $I(x)$, the x-intercept of the tangent line to the upper-half of the ellipse at the point corresponding to x.

    d. Show that $Q(x)$, the function describing the x-intercept of the tangent line to the bottom-half of the ellipse at the point corresponding to x, is the same as $I(x)$.

43. The set of points satisfying the equation

$$y^2 + 4xy + 4x^2 = 1$$

    is a pair of lines.

    a. Use implicit differentiation to find $\dfrac{dy}{dx}$ and $\dfrac{d^2y}{dx^2}$.

    b. Factor the left-hand side and, for each of the two lines, find an explicit equation which defines the line. Show that your answer from part (a) makes sense.

44. The *Kampyle of Eudoxus* is defined to be the set of points satisfying the equation

$$a^2 x^4 = b^4(x^2 + y^2),$$

    where $a \neq 0$ and $b \neq 0$.

    a. Show that for all a, b, the point $(0,0)$ satisfies the equation.

    b. For $a = b = 1$, show that $(0,0)$ is the only point on the graph whose x-coordinate has absolute value less than 1.

    c. Find the points of vertical tangency to the graph.

    d. Show that there are no points of horizontal tangency to the graph.

45. A circular metal plate of radius 10 centimeters is heated in such a way that, if the center of the circle is placed at the origin, then, at the point (x, y), the temperature is given by the function

$$T(x, y) = 20 \left(\sin(5x) + \cos y \right) + 40,$$

measured in degrees Fahrenheit.

    a. An *isotherm* is a curve so that every point (x, y) on this curve is at the same temperature. Find an equation for the isotherm of $40°$F on this plate.

    b. Verify that the point $(\pi/4, \pi/4)$ lies on this isotherm.

c. It can be shown that the temperature changes most rapidly when you move in directions which are orthogonal (perpendicular) to the isotherms. Find an equation for the line orthogonal to the isotherm at the point $(\pi/4, \pi/4)$.

Recall from Section 2.2, Exercise 54, that the *curvature* $\kappa(x)$ of a function $f(x)$ at a point is given by

$$\kappa(x) = \frac{f''(x)}{(1 + [f'(x)]^2)^{3/2}}.$$

In Exercises 46 through 50, find the curvature of the implicitly defined curve. You may have calculated the derivatives of some of these functions in earlier problems.

46. Circle: $x^2 + y^2 = r^2$.

47. "X" shape: $y^2 = x^2$.

48. Ellipse: $ax^2 + by^2 = 1$, $a > 0, b > 0$.

49. Hyperbola: $y^2 - x^2 = 1$.

50. Parabola: $y - x^2 = 0$.

51. Find $\kappa'(x)$, where $\kappa(x)$ is the curvature of the implicit function y defined by $(x - h)^2 + (y - k)^2 = r^2$, i.e., in the case of a general circle.

52. In economics, it is often necessary to calculate the usefulness, called *utility*, U, of a collection (called a *basket*) of products (called *goods*) to a group of consumers. Economists measure utility in abstract units called *utils*; a useful way to think of utility and utils is that U should be proportional to how much money the average person would be willing to spend to buy the basket.

Suppose that a company wants to sell a basket consisting of some macaroni and cheese (mac & cheese) and some steak. Let's suppose then that we have a utility function $U(x, y)$, where x is the number of ounces of mac & cheese in the basket, and y is the number of ounces of steak in the basket; the value of $U(x, y)$ is the utility for the college students to have that (x, y) combination of mac & cheese and steak.

Suppose that market research is conducted, and shows that $U(x, y)$ is of the form:

$$U(x, y) = e^{-0.1(x+y)} + xy - 1.$$

Note that our utility function is symmetry with respect to x and y. This means that consumers value a basket of 10 ounces of mac & cheese and 0 ounces of steak just as much as they value a basket of 0 ounces of mac & cheese and 10 ounces of steak. However, this does **not** mean that they value all combinations of steak and mac & cheese, which total 10 ounces, equally. For instance, $U(5,5) = U(0,10) + 25$, i.e., the utility of 5 ounces of mac & cheese and 5 ounces of steak is 25 utils greater than the utility of 0 ounces of mac & cheese and 10 ounces of steak. This would be a reflection, perhaps, of the fact that many people like to have steak together with mac & cheese more than they like to have steak (even more steak) all by itself.

a. What equation yields an implicitly defined function which gives the set of pairs of amounts (x, y) so that the utility to the college student is $e^{-1.2} + 31$ utils? (Such a curve is called an *indifference curve*, because the consumer is just as happy with any portions of the two foods lying on that curve.)

b. Find $\dfrac{dy}{dx}$ for this indifference curve.

c. Verify that $(4, 8)$ is on this curve and calculate $\dfrac{dy}{dx}$ and $\dfrac{d^2y}{dx^2}$ at this point.

Economists call the absolute value of dy/dx the *marginal rate of substitution of y for x*, since this quantity essentially measures how much steak would be needed to entice the student to substitute the steak for an ounce of mac & cheese.

Assume $G(x, y) = 0$, that G is differentiable, that (a, b) is a specific solution to the equation, and assume $0 \neq K = \dfrac{dG}{dy}\bigg|_{(x,y)=(a,b)}$. Define an iterative process as follows.

Let $f_0(x) = b$ and, for all integers $n \geq 0$, let

$$f_{n+1}(x) = f_n(x) - \frac{G(x, f_n(x))}{K}.$$

One way to prove the Implicit Function Theorem is to show that for x sufficiently close to a, the sequence $\{f_n\}$ converges to a function f such that $G(x, f(x)) = 0$. We will look at this process in Exercises 53 and 54.

53. The equation $G(x, y) = 0$, where $G(x, y) = y - x^2$, is equivalent, of course, to the equation $y = x^2$. In this case we know that $y = f(x) = x^2$. We use this example to demonstrate the iterative technique.

a. Let $(a, b) = (3, 9)$. Show that $1 = K = \dfrac{dG}{dy}\bigg|_{(x,y)=(3,9)}$. In fact, $K = 1$ regardless of the particular point chosen.

    b. What is $f_0(x)$? This is trivially constant.

    c. Show that $f_1(x) = x^2$.

    d. Show that, for all $n \geq 1$, $f_n(x) = x^2$. This verifies the iterative method arrives at the correct answer.

54. Let $G(x, y) = xy - 1$. The set of points (x, y) such that $G(x, y) = 0$ is the same as the set of points satisfying $y = 1/x$. Once again we already know our target and verify that the iterative method converges to $f(x) = 1/x$.

    a. Let $(a, b) = (1, 1)$. Show that $1 = K = \dfrac{dG}{dy}\bigg|_{(x,y)=(1,1)}$.

    b. Show that $f_0(x) = 1$ and $f_1(x) = 2 - x$.

    c. Show that $f_2(x) = x^2 - 3x + 3$, and that $f_3(x) = -x^3 + 4x^2 - 6x + 4$.

    d. If you're familiar with the Binomial Theorem, the coefficients may look familiar. Prove inductively that $f_n(x) = \dfrac{1 - (1 - x)^{n+1}}{x}$.

    e. Show that $\lim_{n \to \infty} f_n(x) = 1/x$ *for x sufficiently close to $a = 1$.* For what values of x does the sequence converge?

This sequence is closely related to Newton's Method, which is covered in Section 3.4. When an existence theorem is proved by demonstrating an explicit solution, the proof is said to be *constructive*. The Implicit Function Theorem is usually proved via constructive methods, like the one above, but non-constructive methods also exist.

Appendix 2.A   Technical Matters

For real functions $f : A \to B$ and $g : C \to D$, the domains and codomains for $f \cdot g$ and f/g are taken to be the "natural" ones. The domain of $f \cdot g$ is $A \cap C$, and the codomain is the set $B \cdot D := \{b \cdot d \mid b \in B, d \in D\}$. The domain of f/g is $\{x \in A \cap C \mid g(x) \neq 0\}$ and the codomain is the set $B/D := \{b/d \mid b \in B, d \in D, d \neq 0\}$.

Theorem 2.A.1. (The Chain Rule) *Suppose that $f : B \to C$ and $g : A \to B$ are real functions, and that x is a point in A such that g is differentiable at x, and f is differentiable at $g(x)$.*

Then, $f \circ g$ is differentiable at x, and

$$(f \circ g)'(x) = f'(g(x)) \cdot g'(x).$$

Proof. Note that x is fixed throughout this proof.

Let I be an open interval around 0 such that, for all $h \in I$, $x + h$ is contained in A. Define the function $u : I \to \mathbb{R}$ by $u(h) = g(x + h) - g(x)$. As g is differentiable at x, g is continuous at x, and so $\lim_{h \to 0} u(h) = 0$.

Define $m : I \to \mathbb{R}$ by

$$m(h) = \begin{cases} \dfrac{f(g(x+h)) - f(g(x))}{g(x+h) - g(x)}, & \text{if } g(x+h) \neq g(x); \\[3mm] f'(g(x)), & \text{if } g(x+h) = g(x). \end{cases}$$

Suppose that $h \neq 0$, and note, then, that

$$m(h) \cdot \frac{g(x+h) - g(x)}{h} = \begin{cases} \dfrac{f(g(x+h)) - f(g(x))}{h}, & \text{if } g(x+h) \neq g(x); \\[3mm] f'(g(x)) \cdot \dfrac{g(x+h) - g(x)}{h}, & \text{if } g(x+h) = g(x). \end{cases}$$

As both of the expressions on the right are equal to zero when $g(x + h) = g(x)$, we find that, for $h \neq 0$,

$$m(h) \cdot \frac{g(x+h) - g(x)}{h} = \frac{f(g(x+h)) - f(g(x))}{h}. \tag{2.20}$$

Therefore, if we can show that $\lim_{h \to 0} m(h) = f'(g(x))$, we will be finished.

Define a function $Q : \mathbb{R} \to \mathbb{R}$ by

$$
Q(y) = \begin{cases} \dfrac{f(g(x) + y) - f(g(x))}{y}, & \text{if } y \neq 0; \\[2ex] f'(g(x)), & \text{if } y = 0. \end{cases}
$$

Then, Q is continuous at 0, and $\lim_{y \to 0} Q(y) = f'(g(x))$. In addition, $Q(u(h)) = m(h)$.

Now, by Theorem 1.A.12, using the Item 3a case, we obtain that

$$
\lim_{h \to 0} m(h) \;=\; \lim_{h \to 0} Q(u(h)) \;=\; f'(g(x)).
$$

$\square$

Theorem 2.A.2. *Suppose that I is an interval, that $f : I \to J$ is a continuous, one-to-one and onto function, that x is a point in I at which f is differentiable, and that $f'(x) \neq 0$. Let $y = f(x)$.*

Then, f^{-1} is differentiable at y, and

$$
(f^{-1})'(y) \;=\; \frac{1}{f'(x)} \;=\; \frac{1}{f'(f^{-1}(y))}.
$$

Proof. As f is differentiable at x, x must be an interior point of I. By restricting, we may assume that I is an open interval.

Then, we know from Theorem 1.A.19 that f^{-1} is continuous, and that J is open. Thus, there exists a $\delta > 0$ such that $(y - \delta, y + \delta) \subseteq J$. Define the function $m : (-\delta, \delta) \setminus \{0\} \to \mathbb{R}$ by

$$
m(u) \;=\; \frac{f^{-1}(y + u) - f^{-1}(y)}{u}.
$$

We wish to show that $\lim_{u \to 0} m(u) = 1/f'(x)$.

Define $q : I \to \mathbb{R}$ by $q(t) = f(t) - f(x)$. Then q is continuous, one-to-one, and $q(x) = 0$; let us continue to denote by q the restriction of this previous q to its range. By Corollary 1.A.20, there exists a deleted open interval B around x such that $q(B) \subseteq (-\delta, \delta) \setminus \{0\}$, and $\lim_{u \to 0} m(u) = L$ if and only if $\lim_{t \to x} m(q(t)) = L$.

It remains for us to show that $\lim_{t \to x} m(q(t))$ equals $1/f'(x)$. However, this is easy:

$$\lim_{t \to x} m(q(t)) \;=\; \lim_{t \to x} \frac{t - x}{f(t) - f(x)} \;=\; \frac{1}{f'(x)}.$$

$\square$

The Inverse Function Theorem follows quickly as a corollary.

Theorem 2.A.3. (Inverse Function Theorem) *Suppose that f is differentiable on an open interval I, and that, for all x in I, $f'(x) \neq 0$.*

Then, the range of the restriction of f to I is an open interval J, and the restricted function $f : I \to J$ is one-to-one and onto. The inverse $f^{-1} : J \to I$ is differentiable and, for all $b \in J$,

$$(f^{-1})'(b) \;=\; \frac{1}{f'(f^{-1}(b))}.$$

Proof. The Intermediate Value Theorem for Derivatives, Theorem 1.A.25, implies that f' is either always positive or always negative at each point of I. In either case, Corollary 1.5.15 implies that f is strictly monotonic on I. It follows that $f : I \to J$ is one-to-one and that J is an open interval; see Theorem 1.A.19. Now, apply Theorem 2.3.8. $\square$

For a proof of the following theorem, see [3] or [2]. Recall, from Definition 2.10.3, what it means for an equation to implicitly define y as a function of x, and recall the "definition" of partial derivatives from Exercises 18 through 22 of Section 2.10. For careful definitions of partial derivatives and continuously differentiable functions of more than one variable, we once again refer you to [3] or [2].

Theorem 2.A.4. (Implicit Function Theorem) *Let $(a, b) \in \mathbb{R}^2$, and let $\mathcal{U}$ be an open subset of $\mathbb{R}^2$, containing the point (a, b). Let $H : \mathcal{U} \to \mathbb{R}$ be continuously differentiable, and let $C = H(a, b)$.*

If $\left. \dfrac{\partial H}{\partial y} \right|_{(a,b)} \neq 0$, then the equation $H(x, y) = C$ implicitly defines y as a differentiable function $f(x)$, near (a, b), and

$$f'(a) \;=\; \left. \frac{dy}{dx} \right|_{(a,b)} \;=\; \left(-\frac{\partial H}{\partial x} \middle/ \frac{\partial H}{\partial y} \right)_{\bigg|_{(a,b)}}.$$

Remark 2.A.5. We need to explain why Theorem 2.A.4 implies the theorem as stated in Theorem 2.10.5. All that we need to show to conclude Theorem 2.10.5 is that $A(a,b) \neq 0$ implies that $F(x,y) = G(x,y)$ implicitly defines y as a function of x near (a,b); the remainder of Theorem 2.10.5 follows easily.

Suppose that $F(x,y)$ and $G(x,y)$ are continuously differentiable on some open subset $\mathcal{U}$ in $\mathbb{R}^2$; for instance, F and G could be elementary functions which yield functions $A(x,y)$ and $B(x,y)$ which satisfy Items 1, 2, and 3 of Theorem 2.10.5. Define $H(x,y) = F(x,y) - G(x,y)$.

Now, suppose that we take any differentiable function $y = f(x)$ on an open subset $\mathcal{W}$ of $\mathbb{R}$ such that, for all $x \in \mathcal{W}$, $(x, f(x)) \in \mathcal{U}$. We are **not** necessarily assuming that $H(x, f(x)) \equiv 0$, i.e., we are not necessarily assuming that $y = f(x)$ satisfies $F(x,y) = G(x,y)$.

We have two ways of calculating $\dfrac{d}{dx} H(x, f(x))$.

One way is to use rules from one-variable Calculus and the one-variable Chain Rule, as described in Theorem 2.10.5, to produce functions $A(x,y)$ and $B(x,y)$ such that

$$\frac{d}{dx} H(x, f(x)) = A(x, f(x)) f'(x) - B(x, f(x)). \qquad (2.21)$$

It is important now that the rules that are used to derive this equality imply that the equality holds, **regardless** of whether or not $y = f(x)$ satisfies $F(x,y) = G(x,y)$. Of course, if, in fact, $F(x, f(x)) \equiv G(x, f(x))$, then $H(x, f(x)) \equiv 0$ and the above equality yields

$$A(x,y) \frac{dy}{dx} = B(x,y),$$

as in Theorem 2.10.5.

Our second way of calculating $\dfrac{d}{dx} H(x, f(x))$ is by using the multi-variable Chain Rule to obtain

$$\frac{d}{dx} H(x, f(x)) = \frac{\partial H}{\partial y}\bigg|_{(x, f(x))} \cdot f'(x) + \frac{\partial H}{\partial x}\bigg|_{(x, f(x))}. \qquad (2.22)$$

We claim that, together, Formula 2.21 and Formula 2.22, imply that $A(x,y) = \partial H/\partial y$ and that $-B(x,y) = \partial H/\partial x$; the former equality is what need, so that $A(a,b) \neq 0$ implies $\partial H/\partial y \neq 0$ at (a,b).

By subtracting Formula 2.21 from Formula 2.22 (or vice-versa), we are reduced to needing to prove that, if we have continuously differentiable functions $P(x,y)$ and $Q(x,y)$ on an open set $\mathcal{U}$ such that, for all differentiable $f(x)$,

$$P(x, f(x)) \cdot f'(x) + Q(x, f(x)) \equiv 0 \qquad (2.23)$$

then $P(x, y)$ and $Q(x, y)$ are always zero.

This is actually easy. Suppose that $(x_0, y_0) \in \mathcal{U}$. Consider, first, the function $f(x) \equiv y_0$. Then, $f(x_0) = y_0$ and $f'(x_0) = 0$, and so Formula 2.23 tells us that $P(x_0, y_0) \cdot 0 + Q(x_0, y_0) = 0$; hence, $Q(x_0, y_0) = 0$. Therefore, Q is always zero, which tells us that, for all continuously differentiable $f(x)$,

$$P(x, f(x)) \cdot f'(x) \equiv 0 \qquad (2.24)$$

But, now, let $f(x) = x + y_0 - x_0$. Then $f'(x_0) = 1$ and $f(x_0) = y_0$, , and so Formula 2.24 tells us that $P(x_0, y_0) \cdot 1 = 0$, and we are finished.

Chapter 3

Applications of Differentiation

Throughout Chapters 1 and 2, we provided numerous applications of the derivative. It is hardly surprising that the derivative has many, many applications; after all, the derivative is the mathematically precise notion of the instantaneous rate of change of one quantity, with respect to another. The study of such rates of change are fundamentally important in a wide range of areas, including mathematics, physics, chemistry, biology, engineering, medicine, economics, and many other areas.

In this chapter, we group together various problems according to the types of data given, and the types of answers that are asked for. In contrast with Chapters 1 and 2, this chapter consists mainly of examples and remarks, with few definitions and theorems.

3.1 Related Rates

Related rates problems arise from the fact that all physical quantities are functions of time. Thus, if you have a number of variables (two, at least), representing physical quantities, and they are related by one or more equations, which are valid at all times t (at least, at all times t in some open interval), then you may differentiate both sides of all the equations, with respect to t, using the Chain Rule, and still obtain equalities. These new equalities relate the rates of change, with respect to time, of the various quantities involved. We already looked at such problems in Example 2.2.10, Example 2.3.5, Example 2.3.6, Example 2.8.5, and Example 2.9.11.

Throughout this section, we will assume that the derivatives, with respect to time, of the various functions exist, at least at the times at which we are interested.

Example 3.1.1. Consider a barge, which is being pulled towards a dock. The dock is 5 feet above the water, and a cable runs from a 3 foot tall winch, on the dock, to the front (the bow) of the barge, right at the waterline. See Figure 3.1.

If the cable is being reeled in at 0.5 ft/s (and no slack ever exists in the cable), how fast is the barge approaching the dock when the barge is 6 feet from the dock?

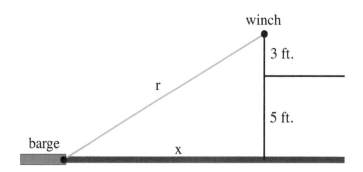

Figure 3.1: A barge being pulled towards a dock.

Solution:

Let r denote the length of the cable, in feet, and let x denote the distance from the bow of the barge to the dock (at the closest point on the dock, the point right at the waterline). Let t denote the time, in seconds. We are given that dr/dt is constantly -0.5 ft/s, where the negative sign reflects the fact that the length of the cable is decreasing. We are asked for $-dx/dt$, where, again, the negative sign is present because dx/dt itself will be negative, since x is decreasing.

To relate dx/dt and dr/dt, we need an equation which relates x and r, an equation that holds at **all** times t in some open interval that contains all of the times in which we are interested. We need to have an equality at all times in some open interval so that we have an equality of **functions of** t, which we may differentiate in order to obtain an equality of the derivatives.

By the Pythagorean Theorem, we know that

$$x^2 + 8^2 \; = \; r^2, \tag{3.1}$$

at all of the times in which we're interested; thus, this is an equality of functions of t. Differentiating both sides of this equation, with respect to t, and using the Chain Rule, we obtain

$$2x\frac{dx}{dt} + 0 \; = \; 2r\frac{dr}{dt},$$

or, equivalently,

$$\frac{dx}{dt} = \frac{r}{x}\frac{dr}{dt}.$$

We are asked for $-dx/dt$ when $x = 6$ feet. Using Formula 3.1, when $x = 6$, we find that $r = 10$ feet. Therefore, we obtain the answer:

$$-\frac{dx}{dt}\Big|_{x=6} = -\left(\frac{r}{x}\frac{dr}{dt}\right)\Big|_{x=6} = -\left(\frac{10}{6}\right)(-0.5) = \frac{5}{6} \text{ ft/s}.$$

Note that, in the midst of this problem, we used that $6^2 + 8^2 = r^2$, **at the specific time that we were asked about.** This is **NOT** an equality of functions of t. It does not hold for all t in some open interval; it holds at one particular point in time. You may **NOT** differentiate both sides and obtain an equality.

Example 3.1.2. Sand is falling in an hourglass at a constant rate of 1 cubic inch every 5 minutes.

Figure 3.2: Sand, falling in an hourglass.

Assume that the sand always forms a conical pile (a right, circular, conical pile), whose radius always equals the height. See Figure 3.2. At what rate is the height of the pile increasing when the height of the pile is 1 inch?

Solution:

The volume of a right, circular cone is

$$V \;=\; \frac{1}{3} \cdot (\text{area of the base}) \cdot (\text{height}) \;=\; \frac{1}{3}\pi r^2 h.$$

This equation holds at all times under consideration.

Also, at all times, we are told that $h = r$. Therefore, we have an equality of functions of the time t, in minutes:

$$V \;=\; \frac{1}{3}\pi h^3.$$

Differentiating, we find:

$$\frac{dV}{dt} \;=\; \frac{1}{3}\pi \cdot 3h^2 \frac{dh}{dt} \quad \text{in}^3/\text{min}.$$

We are told that dV/dt is constantly $1/5$ in^3/min, and so

$$\frac{dh}{dt} \;=\; \frac{1}{5\pi h^2} \quad \text{in}/\text{min}.$$

Therefore, when the height of the pile of sand is 1 inch, we have:

$$\left.\frac{dh}{dt}\right|_{h=1} \;=\; \frac{1}{5\pi} \;\approx\; 0.06366 \quad \text{in}/\text{min}.$$

Example 3.1.3. Two airplanes are flying on perpendicular courses (at the same altitude). We set up x- and y-axes in such a way that plane A is moving along the positive x-axis, in the positive direction, and plane B is moving along the positive y-axis, in the positive direction. See Figure 3.3.

Suppose that plane A is moving at a constant speed of 300 mph and plane B is moving at a constant speed of 200 mph. When $x = 30$ miles and $y = 40$ miles, how fast is the distance r between the planes increasing, and at what rate is the angle θ changing?

Solution:

We are given that dx/dt is constantly 300 mph and dy/dt is constantly 200 mph. Looking at Figure 3.3, we see that what we are after is dr/dt and $d\theta/dt$, when $x = 30$ miles and $y = 40$ miles.

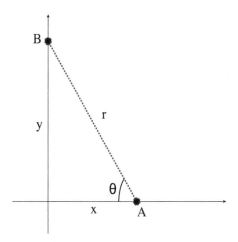

Figure 3.3: Two planes, on perpendicular courses.

By the Pythagorean Theorem and trigonometry, at all times t, in hours, we have

$$x^2 + y^2 \; = \; r^2$$

and

$$\tan \theta \; = \; y/x.$$

Differentiating each of these equations, with respect to t, we obtain

$$2x \frac{dx}{dt} + 2y \frac{dy}{dt} \; = \; 2r \frac{dr}{dt} \tag{3.2}$$

and

$$(\sec^2 \theta) \frac{d\theta}{dt} \; = \; \frac{x \dfrac{dy}{dt} - y \dfrac{dx}{dt}}{x^2}. \tag{3.3}$$

Now, since $\tan^2 \theta + 1 = \sec^2 \theta$ and $\tan \theta = y/x$, we conclude that

$$\sec^2 \theta \; = \; (y/x)^2 + 1 \; = \; \frac{x^2 + y^2}{x^2} \; = \; \frac{r^2}{x^2}.$$

Thus, Formula 3.3 becomes

$$\frac{r^2}{x^2} \frac{d\theta}{dt} \; = \; \frac{x \dfrac{dy}{dt} - y \dfrac{dx}{dt}}{x^2},$$

that is,

$$\frac{d\theta}{dt} = \frac{x\dfrac{dy}{dt} - y\dfrac{dx}{dt}}{r^2}.$$

In addition, dividing each side of Formula 3.2 by $2r$, we find

$$\frac{dr}{dt} = \frac{x\dfrac{dx}{dt} + y\dfrac{dy}{dt}}{r}$$

When $x = 30$ miles and $y = 40$ miles, the Pythagorean Theorem tells us that $r = 50$ miles, and so, at this point in time, the last two equations above tell us that

$$\frac{d\theta}{dt} = \frac{x\dfrac{dy}{dt} - y\dfrac{dx}{dt}}{r^2} = \frac{30(200) - 40(300)}{2500} = -2.4 \text{ rad/hr}$$

and

$$\frac{dr}{dt} = \frac{x\dfrac{dx}{dt} + y\dfrac{dy}{dt}}{r} = \frac{30(300) + 40(200)}{50} = 340 \text{ mph.}$$

3.1.1   Exercises

In each of Exercises 1 through 10, assume that the given equation relates functions of time t, and that the equation is valid at all times. Use the given information to find $\dfrac{dy}{dt}$.

1. $y = \sqrt{1 + 2x^2}$, when $\dfrac{dx}{dt} = 3$ and $x = 1$.

2. $x^2 + y^2 = 25$, when $\dfrac{dx}{dt} = 1$, $x = 3$, and $y > 0$.

3. $y^2 = x^3$, when $\dfrac{dx}{dt} = 3$, $x = 1$, and $y < 0$.

4. $y = \tan x$, when $\dfrac{dx}{dt} = \dfrac{1}{\sqrt{2}}$ and $x = \dfrac{\pi}{4}$.

5. $xy = 1$, when $\dfrac{dx}{dt} = -5$ and $x = \dfrac{1}{3}$.

6. $x^2 - y^2 = 80$, when $\dfrac{dx}{dt} = 12$, $x = 10$, and $y > 0$.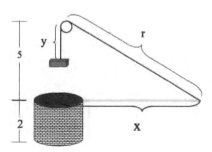

7. $y = xe^{-x}$, when $\dfrac{dx}{dt} = -2$, $x = -5$.

8. $y \cos x = 1$, when $\dfrac{dx}{dt} = 3$, $x = \pi/4$.

9. $y = 4 \ln \sin x$, when $\dfrac{dx}{dt} = 5$, $x = \pi/2$.

10. $y = \tan^{-1} x$, when $\dfrac{dx}{dt} = 4$, $x = 20$.

11. If $z = x^2 + y^2$, find $\dfrac{dz}{dt}$, when $\dfrac{dx}{dt} = \dfrac{1}{2}$, $\dfrac{dy}{dt} = \dfrac{2}{3}$, $x = -6$, and $y = 12$.

12. If $z^2 = x^2 + y^2$, find $\dfrac{dz}{dt}$, when $\dfrac{dx}{dt} = 5$, $\dfrac{dy}{dt} = -7$, $x = 5$, $y = 12$ and $z > 0$.

13. If $z = e^{x+y}$, find $\dfrac{dz}{dt}$, when $\dfrac{dx}{dt} = 3$. $\dfrac{dy}{dt} = 0.6$, $x = \ln 5$ and $y = 0$.

In Exercises 14 through 16, assume a particle is moving along the graph of $y = f(x)$. Determine the point(s) on the curve where the x-coordinate and y-coordinate are changing at the same rate with respect to time.

14. $f(x) = 5x^2$.

15. $f(x) = \sqrt{x}$.

16. $f(x) = \sin x$.

17. Suppose a particle moves along the curve $y = e^{-x}$. At the point $(0, 1)$, the rate of change of the x-coordinate, with respect to time, is 2 units per second. At what rate is the distance between the particle and the point $(1, 1)$ changing, with respect to time?

18. By walking towards a well, a farmer uses a rope that is 13 feet long to lower a small brick into the 2-foot-deep well, via a pulley 5 feet above the mouth of the well.

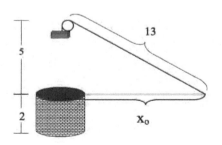

    a. How far is the farmer from the well when the brick is at the pulley as in the first diagram?

    b. How far is the farmer from the well when the rope reaches the bottom of the well?

    c. If the farmer is walking at 1 foot per second towards the well, how quickly is the brick descending when the farmer is 6 feet from the center of the well?

19. A spherical balloon is being inflated at a rate of 10 cubic centimeters per second. How fast is the radius changing when the radius is equal to 4 centimeters?

20. Suppose a right triangle has a constant perimeter of 60 feet but that its hypotenuse is increasing. What is the rate of change in the length of the other two sides when the hypotenuse is increasing at a rate of 2 meters per second and the side lengths are 10 feet, 24 feet and 26 feet?

21. Suppose that water is being pumped into a trough. The trough is 15 feet long, and every cross section of the trough is an equilateral triangle, with a vertex of the triangle pointing towards the ground. Find the rate of change of the height of the water when the height is 4 feet, if the volume of water is increasing at a rate of 8 cubic feet per second.

22. A rock is dropped into a pond, causing a wave to ripple out in a circular pattern. Suppose that the wave moves outward from the center at a constant rate of 0.879 meters per second. Find the rate of change of the area enclosed by the wave, with respect to time, when the radius is 10 meters.

23. A large spherical piece of ice is melting in such a way that it always remains perfectly spherical. Suppose that the surface area of the ice is decreasing at a rate of 2 square meters per second.

    a. What is the volume, V, of a sphere as a function of its surface area, S?

    b. What is dV/dS?

    c. At what rate is the volume of the ice decreasing when the surface area is 18 square meters?

24. Recall that if n capacitors with capacitances C_1, C_2,..., C_n, in farads, are wired in series then the total capacitance of the circuit is given by

$$\frac{1}{\frac{1}{C_1} + \frac{1}{C_2} + \cdots + \frac{1}{C_n}}.$$

Suppose a circuit with two capacitors in series have variable capacitance. The capacitance of the first capacitor is increasing at a rate of 3 farads per second while the capacitance of the second capacitor is decreasing at a rate of 2 farads per second. At what rate, in farads per second, is the capacitance of the circuit changing, when the first capacitor has capacitance 6 farads and the second capacitor has capacitance 4 farads?

Consider a particle moving in a circular path of radius r. Let $s(t)$ be the distance the particle has traveled after t seconds and let $\theta(t)$ be the quantity of radians the particle has traversed after t seconds. We define the *linear speed* v of the particle as ds/dt and the angular speed ω as $d\theta/dt$. Use this information to solve Exercises 25 through 29.

25. Prove that these two rates are related by the equation

$$v = r\omega \quad \text{or} \quad \frac{ds}{dt} = r\,\frac{d\theta}{dt}.$$

26. Calculate the linear speed of a particle traveling in a circular orbit of radius 12 feet that has $\theta(t) = \dfrac{\pi t}{1+t}$. Assume t is measured in seconds.

27. Calculate the linear speed of a person standing on the equator of the Earth. Assume that the radius of the Earth is 4000 miles.

28. Consider a clock where the minute hand has length k inches. What is the linear speed of the tip of the minute hand?

29. What is the angular speed of a particle moving in a circular path of radius 15 miles with distance function $s(t) = \dfrac{t^2+1}{t+1}$?

30. Suppose that two sides of a triangle have lengths $a = 10$ feet and $b = 12$ feet and that the angle θ between them is increasing at a rate of $\pi/3$ radians per minute. What is the rate of change of the length c of the third side, when the angle between the sides of fixed length is $\pi/4$? Hint: Use the Law of Cosines, $c^2 = a^2 + b^2 - 2ab\cos\theta$. ▶

31. If the edge length of a cube is increasing at a constant rate of 3 meters per second, how fast is the volume increasing, when the edge length is 50 meters?

32. If the surface area of a cube is increasing at a constant rate of 12 square meters per second, how fast is the volume of the cube increasing, when the edge length is 4 meters?

33. A particle moves along a differentiable curve $y = f(x)$. If, at some point, the slope of the tangent line to the curve is 12 and the x-coordinate is increasing at a rate of 5 units per second, what is the rate of change of the y-coordinate?

The *kinetic energy* of an object, measured in joules, of mass m, moving with velocity v, is given by the equation $E = \frac{1}{2}mv^2$. **Use this formula to solve Exercises 34 through 36.**

34. Suppose that the mass of an object is increasing at a rate of 5 kilograms per second. At what rate is the kinetic energy changing, if the velocity remains constant at 12 meters per second?

35. Assume a constant mass of 15 kilograms and a constant acceleration of 4 meters per second per second. What is the rate of change of the kinetic energy, when the velocity is 25 meters per second?

36. Assume that the mass and velocity of an object are both functions of time with $m(t) = 2t^2 + 1$ and $v(t) = 4t$.

    a. Express E as a function of t and calculate dE/dt when $t = 5$.

    b. Recall the definition of *partial derivatives* from Exercises 18 through 22 of Section 2.10. Verify that your answer in part (a) agrees with the formula

    $$\frac{dE}{dt} = \frac{\partial E}{\partial m}\frac{dm}{dt} + \frac{\partial E}{\partial v}\frac{dv}{dt},$$

    where $\partial E/\partial m$ and $\partial E/\partial v$ are the partial derivatives of E with respect to m and v, respectively. This is an example of the *multivariable Chain Rule* from multivariable Calculus.

37. A ladder that is 10 meters long is leaning against a wall. Atop the wall, a worker hoists the ladder upwards at a constant rate of 0.1 meters per second in such a way that the bottom of each leg continues to make contact with the ground, and the top continues to

make contact with the wall as shown in the diagram.

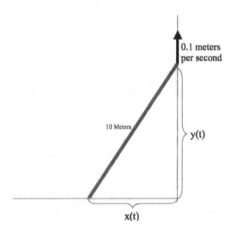

a. Find an equation relating $y(t)$, the height of the end of the ladder, and $x(t)$ the distance of the bottom of the ladder from the wall.

b. What are the velocity and acceleration of the bottom of the ladder, when the bottom of the ladder is 6 meters from the wall?

c. What is the problem as x tends to 0?

38. A block of dry ice (frozen carbon dioxide), which is always in the shape of a right circular cylinder, is sublimating (changing directly from a solid into a gas) in such a way that both the height h and radius r are decreasing with time.

Recall that the volume of a cylinder is the area of the base times the height, while the surface area is the area of the top and bottom disks (filled-in circles) plus the area of the side(s), which is the circumference of the base times the height.

Suppose that the rate of change of the volume, V, with respect to time, is proportional to the surface area, A. (This is reasonable since the surface of the dry ice is what is in contact with the warmer surroundings.) Suppose that the proportionality constant is -2 cubic length units per square length unit.

a. Using the formula for V in terms of r and h, calculate dV/dt, in terms of r, h, dr/dt, and dh/dt.

b. On the other hand, the physical set-up tells us that dV/dt equals what, in terms of r and h?

c. Combine parts (a) and (b), and find a formula for $\dfrac{dr}{dt}$, in terms of h, r, and $\dfrac{dh}{dt}$.

d. If, at time t_0, $h = 4$, $r = 3$, and $\dfrac{dh}{dt} = -0.5$, then what is $\dfrac{dr}{dt}$ at time t_0?

e. Assuming that the sublimation takes place in such a way that the rate of change of the height, with respect to time, is a constant, b, what is $\dfrac{d^2r}{dt^2}$, in terms of b, r, and h?

39. Two cars leave the same point at the same time. The first car travels north at a constant speed of 30 mph and the second car heads west at a constant speed of 55 mph. At what rate is the distance between the two cars increasing 3 hours after they depart?

40. Suppose that two cars leave the same point at the same time. The first car travels north at a constant speed of A mph and the second car travels west at a constant speed of B mph. At what rate is the distance between the two cars increasing t hours after they depart?

41. Water is being pumped out of a container whose shape is an inverted right circular cone, which has base radius 5 meters and height 20 meters, as shown in the diagram below. Recall that the volume of a cone is 1/3 of the area of the base times the height.

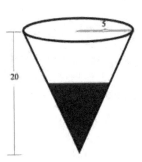

a. If the tank starts out full, and water is pumped out at a constant rate of 0.2 cubic meters per second, how long will it take for the tank to be empty?

b. How quickly is the height of the water decreasing, with respect to time, when the height of the remaining water is 10 meters? Note that you are asked for the rate of **decrease**.

c. How quickly is the radius of the cone of water decreasing, with respect to time, at the same moment? Note that you are again asked for a rate of **decrease**.

42. The (Euclidean) distance between two points $p_0 = (x_0, y_0)$ and $p_1 = (x_1, y_1)$ is given by the formula:
$$E(p_0, p_1) = \sqrt{(x_0 - x_1)^2 + (y_0 - y_1)^2}.$$

a. Suppose that a bug walks on an xy-plane, so that the bug's $x = x(t)$ position and $y = y(t)$ position are functions of time. What is the instantaneous rate of change of the distance between the bug and the origin, in terms of x, y, dx/dt, and dy/dt? (Hint: You may find it easier to differentiate the equation for E^2.)

b. If the bug walks along the path $y = \sin x$ and in such a way that its x-coordinate at time t is $x(t) = 3t^3 - 10$, for $0 \le t \le 3$, what are the instantaneous rates of change of the distance between the bug and the point $(0, 1)$, with respect to time, at times $t = 1$ and $t = 2$?

c. At what time is the bug at the point $(0, 1)$?

43. Jane and Dave are running on two circular tracks of radius 10 and 20 meters, respectively, whose centers are separated by 100 meters.

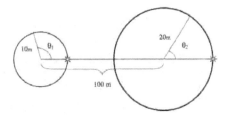

a. Write a function that describes the distance between the two runners as a function of θ_1 and θ_2, the angles labeled in the diagram.

b. Suppose that both run at a constant rate of 4 meters per second. What are $\theta_1(t)$ and $\theta_2(t)$ (assume that, at time 0, both θ_1 and θ_2 are 0 as shown in the diagram)?

c. What is the instantaneous rate of change of the distance between the runners, with respect to time, when $t = 3$ seconds?

44. An inverted right circular cone (as in Exercise 41) of height h_0 meters and base radius r_0 meters is filled with a liquid and put in a warm room. The liquid evaporates. Let $r = r(t)$ denote the radius of the exposed area of the remaining liquid, at time t hours, and let $h = h(t)$ denote the height of the remaining liquid (both in meters).

Suppose that the rate of change of the volume, V, of the liquid in the container, with respect to time, is proportional to the surface area of liquid that is exposed. Let k denote the (negative) proportionality constant, in meters per second.

a. The physical set-up immediately implies that dV/dt equals what, in terms of k, r, and h (though you may not need all of these)?

b. Use similar triangles to derive an equation relating r and h.

    c. Write a formula for V, in terms of r.

    d. Use your answer to part (c) to express dV/dt, in terms of r and dr/dt.

    e. Use your answers to part (a) and part (d) to determine the rate of change of r, with respect to time.

    f. Show that k is equal to the rate of change of h, with respect to time.

    g. If $h_0 = 0.5$ meters, and the last bit of liquid evaporates at $t = 24$ hours, what is the value of k?

45. A dog, whose height is 0.75 meters, is between a light, which is sitting on the ground, and a very tall wall, which is 20 meters from the light.

    a. How tall is the shadow cast by the dog against the wall, resulting from the light, when the dog is 5 meters from the wall?

    b. How quickly is the height of the shadow growing, when the dog is 10 meters from the wall and walking towards the light at a rate of 1 meter per second?

    c. How quickly is the height of the shadow shrinking, when the dog is 6 meters from the wall and walking away from the light at a rate of 2 meters per second?

3.2 Graphing

In this section, we discuss how derivatives and limits help you draw graphs, and help you find the places where the graph has "interesting" features. To be honest, with the proliferation of graphing calculators and computers, being able to graph by hand is not nearly as important as it used to be.

Nonetheless, graphing programs on a calculator or computer usually require you to give a "window" for the graph: a range of x values and y values that it will display for you. The choice of such a window is not always easy; you have to have some idea of what part of the graph you want to see **before** you've seen the graph. Calculus can certainly help with that.

In addition, calculators and computers always use one scale on each axis, which may seem reasonable, until you realize that, in order to see the interesting activity at various points on the graph, you need to use different scales in different regions. This means selecting multiple windows and scales in order to see what you want. By hand, we can alter our scales as we want, in various portions of the xy-plane, according to where our analysis tells us there are important features.

When graphing $y = f(x)$ by hand, we consider the following:

1. **The Domain of f**: If a domain is not explicitly given, we need to determine the set of points where f is actually defined. The range usually cannot be determined as a first step, but follows from the calculations below.

2. **Monotonicity and Extrema**: We look at the sign of the first derivative of f, i.e, where $f'(x)$ is positive and where $f'(x)$ is negative, in order to find intervals on which f is monotonic. (Recall Corollary 1.5.15.) On intervals where f is increasing, the graph of f goes, roughly, from the lower-left to the upper-right, i.e., has positive slope. On intervals where f is decreasing, the graph of f goes, roughly, from the upper-left to the lower-right, i.e., has negative slope. At points in the domain of f where the sign of f' changes from positive to negative, or vice-versa, f attains a local extreme value.

3. **Concavity and Inflection Points**: We look at the sign of the second derivative of f, i.e, where $f''(x)$ is positive and where $f''(x)$ is negative, in order to find where the graph of f is concave up ("curves upward") and where the graph of f is concave down ("curves downward"), respectively. At points in the domain of f where the sign of f'' changes from positive to negative, or vice-versa, the graph of f has an inflection point.

4. **Asymptotes and Activity "at" Infinity**: An **asymptote** is a line which is approached by a graph as the graph "goes out to infinity". We usually consider two types of asymptotes: **horizontal asymptotes** and **vertical asymptotes**. We discussed asymptotes for rational functions back in Subsection 1.3.3.

   A horizontal asymptote is a horizontal line that the graph approaches as $x \to \pm\infty$. If $\lim_{x\to\infty} f(x)$ exists and equals L, then the graph of $y = f(x)$ has a horizontal asymptote of $y = L$. Similarly, if $\lim_{x\to-\infty} f(x)$ exists and equals L, then the graph of $y = f(x)$ has a horizontal asymptote of $y = L$. If $\lim_{x\to\pm\infty} f(x) = \pm\infty$, there is no corresponding horizontal asymptote, but this is nonetheless important information about the graph.

   Vertical asymptotes are vertical lines $x = b$ such that either (or both) of the one-sided limits $\lim_{x\to b^-} f(x)$ or $\lim_{x\to b^+} f(x)$ equals $\pm\infty$. If f is continuous, and its graph has a vertical asymptote at $x = b$, then b is necessarily **not** in the domain of f.

5. **Symmetry: Even- and Oddness**: Recall that a function f is *even* if, for all x in the domain of f, $-x$ is also in the domain of f, and $f(-x) = f(x)$. If n is an even integer, then $f(x) = x^n$ is an even function; hence, the term "even". If f is even, then (a, b) is a point on the graph of f if and only if $(-a, b)$ is a point on the graph; this means that the graph of an even function is *symmetric about the y-axis*. This means that if you graph $y = f(x)$ for $x \geq 0$, you can obtain the part of the graph where $x \leq 0$ by flipping/rotating the positive x portion around the y-axis. See Figure 3.4.

   A function f is *odd* if, for all x in the domain of f, $-x$ is also in the domain of f, and $f(-x) = -f(x)$. If n is an odd integer, then $f(x) = x^n$ is an odd function; hence, the term "odd". If f is odd, then (a, b) is a point on the graph of f if and only if $(-a, -b)$ is a point on the graph; this means that the graph of an odd function is *symmetric about the origin*. This means that if you graph $y = f(x)$ for $x \geq 0$, you can obtain the part of the graph where $x \leq 0$ by rotating the positive x portion around the origin by $180°$ (π radians). See Figure 3.5.

   Note that, while an integer must be even or odd, it is unusual for a function to have either of these properties; most functions are neither even nor odd. Still, enough functions are even or odd, and it's easy enough to check, that it is worth quickly checking before graphing.

6. **Special Points**: It can be very helpful to indicate some special points on the graph, for instance, where the graph intersects the coordinate axes (if it does). The y-intercept is the point $(0, f(0))$, if 0 is in the domain of f. The x-intercept(s) are all of the points $(a, 0)$ such that $f(a) = 0$, if there are any such points.

   Other special points, other than the axis-intercepts can also be helpful to graph.

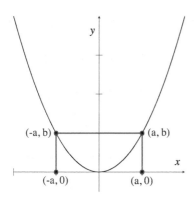

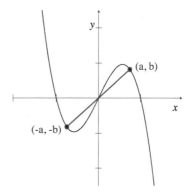

Figure 3.4: An even function.   Figure 3.5: An odd function.

Before we look at some examples, we need to comment on how you tell where f' is positive or negative, and where f'' is positive or negative.

The Intermediate Value Theorem for Derivatives, Theorem 1.5.19 or Theorem 1.A.25, tells us that, if f' (or f'') is defined on an interval I, and is negative at one point of I and positive at another, then f' (resp., f'') must hit zero somewhere in-between.

Therefore, if f' (or f'') is defined on an interval I, and is not equal to zero at any point in I, then f' (resp., f'') must either be positive at every point of I or must be negative at every point of I. How do you tell which sign you have? Pick a point in I and evaluate f' (resp., f'') there, to see whether it's positive or negative.

Okay. Fine. So, how do you find intervals I on which f' (or f'') is defined, but is never equal to zero? Simple. You first find all of places where f' (resp., f'') **does** equal zero or is undefined; call these *unsigned points*, and call the other points *signed points*. After determining the unsigned points, you look at maximum intervals that do not contain any of these unsigned points, i.e., you look at maximum intervals of signed points. For instance, if a is an unsigned point, and there exists a next biggest unsigned point b, then the interval (a, b) would consist entirely of signed points, on which f' or f'' could not switch signs. As another example, if there is a largest (resp., smallest) unsigned point, b, then (b, ∞) (resp., $(-\infty, b)$) would be an interval consisting entirely of signed points, on which f' or f'' could not switch signs.

Now let's see how well all of this enables us to sketch graphs.

Example 3.2.1. Consider the function $y = f(x) = \dfrac{x^2}{x^2 - 1}$. Note that the domain is all x other than ± 1.

For calculating the first derivative, it will help to algebraically simplify the function. We accomplish this by "adding zero in a clever way" in the numerator; it is also what you would obtain by long-dividing the numerator by the denominator (although the division would not be very "long"). We find

$$y \; = \; f(x) \; = \; \frac{x^2}{x^2-1} \; = \; \frac{(x^2-1)+1}{x^2-1} \; = \; 1 + \frac{1}{x^2-1} \; = \; 1 + (x^2-1)^{-1},$$

and so

$$y' \; = \; (-1)(x^2-1)^{-2}(2x) \; = \; -2 \cdot \frac{x}{(x^2-1)^2}.$$

For the second derivative, we find

$$y'' \; = \; -2 \cdot \left[\frac{(x^2-1)^2 \cdot 1 - x \cdot 2(x^2-1)(2x)}{(x^2-1)^4} \right].$$

Canceling a factor of x^2-1 in the numerator and denominator, we obtain

$$y'' \; = \; -2 \cdot \left[\frac{(x^2-1) \cdot 1 - x \cdot 2(2x)}{(x^2-1)^3} \right] \; = \; 2 \cdot \left[\frac{3x^2+1}{(x^2-1)^3} \right].$$

The unsigned points of y' are $x=0$ and $x=\pm 1$. Since the denominator of y' is always ≥ 0, we easily find that $y'<0$ where $x>0$ and $\neq 1$, and $y'>0$ where $x<0$ and $\neq -1$. We indicate this on a number line in Figure 3.6.

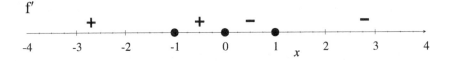

Figure 3.6: The sign of f'.

In Figure 3.6, we have indicated the three unsigned points for f' with black dots, and have placed plus or minus signs over the four intervals into which the real line is divided by the unsigned points; these signs over the intervals indicate the sign of f' everywhere on that interval.

Thus, we see that f is strictly increasing on the open interval $(-\infty, -1)$, and strictly decreasing on the interval $(1, \infty)$. We also see that, since f is defined and continuous at 0, f is strictly increasing on the half-open interval $(-1, 0]$ and is strictly decreasing on the half-open interval

$[0, 1)$; consequently, the First Derivative Test, Theorem 1.5.20, tells us that the restriction of f to the open interval $(-1, 1)$ attains a global maximum value at $x = 0$.

Note that we are absolutely **not** claiming that f, considering **all** the points in its domain, attains a global maximum at $x = 0$; we are merely claiming that, if you look only at x's in the interval $(-1, 1)$, then $f(0)$ is the maximum value of $f(x)$ among all of those $f(x)$'s.

Now let's look at

$$y'' = 2 \cdot \left[\frac{3x^2 + 1}{(x^2 - 1)^3} \right].$$

The unsigned points of f'' are exactly $x = \pm 1$. As the numerator of y'' is always positive, and the denominator is cubed, we see that y'' always has the same sign as $x^2 - 1$. Thus, $y'' > 0$ when $x^2 - 1 > 0$, i.e., when $x > 1$ or $x < -1$, and $y'' < 0$ when $x^2 - 1 < 0$, i.e., when $-1 < x < 1$. We display this information on a number line as follows:

Figure 3.7: The sign of f''.

Now, let's look for asymptotes. As f is an elementary function, it can have vertical asymptotes only at x values not in its domain, i.e., at $x = \pm 1$. Does f actually have vertical asymptotes at these x values? We need to examine limits from the left and right as x approaches -1 and 1.

As $x^2 - 1$ is positive for $x < -1$, we find that $x^2 - 1$ approaches 0 from the right as $x \to -1^-$ and certainly, $\lim_{x \to -1^-} x^2 = 1$; thus,

$$\lim_{x \to -1^-} \frac{x^2}{x^2 - 1} = +\infty.$$

Therefore, the graph of $y = f(x)$ has a vertical asymptote at $x = -1$.

In a similar fashion, we find that

$$\lim_{x \to -1^+} \frac{x^2}{x^2 - 1} = -\infty, \quad \lim_{x \to 1^-} \frac{x^2}{x^2 - 1} = -\infty, \quad \text{and} \quad \lim_{x \to 1^+} \frac{x^2}{x^2 - 1} = +\infty.$$

Hence, the graph of $y = f(x)$ also has a vertical asymptote at $x = 1$.

What about horizontal asymptotes? As $y = 1 + 1/(x^2 - 1)$, we find easily $\lim_{x \to -\infty} y =$

$1+0 = 1$ and $\lim_{x\to\infty} y = 1+0 = 1$. Therefore, there is a horizontal asymptote of $y = 1$, which the graph of $y = f(x)$ approaches as $x \to \pm\infty$.

We should check for symmetry. We see that f is even:

$$f(-x) \;=\; \frac{(-x)^2}{(-x)^2 - 1} \;=\; \frac{x^2}{x^2 - 1} \;=\; f(x).$$

Thus, the graph of f is symmetric about the y-axis.

Finally, the y-intercept is also the only x-intercept: $(0,0)$.

Putting all of the above data together, we can sketch a "qualitatively accurate" graph of $y = f(x)$ by hand. We also include the computer-generated graph. You'll have to believe that we didn't "cheat" and just copy the computer graph.

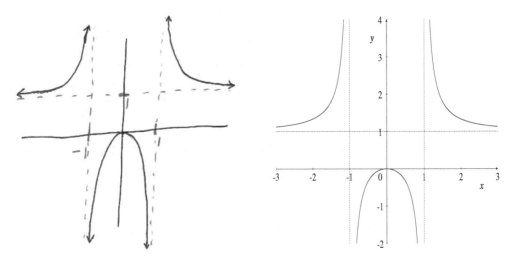

Figure 3.8: By Hand: $y = x^2/(x^2 - 1)$.   Figure 3.9: Computer: $y = x^2/(x^2 - 1)$.

Example 3.2.2. Let's look at another example: $y = g(x) = xe^{-x}$. The domain of this function is all of $(-\infty, \infty)$.

We quickly calculate

$$y' \;=\; xe^{-x}(-1) + e^{-x} \;=\; (-x+1)e^{-x}$$

and

$$y'' \;=\; (-x+1)e^{-x}(-1) + e^{-x}(-1) \;=\; (x-2)e^{-x}.$$

As e^{-x} is always > 0, the unsigned points of y' and y'' are $x = 1$ and $x = 2$, respectively, and the number lines indicating their signs are:

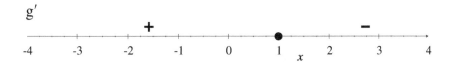

Figure 3.10: The sign of g'.

Figure 3.11: The sign of g''.

We see from the First Derivative Test that g attains a global maximum value at $x = 1$, and we see that the concavity of the graph of $y = g(x)$ switches at $x = 2$, i.e., the graph of $y = g(x)$ has an inflection point at $x = 2$.

The function g is an elementary function and its domain is all of $(-\infty, \infty)$, so there are no vertical asymptotes. What about horizontal asymptotes? We find

$$\lim_{x \to -\infty} xe^{-x} \;=\; -\infty \cdot \infty \;=\; \infty.$$

The limit of $g(x)$ as $x \to \infty$ is more difficult. We need to use Exercise 54 from Subsection 2.4.1 (or, look ahead to Example 3.5.3) to conclude that $\lim_{x \to \infty} xe^{-x} = 0$. So, $y = 0$ is a horizontal asymptote, which the graph of g approaches as $x \to \infty$, but not as $x \to -\infty$.

The function g is neither even nor odd. There are at least three nice special points to indicate on the graph:

1. the x- and y-intercepts are both at the origin,

2. the absolute maximum value occurs at $x = 1$, and is equal to $g(1) = 1 \cdot e^{-1} = 1/e$, and

3. the inflection point occurs where $x = 2$, and $y = 2 \cdot e^{-2} = 2/e^2$.

Putting this all together, we sketch the graph in Figure 3.12, and compare it with the computer-generated graph in Figure 3.13.

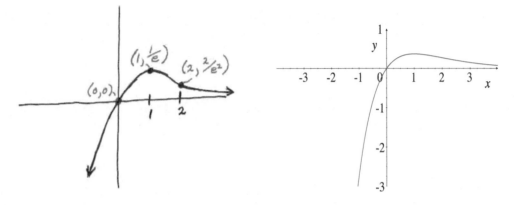

Figure 3.12: By hand: $y = xe^{-x}$.   Figure 3.13: Computer: $y = xe^{-x}$.

Example 3.2.3. Now let's try graphing something fairly complicated:

$$y \; = \; h(x) \; = \; e^{-x} \sec x \; = \; \frac{1}{e^x \cos x}.$$

The function is undefined precisely when $\cos x = 0$, i.e., the domain of h is all x other than $x = \pi/2 + k\pi$, where k is an integer. Note that the sign of y itself is the same as the sign of $\cos x$. This means that the sign of y, for x in the interval $(\pi/2 + (k-1)\pi, \pi/2 + k\pi)$, is the same as the sign of $(-1)^k$, i.e., positive if k is even, and negative if k is odd. For instance, when $k = 0$, we are saying that y is positive if x is in the interval $(\pi/2 + (0-1)\pi, \pi/2 + 0 \cdot \pi) = (-\pi/2, \pi/2)$. What this means is that the sign of y alternates on the intervals on which h is defined. A number line indicating this is given in Figure 3.14.

Figure 3.14: The sign of h.

We found the derivative of h in Example 2.8.4 (though, there, we used different variable names); what we found was:

$$y' = h'(x) = e^{-x}\sec x(\tan x - 1) = y(\tan x - 1).$$

The unsigned points for h' are $x = \pi/2 + k\pi$, where h' is undefined, and $x = \pi/4 + j\pi$, for integers j, where $h' = 0$.

On the interval $(\pi/2 + (k-1)\pi, \pi/2 + k\pi)$, $\tan x - 1$ is positive precisely when x is in the interval $(\pi/4 + k\pi, \pi/2 + k\pi)$. Since we know where y is positive or negative from Figure 3.14, the sign of h' is given in Figure 3.15.

Figure 3.15: The sign of h'.

From this, and the First Derivative Test, we see that, restricted to each interval $(\pi/2 + (k-1)\pi, \pi/2 + k\pi)$, the function h attains an extreme value at $\pi/4 + k\pi$: a minimum value when k is even (and so, h itself is positive), and a maximum value when k is odd (and so, h itself is negative).

Thus, the extreme value of $h(x)$ on the interval $(\pi/2 + (k-1)\pi, \pi/2 + k\pi)$ is

$$e^{-(\pi/4+k\pi)}\sec(\pi/4 + k\pi) = \frac{e^{-(\pi/4+k\pi)}}{\cos(\pi/4 + k\pi)} =$$

$$\frac{e^{-(\pi/4+k\pi)}}{\cos(\pi/4 + k\pi)} = \frac{e^{-(\pi/4+k\pi)}}{(-1)^k \cdot 1/\sqrt{2}} = (-1)^k\sqrt{2}\,e^{-(\pi/4+k\pi)}.$$

We now wish to look at y''. While we could simply differentiate $e^{-x}\sec x(\tan x - 1)$, it is easier to use (twice) that $y' = y(\tan x - 1)$. Using the Product Rule, we find

$$y'' = y\sec^2 x + (\tan x - 1)y' = y\sec^2 x + (\tan x - 1)y(\tan x - 1) =$$

$$y\big(\sec^2 x + (\tan x - 1)^2\big).$$

As the expression in the parentheses is always positive (where it's defined), we see that the sign of y'' is exactly the same as the sign of y itself. see Figure 3.16.

Figure 3.16: The sign of h''.

Thus, the graph of $y = h(x)$ is concave up where $h(x)$ is positive, and is concave down where $h(x)$ is negative.

The function $h(x) = e^{-x} \sec x$ certainly approaches $\pm\infty$ as x approaches any value of the form $\pi/2 + k\pi$. Thus, we have vertical asymptotes at these points. The easiest way to express where the graph approaches ∞ or $-\infty$ is simply to say that, if x approaches a vertical asymptote from a side where y is always positive, then y must approach ∞; if x approaches a vertical asymptote from a side where y is always negative, then y must approach $-\infty$.

As there are vertical asymptotes for positive and negative x values that are arbitrarily large in absolute value, there are no horizontal asymptotes. We will say more about what happens for large $|x|$ after we sketch the graph.

While $\sec x$ is even, e^{-x} is neither even nor odd and, hence, h is neither even nor odd.

As $h(0) = e^0 \sec 0 = 1$, $(0, 1)$ is a point on the graph. The other "interesting points" on the graph are the points where the extrema occur.

Finally, we put all of our work together, and draw the sketch in Figure 3.17.

Note that we have not indicated any y-coordinates along the y-axis. We have also made no attempt to draw the local extreme points in a way consistent with any particular scale on the y-axis. We were merely trying to capture the correct qualitative aspects of the graph.

The computer-generated graph is fine, but, in fact, we "cheated" to produce it; to give the computer a window that showed the local extreme value which occurs where $x = -3\pi/4$, we first did the Calculus problem and determined what the local extreme value was.

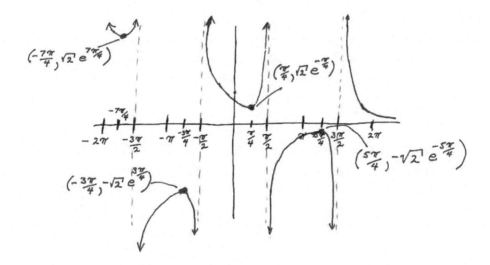

Figure 3.17: By hand: $y = e^{-x} \sec x$.

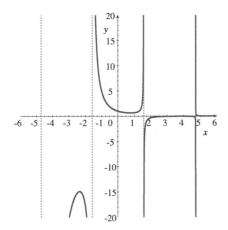

Figure 3.18: Computer: $y = e^{-x} \sec x$.

Notice that, for arbitrarily large x, the graph of $y = h(x)$ gets arbitrarily close to the x-axis at the local extrema, but also gets arbitrarily far away from the x-axis near the vertical asymptotes. Therefore, not only does $\lim_{x \to \infty} h(x)$ not exist, it fails to exist in a fairly complex manner, as opposed to being $\pm\infty$.

As $x \to -\infty$, the points on the graph move arbitrarily far away from the x-axis, but alternate having positive and negative y-coordinates. Thus, $\lim_{x \to -\infty} h(x)$ does not exist, and is also unequal to $\pm\infty$.

A final comment:

We wrote above that, before having a computer generate the graph in Figure 3.18, we first used Calculus to find the "interesting" points, and then used that knowledge to specify for the computer a good window for the graph.

In the modern age of computers and graphing calculators, this hybrid approach to graphing is probably the best: use Calculus to determine the best graphing parameters to give to a calculator or computer, and then let the software take over.

3.2.1   Exercises

Using methods from this section, graph the functions in Exercises 1 through 12.

1. $g(t) = t(t-1)(t+2) = t^3 + t^2 - 2t$.

2. $g(t) = t^2(t-1)(t+2) = t^4 + t^3 - 2t^2$.

3. $f(x) = \dfrac{x}{x-2}$.

4. $f(x) = \dfrac{x^2}{x-2}$.

5. $f(x) = \dfrac{x}{x^2-2}$.

6. $f(x) = e^x \cos x$.

7. $g(\theta) = \dfrac{\sin\theta}{1-\cos\theta}$.

8. $h(x) = \sin\left(\dfrac{\pi x}{2x-9}\right)$.

9. $k(t) = \dfrac{t^3 + 4t^2 - 4t - 8}{t-2}$.

10. $f(x) = \dfrac{\sin x}{x}$.

11. $f(x) = x + \sin x$.

12. $f(x) = \sqrt{x^3 + x^2} = |x|\sqrt{x+1}$. (Hint: consider where $x \geq 0$ and where $x \leq 0$ separately.)

13. Recall that the *demand* $D = D(p)$ for a product (a *good*) is a function which predicts the quantity of an item sold as a function of the price p charged for each item. Many models of demand for many goods follow these rules: $D(p)$ for $p \geq 0$ is decreasing, positive, and approaches 0 as p tends toward infinity. (Economists will occasionally use a piecewise linear function.)

    a. Sketch a differentiable demand curve which has no inflection points. If the demand curve has no inflection points, what must its concavity be?

    b. Sketch a differentiable demand curve which has exactly one inflection point.

14. Graph the graph of a function $f(x)$, whose domain is the interval $[0, \infty)$, and which has the following properties:

    i. $\lim\limits_{x \to \infty} f(x) = 0$;

    ii. $-1 < f(x) < 1$, for all x in $[0, \infty)$; and

    iii. $f''(x) > 0$, if there exists an integer n such that $2n\pi < x < (2n+1)\pi$, and

    iv. $f''(x) < 0$, if there exists an integer n such that $(2n+1) < x < 2n\pi$.

In Exercises 15 through 22, match the functions with their corresponding graphs (a) through (h).

15. $y = \tan^{-1}\left(\dfrac{6}{x^2}\right)$.

16. $y = e^{0.1\left(\frac{1}{3}x^3 - 2x^2\right)}$.

17. $y = \sqrt[3]{x^2 - 4x + 3}$.

18. $y = \dfrac{4\sin x}{x - 4}$.

19. $y = \dfrac{(x - 10)(x^2 - 20x + 100)}{(x^2 - 10x + 25)(x^2 - 30x + 225)}$.

20. $y = 10\sin(0.1x^2 - 0.9)$.

21. $y = |x^2 - 3x - 4| - 1$.

22. $y = \dfrac{3x + 3}{x^2 + 1}$.

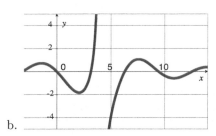

a.

b.

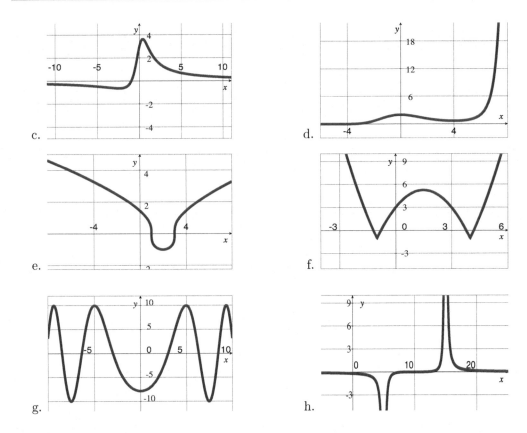

c.

d.

e.

f.

g.

h.

3.3 Optimization

The term *optimization* means the determining of how to maximize or minimize quantities in which we are interested. For instance, you might want to find the dimensions that minimize the amount of material that you need to use to construct a rectangular box that satisfies certain conditions, such as having no top, a square bottom, and a given volume. Or, you might want to determine what price to charge for product in order to maximize your profit. There are many, many problems where we wish to find the value of one quantity that will maximize or minimize another quantity.

Of course, in Section 3.2, we discussed how to use Calculus to graph functions. If you can graph a function $y = f(x)$ accurately, then you can certainly answer any question about what value of x produces a maximum or minimum value of y. However, graphing is time-consuming, and we would like a more streamlined process; in other words, we would like to optimize our optimization method.

We shall first discuss the theory, give some purely mathematical examples, and then look at some more serious examples. Optimization problems are essentially always "word problems", which are also called "modeling problems", since we are taking a real-world problem and modeling it by a pure mathematics problem.

There are three types of optimization problems that we deal with. In all of them, **we assume that we have a continuous function f, whose domain is an interval I, and that f has a finite number of critical points, where f' is zero or undefined, on the interior of the interval I.**

Optimization Case 1: The interval I is a closed, bounded interval $[a, b]$.

In this case, we have a theorem which tells us how to proceed.

Theorem 3.3.1. *The function f attains a global maximum and a global minimum value on I. These extreme values are attained at critical points of f (on the interval I); thus, they are attained either when $x = a$, $x = b$, or when x is in the open interval (a, b) and $f'(x) = 0$ or $f'(x)$ is undefined.*

Therefore, we can find the global extreme values of f by making a table of x and $f(x)$ values, where the x values are a, b, and all of the x-coordinates in (a, b) where $f'(x) = 0$ or $f'(x)$ is undefined. Then, we simply look at the $f(x)$ values and select the largest and smallest ones.

Proof. This follows immediately from the Extreme Value Theorem, Theorem 1.3.30, and Theorem 1.5.4. □

Remark 3.3.2. While there are unique maximum and minimum **values** of f on I, the maximum and/or minimum value could occur at multiple x values.

> You need to be careful when looking at critical points where $f'(x) = 0$. Frequently, we have a function F that is defined and differentiable on a larger set than the interval $[a, b]$, but physical constraints require that x be in the closed interval $[a, b]$; so, our function f is actually the function F, restricted to the interval $[a, b]$. This function f is technically a different function from F, even though the rule specifying them is the same.
>
> The danger is that the formula you derive for the derivative will be $F'(x)$, which may be defined and equal to zero **outside of the interval** $[a, b]$. The points where $F'(x) = 0$ that are outside $[a, b]$ are **not** critical points of f, and do not belong in the table of x and $f(x)$ values that you use to find the maximum and minimum values of f on $[a, b]$.

Example 3.3.3. Find the extreme values of $f(x) = 2x^3 - 3x^2 - 12x$ on the interval $[0, 3]$.

Solution:

We find
$$f'(x) = 6x^2 - 6x - 12 = 6(x^2 - x - 2) = 6(x - 2)(x + 1).$$

When does $f'(x) = 0$? When $x = 2$ and when $x = -1$. However, $x = -1$ is **outside the interval under consideration**, and so we ignore it. We find one critical point, $x = 2$, of f, inside the open interval $(0, 3)$, and we also have to check the endpoints of the closed interval. Thus, we make the small table

x	0	2	3
$f(x)$	0	-20	-9

Therefore, the maximum value of f on the interval $[0, 3]$ is 0, which occurs at the single x-coordinate $x = 0$, and the minimum value of f on the interval $[0, 3]$ is -20, which occurs at the single x-coordinate $x = 2$.

Optimization Case 2: The interval I is not closed and bounded, but the First or Second Derivative Test applies.

By "the First Derivative Test applies", we mean that there exists a point c in the interior of I such that, for all x in the interior of I, if $x < c$, then $f'(x) > 0$ (resp., < 0), and if $x > c$, then $f'(x) < 0$ (resp., > 0). In other words, the derivative switches signs exactly once in the interior of I.

In this case, the First Derivative Test, Theorem 1.5.20, tells us that, if the sign of f' switches from positive to negative at $x = c$, then f attains a global maximum value at c; if the sign of f' switches from negative to positive at $x = c$, then f attains a global minimum value at c.

In a similar fashion, the Second Derivative Test, Theorem 1.6.9, applies if f has a single critical point c in the interior of I, where $f'(c) = 0$ and $f''(c) \neq 0$.

Example 3.3.4. Find the maximum value of $y = f(x) = xe^{-x}$ on the interval $(-\infty, \infty)$, if a maximum value is actually attained.

Solution:

We already looked at this function back in Example 3.2.2, and found

$$y' = xe^{-x}(-1) + e^{-x} = (-x + 1)e^{-x}.$$

As we discussed before, $y' > 0$ when $x < 1$, and $y' < 0$ when $x > 1$. Thus, the First Derivative Test tells us that f attains a global maximum value at $x = 1$, and that value is $f(1) = 1 \cdot e^{-1} = 1/e$.

Alternatively, as f has the single critical point $x = 1$ where $y' = 0$, we could have used the Second Derivative Test. In Example 3.2.2, we found

$$y'' = (x - 2)e^{-x}.$$

Thus, $y''(1) = -1 \cdot e^{-1} < 0$, and we again conclude that f attains a global maximum value at $x = 1$.

Optimization Case 3: We are told, or physical reasons imply, that an extreme value is attained, and there is only one critical point.

If we know that an extreme value is attained, then Theorem 1.5.4 tells us that it must occur at a critical point; if there's only one critical point, then that must be where the extreme value is attained. This critical point could be at an endpoint of a closed, or half-open, interval, or could be at an interior point of the interval.

Example 3.3.5. The function $y = e^{-x} \sec x$ attains a maximum value on the interval $(\pi/2, 3\pi/2)$. Determine that maximum value.

Solution:

We looked at this function in Example 3.2.3, where we found that

$$y' = e^{-x} \sec x (\tan x - 1).$$

The only critical point of y, on the given interval, occurs where $\tan x - 1 = 0$, which is at $x = \pi/4 + \pi = 5\pi/4$. Thus, the only possible place where y could attain a maximum value, on the given interval, is at $5\pi/4$, and we are told that, in fact, y does attain a maximum value. Therefore, the maximum value must occur at $x = 5\pi/4$, and that value is

$$y(5\pi/4) = e^{-5\pi/4} \sec(5\pi/4) = -\sqrt{2} e^{-5\pi/4}.$$

Now, let's look at a few more-interesting word/modeling problems.

Example 3.3.6. A tool company is producing wrenches. The initial set-up to produce the wrenches costs the company $10,000$ (the *fixed cost*), and then it costs them 5 (the *variable cost* per wrench) for each wrench that they produce. A market study indicates that, if they charge a price of 5 for the wrench (which, of course, would be a bad idea for the company), then they will sell $200,000$ wrenches, but, for every 10 that they raise the price, only half as many people will buy a wrench.

What price should the company charge for their wrenches in order to maximize their profit?

Solution:

Let $N = N(p)$ denote the number of wrenches sold (the demand) when the price per wrench is p dollars. We need to find a formula for $N(p)$.

The data that we are given about N is just like the data we discussed for radioactive decay in Example 2.6.8 or drug decay in a person in Example 2.6.9; we are given a "half-life" or, perhaps, here, it should be call a "halving-price": a fixed change in the price that halves the number of purchasers.

As in Example 2.6.9, $N(p)$ is easiest to describe using base $1/2$. We are given "initial" data that $N = 200,000$, when $p = 5$, i.e., when $p - 5 = 0$. Every time that p goes up by \$10, $p - 5$ goes up by \$10. Thus, $N = N(p)$ is a function whose value is $200,000$ when $p - 5 = 0$, and which gets multiplied by another $1/2$ every time $p - 5$ goes up by \$10. Thus,

$$N = 200,000 \cdot \left(\frac{1}{2}\right)^{\frac{p-5}{10}} = 200,000\sqrt{2} \cdot 2^{-p/10}.$$

The total revenue, R, the total money taken in from wrench sales, is the price per wrench times the number of wrenches sold, i.e., $N \cdot p$. Therefore, in dollars,

$$R = N \cdot p = 200,000\sqrt{2} \cdot 2^{-p/10}p.$$

The total cost C, in dollars, of producing N wrenches, is the fixed cost plus \$5 per wrench, i.e.,

$$C = 10,000 + 5N = 10,000 + 1,000,000\sqrt{2} \cdot 2^{-p/10}.$$

The total profit, T, in dollars, that the company makes from wrench sales, when they charge p dollars, is

$$T = R - C = 200,000\sqrt{2} \cdot 2^{-p/10}p - 10,000 - 1,000,000\sqrt{2} \cdot 2^{-p/10} =$$

$$(200,000\sqrt{2} \cdot 2^{-p/10})(p - 5) - 10,000.$$

The price p could be any real number, even a negative one (the company **could** pay you to take a wrench); of course, you intuitively know that the maximum profit, if there is one, must be for some $p \geq 5$.

The function T is differentiable everywhere, and its derivative, the marginal profit per dollar of the price per wrench, is:

$$\frac{dT}{dp} = (200,000\sqrt{2})\left[2^{-p/10} \cdot 1 + (p-5)2^{-p/10}(\ln 2)(-1/10)\right] =$$

$$(200,000\sqrt{2})2^{-p/10}\left[1 + (p-5)(\ln 2)(-1/10)\right],$$

which equals 0 for exactly one p value, when $1 + (p-5)(\ln 2)(-1/10) = 0$, i.e., when the price, in dollars, is

$$p = 5 + 10/\ln 2 \approx 19.43.$$

When $p > 5 + 10/\ln 2$, $dT/dp < 0$, and when $p < 5 + 10/\ln 2$, $dT/dp > 0$.

Therefore, the First Derivative Test tells us that the profit T attains a global maximum value when the price per wrench is $p = 5 + 10/\ln 2 \approx \$19.43$.

Example 3.3.7. A particle is moving in the xy-plane. The x- and y-coordinates, in feet, at time t seconds, are given by

$$x = (t-1)e^{-t} \qquad \text{and} \qquad y = te^{-t}.$$

For $0 \leq t \leq 2$, what is the closest that the particle comes to the origin, and what is the farthest that the particle gets from the origin?

Solution:

The distance D of a point (x,y) from the origin is $D = \sqrt{x^2 + y^2}$. Thus, the distance, in feet, between the particle and the origin at time t seconds is

$$D = \sqrt{\left((t-1)e^{-t}\right)^2 + \left(te^{-t}\right)^2} = \sqrt{(t-1)^2 e^{-2t} + t^2 e^{-2t}} =$$

$$e^{-t}\sqrt{t^2 - 2t + 1 + t^2} = e^{-t}\sqrt{2t^2 - 2t + 1}.$$

This continuous function does, in fact, attain both a maximum and minimum value on the closed interval $[0, 2]$. However, because the square root in the formula for D will lead to a slightly ugly derivative, we will be slightly tricky: instead of finding where the distance from the origin attains a maximum and a minimum value, we will find where the distance **squared** from the origin attains a maximum and minimum value. Intuitively, the extreme values of D and of D^2 should occur at the same values of t; mathematically, this is correct because the range of D is contained in the interval $[0, \infty)$ and the squaring function is strictly increasing on $[0, \infty)$.

So, let

$$E = D^2 = e^{-2t}(2t^2 - 2t + 1).$$

We calculate

$$E' = e^{-2t}(4t - 2) + (2t^2 - 2t + 1)e^{-2t}(-2) = e^{-2t}(-4t^2 + 8t - 4) =$$

$$-4e^{-2t}(t^2 - 2t + 1) = -4e^{-2t}(t - 1)^2.$$

Thus, the only critical point of E in the interval $(0, 2)$ is at $t = 1$, and so we make the following table to look for maximum and minimum values of E on $[0, 2]$:

t	0	1	2
E	1	e^{-2}	$5 \cdot e^{-4}$

Thus, on the interval $[0, 2]$, E is largest when $t = 0$ and is smallest when $t = 5 \cdot e^{-4}$. As D is the square root of E, we find that, for $0 \leq t \leq 2$ seconds, the particle is farthest from the origin at $t = 0$ seconds, when it is 1 foot from the origin; the particle is closest to the origin at $t = 2$ seconds, when it is

$$\sqrt{5e^{-4}} = \sqrt{5}\,e^{-2} \approx 0.30261889307 \text{ feet}$$

from the origin.

In fact, we can say even more in this example. Note that E' is always ≤ 0, and equals 0 at exactly one point. It follows that E is strictly decreasing and, therefore, so is D.

Example 3.3.8. Suppose that we need to construct an aluminum can that holds 500 cubic centimeters (0.5 liters).

The can is required to have the shape of a right circular cylinder. We assume that the aluminum has some uniform small thickness, so that it is reasonable to describe the amount

Figure 3.19: An aluminum can.

of aluminum using area units. Suppose that the cost per square centimeter of aluminum for the sides of the can is some constant $c > 0$ dollars, while the top and bottom of the can use aluminum that costs $2c$ dollars per square centimeter. Find the dimensions (radius and height) of the can that minimizes the cost.

Solution:

Let r denote the radius of the can, and h the height, both in centimeters. The volume of the can is the area of the base times the height, i.e., $\pi r^2 h$; this is required to equal 500. Therefore, we have

$$\pi r^2 h \;=\; 500.$$

This type of equation is known as a *constraint*, because it implies that r and h may not vary independently; the allowable (r, h) pairs are *constrained* by the equation.

We are trying to minimize T, the total cost of the can in dollars. The cost of the sides of the can is the area of the sides times the cost per unit area, i.e., $2\pi r h \cdot c$. The cost of the top and bottom of the can is the area of the top and bottom times the cost per unit area, i.e., $2\pi r^2 \cdot 2c$.

Thus, the total cost of the can is

$$T \;=\; 2c\pi r h + 4c\pi r^2.$$

We need a function of one variable to apply our methods; that is, we need to write T as a function of either r or h, but not both. As we mentioned above, r and h are not allowed to vary independently. So, at this point, we solve the constraint equation for h, and write T as a function of the single variable r. We find

$$h \;=\; \frac{500}{\pi r^2}$$

and
$$T = 2c\pi r \cdot \frac{500}{\pi r^2} + 4c\pi r^2 = \frac{1000c}{r} + 4c\pi r^2 = 4c\left(250r^{-1} + \pi r^2\right).$$
The domain of this function is all $r > 0$.

We calculate the derivative

$$\frac{dT}{dr} = 4c(-250r^{-2} + 2\pi r),$$

and see that it equals 0 when $-250r^{-2} + 2\pi r = 0$, i.e., when $2\pi r = 250r^{-2}$; this means that $r^3 = 125/\pi$, i.e., $r = 5/\sqrt[3]{\pi}$.

Hence, T has a single critical point at $r = 5/\sqrt[3]{\pi}$ on the interval $(0, \infty)$. The 2nd derivative of T is
$$\frac{d^2T}{dr^2} = 4c(500r^{-3} + 2\pi),$$
which is positive for $r > 0$. Thus, the Second Derivative Test tells us that T attains a global minimum value at $r = 5/\sqrt[3]{\pi}$. To find the corresponding height, we go back to

$$h = \frac{500}{\pi r^2} = \frac{500}{\pi(5/\sqrt[3]{\pi})^2} = \frac{20}{\sqrt[3]{\pi}} = 4r.$$

Finally, we conclude that the dimensions of the can that minimize the total cost of the can are $r = 5/\sqrt[3]{\pi}$ centimeters $h = 20/\sqrt[3]{\pi}$ centimeters, which means that the height of the can should be twice the diameter of the can.

3.3.1 Exercises

In Exercises 1 through 5, use the technique outlined in "Optimization Case 1" to find the global extreme values of the function on the interval.

1. $g(x) = 3x^4 + 4x^3 - 12x^2$, $I = [-3, 2]$.

2. $h(x) = x^3 - 6x^2 - 9x + 8$, $I = [0, 4]$.

3. $r(\theta) = \sin 2\theta + \cos \theta$, $I = \left[-\pi, \frac{3\pi}{4}\right]$.

4. $m(t) = \frac{t^2 + 1}{t + 1}$, $I = \left[-\frac{1}{2}, 3\right]$.

5. $g(p) = \sec p$, $I = \left[-\frac{1}{2}, \frac{3}{2}\right]$.

6. Consider the function defined below. $f(x) = \begin{cases} x \sin \frac{1}{x} & x \neq 0 \\ 0 & x = 0. \end{cases}$

   a. Show that f is continuous on $[-1, 1]$.

   b. Identify the critical points of f. Note that, contrary to the hypothesis preceding Theorem 3.3.1, f has infinitely many critical points.

   c. Identify the global extreme values of f.

7. Find a pair of real numbers a, b, whose sum is 200, such that the product ab is as large as possible.

8. Find a, b, the pair of positive real numbers with the maximum square of their sum, given that the sum of their squares must be 100.

9. A particle travels along the y-axis with position function $y(t) = te^{-t}$, for $0 \leq t \leq 10$.

   a. Find the velocity of the particle for $0 < t < 10$.

   b. At what point in time does the particle have the smallest y-coordinate? At what point in time does the particle have the largest y-coordinate?

   c. At what point in time is the particle moving most quickly downward? At what point is it moving most quickly upwards?

10. What is the greatest possible area of a triangular region with one vertex at the center of a circle of radius 1 and the other two vertices on the circle?

11. Why *can't* we find the greatest possible perimeter of a triangular region with one vertex at the center of a circle of radius 1 and the other two vertices on the circle?

Suppose a simple circuit contains two resistors with resistances R_1 and R_2 in ohms. The resistors are wired in parallel, which implies that the total resistance of the circuit is given by

$$R_{\text{total}} = \frac{1}{\frac{1}{R_1} + \frac{1}{R_2}}.$$

In Exercises 12 through 14, find the minimum total resistance of the circuit given the constraint. Note that in problems like this the domain of the function is implied by the physical reality of the situation. Specifically, $R_1 > 0$ and $R_2 > 0$ is implied if not explicitly stated. A constraint is used to eliminate either R_1 or R_2 from the problem, leaving you with a function of a single variable.

12. The sum of the resistances of the two resistors is 30 ohms.

13. The sum of the resistances of the two resistors is Ω ohms, $\Omega > 0$.

14. The resistance of one resistor is twice the resistance of the second resistor.

15. More generally, a parallel circuit consisting of n resistors has total resistance

$$R_{\text{total}} = \frac{1}{\frac{1}{R_1} + \frac{1}{R_2} + \cdots + \frac{1}{R_n}}$$

where R_k is the resistance of the kth resistor. Suppose that the sum of the resistances in three resistors is 120 ohms and that the resistance of one of the resistors is twice the resistance of one of the others. Calculate the minimum total resistance in the circuit.

16. Suppose we want to build a rectangular storage container with a volume of 12 cubic meters. Assume that the cost of materials for the base is \$12 per square meter, and the cost of materials for the sides is \$8 per square meter. The height of the box is three times the width of the base. What's the least amount of money we can spend to build such a container?

17. Suppose you wish to construct a right, circular, cylindrical cup (without a top) that will hold 1000 cubic centimeters of liquid. What dimensions should the cup have in order to minimize the total amount of material used to construct the cup? Assume that the cup is arbitrarily thin, so that it is the surface area that we wish to minimize. ⊙

18. As in the previous problem, suppose you wish to construct a right, circular, cylindrical cup with no top that will hold 1000 cubic centimeters of liquid. Assume that the base needs to be built out of higher quality material than the rest of the cup, perhaps because the cup will hold highly acidic liquid. Material for the base therefore costs \$8 per square centimeter while material for the rest of the cup costs \$6 per square centimeter. Find the dimensions of the cup that minimize the cost of materials.

19. A rectangle is inscribed in a circle of radius r. What is the maximum area of the rectangle?

20. Show that the conclusion in Example 3.3.8, that the height of the can is twice the diameter, is independent of the specified volume of the can, i.e., re-do the problem with the volume being some constant V_0, and show that the ratio of the height to the diameter of the minimal-cost can remains the same (though, of course, the individual dimensions change).

21. Suppose that 100 meters of fencing are used to build a rectangular enclosure against a very long wall (so that no fencing needs to be used along the wall).

    a. What is the largest area of the enclosure that can be constructed using the fence?

b. What are the dimensions of the maximum-area enclosure?

22. A farmer is building a rectangular enclosure for 4000 square yards of land. The enclosure is to be split lengthwise into three rectangular pens of equal area. The outer fencing for the enclosure costs $10 per yard and the inner partitions cost $5 per yard.

   a. Find the dimensions of the enclosure that would minimize the total cost.

   b. Find the dimensions of the enclosure that would minimize the total fencing used.

23. Suppose more generally that F feet of fencing is available to form three sides of rectangular enclosure against a wall. What is the maximum possible area of the yard in terms of F?

24. It can be shown that the amount of liquid needed to fill a sphere of radius r to a height $h \le 2r$ is given by the formula

$$V(h) = \pi \cdot \left(rh^2 - \frac{h^3}{3} \right).$$

   a. What is the rate of change of V with respect to h, when $h = \frac{1}{4}r$?

   b. For which h value(s) (for fixed radius) is $\dfrac{dV}{dh}$ at a maximum?

   c. For which h value(s) is this rate of change at a minimum?

25. The distance function generalizes to higher dimensions. For example, the distance between two point (x_0, y_0, z_0) and (x_1, y_1, z_1) is $\sqrt{(x_0 - x_1)^2 + (y_0 - y_1)^2 + (z_0 - z_1)^2}$. Suppose that the x, y and z coordinates of a particle in space at time t are given by the equations: $x(t) = \cos t$, $y(t) = \sin t$, $z(t) = t$. At what time is the particle closest to the point $(0, 0, 1)$. As usual, it's easier to work with the square of the distance function.

26. An isosceles triangle is inscribed in a circle of radius 2 as shown in the diagram below. Find the dimensions of the triangle with largest area inside the circle. (Hint: Let x denote half the length of the base, and show that the height of the triangle is then $2 + \sqrt{4 - x^2}$.)

27. A rectangle is inscribed in the upper half of the ellipse

$$\frac{x^2}{4} + y^2 = 1$$

as shown in the diagram. Find the dimensions of the rectangle with maximal area drawn in this fashion.

28. Suppose, more generally, that a rectangle is inscribed in the upper half of the ellipse

$$\frac{x^2}{a^2} + \frac{y^2}{b^2} = 1.$$

Find the dimensions of the rectangle with maximal area in terms of a and b.

29. Find the maximum perimeter of a rectangle that can be inscribed in the ellipse

$$\frac{x^2}{a^2} + \frac{y^2}{b^2} = 1.$$

In Exercises 30 through 32, find the shortest distance between the graph of the equation and the point. As noted in the text, you may find it easier to minimize the square of the distance between the two.

30. $y = 3x - 5$, $(0, 0)$.

31. $y = 9x^2$, $(0, 1)$.

32. $y = \sqrt{1 - x^2}$, $(20, 20)$.

33. Show that the point on the circle $x^2 + y^2 = 1$ closest to the point (p, q), where (p, q) is not the origin, lies on the line containing the origin and (p, q).

34. Suppose a box with an open top (a.k.a., no top) and a square base is to be built and 2000 square centimeters of cardboard is available. Find the maximum volume of the box under these constraints.

35. a. Show that the maximum product of two numbers x and y subject to the constraints (i) $x + y = 1$ and (ii) $0 \leq x \leq 1$ is 1/4.

    b. Let p and q be positive numbers, and apply part (a) to $p/(p + q)$ and $q/(p + q)$ to conclude that
$$\sqrt{pq} \leq \frac{1}{2}(p + q).$$
This shows the arithmetic mean of two positive numbers is always greater than or equal to the geometric mean. This can also be deduced from elementary algebra.

36. A trapezoid is inscribed in the upper half of a circle of radius 1 in the following way: the points at an angle of θ and $\pi - \theta$ are marked and connected to the points $(1, 0)$ and $(-1, 0)$. Find the area of the trapezoid of largest area arising in this way.

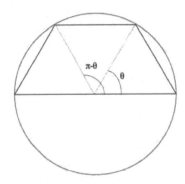

37. Extend Theorem 3.3.1 by showing that, if f is continuous on the domain $D = I_1 \cup I_2 \cup ... \cup I_n$ (the union of the intervals), where $I_k = [a_k, b_k]$, then f attains a global maximum and minimum on D. Furthermore, the extreme values are attained at critical points of f.

38. Show by example that the conclusion of Theorem 3.3.1 need not hold if the domain is the union of infinitely many closed and bounded intervals.

39. Suppose that to make n sunglasses, it costs a company $f(n)$ dollars, where $f(n) = n^3 + 2n + 2000$. ▶

a. What are the fixed costs associated with production of the sunglasses, i.e., what are the costs that the company incurs for setting up to produce the sunglasses, even if they don't actually produce any?

b. Define $A(n)$ to be the function giving the average cost per pair of sunglasses to produce n sunglasses. Find an expression for $A(n)$, and determine the interval on which it should be defined.

c. How many sunglasses should be made to minimize the average cost per pair of sunglasses? What is the minimum average cost?

40. A certain company makes the packaging for pre-packaged ice cream cones. The packaging is in the shape of a right circular cone, and is made from paper on the sides, with an aluminum lid. Each cone is supposed to hold 500 cubic centimeters of ice cream.

Recall that the volume of a cone is 1/3 of the area of the base times the height. The area of the conical sides is 1/2 of the circumference of the base times the slanted length from the vertex to the edge of the base.

a. If special lactose-proof paper costs 0.25 dollars per square centimeter, and aluminum costs 0.40 dollars per square centimeter, what dimensions of the conical container would minimize the cost of production?

b. What is the minimum cost to produce these packages?

41. Suppose that the marketing division of a hat company conducts a study that suggests that the demand for a certain hat can be modeled by the function $D(p) = 10000e^{-2p}$, where p is in dollars and $D(p)$ is an approximation of the number of hats that will be sold at p dollars per hat.

a. What function describes $R(p)$, the net revenue (total amount earned), if the price of each hat is p dollars?

b. What price would maximize revenue, and what would be the revenue at this price?

c. Data from the finance division says that the start-up cost for the business is $5000, and that each hat costs $2.00 to produce. Find a function relating cost to the company and price charged (assume that the company will produce exactly as many hats as it would be able to sell at the price of p dollars).

d. Profit is defined to be the difference between revenue and cost, i.e, it is the income after paying for the goods that are produced and the start-up costs. Find $P(p)$ profit as a function of price charged. What price would maximize profit?

42. A fire hydrant is to be built next to a street. The nearest junction in the water line lies 30 feet across the street and down the street another 40 feet as in the diagram. (We assume

that the street extends arbitrarily far off the top and bottom of the diagram.)

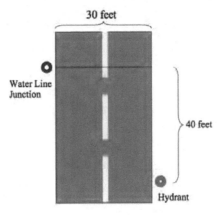

a. Find the three (reasonable) *extreme* piecewise linear paths for the pipeline, i.e., the longest pipelines and shortest pipelines within reason.

b. For materials and labor, it costs $40 per foot for pipe buried under the street, and $20 per foot for pipe buried under the ground. Find the path of the pipeline that would minimize the cost for the operation, assuming that the pipe first goes straight from the hydrant under the ground (up the page, just to the right of the street, in the diagram), and then cuts straight under the street to the junction.

c. Without doing any Calculus, explain why you would get the same answer, as what you got in part (b), if you assumed that the pipe first cuts straight across the street to the other side, and then goes straight under the ground to the junction.

d. Calculate the minimum total cost of the pipeline, using your answers to (a), (b), and (c).

43. A ship that carries supplies from a factory to an offshore drilling site has an inconvenient route due to poor factory design. The ship must first sail 50 miles up a man-made waterway, in calm waters, then set to sea, with much more rough waters, to finish the 50 mile trip to the drilling site as shown in the diagram. The planning division decides that in order to economize, it would be wise to dig a short canal in the embankment on the

seaward side of the channel, so that the ship can reach the open waters sooner.

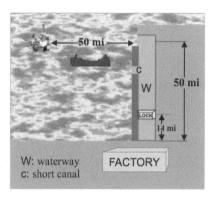

a. Assuming that it costs $400 per mile to travel in the waterway and $800 per mile to travel on the open sea, find the optimal location for the canal.

b. What is the cost for each round trip following this route?

c. The owner of the waterway, realizing that he will lose out on use-of-waters revenues in the long run, decides to start charging $100 for use of the lock that he owns that is 14 miles from the factory. Does it still behoove the shipping company to dig the canal in the same place? Explain.

Suppose that the domain of f is an open interval I. We say that f is an *open mapping* if for every open interval J contained in I, $f(J)$ is also open. For example, Corollary 2.3.8 implies that a differentiable function, on an open interval, whose derivative is never zero, is an open mapping. Use this definition in answering the next two problems.

44. Say whether or not the functions below are open mappings. Assume $I = (-\infty, \infty)$.

    a. $f(x) = c$.

    b. $f(x) = x$.

    c. $f(x) = x^2$.

    d. $f(x) = \sin x$.

45. Show that, if f is an open mapping, then f does not attain local maximum or minimum values.

3.4   Linear Approximation, Differentials, and Newton's Method

In this section, we come full circle. At the beginning of the book, in Section 1.1 and Section 1.4, we introduced the derivative, the instantaneous rate of change, by first looking at average rates of change, and then saying that, when the change in the independent variable was "small", this average rate of change of change should be a reasonable approximation of the instantaneous rate of change. Now that we have some many nice, quick formulas for calculating derivatives, we can use derivatives to approximate average rates of change, instead of the other way around.

Suppose that we have $y = f(x)$ and f is differentiable at $x = a$. Then,

$$f'(a) \ = \ \lim_{x \to a} \frac{f(x) - f(a)}{x - a}. \tag{3.4}$$

Therefore, when x is "close" to a, but unequal to a, $f'(a)$ is approximately equal to $(f(x) - f(a))/(x - a)$; multiplying by $x - a$, we obtain that $f(x) - f(a)$ is approximately $f'(a)(x - a)$. Adding $f(a)$ to each side of this approximation, we find that, if x is close to a, then

$$f(x) \ \approx \ f(a) + f'(a)(x - a). \tag{3.5}$$

Note that, in the above approximation, we do not need to disallow that $x = a$; when $x = a$, the two sides of the approximation are equal, namely $f(a)$. Also note that, since a is a fixed constant, the right-hand side of the approximation is a *linear function*, i.e., it is of the form $mx + b$, where $m = f'(a)$ and $b = f(a) - af'(a)$.

Thus, when x is close to a, the conceivably very complicated function $f(x)$ can be approximated by a simple function, a linear function. Thus, we make the following definition.

Definition 3.4.1. *Given a function f and a point a at which f is differentiable, we define the* **linearization of f at a,** *$L_f(x; a)$, to be the linear function of x given by*

$$L_f(x; a) \ = \ f(a) + f'(a)(x - a).$$

If f and a are clearly specified, we usually simply write $L(x)$ for the linearization of f at a.

Note that the graph of the linearization of f at a is a line with slope $f'(a)$ and which contains the point $(a, f(a))$. Therefore, :

Proposition 3.4.2. *The graph of the linearization of f at a is the tangent line to the graph of f where $x = a$.*

Definition 3.4.3. *Given a function f and a point a at which f is differentiable, the approximation of Formula 3.5,*

$$f(x) \approx f(a) + f'(a)(x - a) = L_f(x; a),$$

when x is close to a, is called the **linear approximation of f at a** *or the* **tangent line approximation of f at a.**

Example 3.4.4. Use linear approximation to approximate the function $\ln x$, when x is close to 1. What approximation do you obtain for $\ln(0.9)$? How does this compare with the result for $\ln(0.9)$ from your calculator or computer? Compare the graph of $y = \ln x$ with the graph of $y = L(x)$, when x is close to 1.

Solution:

The linearization of $\ln x$ at 1, $L(x) = L_{\ln x}(x; 1)$, is

$$L(x) = \ln(1) + \ln'(1) \cdot (x - 1) = 0 + (1/1)(x - 1) = x - 1.$$

Thus, linear approximation tells us that $\ln(0.9) \approx 0.9 - 1 = -0.1$. From a calculator, we obtain $\ln(0.9) \approx -0.10536051566$; so, -0.1 seems like a reasonable approximation.

The graphs of $y = \ln x$ and $y = x - 1$ are shown together in Figure 3.20. In Figure 3.21, we "zoomed in" on what happens near $(1, 0)$ by graphing in a window where the x's were closer to 1, and the y's were closer to 0, than in Figure 3.20.

Notice that, when you zoom in, the graph of $y = \ln x$ looks straighter and, thus, harder to distinguish from the tangent line.

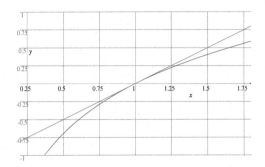   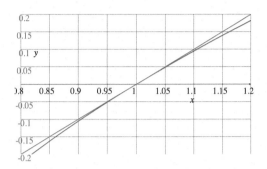

Figure 3.20: $y = \ln x$ in blue. $y = x - 1$ in red.   Figure 3.21: $y = \ln x$ in blue. $y = x - 1$ in red.

Example 3.4.5. Let $f(x) = (1 + x)^p$, where p is a constant. Find the linearization of f at 0.

Solution: Using the Power Rule, and the Chain Rule, we find

$$f'(x) \;=\; p(1 + x)^{p-1}.$$

Therefore, $L(x) = f(0) + f'(0) \cdot (x - 0) = 1 + px$.

Hence, linear approximation tells us that, for x close to 0,

$$(1 + x)^p \;\approx\; 1 + px.$$

For instance, $\sqrt{0.99} = (1 - 0.01)^{1/2} \approx 1 + (1/2)(-0.01) = 0.995$. (The calculator result is that $\sqrt{0.99} \approx 0.99498743711$.)

Remark 3.4.6. You may be saying to yourself something along the lines of "Yes, $\sqrt{0.99}$ is approximately 0.995. It's also approximately 3451; this second approximation is just a lot worse".

Moreover, if you don't know the exact value of a quantity, and you want to approximate it, you can't know how good your approximation is: if you knew exactly how accurate your approximation was, then you'd know the exact value of the quantity in question.

So, what does it mean to write something like "when x is close to a, $f(x)$ is approximately equal to $L_f(x; a)$"?

To answer this, it will help to first make another definition.

Definition 3.4.7. *Let*
$$R_f(x; a) = f(x) - L_f(x; a).$$
*We call $R_f(x; a)$ the **remainder** when approximating f by $L_f(x; a)$.*

Now, saying that "when x is close to a, $f(x)$ is approximately equal to $L_f(x; a)$" is the same as saying "when x is close to a, $R_f(x; a)$ is close to 0". So, one rigorous mathematical thing that we could mean by "when x is close to a, $f(x)$ is approximately equal to $L_f(x; a)$" is that $\lim_{x \to a} R_f(x; a) = 0$. Is this true?

In fact, more than this is true. As we showed in Lemma 1.A.23, there exists a function $E_f(x; a)$, defined on some open interval I around a, such that $E_f(x; a)$ is continuous at a, $E_f(a; a) = 0$, and such that, for all $x \in I$,

$$f(x) = f(a) + f'(a)(x - a) + E_f(x; a)(x - a).$$

Therefore,

$$R_f(x; a) \;=\; f(x) - L_f(x; a) \;=\; E_f(x; a)(x - a),$$

and so we see that, as $x \to a$, $R_f(x; a)$ approaches 0 in a two-fold sense; first, the factor $E_f(x; a)$ approaches 0 and, second, the factor $x - a$ approaches 0.

If f possesses a second derivative near a, it is possible to give an interesting characterization of the remainder above in terms of the second derivative.

Theorem 3.4.8. *Suppose that f is second-differentiable on an open interval I around a point a. Then, for all $x \neq a$ in I, there exists a c in the open interval between x and a (i.e., in (x, a) if $x < a$, or in (a, x) if $a < x$) such that*

$$f(x) \;=\; f(a) + f'(a)(x - a) + \frac{f''(c)}{2}(x - a)^2,$$

i.e., such that $R_f(x; a) = (f''(c)/2)(x - a)^2$.

We defer the proof of this result until Theorem 3.A.1 in Section 3.A (where we actually prove a much more general result), but note here that the theorem looks like some sort of Mean Value Theorem, which is, in fact, at the heart of the proof.

Let's look at an example of how Theorem 3.4.8 is used in practice.

Example 3.4.9. Suppose that we use linear approximation to approximate $\sqrt{1+x}$, for $-0.1 \leq x \leq 0.1$. Can we put some upper bound on how bad this approximation can be, i.e., can we come up with a "small" value $M > 0$ such that the difference between $\sqrt{1+x}$ and the approximation has absolute value less than or equal to M?

Solution:

As we saw in Example 3.4.5, the linear approximation of the function $f(x) = (1+x)^{1/2}$ at 0 is $L(x) = 1 + (1/2)x$. We want to know how big

$$\left| (1+x)^{1/2} - \left(1 + \frac{1}{2}x \right) \right| \; = \; |R_f(x; 0)|$$

can be, given that $|x| \leq 0.1$.

To apply Theorem 3.4.8, we need f''. We easily find that

$$f''(x) = -\frac{1}{4}(1+x)^{-3/2}.$$

Now, for any x in the interval under consideration, any value of c in Theorem 3.4.8 is between x and 0, and so certainly $-0.1 < c < 0.1$. In addition, the derivative of f'' is positive, which tells us that f'' is a strictly increasing function; hence,

$$f''(-0.1) < f''(c) < f''(0.1) < 0.$$

This tells us that, for all x under consideration, any corresponding c from Theorem 3.4.8 satisfies

$$|f''(c)| \; < \; |f(-0.1)| = \frac{1}{4(0.99)^{3/2}},$$

and so

$$|R_f(x;0)| = \left| \frac{f''(c)}{2} x^2 \right| < \frac{1}{8(0.99)^{3/2}} (0.1)^2.$$

This tells us that, if $|x| \leq 0.1$, and we use $1 + (1/2)x$ to approximate $\sqrt{1+x}$, then our approximation will be within $\pm \frac{1}{8(0.99)^{3/2}} (0.1)^2$ of the exact value.

If you object to this as answer, saying that "we're approximating square roots, so how do we know $(0.99)^{3/2}$?", there are at least two responses to this.

First, we could use a worse (bigger) value for the error, but one that we can calculate by hand. For instance, $0.99 > 0.81 = (0.9)^2$, and raising to the $3/2$ power and taking reciprocals reverses the inequality sign to yield:

$$|R_f(x;0)| < \frac{1}{8(0.99)^{3/2}} (0.1)^2 < \frac{1}{8((0.9)^2)^{3/2}} (0.1)^2 = \frac{10}{8 \cdot 9^3}.$$

You can easily check **by hand** that this last quantity is less than 0.002. So, without using a calculator, we can say that the error in the approximation is less than 0.002.

Second, depending on the application, we may not mind using a calculator **once** to find that $\frac{1}{8(0.99)^{3/2}} (0.1)^2 < 0.001269$. This would be reasonable if we don't actually have just a few x values for which we want to calculate $\sqrt{1+x}$, but rather have $\sqrt{1+x}$ in a complicated formula, where we are happy to simplify the formula by replacing $\sqrt{1+x}$ with $1 + (1/2)x$, even though it only yields an approximation, once we know that the approximation is pretty good. Another sample application, where using a calculator once would be reasonable, would be if we record somewhere, once and for all, that $1 + (1/2)x$ approximates $\sqrt{1+x}$ to within 0.001269, as long as $|x| < 0.1$. Then, we use this approximation in the future, without ever getting out a calculator again.

Now we wish to look at linear approximation in a different way. Suppose we have $y = f(x)$, and f is differentiable at a. Then, Formula 3.4 says that

$$f'(a) = \lim_{x \to a} \frac{f(x) - f(a)}{x - a} = \lim_{\Delta x \to 0} \frac{\Delta y}{\Delta x}.$$

Thus, another way to write linear approximation is

$$\Delta y \approx f'(a)\Delta x,$$

provided that Δx is close to 0. If we treat Δx and Δy as new variables (whose names use two symbols each), then there is no reference to just the variable x in the inequality above and, instead of writing $f'(a)$, it is common to write $f'(x)$ at an arbitrary x. Hence, we have

$$\Delta y \approx f'(x)\Delta x \quad \text{or} \quad \Delta y \approx \frac{dy}{dx}\Delta x,$$

where we expect this approximation to be better when Δx is closer to 0. This leads us to think intuitively, but non-rigorously, that if Δx were an **infinitesimally small** change in x, then $\frac{dy}{dx}\Delta x$ would be **exactly** the corresponding infinitesimal change in y.

This leads us to introduce a new variable dx, which we think of as ideally representing an infinitesimal change in x. We think of dx as the denominator of dy/dx, even though dy/dx **is not a fraction**. To make things even more confusing, we write dy for a new variable, whose value is precisely $(dy/dx)dx$.

Understand what's going on. Given a function $y = f(x)$, at any particular x value, dy/dx is a number which is **not** defined as a quotient. However, we can certainly say, given any real number u, we'll use v to denote the $(dy/dx) \cdot u$. If we do this, and $u \neq 0$, then, of course, $v/u = dy/dx$, where the left-hand expression indicates the quotient of two numbers, while the right-hand expression does not. What we're doing is using the variables dx and dy in place of the variables u and v, respectively.

Using dx and dy in this manner, we obtain the seemingly obvious equation

$$dy = \frac{dy}{dx}\,dx,$$

which looks like we simply cancelled the dx's, but we did not. The dx in the derivative is part of the differentiation symbol, and does **not** denote a number; the dx on the right is an independent variable, which **does** denote a number. The equation above really just gives the definition of dy.

As a quick example, suppose that $y = x^2$. Then, $dy/dx = 2x$, and we write $dy = 2x\,dx$, where x and dx can be any real numbers, and dy is the real number that results from the multiplication. So, for instance, when $x = 3$ and $dx = 7$, we obtain $dy = (2 \cdot 3) \cdot 7 = 42$.

However, the example above does not give you a feel for how the equation $dy = (dy/dx)dx$ is really used. As we discussed, linear approximation can be written in the form $\Delta y \approx (dy/dx)\Delta x$. Thus, linear approximation is equivalent to:

Theorem 3.4.10. *If $dx = \Delta x$, and dx is close to 0, then*

$$dy \;=\; \frac{dy}{dx}\,dx$$

is approximately equal to the corresponding Δy, i.e., $dy \approx \Delta y$.

Definition 3.4.11. *The notation above is referred to as* **differential notation**. *dx is the* **differential of** *x and dy is the* **differential of** *y.*

Given $y = f(x)$, to **calculate** *dy means to calculate dy/dx, and write $dy = (dy/dx)dx$.*

Differential approximation *refers to the approximation $dy \approx \Delta y$, when dx is a change in x which is close to 0.*

The quantities dx/x and dy/y are called the **relative differentials of** *x and y, respectively.*

Example 3.4.12. Back in Example 1.1.7, in Section 1.1, we looked at the change in the area of a wide-screen television, with respect to changes in the diagonal length.

We derived the formula that the area of the television screen A, in square inches, was related to the diagonal length, in inches, by $A = 144d^2/337$. Using this, we found that a fixed change in d produced a larger change in A when d was larger. However, we found that a fixed change in d produced a smaller **relative** change in A when d was larger.

We wish to consider this example again with our new notation and results. We have a mild notational problem; d is a variable name, and will also indicate a differential. So, let's change the name of the diagonal length from d to x, so that

$$A = 144x^2/337.$$

Now, we easily calculate

$$dA \;=\; \frac{288}{337}\,x\,dx,$$

and the relative differentials are related by

$$\frac{dA}{A} = \frac{(288/337)x\,dx}{(144/337)x^2} = 2 \cdot \frac{dx}{x}.$$

The two equations above tell us three different things:

1. that a fixed small change in x produces an approximate change in A that is $288x\,dx/337$, and so a fixed small change in the diagonal length of the television produces a bigger (approximate) change in the area for bigger televisions;

2. that the approximate relative change in the area of the television is 2 times the relative change in the diagonal length, for small changes in the diagonal length, **regardless of the size of the television**; and

3. that the approximate relative change in the area of the television is $2/x$ times the (non-relative) change in the diagonal length, for small changes in the diagonal length, and so a fixed small change in the diagonal length of the television produces a bigger (approximate) change in the area for **smaller** televisions.

Differential approximation is common in *error analysis*, which is best demonstrated by an example.

Example 3.4.13. The radius of a spherical container is measured to be 1 meter, with an error of at most ± 0.001 meters (1 millimeter). What is the differential approximation for the possible error in the calculation of the volume of the container?

More generally, if the radius is measured to be r meters, what does differential approximation tell us about the relationship between the relative change in the volume V, in cubic meters, and the relative change in r, for small changes in r?

Solution:

Though the question is phrased in terms of errors, you shouldn't be confused. In the first question, you are given that $|\Delta r| \leq 0.001$, and you are asked to approximate ΔV via the differential approximation. In the second question, you are asked to compare dV/V and dr/r, using differential approximation.

The volume is related to the radius by

$$V = \frac{4}{3}\pi r^3,$$

and, hence, we easily find

$$dV = 4\pi r^2\, dr.$$

Thus, when $r = 1$ and $|dr| \leq 0.001$, we find

$$|\Delta V| \approx |dV| = \left|4\pi (1)^2 dr\right| = 4\pi |dr| \leq 4\pi(0.001) \approx 0.013\ \text{m}^3;$$

thus, we approximate that the volume of the container is $4\pi(1)^3/3$ m^3, to within ± 0.013 m^3.

The relative differentials are related by

$$\frac{dV}{V} = \frac{4\pi r^2\, dr}{4\pi r^3/3} = 3 \cdot \frac{dr}{r}.$$

Hence, the relative error in calculating V is approximately 3 times the relative error in measuring r, for small amounts of error in the measurement of r.

Example 3.4.14. In business and economics, *marginal analysis*, the use of derivatives of cost, revenue, and profit functions, is very important. Much of this importance stems from an implicit use of the differential approximation, in which dy/dx is calculated and the differential $dx = \Delta x =$ a change of one unit in x; then,

$$\Delta y \approx dy = \frac{dy}{dx}\, dx$$

has the same numeric value as dy/dx, but now gives an approximation of the change in y, given a unit change in x.

For instance, suppose that a company produces x toasters, and their total cost $C = C(x)$, in dollars, for producing the x toasters (the fixed cost plus the variable cost) is

$$C(x) = 5000 + 0.01x^2 + 3x.$$

Then, the marginal cost per toaster, when the company is producing $x = 1000$ toasters, is dC/dx at $x = 1000$. We calculate $dC/dx = 0.02x + 3$, which equals 23 dollars per toaster, when $x = 1000$. This is, of course, an instantaneous rate of change.

However, it is usually interpreted as an approximation, the differential approximation with $dx = 1$, which says that, when the company is producing $x = 1000$ toasters, the cost of producing each new toaster is approximately

$$ dC \;=\; \frac{dC}{dx}\, dx \;=\; (0.02 \cdot 1000 + 3) \cdot 1 \;=\; \$23. $$

How good is this approximation? The actual cost of producing the 1001th toaster is

$$ C(1001) - C(1000) \;=\; \$23.01. $$

So, the differential approximation was off by 1 cent.

———

As our final topic related to linear approximation, we wish to show how it is used to find/approximate roots of functions, i.e., x values where $f(x) = 0$.

Suppose you have a differentiable (and, hence, continuous) function $y = f(x)$ and you wish to find an x value such that $f(x) = 0$. What do you do? Well...these days, you'd probably just ask your calculator to solve the equation for you. However, what if you had to approximate a root by hand or, possibly, what if you had to be the one to program the calculator or computer to find roots? Then what would/could you do?

What we are about to describe is called **Newton's Method** for approximating roots. It is an iterative process, that is, you start with an approximate root x_0, then use that to produce a (hopefully) better approximation x_1, then use x_1 to produce a "better" x_2, and so on. In general, you assume you have a approximation x_n and use that to produce what you hope is a better approximation x_{n+1}.

The steps of Newton's Method are as follows:

Step 1: Guess a root. That's not a joke. Guess some value x_0 that you hope is a root, or is close to a root. One way that you might make this initial choice of x_0 is to plug a few x values into f until you find one (call it a) where f is positive and one (call it b) where f is negative;

then, the Intermediate Value Theorem tells you that, somewhere in-between those two x values, there is a root of $f(x)$. In this case, you could take, as a reasonable guess, the midpoint $(a+b)/2$ as x_0.

As we will refer to x_n below (for our general iteration), set n equal to 0 at this point.

Step 2: Evaluate $f(x_n)$. If it equals 0, then great, x_n is a root of f. You're finished. If $f(x_n)$ does not equal 0, but is close enough to 0 for your purposes, then you're finished in that case too.

Step 3: Assuming that $f(x_n)$ is not close enough to 0 for your purposes, if $f'(x_n) = 0$, then Newton's method cannot proceed. You need to go back to Step 1 and pick a new x_0.

Step 4: Assuming that $f(x_n)$ is not close enough to 0 for your purposes, and that $f'(x_n) \neq 0$, you proceed as follows.

For x values close to x_n, $f(x)$ is approximately equal to its linearization $L_f(x; x_n)$ at x_n. Even though x_n is not close enough to a root of f for you to stop the process, you hope that x_n is fairly close to a root of f, close enough so that you ought to be able to better approximate such a root of f, one that's fairly close to x_n, by finding instead a root of $L_f(x; x_n)$. Let x_{n+1} be the root of $L_f(x; x_n)$.

That is, you let x_{n+1} be the x value such that

$$0 = L_f(x; x_n) = f(x_n) + f'(x_n)(x - x_n),$$

i.e., let

$$x_{n+1} = x_n - \frac{f(x_n)}{f'(x_n)}.$$

Step 5: Let the value of $n + 1$ now take the place of the old n, and go back to Step 2.

That's all there is to Newton's Method, and we'll run through an example below, but let us first mention that Newton's Method can fail; you can keep hitting places where $f'(x_n) = 0$ or the sequence x_0, x_1, x_2, ... may not converge to a root of f. See the exercises at the end of this section for examples. Nonetheless, Newton's Method works well in many, many cases, so it's worth knowing.

Example 3.4.15. Let's see how to use Newton's method to approximate square roots. We will use mainly fractions here, rather than decimals, in hopes of emphasizing that we can do this by hand, and because we don't want to insert extra error into our calculation by truncating decimal expansions. However, we will use **some** decimals, so that we can easily compare our results with those from Example 3.4.5.

Let $f(x) = x^2 - 0.99$. So, the roots of f are $\pm\sqrt{0.99}$. There are two obvious "nice" choices for approximate roots of f, namely ± 1. Let's start with $x_0 = 1$, and see what Newton's Method gives us, and compare it with what we obtained by linear approximation in Example 3.4.5. We'll decide ahead of time that we'll stop Newton's Method when we find an x_n such $|f(x_n)| < 0.00001 = 1/100,000$.

It is convenient to write the general iteration formula for Newton's Method, with $f(x) = x^2 - 0.99$:

$$x_{n+1} \;=\; x_n \;-\; \frac{f(x_n)}{f'(x_n)} \;=\; x_n \;-\; \frac{x_n^2 - 0.99}{2x_n}.$$

Now, we find that $f(x_0) = f(1) = 0.01 = 1/100$, which is not close enough to 0 for us. We note that $f'(x) = 2x$, and so $f'(x_0) = f'(1) = 2 \neq 0$; thus, we may proceed with Newton's Method. We calculate x_1:

$$x_1 \;=\; x_0 - \frac{f(x_0)}{f'(x_0)} \;=\; 1 - \frac{1/100}{2} \;=\; 199/200 \;=\; 0.995.$$

Note that this is precisely what we found in Example 3.4.5, where we used linear approximation to estimate $\sqrt{0.99}$. This is **not** surprising; Newton's Method is just an iterative form of linear approximation.

Now, we evaluate $f(x_1)$ and see if it's as close to 0 as we wanted. We find

$$f(x_1) \;=\; f\left(\frac{199}{200}\right) \;=\; \left(\frac{199}{200}\right)^2 - \frac{99}{100} \;=\; \frac{(199)^2 - 99 \cdot 400}{40,000} \;=\; \frac{1}{40,000},$$

which is not as close to 0 as we required. We also find that $f'(x_1) = 2x_1 = 199/100$. So, we continue with Newton's Method and find

$$x_2 \;=\; x_1 - \frac{f(x_1)}{f'(x_1)} \;=\; \frac{199}{200} - \frac{1/40,000}{199/100} \;=\; \frac{79,201}{79,600}.$$

We leave it for you to check that $f(x_2)$ is definitely as close to 0 as we specified; in fact, $|f(x_2)| < 1/1,000,000,000$.

Therefore, we arrive at the approximation

$$\sqrt{0.99} \approx \frac{79,201}{79,600}.$$

3.4.1 Exercises

For each of the functions in Exercises 1 through 5, find the linearization $L_f(x; a)$ of f at $x = a$.

1. $f(x) = 3x^4 - 2x^2$, when $a = -1$.

2. $f(x) = x^{\frac{5}{4}}$, when $a = 81$.

3. $f(x) = \ln(2x)$, when $a = \frac{1}{2}$.

4. $f(x) = \tan x$, when $a = \frac{\pi}{3}$

5. $f(x) = \sec^{-1} x$, when $a = -\sqrt{2}$.

For each of the functions in Exercises 6 through 10, find the differential dy, and then evaluate this differential for the given values of x and dx.

6. $y = e^{\frac{x}{4}}$, $x = 0$, $dx = 0.4$.

7. $y = \frac{1}{(x+1)^2}$, $x = 1$, $dx = -0.1$.

8. $y = \cos x$, $x = \frac{\pi}{4}$, $dx = 0.01$.

9. $y = \ln x$, $x = 2$, $dx = 0.02$.

10. $y = x^{\frac{4}{3}}$, $x = 27$, $dx = 0.025$.

In each of Exercises 11 through 15, use Newton's Method, with the specified x_0, to find x_2, the second approximation of a solution to the given equation.

11. $x^4 - x - 1 = 0$ and $x_0 = 1$.

12. $x^5 + 2 = 0$ and $x_0 = -1$.

13. $x^3 + 2x - 4 = 0$ and $x_0 = 1$.

14. $\sin x - \cos x = 0$ and $x_0 = 0$.

15. $e^x + 2x = 0$ and x_0.

In Exercises 16 through 18, use differentials, linearization, or Newton's Method to approximate the value.

16. $\sqrt[3]{345}$.

17. $\dfrac{1}{19.5}$.

18. $\sin(29°)$.

In Exercises 19 through 22, calculate $R_f(b; a)$.

19. $f(x) = \sqrt{x}$, $a = 144$, $b = 145$.

20. $f(x) = \sqrt[3]{x+1}$, $a = 27$, $b = 28$.

21. $f(x) = \tan x$, $a = \pi/4$, $b = 3/4$.

22. $f(x) = 2e^{3x}$, $a = 0$, $b = 0.5$.

23. Suppose that the total cost of producing n units of a product is given by $C(n) = 2\sqrt{3n} + 20$. Recall that the *average cost* is given by the function $AC(n) = \dfrac{C(n)}{n}$. Suppose that a monthly report shows 75 units were produced with a possible error of ± 1 due to pending sales.

    a. Use differentials to estimate the maximum error in the average cost function.

    b. What is the relative error of the average cost function to units produced?

24. The total dollars accumulated after n years, if P_0 dollars is initially invested at interest rate r and compounded annually, is $P(r, n) = P_0(1 + r)^n$. Suppose that $10,000 is invested for 5 years and the expected interest rate is $r = 5\%$ with an error of $\pm 0.5\%$. Use differentials to estimate the error term. What is the relative differential?

25. Recall that the mass $A(t)$, at time t, of a radioactive substance is given by the equation $A(t) = A_0 e^{-kt}$ where k is a constant and A_0 is the initial mass. Note that A can also be viewed as single variable function of k, if t is fixed. The estimate of k for uranium-238 is $k = 1.55 \cdot 10^{-10}$ with an estimated error of $\pm 10^{-12}$. Assume $A_0 = 1000$ kg and use differentials to estimate the error in the amount of uranium remaining after 3 billion years.

26. The *Poisson distribution* is often used to model the number of insurance claims a driver will experience in a year. The *probability mass function* of the distribution is

$$f(n, \lambda) = \frac{\lambda^n e^{-\lambda}}{n!}.$$

The value $f(n, \lambda)$ is the probability that a driver will file n claims, and λ is the expected number of claims per year. *Parameter error* refers to the possibility that λ is incorrectly specified. Suppose an insurer sets $\lambda = 2$, with a possible error of ± 0.1.

   a. Use differentials to estimate the error in the probability that $n = 3$.

   b. What is the relative error?

Recall that the height, h, of a projectile fired from an initial height h_0 with initial velocity v_0 is given by the equation

$$h(t) = -\frac{1}{2}gt^2 + v_0 t + h_0,$$

where $g = 9.8$ m/s^2 is the acceleration due to gravity. You will use this in Exercises 27 through 30.

27. Show that, if we consider h as a function of t, with all other variables fixed, then $dh = (-gt + v_0)\, dt$. On the other hand, show that, if we consider h as a function of g, with all other variables fixed, then $dh = (-t^2/2))\, dg$.

28. Suppose the error in calculating the acceleration due to gravity is within 0.1 m/s^2 and that a ball is dropped from a height of 40 meters ($v_0 = 0$, $h_0 = 40$). Use differentials to estimate the maximum error in determining the height of the ball after 2 seconds (by using the formula from the previous exercise). What is your estimate for the maximum relative error?

29. Suppose that $h_0 = 4$ meters, $v_0 = 0$ m/s and that the height is measured at $t = 3$ seconds. If the time measurement error (for example, a measurer's finger on a stopwatch) is within 0.1 seconds, what is the approximate maximum error in the height? Use differentials.

30. Suppose that $h_0 = 0$ and the time it takes for a projectile to hit the ground, 5 seconds, is used to estimate v_0. If the measurement error is within 0.1 seconds, what is the approximate maximum error of the initial velocity? The approximate maximum relative error?

31. Let $f(x) = x^3 + 1$ and $x_0 = 0$. Why can't Newton's Method be used to approximate a zero of f?

32. Let $f(x) = \sqrt[3]{x}$ and let $x_0 \neq 0$. Show that Newton's Method fails to converge to zero.

33. Let f be a differentiable function such that $f(1) = 1$, $f(-1) = -1$, $f'(1) = f'(-1) = 1/2$. Show that Newton's Method fails to converge when $x_0 = 1$.

34. A function h defined on the domain $[a, b]$ is a *contraction mapping* if:

    1. there exists a constant $k < 1$ such that, for all x and y in $[a, b]$,

    $$|h(x) - h(y)| \leq k|x - y|, \text{ and}$$

    2. the range of f is contained in $[a, b]$.

    Suppose that f is differentiable on an open interval which contains the closed interval $[a, b]$, and that f' is continuous on $[a, b]$. Assume that $f(a) < 0 < f(b)$, and that $f'(x) > 0$, for all $x \in [a, b]$. Let M be the maximum value of $f'(x)$ on $[a, b]$, and set $\phi(x) = x - f(x)/M$.

    a. Show that $0 \leq \phi'(x) < 1$. Conclude that ϕ' attains a maximum value k on $[a, b]$, and that $k < 1$.

    b. Use the Mean Value Theorem, Theorem 1.5.10, to show that ϕ satisfies the first contraction mapping condition.

    c. Show that ϕ satisfies the second contraction mapping condition. Hint: use the fact that ϕ is non-decreasing.

35. Referring to the previous exercise, the function $\phi(x) = x - f(x)/M$ defines a sequence $x_{n+1} = x_n - f(x_n)/M$. Let $f(x) = x^4 - x - 1$ and set $x_0 = 1$.

    a. Find x_1 and x_2 using Newton's Method. You may have done this already in Exercise 11.

    b. Find x_1 and x_2 using the ϕ-defined sequence above. Use the interval $[1, 2]$ in your calculation of M.

    c. Which sequence appears to converge more quickly?

36. Let $f(x) = x^5 - 2.25x^3 + 2.25x$ and let $x_0 = 1$. Show that Newton's Method fails to converge. Hint: show that $x_0 = x_2 = \dots$ and $x_1 = x_3 = \dots$

37. Let $f(x) = x^5 - 2.25x^3 + 2.25x$ and let $x_0 = 1$.

    a. Let $M = \max |f'(x)|$ where $x \in [a, b]$. Calculate M.

b. Calculate x_1 and x_2 using the algorithm $x_{n+1} = x_n - \dfrac{f(x_n)}{M}$.

c. Does the sequence appear to converge?

The Secant Line Method is an alternative to Newton's Method. Its disadvantage is it converges more slowly than Newton's Method. Its advantage is that no derivatives need to be calculated. Exercises 38 through 40 explore this method.

38.   a. Suppose a root of f is known to lie in the interval $[x_0, x_1]$. Let x_2 be the x-intercept of the secant line of f through the points $f(x_0)$ and $f(x_1)$. Show that

$$x_2 = x_1 - \frac{x_1 - x_0}{f(x_1) - f(x_0)} f(x_1).$$

    b. Given approximations $x_0, x_1, ...x_n$, let x_{n+1} be the x-intercept of the secant line through $f(x_n)$ and $f(x_{n-1})$. Show that

$$x_{n+1} = x_n - \frac{x_n - x_{n-1}}{f(x_n) - f(x_{n-1})} f(x_n).$$

39. Let $f(x) = x^3 - \cos(x)$, $x_0 = 0$ and $x_1 = 1$. Find x_2 and x_3 using the Secant Line Method.

40. Use the Secant Line Method to approximate $\sqrt[3]{50}$. Let $x_0 = 3$, $x_1 = 4$ and calculate x_2 and x_3.

41. Newton's Method is often used by computers to quickly calculate reciprocals. To calculate $1/a$, let $h(x) = \dfrac{1}{x} - a$. Show that $x_{n+1} = 2x_n - ax_n^2$.

In Exercises 42 through 44 use the method outlined in the previous problem to approximate $1/a$. Calculate x_1, x_2 and x_3.

42. $a = 8$, $x_0 = 1$.

43. $a = 0.23$, $x_0 = 5$.

44. $a = -0.1753$, $x_0 = -5$.

45. A thin rectangular piece of foil is wrapped tightly around the side of a solid right-cylindrical wooden peg (and not wrapped around the top or bottom).

a. By approximately how much does the surface area of the peg grow if the height is 8 centimeters, the radius is 3 centimeters, and the aluminum is approximately 0.0002 centimeters thick?

b. What is the relationship between the relative differential of the surface area and the relative differential of r?

c. If r grows by 0.025%n (and the height remains constant), by approximately what proportion does the surface area grow?

d. Answer (a)-(c) replacing "surface area" by "volume".

46. A sphere of radius r is painted, and so its radius increases by a small amount.

a. If the radius is 2 meters, and 0.0001 cubic meters of paint are evenly spread on the sphere, what is the differential approximation for the change in radius?

b. Using the answer from part (a), what is the approximation for the new radius?

c. What is the actual new radius?

d. For an arbitrary radius r, what is the relationship between the relative differential of V and the relative differential of r?

47. A cup in the shape of an upside-down right circular cone, of height 5 centimeters and base-radius 2 centimeters, is filled to the brim with water.

a. Use differential approximation to estimate the decrease in the height of water in the cup, if 0.02 cubic centimeters of water evaporate.

b. The owner of the cup accidentally pours some of the water on the floor so that the water no longer entirely fills the cup. If 0.5% of the water now evaporates, by approximately what proportion will the height of the water drop?

48. Consider the functions $f(x) = \sin x$ and $g(x) = x + 1$.

a. Graph both curves $y = f(x)$ and $y = g(x)$ on the same coordinate plane.

b. Pick an initial x_0, and use x_3 from Newton's Method to approximate the point at which $f(x) = g(x)$.

49. Use x_2 from Newton's Method to approximate the value of $\sqrt[3]{10}$, with an initial approximation $x_0 = 2$.

50. Use Newton's Method to approximate the unique critical point of $f(x) = \sin x + \frac{1}{2}x^2 + 1$.

51. Consider the function $f(x) = \sin x$.

a. Find the linearization of f at $\dfrac{\pi}{6}$, i.e., find $L_f\left(x; \frac{\pi}{6}\right)$.

    b. Use the linearization to approximate $f\left(\dfrac{\pi}{3}\right)$.

    c. What is $R_f\left(\dfrac{\pi}{3}; \dfrac{\pi}{6}\right)$?

    d. What is an upper bound on the error in the approximation on the interval $I = \left[0, \dfrac{\pi}{3}\right]$?

52. Find the linearization of $f(t) = 2e^{2t}$ at $t = 0$. What is an upper bound on the error in the linear approximation on the interval $-0.5 < t < 0.5$?

3.5   Indeterminate Forms and l'Hôpital's Rule

In this section, as in the previous one, we use our ability to find derivatives easily to help us calculate something involved in the definition of the derivative. In the previous section, that "something" was the average rate of change (which we calculated approximately); in this section, that "something" is certain kind of limits.

Consider the limit that defines the derivative:

$$f'(x) \;=\; \lim_{h \to 0} \frac{f(x+h) - f(x)}{h}.$$

If we simply "plug in" $h = 0$ into the numerator, we get $f(x) - f(x)$, which, of course, is 0, and, certainly if we let $h = 0$, the denominator is 0. In other words, if we simply let h be 0, we arrive at the undefined quantity $0/0$. In Definition 1.3.50, we referred to this as an *indeterminate form*. The limit, itself, may well be defined (if the function is differentiable at x), but what you obtain if you simply let h equal 0 is not defined.

There are other times when limits may exist (or, possibly, be extended real numbers), where, if you simply "plug in", or take separate limits of the numerator and denominator, or take separate limits in the factors in a product, you get one of the indeterminate forms in Definition 1.3.50. Examples are

$$\lim_{x \to \infty} \frac{x}{e^x}, \qquad \text{and} \qquad \lim_{x \to 0^+} x \ln x,$$

where, in the first limit, both the numerator and denominator approach ∞ as x approaches ∞ and, in the second limit, we obtain $0 \cdot -\infty$.

Of course, we could rewrite the second limit above as the limit of a quotient

$$\lim_{x \to 0^+} x \ln x \;=\; \lim_{x \to 0^+} \frac{\ln x}{1/x},$$

and, thus, see that this limit also involves a quotient that looks like $\pm\infty$ divided by $\pm\infty$, if we look at the numerator and denominator separately.

Such limits actually arise in many physical settings, particularly when you are interested in the long-term behavior of some physical quantity x, which depends on time t. For instance, if we look at a block, attached to a spring, sliding on a floor which produces friction, then, if there is relatively little friction (when the mass-spring system is said to be *underdamped*), we get a solution like that in Example 2.7.11.

However, if there is precisely the right amount of friction (when the mass-spring system is said to be *critically damped*), then the position of the block would be given by a function such as $x = x(t) = te^{-t}$. What happens to the block as time goes on, i.e., what is $\lim_{t \to \infty} x(t)$? Intuitively, the block should approach the equilibrium position $x = 0$. We need to be able to show this mathematically.

The following theorem is **very** useful for the evaluation of limits which yield indeterminate forms if you take separate limits of the numerator and denominator of a quotient.

Theorem 3.5.1. (l'Hôpital's Rule) *Suppose that* $-\infty \leq a < b \leq \infty$, *and that* $-\infty \leq L \leq \infty$. *Let* f *and* g *be differentiable on the interval* (a, b), *and suppose that, for all* x *in* (a, b), $g'(x) \neq 0$. *Let* c *be in* (a, b).

Suppose that either $\lim_{x \to c} f(x) = \pm\infty$ *and* $\lim_{x \to c} g(x) = \pm\infty$, *or that* $f(c) = g(c) = 0$. *Also, suppose that*

$$\lim_{x \to c} \frac{f'(x)}{g'(x)} = L.$$

Then,

$$\lim_{x \to c} \frac{f(x)}{g(x)} = \lim_{x \to c} \frac{f'(x)}{g'(x)} = L.$$

In addition, the theorem remains true if all of the limits $\lim_{x \to c}$ *above are replaced by* $\lim_{x \to a^+}$ *or* $\lim_{x \to b^-}$, *and the condition that* $f(c) = g(c) = 0$ *is replaced by* $\lim_{x \to a^+} f(x) = \lim_{x \to a^+} g(x) = 0$ *or* $\lim_{x \to b^-} f(x) = \lim_{x \to b^-} g(x) = 0$, *respectively.*

Proof. We prove here only the easiest case; the general proof is in Theorem 3.A.4.

Suppose that $f(c) = g(c) = 0$, $L \neq \pm\infty$, and

$$\lim_{x \to c} \frac{f'(x)}{g'(x)} = \frac{f'(c)}{g'(c)} = L,$$

which, in particular, implies that $g'(c) \neq 0$.

Then, we find

$$L = \frac{f'(c)}{g'(c)} = \frac{\lim_{x \to c} \frac{f(x) - f(c)}{x - c}}{\lim_{x \to c} \frac{g(x) - g(c)}{x - c}} = \lim_{x \to c} \left(\frac{\frac{f(x) - f(c)}{x - c}}{\frac{g(x) - g(c)}{x - c}} \right) = \lim_{x \to c} \frac{f(x) - f(c)}{g(x) - g(c)} = \lim_{x \to c} \frac{f(x)}{g(x)}.$$

□

Remark 3.5.2. As with many of our theorems on limits, when using l'Hôpital's Rule, we frequently write equalities that we only know really hold after we finish the problem. That is, after verifying that the numerator and denominator both approach 0, or that the denominator approaches $\pm\infty$, we write

$$\lim_{x \to c} \frac{f(x)}{g(x)} = \lim_{x \to c} \frac{f'(x)}{g'(x)},$$

which we only really know is true **after** we prove that the limit on the right exists (as an extended real number). In addition, in some problems, you may need to iterate the use of l'Hôpital's Rule and continue taking derivatives of the numerator and denominator.

Example 3.5.3. Use l'Hôpital's Rule to calculate

$$\lim_{t \to \infty} \frac{t}{e^t} \quad \text{and then, more generally,} \quad \lim_{t \to \infty} \frac{t^n}{e^t},$$

where n is any positive integer. (This second part, via a different method, was given as an exercise, in Exercise 54 in Subsection 2.4.1.)

Solution:

As $t \to \infty$, both t and e^t approach ∞, and so we are in a position to use l'Hôpital's Rule. We find

$$\lim_{t \to \infty} \frac{t}{e^t} = \lim_{t \to \infty} \frac{(t)'}{(e^t)'} = \lim_{t \to \infty} \frac{1}{e^t} = 0,$$

where we really don't know that the first equality holds until after we find that the final limit exists (or is $\pm\infty$).

Suppose that n is a positive integer. As $t \to \infty$, both t^n and e^t approach ∞, and so, we can use l'Hôpital's Rule to find

$$\lim_{t \to \infty} \frac{t^n}{e^t} = \lim_{t \to \infty} \frac{(t^n)'}{(e^t)'} = \lim_{t \to \infty} \frac{nt^{n-1}}{e^t},$$

provided that the limit on the right exists as an extended real number. Assuming that $n \geq 2$, the numerator and denominator still both approach infinity, so we use l'Hôpital's Rule again:

$$\lim_{t \to \infty} \frac{nt^{n-1}}{e^t} = \lim_{t \to \infty} \frac{(nt^{n-1})'}{(e^t)'} = \lim_{t \to \infty} \frac{n(n-1)t^{n-2}}{e^t}.$$

While, strictly speaking, we need to use induction here, it is easy to see what happens if we keep applying l'Hôpital's Rule: the derivative of e^t is always just e^t, and every time we differentiate the numerator, the power of t goes down by 1. As long as the exponent of the t is not 0, both the numerator and denominator approach ∞ as $x \to \infty$, so we can keep applying l'Hôpital's Rule. Eventually, we end up with

$$\lim_{t \to \infty} \frac{n(n-1)(n-2) \cdots (2)t^1}{e^t} = \lim_{t \to \infty} \frac{n(n-1)(n-2) \cdots (2)(1)}{e^t} = 0,$$

since the numerator is constant (though it may be very large), while $e^t \to \infty$.

Therefore, we find that

$$\lim_{t \to \infty} \frac{t^n}{e^t} = 0,$$

which is frequently described by saying that "the exponential function grows faster than any power of t".

Example 3.5.4. Calculate

$$\lim_{x \to 0} \frac{x - \sin x}{x^3}.$$

Solution:

The numerator and denominator are both 0 when $x = 0$. We use l'Hôpital's Rule as long as we continue to get $0/0$ when we let $x = 0$:

$$\lim_{x \to 0} \frac{x - \sin x}{x^3} = \lim_{x \to 0} \frac{1 - \cos x}{3x^2} = \lim_{x \to 0} \frac{\sin x}{6x} = \lim_{x \to 0} \frac{\cos x}{6} = \frac{1}{6}.$$

If you look at the original function involved in the limit, it may be somewhat hard to believe that you get $1/6$; you don't see $1/6 = 0.166\overline{6}$ anywhere. You might want to get a calculator, make sure that it's set to use radians, and plug in something "close to 0" for x in the expression

$(x - \sin x)/x^3$. How close to 0? Try a few numbers, like $x = 0.01$ and $x = 0.00001$, and see what you get.

Example 3.5.5. A particle is moving along the x-axis, and its position, in meters, is given by $x = (1 - t)\ln(1 - t)$, for $0 \le t < 1$. As t approaches 1, what happens to the position of the particle?

Solution:

We need to calculate

$$\lim_{t \to 1^-} x = \lim_{t \to 1^-} \left[(1 - t)\ln(1 - t) \right].$$

In order to use l'Hôpital's Rule on this, we need to write this limit as

$$\lim_{t \to 1^-} \left[(1 - t)\ln(1 - t) \right] = \lim_{t \to 1^-} \frac{\ln(1 - t)}{(1 - t)^{-1}}.$$

Now, as $t \to 1^-$, the numerator in this last limit approaches $-\infty$, while the denominator approaches ∞. Thus, we may apply l'Hôpital's Rule:

$$\lim_{t \to 1^-} \frac{\ln(1 - t)}{(1 - t)^{-1}} = \lim_{t \to 1^-} \frac{\left(\ln(1 - t)\right)'}{\left((1 - t)^{-1}\right)'} =$$

$$\lim_{t \to 1^-} \frac{\left(\frac{1}{1-t}\right)(-1)}{(-1)(1 - t)^{-2}(-1)} = \lim_{t \to 1^-} \left[(1 - t)(-1) \right] = 0,$$

where the next-to-last equality is obtained from algebra, not by applying l'Hôpital's Rule.

Thus, the particle approaches the origin as the time approaches 1 second.

Example 3.5.6. Consider the limit

$$\lim_{p \to \infty} \left[p \ln\left(1 + \frac{x}{p}\right) \right].$$

It is important to understand which "variables" are constant and which are really variables as far as the limit is concerned. It is p that is approaching some value; x is a constant in this problem. Once we set this up to use l'Hôpital's Rule, we will take derivatives with respect to p, not x.

We rewrite the limit expression as a quotient:

$$\lim_{p \to \infty} \left[p \ln \left(1 + \frac{x}{p} \right) \right] = \lim_{p \to \infty} \frac{\ln \left(1 + \frac{x}{p} \right)}{1/p}.$$

Now, written in this form, we see that the numerator and denominator each approach 0, and so we may apply l'Hôpital's Rule:

$$\lim_{p \to \infty} \frac{\ln \left(1 + \frac{x}{p} \right)}{1/p} = \lim_{p \to \infty} \frac{\frac{d}{dp} \left[\ln \left(1 + \frac{x}{p} \right) \right]}{\frac{d}{dp} [1/p]} =$$

$$\lim_{p \to \infty} \frac{\left[1/(1 + x/p) \right] (-x/p^2)}{-1/p^2} = \lim_{p \to \infty} \frac{x}{1 + x/p} = x,$$

where, again, the next-to-last equality is obtained from algebra, not by applying l'Hôpital's Rule.

Therefore, we find that

$$\lim_{p \to \infty} \left[p \ln \left(1 + \frac{x}{p} \right) \right] = x.$$

We can use our limit calculation above to calculate other, related, limits. For instance,

$$\lim_{p \to \infty} \left(1 + \frac{x}{p} \right)^p = \lim_{p \to \infty} e^{p \ln(1+x/p)} = e^{\lim_{p \to \infty} p \ln(1+x/p)} = e^x,$$

which uses that the exponential function is continuous, so that we may move the limit "inside". Note that this agrees with Item 13 of Theorem 2.4.21 and Theorem 2.5.10, where we used limits over exponents which were natural numbers.

3.5.1   Exercises

In Exercises 1 through 21, calculate the limits, if they exist.

1. $\displaystyle\lim_{x\to1}\frac{x^p-1}{x^q-1}$ with $p\neq0$ and $q\neq0$.

2. $\displaystyle\lim_{x\to0}\frac{\sin(4x)}{\tan(5x)}$. $\circledR$

3. $\displaystyle\lim_{t\to0}\frac{e^t-1}{t^3}$. $\circledR$

4. $\displaystyle\lim_{\theta\to0}\frac{1-\cos\theta}{1-\sec\theta}$.

5. $\displaystyle\lim_{x\to\infty}\frac{\ln x}{\sqrt{x}}$.

6. $\displaystyle\lim_{x\to\infty}\frac{x^2+2x-1}{1-x^2}$.

7. $\displaystyle\lim_{x\to0}\frac{x-\sin x}{x-\tan x}$.

8. $\displaystyle\lim_{u\to0}\frac{2^u-4^u}{u}$.

9. $\displaystyle\lim_{x\to0}\frac{\sin^{-1}x}{2x}$.

10. $\displaystyle\lim_{x\to\infty}\frac{(\ln x)^3}{x}$.

11. $\displaystyle\lim_{t\to-\infty}t^2e^t$.

12. $\displaystyle\lim_{x\to\infty}x\tan\left(\frac{1}{x}\right)$.

13. $\displaystyle\lim_{p\to0^+}\frac{\ln p}{p}$. $\circledR$

14. $\displaystyle\lim_{x\to\infty}x^{\frac{1}{x}}$.

15. $\displaystyle\lim_{x\to\infty}\left(1+\frac{2}{x}\right)^{-3x}$.

16. $\displaystyle\lim_{x\to0}\frac{\cos(2x)-\cos x}{x^3}$.

17. $\displaystyle\lim_{x\to0}\frac{e^{5x}-1}{x}$.

18. $\lim\limits_{t \to \infty} \ln t + t^2 - t$.

19. $\lim\limits_{x \to \infty} e^{-x^2}$.

20. $\lim\limits_{x \to 0} \dfrac{\sin^3 x}{\sin x^3}$.

21. $\lim\limits_{x \to 1} \left(\dfrac{x}{x-1} - \dfrac{1}{\ln x} \right)$. Hint: rewrite as a single fraction over a common denominator.

The two limits below were calculated using different methods in Section 2.7. Use l'Hôpital's Rule to calculate them. Think about why we couldn't we have used this method to calculate these limits back in Section 2.7, even if we had known l'Hôpital's Rule then.

22. $\lim\limits_{x \to 0} \dfrac{\sin x}{x}$.

23. $\lim\limits_{x \to 0} \dfrac{1 - \cos x}{x}$.

24. Two particles leave the origin at the same time and travel along the x-axis. The distance functions of the two particles are $d_1(t) = 3t^2 + 2t + 7$ and $d_2(t) = 7t^2 + 3t + 9$. Show that the limit of the ratio of the distances traveled, as t goes to infinity, is equal to the ratio of the accelerations of the particles.

25. Evaluate $\lim\limits_{x \to \infty} \dfrac{x - \sin x \cos x}{x}$. Does l'Hôpital's Rule apply to this limit?

26. Calculate $\lim\limits_{x \to \infty} \dfrac{(x-a)^n}{(x-b)^n}$ where a and b are arbitrary constants and $n > 0$ is an integer.

27. Let $p(x)$ be any differentiable function. Show that $\lim\limits_{x \to 0} \dfrac{p(x) \sin x}{x} = p(0)$. Do this two different ways, by l'Hôpital's Rule, and in a different way that requires only that p is continuous at 0.

28. Suppose we want to apply l'Hôpital's rule to the function $\phi(x) = \dfrac{f(x)}{g(x)}$ where $f(c) = g(c) = 0$. Assume that both f and g are differentiable. Suppose that neither $f'(c)$ nor $g'(c)$ exists, but that $g'(x)$ exists and is non-zero in an open interval containing c and that $\lim\limits_{x \to c} \dfrac{f'(x)}{g'(x)} = L$. Can we conclude that $\lim\limits_{x \to c} \phi(x) = L$?

29. Use l'Hôpital's Rule to calculate $\lim\limits_{x \to a} \dfrac{x^n - a^n}{x - a}$, where $n > 0$ is an integer.

In some circumstances, l'Hôpital's Rule can be used to evaluate limits of the form $f(x)^{g(x)}$ when the limit is in the indeterminate form 0^0 or 1^∞. In the first problem,

we outline the steps in this method. **Exercises 30 - 37** give you a chance to practice applying the technique. Finally, **Exercise 38** suggests a mathematical justification for the method.

30. $\lim_{x \to 0^+} (\tan x)^{\sin x}$.

    a. Let $\phi(x) = (\tan x)^{\sin x}$. Show that $\ln \phi(x) = \dfrac{\ln(\tan x)}{\csc x}$. Note that for $x \in (0, \pi/2)$, $\tan(x) > 0$ so we're justified in taking a logarithm. There's nothing special about the right endpoint $\pi/2$; we could have chosen any number smaller than $\pi/2$ and greater than 0.

    b. Use l'Hôpital's Rule to prove that $\lim_{x \to 0^+} \ln \phi(x) = 0$.

    c. Conclude that $\lim_{x \to 0^+} \phi(x) = e^0 = 1$.

31. $\lim_{x \to \pi^-} (\sin x)^{\tan x}$.

32. $\lim_{x \to 0^+} (\sin x)^{\tan x}$.

33. $\lim_{x \to 0^+} (x^2 + 3x + 1)^{\cot x}$.

34. $\lim_{x \to 0^+} (\sin 37x)^{\sin 51x}$. Hint: it may be easier to calculate the limit of the general case $(\sin ax)^{\sin bx}$.

35. Calculate $\lim_{x \to 0} (\cos x)^{\csc^2 x}$.

36. Let $a > 0$ and $b > 0$. Calculate $\lim_{x \to 0} (\cos ax)^{\csc^2 bx}$.

37. $\lim_{x \to 1^+} \left(\tan \dfrac{\pi x}{4} \right)^{3/(x-1)}$.

38. The method used in the past several exercises is justified by parts (a) and (b) of this problem.

    a. Prove that if $\lim_{x \to b} h(x) = M$ then $\lim_{x \to b} e^{h(x)} = e^M$. Assume M is finite, but that b may be finite or infinite.

    b. What does statement (a) say if $h(x) = \ln j(x)$? Note that part (a) still holds if only a left or right hand limit is being taken.

39. Suppose f has a continuous derivative, that $f(3) = 0$ and $f'(3) = 4$. Find $\lim_{x \to 0} \dfrac{f(3 + 3x) + f(3 + 7x)}{x}$.

40. Use l'Hôpital's Rule to prove that if f has a continuous derivative, then

$$\lim_{h \to 0} \frac{f(x+h) - f(x-h)}{2h} = f'(x).$$

41. Prove that if $f''(x)$ is continuous, then $\lim\limits_{h\to 0} \dfrac{f(x+h) + f(x-h) - 2f(x)}{h^2} = f''(x)$.

42. Recall that if \$1 is invested at a rate of r and compounded k times a year, then the amount in the account after n years is $A(k) = (1 + \dfrac{r}{k})^{kn}$. Calculate $\lim\limits_{k\to\infty} A(k)$.

43. Show that, if $a > 0$ and $b > 0$, then $\lim\limits_{x\to 0} \dfrac{a^x - b^x}{x} = \ln\left(\dfrac{a}{b}\right)$. Note that this is a generalization of the formula $\lim\limits_{x\to 0} \dfrac{e^x - 1}{x} = 1$.

44. Show that, if $\phi(x) = \dfrac{f(x)}{g(x)}$, where f and g are polynomials with degrees m and n, respectively, and $m > n$, then $\lim\limits_{x\to\infty} |\phi(x)| = \infty$. Assume m and n are integers.

45. Show that if $\phi(x) = \dfrac{f(x)}{g(x)}$ where f and g are polynomials with degrees m and n, respectively, with $m < n$ then $\lim\limits_{x\to\infty} \phi(x) = 0$. Assume m and n are integers.

46. What is $\lim\limits_{x\to 0} \dfrac{x\tan x}{\sqrt{1 - x^2} - 1}$?

47. What happens when you try to apply l'Hôpital's Rule to $\lim\limits_{x\to\infty} \dfrac{x - \sin x}{x + \sin x}$? Can you find a better way to evaluate this limit?

48. Use l'Hôpital's Rule to calculate $\lim\limits_{x\to 0} \dfrac{\tan x - \sin x}{(\sin^{-1} x) - x}$.

49. Let $p(x)$ be a polynomial function. Prove that $\lim\limits_{x\to\infty} \dfrac{p(x)}{e^x} = 0$.

50. a. Let $p(1/x)$ be a polynomial in $1/x$. Show that $\lim\limits_{x\to 0} \dfrac{p(1/x)}{e^{1/x^2}} = 0$.

   b. Show that if $H(x) = e^{-1/x^2}$, then $H^{(n)}(x) = p_n(1/x)H(x)$ where each $p_n(1/x)$ is a polynomial in $1/x$ and $n = 1, 2, 3, \dots$.

   c. Conclude that the function below has continuous derivatives of all orders.

$$G(x) = \begin{cases} H(x) & x > 0 \\ 0 & x \le 0. \end{cases}$$

51. Consider the differential equation (covered in more depth in the next chapter) given by

$$\frac{dy}{dt} = \frac{t}{e^t}.$$

   a. Find $\lim\limits_{t\to\infty} \dfrac{dy}{dt}$.

b. Show that $f(t) = te^{-t} + e^{-t} + 5$ is a solution of the differential equation.

c. What is $\lim_{t \to \infty} f(t)$?

52. Two particles are traveling toward the star in the diagram so that, at time t, where $0 \le t \le 1$, their relative positions are at the other two vertices of the triangle shown in the diagram. That is, if the star is at the origin, then the $x-$ and y-coordinates of one particle, at time t, are $(x, y) = (1 - t, 0)$, and the x- and y-coordinates of the other particle are $(x, y) = (1 - t, 1 - e^{t-1})$.

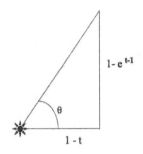

a. Write θ as a function of t.

b. Notice that, at time $t = 1$, both particles reach the star. At what angle do the particles collide with each other?

53. The picture below shows the solid formed by revolving, around the x-axis, the region bounded above by the graph of $f(x) = xe^{-x}$, bounded below by the x-axis, and bounded on the left by the y-axis. It can be shown that, if we truncate the solid at $x = a$, then the volume, V, of the resulting solid, to the left of $x = a$, will be

$$V = \frac{-1}{2}a^2e^{-2a} - \frac{1}{2}ae^{-2a} - \frac{1}{4}e^{-2a} + \frac{1}{4}.$$

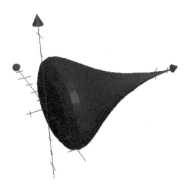

a. Find the volume of the solid resulting from truncating the graph at $x = 2$.

b. Find the volume, if it is finite, of the entire solid, by letting a tend towards infinity.

54. The graph below is a segment of the so-called the *unnormalized sinc function*, often encountered in data processing. The function is

$$\text{sinc}(x) = \begin{cases} \frac{\sin(x)}{x} & \text{if } x \neq 0; \\ 1 & \text{if } x = 0. \end{cases}$$

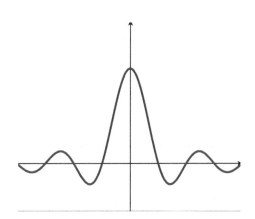

a. Show that this function is continuous at the origin.

b. Does the derivative exist at $x = 0$? If so, what is its value?

c. Show that the graph of the function is concave down at $x = 0$ (show this using Calculus, not simply by looking at the picture).

55. In the diagram below, the point $(0, 1)$ is connected to the point on the unit circle at an

angle of θ to the positive x-axis. For θ in the open interval $(0, 2\pi)$, define $T(\theta)$ to be the y-intercept of this line. (This is one way of carrying out *stereographic projection*, an identification of a line or plane with a circle or sphere with one point removed.)

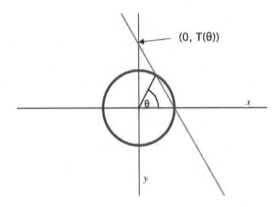

a. Find an explicit description of $T(\theta)$.

b. What is $T'(\theta)$? What is $T'\left(\dfrac{\pi}{2}\right)$?

c. Does $T(\theta)$ have any critical points? Does the graph of $T(\theta)$ have any inflection points?

d. Recreate the drawing with θ very close to 0. What does it seem that $\displaystyle\lim_{\theta \to 0^+} T(\theta)$ should be? Find the limit explicitly.

Appendix 3.A   Technical Matters

The following theorem is a generalization of the Mean Value Theorem. Various forms of this result are known as *Taylor's Theorem*. See below.

Theorem 3.A.1. *Suppose that the $(n+1)$-st derivative of f exists on an open interval I around a point a. Then, for all $x \neq a$ in I, there exists a c in the open interval between x and a (i.e., in (x, a) if $x < a$, or in (a, x) if $a < x$) such that*

$$f(x) \;=\; f(a) + f'(a)(x-a) + \frac{f''(a)}{2!}\,(x-a)^2 + \frac{f'''(a)}{3!}\,(x-a)^3 + \cdots$$

$$+ \frac{f^{(n)}(a)}{n!}\,(x-a)^n + \frac{f^{(n+1)}(c)}{(n+1)!}\,(x-a)^{n+1}.$$

Proof. Define

$$R_f^n(x; a) = f(x) - \sum_{k=0}^{n} \frac{f^{(k)}(a)}{k!}(x-a)^k.$$

Assume that $x \neq a$. Define the function $g(t)$, with domain I, by

$$g(t) = f(x) - R_f^n(x; a)\frac{(x-t)^{n+1}}{(x-a)^{n+1}} - \sum_{k=0}^{n} \frac{f^{(k)}(t)}{k!}(x-t)^k.$$

Since f is $(n+1)$-differentiable on I, g is differentiable on I, and so is certainly continuous on the closed interval between x and a. One easily sees that $g(x) = g(a) = 0$. By Rolle's Theorem, Lemma 1.5.9, there exists c in the open interval between x and a such that $g'(c) = 0$.

We claim that this shows that $R_f^n(x; a) = \frac{f^{(n+1)}(c)}{(n+1)!}(x-a)^{n+1}$, as desired, but we need to calculate $g'(t)$ in order to see this.

We find, using the Product Rule inside the summation,

$$g'(t) \;=\; -R_f^n(x; a)\frac{(n+1)(x-t)^n(-1)}{(x-a)^{n+1}} -$$

$$\sum_{k=0}^{n} \frac{1}{k!}\cdot\left[f^{(k+1)}(t)(x-t)^k + f^{(k)}(t)k(x-t)^{k-1}(-1)\right] =$$

$$(n+1)R_f^n(x;a)\frac{(x-t)^n}{(x-a)^{n+1}} - \sum_{k=0}^{n}\frac{f^{(k+1)}(t)}{k!}(x-t)^k + \sum_{k=1}^{n}\frac{f^{(k)}(t)}{(k-1)!}(x-t)^{k-1}.$$

Let $j = k - 1$, so that $k = j + 1$, and use j to re-index the last summation above. We obtain that $g'(t) =$

$$(n+1)R_f^n(x;a)\frac{(x-t)^n}{(x-a)^{n+1}} - \sum_{k=0}^{n}\frac{f^{(k+1)}(t)}{k!}(x-t)^k + \sum_{j=1}^{n-1}\frac{f^{(j+1)}(t)}{j!}(x-t)^j.$$

Now, in these two summations above, one with a plus sign, one with a minus sign, all of the terms cancel out, except for one: $-f^{(n+1)}(t)(x-t)^n/(n!)$.

Therefore, we find that

$$g'(t) = \frac{(n+1)(x-t)^n}{(x-a)^{n+1}} \cdot \left[R_f^n(x;a) - \frac{f^{(n+1)}(t)}{(n+1)!}(x-a)^{n+1} \right].$$

Since $g'(c) = 0$ and c is in the **open** interval between x and a, we know that $x - c \neq 0$ and so, we conclude that

$$R_f^n(x;a) - \frac{f^{(n+1)}(c)}{(n+1)!}(x-a)^{n+1} = 0,$$

which is what we needed to show. $\square$

The degree n polynomial, using powers of $(x - a)$, which appears above is very important in the study of infinite series and in estimating the values of functions.

Definition 3.A.2. *Suppose that the n-th derivative of f exists on an open interval I around a point a. Then, the polynomial, $T_f^n(x;a)$, given by*

$$T_f^n(x;a) = \sum_{k=0}^{n}\frac{f^{(k)}(a)}{k!}(x-a)^k =$$

$$f(a) + f'(a)(x-a) + \frac{f''(a)}{2!}(x-a)^2 + \cdots + \frac{f^{(n)}(a)}{n!}(x-a)^n$$

*is called the n-**th order Taylor polynomial of f centered at a. In the special case where $a = 0$, $T_f^n(x;a)$ is also referred to as the **degree n Maclaurin polynomial** of f. Note that the n-th order Taylor polynomial will have degree n, provided that $f^{(n)}(a) \neq 0$; for this reason, the n-th order Taylor polynomial is sometimes referred to as the **degree n Taylor polynomial**, without worrying about whether or not $f^{(n)}(a)$ is zero.*

The function on I given by $R_f^n(x;a) = f(x) - T_f^n(x;a)$ is called the n-th **Taylor remainder** *of f centered at a.*

The equality $R_f^n(x;a) = \dfrac{f^{(n+1)}(c)}{(n+1)!}(x-a)^{n+1}$ from Theorem 3.A.1 is called the **Lagrange Form of the Remainder**.

Theorem 3.A.3. (**Cauchy's Mean Value Theorem**) *Suppose that $a < b$, and that f and g are functions which are continuous on the closed interval $[a,b]$ and differentiable on the open interval (a,b).*

Then, there exists c in (a,b) such that

$$[f(b) - f(a)]g'(c) = [g(b) - g(a)]f'(c).$$

Proof. Let $h(x) = [f(b) - f(a)]g(x) - [g(b) - g(a)]f(x)$, and apply Rolle's Theorem, Lemma 1.5.9, to h. $\qquad\square$

Theorem 3.A.4. (**l'Hôpital's Rule**) *Suppose that $-\infty \leq a < b \leq \infty$, and that $-\infty \leq L \leq \infty$. Let f and g be differentiable on the interval (a,b), and suppose that, for all x in (a,b), $g'(x) \neq 0$. Let c be in (a,b).*

Suppose that either $\lim_{x \to c} f(x) = \pm\infty$ and $\lim_{x \to c} g(x) = \pm\infty$, or that $f(c) = g(c) = 0$. Also, suppose that

$$\lim_{x \to c} \frac{f'(x)}{g'(x)} = L.$$

Then,

$$\lim_{x \to c} \frac{f(x)}{g(x)} = \lim_{x \to c} \frac{f'(x)}{g'(x)} = L.$$

In addition, the theorem remains true if all of the limits $\lim_{x \to c}$ above are replaced by $\lim_{x \to a^+}$ or $\lim_{x \to b^-}$, and the condition that $f(c) = g(c) = 0$ is replaced by $\lim_{x \to a^+} f(x) = \lim_{x \to a^+} g(x) = 0$ or $\lim_{x \to b^-} f(x) = \lim_{x \to b^-} g(x) = 0$, respectively.

Proof. The conclusion as $x \to c$ clearly follows from applying the statements involving $\lim_{x \to a^+}$ and $\lim_{x \to b^-}$. We will prove the cases where L is finite, involving limits $\lim_{x \to a^+}$, and leave the remaining cases as an exercise, or you may consult [3] or [2].

Case 1: Suppose that $\lim_{x \to a^+} f(x) = \lim_{x \to a^+} g(x) = 0$. If necessary, redefine f and g at $x = a$, so that $f(a) = g(a) = 0$; this makes f and g continuous on the interval $[a,b)$. Note that this implies that, for all x in (a,b), $g(x) \neq 0$ for, otherwise, we could apply Rolle's Theorem, Lemma 1.5.9, to g on the interval $[a,x]$ to conclude that $g'(t) = 0$, for some t in (a,x); this would contradict the assumption that g' is never equal to 0 on the interval (a,b).

Thus, for each x in (a,b), f and g are continuous on $[a,x]$ and differentiable on (a,x). We may apply Cauchy's Mean Value Theorem, Theorem 3.A.3, to conclude that there exists t_x in (a,x) such that

$$f(x)g'(t_x) \;=\; g(x)f'(t_x), \quad \text{or, equivalently,} \quad \frac{f(x)}{g(x)} \;=\; \frac{f'(t_x)}{g'(t_x)},$$

where we have used that neither $g(x)$ nor $g'(t_x)$ is equal to 0. Now, as $x \to a^+$, $t_x \to a^+$, the desired conclusion follows.

Case 2: Suppose that $\lim_{x \to a^+} f(x) = \pm\infty$ and $\lim_{x \to a^+} g(x) = \pm\infty$. This case is more delicate, but still similar to Case 1. As one may multiply f or g by -1, and simply negate the result, it is enough to prove the case where $\lim_{x \to a^+} f(x) = \infty$ and $\lim_{x \to a^+} g(x) = \infty$.

Let $\epsilon > 0$. Let $\delta_1 > 0$ be such that $a + \delta_1 < b$ and such that, if $a < x < a + \delta_1$, then

$$\left| \frac{f'(x)}{g'(x)} - L \right| < \frac{\epsilon}{2}.$$

Let δ_2 be such that $0 < \delta_2 < \delta_1$ and such that, if $a < x < a+\delta_2$, then $f(x) > 0$ and $g(x) > 0$. Let δ_3 be such that $0 < \delta_3 < \delta_2$ and such that, if $a < x < a + \delta_3$, then $f(x) > f(a + \delta_2)$ and $g(x) > g(a+\delta_2)$. (We can produce such δ_2 and δ_3 because $f(x) \to \infty$ and $g(x) \to \infty$ as $x \to a^+$.)

For x in the open interval $(a, a+\delta_3)$, Cauchy's Mean Value Theorem, applied on the interval closed interval $[x, a + \delta_2]$, tells us that there exists t_x in the open interval $(x, a + \delta_2)$ such that

$$\frac{f'(t_x)}{g'(t_x)} \;=\; \frac{f(x) - f(a+\delta_2)}{g(x) - g(a+\delta_2)} \;=\; \frac{f(x)}{g(x)} \cdot \frac{1 - f(a+\delta_2)/f(x)}{1 - g(a+\delta_2)/g(x)}.$$

For x in the interval $(a, a + \delta_3)$, define

$$H(x) \;=\; \frac{1 - g(a+\delta_2)/g(x)}{1 - f(a+\delta_2)/f(x)}.$$

Then, for x in the interval $(a, a + \delta_3)$,

$$\frac{f'(t_x)}{g'(t_x)} \cdot H(x) = \frac{f(x)}{g(x)},$$

and, as $f(x) \to \infty$ and $g(x) \to \infty$ as $x \to a^+$, $\lim_{x \to a^+} H(x) = 1$.

We would like to take the equation above, take the limit as $x \to a^+$, and claim that we are finished. Unfortunately, as $x \to a^+$, t_x is not forced to approach a; t_x is just somewhere between a and $a + \delta_2$.

We must do more work. We find, for x in the interval $(a, a + \delta_3)$,

$$\left| \frac{f(x)}{g(x)} - L \right| = \left| \frac{f'(t_x)}{g'(t_x)} \cdot H(x) - L \right| \le \left| \frac{f'(t_x)}{g'(t_x)} \cdot H(x) - \frac{f'(t_x)}{g'(t_x)} \right| + \left| \frac{f'(t_x)}{g'(t_x)} - L \right| =$$

$$\left| \frac{f'(t_x)}{g'(t_x)} \right| \cdot |H(x) - 1| + \left| \frac{f'(t_x)}{g'(t_x)} - L \right|,$$

where

$$\left| \frac{f'(t_x)}{g'(t_x)} - L \right| < \frac{\epsilon}{2},$$

which implies that $|f(t_x)/g(t_x)| < |L| + \epsilon/2$. If we now take x close enough to a so that

$$|H(x) - 1| < \frac{\epsilon}{2(|L| + \epsilon/2)},$$

we find that

$$\left| \frac{f(x)}{g(x)} - L \right| < \epsilon,$$

as we desired. $\qquad\qquad\square$

Chapter 4

Anti-differentiation & Differential Equations

In many physical problems, we know that the rate of change of a function must satisfy certain relationships, or equations, at all times, or for all of the values of some variable in some interval; given these equations and some "initial data", we would like to be able to determine the function in question.

Such equations are known as *differential equations*, and a differential equation, together with initial data, is known as an *initial value problem*.

The most basic type of differential equation is

$$\frac{dy}{dx} \;=\; f(x),$$

in which $f(x)$ is specified, and we want to find $y = F(x)$ such that $F'(x) = f(x)$. This is the problem of *anti-differentiating*, not calculating a derivative, but rather finding a function whose derivative has been specified.

In this chapter, we will introduce (or, reintroduce) you to anti-differentiation and to another class of basic differential equations, *separable differential equations*; we will look at initial value problems, and give a number of applications.

4.1 What is a Differential Equation?

We have actually already looked at several differential equations problems in earlier chapters: in Example 1.5.13, Example 2.2.9, Theorem 2.4.14, Example 2.4.15, Example 2.4.24, Example 2.6.8, and in a number of theorems, corollaries, and examples in the second half of Section 2.7.

In this section, we will begin a more careful treatment of differential equations. We will define terminology and discuss some important issues surrounding the solving of differential equations. In this section, we are just trying to familiarize you with the types of questions you might be asked, and what an "answer" means; we will not actually explain how to **produce** answers until later in this chapter.

Suppose $A = A(t)$ is a function which gives the amount, in kilograms, of some substance present at time t seconds. Suppose, for some physical reasons, we know that, for all t in some open interval I, the following equation is satisfied:

$$\frac{dA}{dt} = tA^2. \tag{4.1}$$

This means that we know that the instantaneous rate of change of A, with respect to t, is related to t and A by Formula 4.1, for all t in I. This is called a *1st-order differential equation*; the *order* refers to the highest derivative that appears in the equation.

The standard question is:

Can we determine the function $A = A(t)$, given that Formula 4.1 must be satisfied?

Strictly speaking, given how the question is asked, the answer is "no"; we cannot determine **THE** function which satisfies Formula 4.1 because there are an infinite number of functions $A = A(t)$ which satisfy Formula 4.1. We can show that the functions

$$A(t) = 0 \quad \text{and} \quad A(t) = \frac{2}{C - t^2} \tag{4.2}$$

are solutions, where C can be any constant.

What do we mean when we write that the functions in Formula 4.2 are solutions of Formula 4.1? We mean that, for any one of the functions in Formula 4.2, if you take that A, and calculate dA/dt, and calculate tA^2, then the two functions of t that you get will be equal. Let's verify this.

If $A = A(t) = 0$, then $dA/dt = 0$ and $tA^2 = 0$, so $dA/dt = tA^2$, for all t. Thus, $A = 0$ is a solution of Formula 4.1. Such a solution, one where the function is constant, is called an *equilibrium solution*.

Suppose that $A = 2/(C - t^2)$, for some constant C. Then, we find

$$\frac{dA}{dt} = \left[2(C - t^2)^{-1}\right]' = 2(-1)(C - t^2)^{-2}(-2t) = \frac{4t}{(C - t^2)^2}.$$

Still assuming that $A = 2/(C - t^2)$, we also find

$$tA^2 = t\left(\frac{2}{C - t^2}\right)^2 = \frac{4t}{(C - t^2)^2},$$

and so, $A = 2/(C - t^2)$ is a solution to Formula 4.1 on any interval where $C - t^2 \neq 0$, that is, if $C < 0$, $A = 2/(C - t^2)$ is a solution to Formula 4.1, for all t, and, if $C \geq 0$, then $A = 2/(C - t^2)$ is a solution to Formula 4.1, for all $t \neq \pm\sqrt{C}$.

Later, in Section 4.3, we will be able to show that every solution to Formula 4.1, on an open interval, is one of the solutions given in Formula 4.2; there are no others. The entire collection of all solutions to a given differential equation is referred to as the *general solution to the differential equation*. Thus, Formula 4.2 gives the general solution to the differential equation given in Formula 4.1. The general solution is what you are supposed to find if the instructions are simply "solve the following differential equation", and no other data is given.

It is a technical, but nonetheless important, point that, if you allow your solutions to be defined on (open) sets $\mathcal{U}$ which are **not** intervals, then you might choose different-looking functions on non-intersecting open intervals contained in $\mathcal{U}$. For instance, a solution to Formula 4.1, for $t \neq \pm 2$, would be the piecewise-defined function $A = 2/(4 - t^2)$, if $-2 < t < 2$, and $A = 2/(1 - t^2)$, if $t < -2$ or $t > 2$. In other words, we can pick different constants C in Formula 4.2 on different non-intersecting open intervals. For that matter, we could also have picked the $A = 0$ solution on an interval. Because we can always piece together solutions on non-intersecting open intervals, **it is standard to describe the general solution to a differential equation in terms of functions defined on open intervals**. This is the practice that we shall follow.

Okay. So, the general solution to Formula 4.1 is given in Formula 4.2. In a true physical problem, we don't (usually) want to know that one of an infinite number of things will happen, we want to have posed the question/problem in such a way that there's exactly one solution, that is, we want our solution to be **unique**. What more needs to be given in order that we have a unique solution?

Typically, we are also given *initial data*; that is, for a particular value t_0 of t, we are given additional information about A, usually the value A_0 of A at time t_0. In physical problems, $t = 0$ is frequently taken as the time at which the whole situation under consideration starts; in this case, it is common that we would know the value $A(0)$, i.e., the initial value of A. However, even if $t_0 \neq 0$, if we are given values t_0 and A_0 and told that $A_0 = A(t_0)$, this is referred to as *initial data* (think of it as data that we are given initially, not necessarily data about what's true at time 0). A differential equation, together with initial data, is called an *initial value problem*. What we would like is that such an initial value problem has a unique solution on some open interval that contains t_0.

For example, suppose we are given the initial value problem

$$\frac{dA}{dt} = tA^2, \quad \text{and} \quad A(0) = 1/2. \tag{4.3}$$

Is there a unique solution to this initial value problem on some open interval containing $t = 0$?

Looking back at the general solution in Formula 4.2, we see that the $A = 0$ won't work, for then the initial data would have to be different from what we're given. So, any solution to this initial value problem on an open interval containing 0 must be of the form

$$A = \frac{2}{C - t^2},$$

and we also need that $A(0) = 1/2$. Hence, we need

$$\frac{1}{2} = \frac{2}{C - 0^2},$$

i.e., we need for C to equal 4. Therefore, we find that

$$A = \frac{2}{4 - t^2}$$

is the unique solution to the initial value problem in Formula 4.3 on any open interval J containing 0, as long as J is contained in the open interval $(-2, 2)$.

The last part of what's written above "as long as J is contained in the open interval $(-2, 2)$" may seem annoyingly precise or technical, however, there is an important point here: if you are given the initial value problem in Formula 4.3 and are asked a question such as "what is the value of A when $t = 3$?", then you need to realize that you are not given enough data to answer this question. The initial value problem gives you enough data to determine the value of $A(t)$ for every t value in the open interval $(-2, 2)$, but not for t values outside that interval. As we discussed above, on an open interval K around 3, if K doesn't intersect the interval $(-2, 2)$, then we are free to choose, on the interval K, a function other than $A = 2/(4 - t^2)$ from among the functions in our general solution; each different choice could/would lead to a different value of $A(3)$ for the resulting piecewise definition of A.

A specific solution to a differential equation, as opposed to a collection of solutions (like the general solution), is called a *particular solution* to the differential equation. Thus, to solve an initial value problem, you usually find the general solution to the differential equation, and then among those, you find the particular solution which satisfies the initial data.

We will conclude this extended example/discussion by showing that we could have given the general solution to Formula 4.1 in such a way that $A = 0$ did not appear as a special case, and in such a way that we solve every initial value problem arising from Formula 4.1.

How do we do this? We simply give ourselves completely general initial data, $A_0 = A(t_0)$, and we write the solution to the resulting initial value problem in terms of t_0 and A_0.

Let us temporarily assume that $A_0 \neq 0$, so that $A = 0$ is not a solution to our initial value problem. Then, the given data determines the value of C in $A = 2/(C - t^2)$; we must have

$$A_0 = \frac{2}{C - t_0^2}.$$

Solving for C, we find that $C = t_0^2 + (2/A_0)$, and so

$$A = \frac{2}{t_0^2 + \frac{2}{A_0} - t^2} = \frac{2A_0}{2 + A_0(t_0^2 - t^2)},$$

where, in the last step, we multiplied the numerator and denominator by A_0, and rearranged terms in the denominator. Now, you should see the nice thing, the cool thing, that we just obtained; we found that

$$A = \frac{2A_0}{2 + A_0(t_0^2 - t^2)}, \tag{4.4}$$

when $A_0 \neq 0$, but, written in this final form, the solution yields the correct result, $A(t) = 0$, even when $A_0 = 0$. Therefore, Formula 4.4 is the "best" way to write the general solution to the original differential equation in Formula 4.1.

Unfortunately, for some differential equations and some initial data, it is **not** true that specifying the initial data necessarily yields a **unique** solution, even for a fairly simple-looking differential equation. We will be able to explain what goes "wrong" for this type initial value problem in Section 4.3.

Example 4.1.1. Consider the initial value problem

$$\frac{dy}{dx} = y^{2/3}; \qquad y(0) = 0.$$

Show that $y = f(x) = 0$ and $y = g(x) = x^3/27$ are both solutions to this initial value problem on the entire real line $(-\infty, \infty)$.

Solution:

Certainly, both f and g satisfy the initial data: $f(0) = 0 = g(0)$. Clearly $(0)' = 0^{2/3}$. We need to check that $y = x^3/27$ satisfies the differential equation. We find

$$\left(\frac{x^3}{27}\right)' = \frac{3x^2}{27} = \frac{x^2}{9} = \left(\frac{x^3}{27}\right)^{2/3}.$$

Therefore, $y = f(x) = 0$ and $y = g(x) = x^3/27$ are both solutions to the given initial value problem.

Let's look at an example of how an interesting 1st-order differential equation arises in a relatively easy physical situation, a situation that naturally leads to an initial value problem. We will then see that some of our earlier results allow us to solve the problem.

Example 4.1.2. A mixing tank initially contains 50 gallons of salt water, which has 10 pounds of salt dissolved in it, i.e., the initial salt concentration of the salt water is 1/5 of a pound of salt per gallon. Salt water containing 1 pound of salt per gallon enters the tank at a rate of 2 gallons per second. The water in the tank is continually stirred, so that the salt in the water is uniformly distributed throughout the salt water in the tank. Salt water is also being removed from the tank at 2 gallons per second.

Find the number of pounds, P, of salt in the tank at time t seconds.

Solution:

It might seem hard to get started, but you are given a number of **rates**, and so you write down an equation which describes the rate of change of the pounds of salt in the tank. This will be a differential equation. You are also given initial data that tells you how many pounds of salt are in the tank at time 0. Thus, you obtain an initial value problem to solve.

If you just think of the problem in words, you can easily translate it into mathematics. Throughout this discussion, when we write "the rate of change", we mean with respect to time. Also, we will abbreviate "pounds of salt" by simply writing "salt" in places.

Hopefully, it seems obvious that

rate of change of salt in tank $=$ rate salt enters tank $-$ rate salt leaves tank.

Also, it should seem clear that

rate salt enters tank $=$ (rate gallons enter tank)(salt per gallon in entering water),

and

rate salt leaves tank $=$ (rate gallons leave tank)(salt per gallon in leaving water).

The three "word equations" above practically give us the differential equation that we're after, but we need two more pieces.

First, of course, the rate of change of pounds of salt in the tank is precisely the derivative dP/dt. Second, the salt per gallon in the water that's leaving is the only other data that we need that is not explicitly given to us. However, the salt is uniformly distributed throughout the water in the tank. Thus, the salt concentration in the water that's leaving is the same as the concentration in the entire tank, and that's just the total number of pounds of salt in the tank divided by the total number of gallons of water in the tank. As water is entering and leaving the tank at the same rate, there are always 50 gallons of salt water in the tank, and the total number of pounds of salt in the tank is exactly what P stands for. Therefore, at all times t, the concentration of salt in the water that's leaving is $P/50$ pounds per gallon.

If we put all of this together, and insert the given data, we obtain the differential equation

$$\frac{dP}{dt} = (2 \text{ gal/s})(1 \text{ lb/gal}) - (2 \text{ gal/s})\left(\frac{P}{50} \text{ lb/gal}\right)$$

in pounds per second. Dropping the units, since we know they're consistent, we have

$$\frac{dP}{dt} = 2 - \frac{P}{25},$$

with initial data $P(0) = 10$ lb.

How do we solve this differential equation? Well...in Section 4.3, we'll have a general process for dealing with *separable equations*, like this one. For now, however, we will use Theorem 2.4.14, which tells us (using different notation and variables) that the (general) solution to $du/dt = ku$ is $u = Ce^{kt}$. How do we use this? After all, we don't have $dP/dt = (-1/25)P$; we have an extra 2 on the right. We get clever.

Let $u = dP/dt$, so that our differential equation becomes $u = 2 - P/25$. Differentiate both sides with respect to t to get

$$\frac{du}{dt} = 0 - \frac{1}{25} \cdot \frac{dP}{dt} = -\frac{1}{25}u.$$

Now, our result from Theorem 2.4.14 tells us that $u = Ce^{-t/25}$, for some constant C. Therefore, $dP/dt = Ce^{-t/25}$, but we also know that $dP/dt = 2 - P/25$. Hence,

$$Ce^{-t/25} = 2 - P/25,$$

and we solve for P to obtain

$$P = 50 - 25Ce^{-t/25}.$$

If we use the initial data that $P(0) = 10$, we find that $10 = 50 - 25C$, so $25C = 40$. Therefore, our particular solution to the initial value problem tells us that the number of pounds of salt in the tank at time t seconds is

$$P = 50 - 40e^{-t/25}.$$

Remark 4.1.3. There is a final point that we should address in this introductory section, since it arises in many differential equations problems.

There is the question of whether or not it is reasonable to assume that our original differential equation in Example 4.1.2, $dP/dt = 2 - P/25$, holds when $t = 0$. If the equation does not hold at $t = 0$, then how can we plug the initial data at $t = 0$ into the general solution in order to solve the initial value problem?

There are two ways of addressing this type of recurring problem.

One is to simply assume that the entire physical set-up existed before time $t = 0$, even if it's just for the tiniest amount of time before $t = 0$. Then, we would be able to look for a function P that is defined on an **open** interval containing 0, and require that this P satisfies the differential equation and the given initial condition(s).

The second way around this issue is the one that we will usually adopt, when we are given initial data which involves a value t_0 for the independent variable, and we are looking for a solution P whose domain is an interval which has t_0 as an included endpoint, i.e., an interval I of one of the forms $[t_0, b)$, $[t_0, b]$, $(a, t_0]$, or $[a, t_0]$, where a and b are unequal to t_0. This is very common when the independent variable denotes time, and the physical situation is not assumed to exist before time 0. The problem in such cases is that the differential equation cannot hold at t_0; for dP/dt does not exist at $t = t_0$, if P is not defined on an **open** interval around t_0.

In these cases, we implicitly assume that what we are looking for as a solution to the initial value problem is a function $P = P(t)$ which is **continuous** on the entire interval I, which satisfies the given initial condition(s), and which satisfies the given differential equation on the **interior** of the interval I.

For example, if we followed this approach in Example 4.1.2, we would have obtained that we want P to be a continuous function on the interval $[0, \infty)$ such that $P = 50 - 25Ce^{-t/25}$ on the open interval $(0, \infty)$. Therefore, we find that P must equal $50 - 25Ce^{-t/25}$ on the entire half-open interval $[0, \infty)$, and now we can plug in the initial data to determine C. This is, in fact, exactly what we did in Example 4.1.2 to get our solution to the initial value problem, we simply didn't worry about the possible problem of $t = 0$ being an endpoint of the domain of P. Everything just works out correctly, even if you ignore the issue.

In the remaining sections of this chapter on differential equations, we shall not mention this "endpoint issue" again. We will assume that we have made one of the assumptions above that "fixes" the problem.

4.1.1   Exercises

For each of the differential equations in Exercises 1 through 4, two solutions are proposed. Decide which is a solution, if either one is.

1. $\left(\dfrac{dy}{dx}\right)^2 = 1 - y^2$; potential solutions: $y = \sin^2 x$, or $y = \sin x$.

2. $\dfrac{d^2 y}{dx^2} = y(2 + 4x^2)$; potential solutions: $y = e^x$ or $y = e^{x^2}$.

3. $y - \dfrac{d^2 y}{dx^2} = x$; potential solutions: $y = x + e^x$ or $y = e^{-x}$.

4. $\dfrac{y'}{x} = \dfrac{1}{y}$; potential solutions: $y = \sqrt{x}$ or $y = \sqrt{x^2 + 1}$.

5. Consider the 1st-order differential equation

$$\frac{dy}{dx} + ky = sx,$$

with k and s arbitrary constants.

   a. Take the second derivative of each side, with respect to x, and let $T = d^2 y/dx^2$, in order to yield a well-known differential equation, in terms of T and x.

   b. Solve the differential equation you found in part (a) to yield the general solution for $T = d^2 y/dx^2$.

   c. Find constants a and b so that $y = \dfrac{T}{k} + ax + b$ is a solution to the original differential equation.

In Exercises 6 through 12 and Figures a through f, you are given differential equations and graphs of some solutions. By considering the slopes dy/dx of the solution curves, determine which set of solutions seem to fit each differential equation.

6. $\dfrac{dy}{dx} = y - x^2$.   7. $\dfrac{dy}{dx} + y = e^x$.   8. $\dfrac{dy}{dx} = \dfrac{y^2}{x}$.   9. $\dfrac{dy}{dx} = \dfrac{1 + y^2}{x^2}$.

10. $\dfrac{dy}{dx} = y(x - 1)$.   11. $\dfrac{dy}{dx} = y \cos x$.   12. $\dfrac{dy}{dx} = \dfrac{\sin x}{y^2}$.

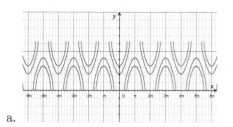

a.

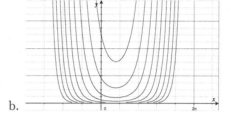

b.

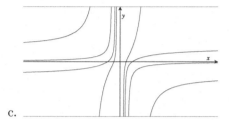

c.

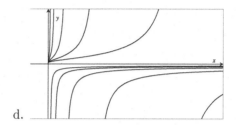

d.

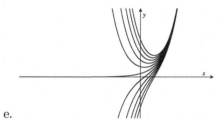

e.

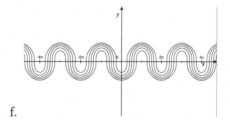

f.

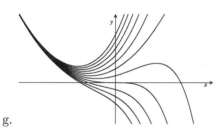

g.

Identify the order of the differential equations in Exercises 13 through 15.

13. $\dfrac{dA}{dx} = x^3$.

14. $y''(x) + 5y^4 y'(x) + 6x^3 y = 0$.

15. $t^3 \dfrac{dA}{dt} - t\dfrac{d^2 A}{dt^2} + 5 = \sin t$.

16. Show the function $y = 1/x$ satisfies the differential equation $y^{(n)}(x) = (-1)^n n! y^{n+1}$, for any positive integer n.

17. Give an explicit solution to $\dfrac{dA}{dt} = tA^2$, if $A(0) = 6$.

18. Define the term *equilibrium solution*.

19. Explain the difference between a *general* and a *particular* solution to a differential equation.

A differential equation of the form $dy/dx = F(y)$ (i.e., where there is no explicit reference to the independent variable on the right-hand side) is said to be *autonomous*. For example, $dy/dx = 3y$ is autonomous but $dy/dx = 3y + x$ is not. In Exercises 20 through 22, determine if the differential equation is autonomous.

20. $\dfrac{dy}{dx} = 3x^2$.

21. $\dfrac{dP}{dt} = kP$.

22. $y'(t) = (k - ay)y$.

23. The equation

$$y''(x) - 2xy'(x) + ky = 0$$

is called the *Hermite Equation* and is used in quantum mechanics. What is the order of the Hermite Equation?

24. An equation in the form $a_n y^{(n)}(x) + a_{n-1} y^{(n-1)}(x) + ... + a_1 y'(x) + a_0 y(x) = 0$ is called a *Homogeneous Linear Differential Equation with Constant Coefficients*. The word "homogeneous" refers to the fact that the right-hand side of the equation is equal to zero.

    (▶)

    a. Show that if y_1 and y_2 are solutions to such an equation, then so is $y_1 + y_2$.

    b. Show that if y is a solution to such an equation, then so is cy, where c is any constant.

In Exercises 25 through 30, show that y_1 and y_2 are solutions to the homogenous second order linear differential equation with constant coefficients.

25. $y''(x) + 7y'(x) + 10y(x) = 0$, $y_1(x) = Ce^{-5x}$, $y_2(x) = Ke^{-2x}$.

26. $y''(x) + y'(x) - 12y(x) = 0$, $y_1(x) = Ce^{-4x}$, $y_2(x) = Ke^{3x}$.

27. $y''(x) + 8y'(x) + 16y(x) = 0$, $y_1(x) = Ce^{-4x}$, $y_2(x) = Kxe^{-4x}$.   (▶)

28. $y''(x) - 14y'(x) + 49y(x) = 0$, $y_1(x) = Ce^{7x}$, $y_2(x) = Kxe^{7x}$.

29. $y''(x) + y'(x) + y(x) = 0$, $y_1(x) = e^{-x/2} \sin\left(\dfrac{\sqrt{3}}{2}x\right)$, $y_2(x) = e^{-x/2} \cos\left(\dfrac{\sqrt{3}}{2}x\right)$.

30. $y''(x) + 9y(x) = 0$, $y_1(x) = \sin(3x)$, $y_2(x) = \cos(3x)$.

31. The *characteristic equation* associated with the differential equation

$$y''(x) + ay'(x) + by = 0$$

   is the algebraic equation

$$r^2 + ar + b = 0.$$

   Somewhat surprisingly, the roots of this algebraic equation give us solutions to the original differential equation.

   a. Show that, if r_0 is a real root of the characteristic equation, then $y(x) = Ce^{r_0 x}$ is a solution of the differential equation.

   b. Show that, if the characteristic equation has a repeated (real) root, r_0, i.e., if $r^2 + ar + b = (r - r_0)^2$, then $y(x) = Cxe^{r_0 x}$ is another solution to the differential equation.

32. Recall that the solutions to a quadratic equation $ar^2 + br + c = 0$ can be written in the form

$$r = \frac{-b \pm \sqrt{b^2 - 4ac}}{2a}.$$

   If the equation has no real roots, the roots take the form $\mu \pm \lambda\sqrt{-1}$.

   Suppose that $y''(x) + by'(x) + cy(x) = 0$ and that the characteristic equation has no real roots. Show that $y_1 = Ce^{\mu x}\sin(\lambda x)$ and $y_2 = Ke^{\mu x}\cos(\lambda x)$ are solutions to the differential equation, where C and K are arbitrary constants.

33. The characteristic equation can be defined for any homogenous linear differential equation with constant coefficients. If

$$a_n y^{(n)}(x) + a_{n-1}y^{(n-1)}(x) + \ldots + a_1 y'(x) + a_0 y(x) = 0$$

   then the associated characteristic equation is

$$a_n r^n + a_{n-1}r^{n-1} + \ldots + a_1 r + a_0 = 0.$$

   Show that if r_0 is a real root of the characteristic equation, then $y = Ce^{r_0 x}$ is a solution to the differential equation.

34. Consider the two differential equations:

$$y''(t) + by'(t) + cy(t) = 0$$
$$y''(t) + by'(t) + cy(t) = h(t)$$

where a and b are constants and h is some function of t. Show that if $y_1(t)$ is a solution to the first equation and y_2 is a solution to the second equation, then $y_1 + y_2$ is also a solution to the second equation.

The general solution of a second order linear differential equation has two free parameters and therefore two initial conditions are needed to determine a particular solution. In Exercises 35 through 37, you are given a differential equation, the general solution, and two initial conditions. Use this information to solve for C and K, and so find the particular solution to the initial value problem.

35. $y''(t) + 13y'(t) + 40y(t) = 0$, $y(t) = Ce^{-5t} + Ke^{-8t}$, $y(0) = 0$, $y'(0) = 10$.

36. $y''(t) + 25y(t) = 0$, $y(t) = C\sin(5t) + K\cos(5t)$, $y(0) = 0$, $y'(0) = 10$.

37. $y''(t) - 12y'(t) + 36y(t) = 0$, $y(t) = Ce^{6t} + Kte^{6t}$, $y(0) = 3$, $y'(0) = 0$.

38. The differential equation $y''(t) + y'(t) - 2y(t) = 0$ has general solution $y(t) = ae^t + be^{2t} + ce^{3t}$. Find a, b and c, if $y(0) = 1$, $y'(0) = -1$ and $y''(0) = -5$.

39. Suppose a mixing tank contains 100 gallons of salt water initially containing 30 pounds of dissolved salt. Salt water containing 4 pounds of salt per gallon enters the tank at a rate of 3 gallons per second. Uniformly concentrated salt water exits the tank at 3 gallons per second.

    a. Find the general equation for the number of pounds of salt in the tank after t seconds.

    b. After how many seconds will there be 50 pounds of salt in the tank?

    c. What is $\lim_{t\to\infty} P(t)$? Why does this make sense physically?

40. Suppose a mixing tank contains G gallons of salt water initially containing P_0 pounds of dissolved salt. Salt water containing a pounds of salt per gallon enters the tank at a rate of r gallons per second. Uniformly concentrated salt water exits the tank at r gallons per second.

    a. Find an expression for the number of pounds of salt in the tank after t seconds in terms of G, P_0, a and r.

   b. What is $\lim_{t \to \infty} P(t)$?

41. Recall Newton's Law of Cooling from Exercise 52 in Section 2.4: if $T = T(t)$ is the temperature of an object, which is placed in surroundings which are kept at a constant (ambient) temperature, T_{amb}, then the rate of change of the temperature of the object, with respect to time, is proportional to the difference between the ambient temperature and the temperature of the object. Assume that t is measured in hours.

This gives us the differential equation

$$\frac{dT}{dt} = k\left(T_{\text{amb}} - T\right),$$

where $k > 0$ is a constant.

In this exercise, the "object" will actually be the **surface** of an ice cube, not the entire ice cube.

   a. Suppose an ice cube with initial surface temperature 32° F is placed in a room with temperature 70° F and that $k = 0.5$. Setup and solve the resulting initial value problem. Hint: use the techniques from the salt-mixing example.

   b. At what time will the surface temperature of the ice be 50° F?

   c. What is $\lim_{t \to \infty} T(t)$? Why does this make sense physically?

42. The *Law of Conservation of Energy* states that the sum of potential and kinetic energy of a moving object is constant.

   a. The *kinetic energy* of an object is defined as $K_E = \frac{1}{2}mv^2$, where m is the mass of the object and v is its velocity. For a falling object, *potential energy* is defined as $P_E = mgh$ where h is the height of the object above the ground, and g is the gravitational constant. Express the Law of Conservation of Energy for a falling object (with only vertical velocity) in the form of a 1st-order differential equation, involving $h = h(t)$ and dh/dt.

   b. Take the derivative with respect to t of your expression from part (a). Show that, during any interval of time in which the velocity is not zero, $\frac{d^2h}{dt^2} = -g$ (so that the acceleration is constant).

   c. Does the height function $h(t) = -\frac{1}{2}gt^2 + v_0t + h_0$ satisfy the latter differential equation from part (b)?

   d. What is the total energy (the sum of kinetic and potential energies) for a falling object whose height function is the one from part (b), in terms of g, m, v_0 and h_0?

43. Suppose that the momentum, in kg·m/s, of an object of mass 10 kilograms moving along the x-axis, marked off in meters, is proportional to the net force, in Newtons, acting on it, with proportionality constant $k = 0.5$ (kg·m/s)/N, where the time t is in seconds.

    a. Set up a differential equation for the velocity of the object.

    b. Solve the differential equation under the assumption that the velocity at time $t = 0$ is 1 meter per second.

    c. Determine the momentum of the object at time $t = 10$ seconds from the start of the experiment.

44. A spherical chunk of ice is melting in such a way that its volume decreases in a rate proportional to the exposed surface area.

    a. Set up a differential equation relating volume and surface area.

    b. Use the formulas for volume and surface area, in terms of the radius, of the sphere to determine a differential equation, involving radius, which models this situation.

    c. Find the volume of the block, as a function of time, if the initial radius of the block is 0.25 meters.

45. A certain city collects data about its fluctuating population. It finds that, on average, each year, 6% of its people move out, and 1000 families, of average size 3.5 people per family, move in for newly created jobs and due to the draw of the educational system. Let t denote the time in years since 1990.

    a. Put the rates of change together to find a differential equation relating P and $\dfrac{dP}{dt}$.

    b. Following Example 4.1.2, solve the differential equation to find a general form for $P(t)$.

    c. If at time $t = 0$, 60,000 people live in this city, at what time will the population reach 150,000?

4.2 Anti-derivatives

In this section, we consider the general solution of the "easiest" kind of differential equation: a differential equation of the form

$$\frac{dy}{dx} = f(x),$$

for a given function $f(x)$. That is, we are given a function $f(x)$ and want to find all of those functions $y = F(x)$ such that $F'(x) = f(x)$.

Example 4.2.1. Find the general solution of the differential equation

$$\frac{dy}{dx} = 3x^2.$$

Solution:

Hopefully, you quickly realize that $y = F(x) = x^3$ is **one** solution of the differential equation. What's the general solution? Corollary 1.5.12 tells us that if $y = G(x)$ is a function whose derivative is $3x^2$, i.e., the same as the derivative of x^3 (on an open interval), then $y = G(x) = x^3 + C$, for some constant C (and, certainly, every such function has derivative equal to $3x^2$).

Therefore, the general solution to the differential equation is $y = x^3 + C$, where C is an arbitrary constant.

Definition 4.2.2. *Given a function $f(x)$, defined on an open interval I, a function $F(x)$, on I, such that $F'(x) = f(x)$ is called an* **anti-derivative of** *$f(x)$, (with respect to x).*

Thus, an anti-derivative $y = F(x)$ of $f(x)$ is a solution to the differential equation $dy/dx = f(x)$.

If $F(x)$ is an anti-derivative of $f(x)$, on an open interval, then every anti-derivative of $f(x)$, on that interval, is given by $y = F(x) + C$, where C is a constant. The collection $y = F(x) + C$ is called the **(general) anti-derivative of** *$f(x)$, (with respect to x); it is the general solution to $dy/dx = f(x)$.*

The notation for the general anti-derivative of $f(x)$, with respect to x, is

$$\int f(x)\, dx.$$

This is also called the **(indefinite) integral of** $f(x)$, *with respect to* x.

Remark 4.2.3. It follows from the definition that the units of $\displaystyle\int f(x)\, dx$ are the units of $f(x)$ times the units of x.

You should think of the anti-differentiation, with respect to x, operator, $\int (\ ) \, dx$, as essentially being the inverse operator of $\dfrac{d}{dx}(\ )$, differentiation with respect to x.

We wrote "essentially" here because, if you first differentiate and then anti-differentiate, you get what you started with, except that there is an additional $+C$; that is, you end up with a collection of functions that all differ by constants, instead of simply the one function that you started with.

We should also comment on the term "indefinite integral". There is another notion, called the *definite integral* of a function over a closed interval. The definite integral is defined in such a way that it agrees with one's intuitive idea of what a "continuous sum of infinitesimal contributions" should mean. This would seem to be unrelated to anti-differentiating. However, there is a theorem, the **Fundamental Theorem of Calculus**, which tells us: i) every continuous function possesses an anti-derivative, and ii) the primary step used to obtain a nice formula for a definite integral is to produce an anti-derivative of the given function.

Hence, anti-differentiation is frequently referred to simply as "integration", and definite integration is also simply referred to as "integration"; the context should always make it clear whether the meaning is anti-differentiation or definite integration. In addition, the symbols for anti-differentiating $\int (\ ) \, dx$ are essentially the same as the symbols used for definite integration; in fact, the $\int$ symbol was introduced by Leibniz, and is supposed to look like a fancy letter "s", for "summa", the Latin word for "sum" or "total".

The definite integral, anti-differentiation, and the Fundamental Theorem of Calculus form the foundations of *Integral Calculus*, the main topic of the book, *Worldwide Integral Calculus*, which is the textbook that follows the one you are currently reading.

All of the differentiation formulas which we have derived up to this point yield corresponding anti-differentiation/integration formulas, it's just a matter of reading things "in reverse", for, if $F'(x) = f(x)$, then the corresponding integration rule is $\int f(x)\, dx = F(x) + C$, where C denotes an arbitrary constant, as it does in all of the integration rules that we will give. In this context, $f(x)$ is frequently referred to as the *integrand*.

For instance, we have a **Power Rule for Integration**:

Theorem 4.2.4. *For all x in an open interval for which the functions involved are defined,*

1. $\int 0\, dx = C$;

2. $\int 1\, dx = x + C$;

3. *if* $p \neq -1$, $\displaystyle\int x^p\, dx = \frac{x^{p+1}}{p+1} + C$; *and*

4. $\displaystyle\int x^{-1}\, dx = \int \frac{1}{x}\, dx = \ln|x| + C.$

Remark 4.2.5. In many books, only the third formula above is referred to as the Power Rule for Integration.

As is frequently the case, you should try to remember this rule not in symbols, but in words; it says that, as long as the exponent is not -1, you obtain the anti-derivative of x raised to a constant exponent by adding one to the exponent, and dividing by the new exponent (and then adding a C).

Remark 4.2.6. There are two fairly common, horrific mistakes associated with the integration rule $\displaystyle\int x^{-1}\, dx = \int \frac{1}{x}\, dx = \ln|x| + C.$

The first big mistake is to treat the $p = -1$ case in the same manner as the cases where $p \neq -1$. If you were to do this, you would obtain

$$\int x^{-1}\, dx \;=\; \frac{x^{-1+1}}{-1+1} + C \;=\; \frac{x^0}{0} + C.$$

The undefined division by 0 should immediately tell you that you've done something wrong, and remind you that you must treat the $p = -1$ case differently.

The second big mistake may come later, when we have more integration rules. It will then be tempting to look at the formula $\int \frac{1}{x}\, dx = \ln|x| + C$ and think that it implies that

$$\int \frac{1}{\text{anything}}\, dx \;=\; \ln|\text{anything}| + C.$$

This is completely wrong (in general); it is not true that the derivative of the expression on the right would be the integrand. The problem is that, if you differentiate the expression on the right, you do, in fact, get a $1/\text{anything}$ factor, but then the Chain Rule tells you that this is multiplied by $d(\text{anything})/dx$.

Example 4.2.7. Solve the initial value problem

$$\frac{dP}{dr} \;=\; \sqrt{r}\;; \qquad P(9) = -7.$$

Solution:

We find that

$$P \;=\; \int r^{1/2}\, dr = \frac{r^{3/2}}{3/2} + C \;=\; \frac{2}{3} r^{3/2} + C.$$

We need to determine C. We have

$$-7 \;=\; P(9) \;=\; \frac{2}{3}\,(9)^{3/2} + C \;=\; 18 + C.$$

Therefore, $C = -25$, and so

$$P \;=\; \frac{2}{3} r^{3/2} - 25.$$

The linearity of the derivative gives us the linearity of the anti-derviative.

Theorem 4.2.8. *If a and b are constants, not both zero, then*

$$\int af(x) + bg(x)\,dx \;=\; a \cdot \int f(x)\,dx \;+\; b \cdot \int g(x)\,dx.$$

Remark 4.2.9. The prohibition against $a = b = 0$ in Theorem 4.2.8 is there for just one reason: we do not want both of the arbitrary constants on the right to be eliminated by multiplying by zero. We will explain this more fully.

Since each indefinite integral actually yields a set, or collection, of functions, there may be some question in your mind about what it means to multiply a set of functions by a constant, like a or b, and what it means to add two such sets. In other words, you may wonder exactly what the right-hand side of the equality in Theorem 4.2.8 means.

For instance, what does it mean to write that

$$\int (5x^3 - 7\sqrt{x})\,dx \;=\; 5 \cdot \int x^3\,dx \;-\; 7 \cdot \int \sqrt{x}\,dx?$$

We know, from the Power Rule for Integration, that $\int x^3\,dx = x^4/4 + C_1$ and

$$\int \sqrt{x}\,dx = \int x^{1/2}\,dx \;=\; \frac{x^{3/2}}{3/2} + C_2 \;=\; 2x^{3/2}/3 + C_2,$$

where we have used C_1 and C_2, in place of using simply C twice, since we don't want to assume that we have to pick the two arbitrary constants to be the same thing.

So, what does $5 \cdot \int x^3\,dx$ mean? It means the collection of functions obtained by taking 5 times any function from the collection of functions $x^4/4 + C_1$; that is, the collection of functions $5x^4/4 + 5C_1$, where C_1 could be any constant. But, if C_1 can be anything, then $5C_1$ can be anything, and we might as well just call it B_1, where B_1 can be any number. Thus, we can write the collection of functions $5 \cdot \int x^3\,dx$ as $5x^4/4 + B_1$.

However, here's the part that can be confusing; instead of using a new constant name, like B_1, it is fairly standard to just use the name C_1 again, i.e., to use C_1 to now denote 5 times the old value of C_1. Assuming that we had not determined some value for the old C_1, there is no harm in doing this, but it certainly can make things look confusing, for you frequently see calculations like

$$5 \cdot \int x^3 \, dx \;=\; 5\bigl(x^4/4 + C_1\bigr) \;=\; 5x^4/4 + C_1,$$

where the C_1 on the far right above is actually 5 times the C_1 in the middle.

Similarly, we write

$$-7 \cdot \int \sqrt{x} \, dx \;=\; \int x^{1/2} \, dx \;=\; -7\bigl(2x^{3/2}/3 + C_2\bigr) \;=\; -14x^{3/2}/3 + C_2.$$

Therefore,

$$5 \cdot \int x^3 \, dx - 7 \cdot \int \sqrt{x} \, dx \;=\; 5x^4/4 + C_1 - 14x^{3/2}/3 + C_2 \;=\;$$

$$5x^4/4 - 14x^{3/2}/3 + (C_1 + C_2) \;=\; 5x^4/4 - 14x^{3/2}/3 + C,$$

where $C = C_1 + C_2$ can be any real number.

Using this example as a guide, you can see what to do more generally: whenever you have a *linear combination* of indefinite integrals, i.e., a sum of constants multiplied times indefinite integrals, you do **not** include an arbitrary constant for each individual indefinite integral; instead, for each indefinite integral you write **one** particular anti-derivative, and then put in a single $+C$ at the end.

The fact that an indefinite integral is actually a collection of functions can lead to seemingly bizarre results. For instance, while it's true that $\int x \, dx = \int x \, dx$, it would nonetheless be bad to write that $\int x \, dx - \int x \, dx = 0$. Why? Because our operations on sets of functions yield that the correct calculation is

$$\int x \, dx - \int x \, dx \; = \; x^2/2 + C_1 - \left(x^2/2 + C_2\right) \; = \; C_1 - C_2 \; = \; C,$$

which is the same as $\int 0 \, dx$.

This agrees with what we wrote above: when you have a linear combination of indefinite integrals, you should use one particular anti-derivative for each integral, and then add a C at the end. Thus, in the above calculation, you should get $x^2/2 - x^2/2 + C$, which is just C.

Example 4.2.10. Calculate the indefinite integral

$$\int \left(\frac{5}{w} - 3 + 7w^3 + 5\sqrt[9]{w} \right) dw.$$

Solution:

We calculate

$$\int \left(\frac{5}{w} - 3 + 7w^3 + 5\sqrt[9]{w} \right) dw \; = \; \int \left(5 \cdot \frac{1}{w} - 3 + 7w^3 + 5w^{1/9} \right) dw \; =$$

$$5 \ln |w| - 3w + 7 \cdot \frac{w^4}{4} + 5 \cdot \frac{w^{(1/9)+1}}{(1/9) + 1} + C \; =$$

$$5 \ln |w| - 3w + \frac{7w^4}{4} + \frac{9w^{10/9}}{2} + C.$$

Example 4.2.11. Suppose that an object moves in a straight line with constant acceleration a meters per second per second. Show that the position of the object $p = p(t)$, in meters, is given by

$$p = at^2/2 + v_0 t + p_0,$$

where p_0 is the initial position of the object (i.e., the position at $t = 0$), in meters, v_0 is the initial velocity in m/s, and t is the time in seconds.

Solution:

Acceleration a is the derivative, with respect to t, of the velocity v, i.e., $a = dv/dt$. This is exactly the same as writing that v is an anti-derivative of a, with respect to t, i.e., $v = \int a\,dt$. Since a is a constant, we find

$$v = \int a\,dt = a \int 1\,dt = at + C \ \text{ m/s},$$

for some constant C. Therefore, $v = at + C$, but we would like to give some physical meaning to the constant C.

How do we do this? We are not given any other data. The answer is that we use *tautological initial data*; that is, we use initial data that is simply obviously true. We use that, when $t = 0$, the velocity v equals v_0. Why is this true? Because it says something that's clearly true: at time 0, the velocity equals the velocity at time 0. It may seem strange, but using this tautological initial data actually gets us somewhere.

We have $v = at + C$. Now, plug in that, when $t = 0$, $v = v_0$. You find $v_0 = a \cdot 0 + C$, i.e., $C = v_0$. Thus, we conclude that $v = at + v_0$. Notice that no one has to give you any extra data to conclude that $C = v_0$; it follows from the equation $v = at + C$.

Now, we have that $v = at + v_0$, and we know that the velocity v equals the rate of change of p, with respect to time, i.e., $v = dp/dt$. This is the same as writing that $p = \int v\,dt$. Therefore, we have

$$p = \int v\,dt = \int (at + v_0)\,dt = a\left(\int t\,dt\right) + v_0\left(\int 1\,dt\right) =$$
$$at^2/2 + v_0 t + C \ \text{ meters},$$

where this C is definitely not the same C that we used in the equation for v.

How do we find this new C? We plug in more tautological initial data, namely that $p = p_0$ when $t = 0$. We find that $p_0 = a(0)^2/2 + v_0(0) + C$, and so $C = p_0$. Thus, we conclude

$$p = at^2/2 + v_0 t + p_0 \ \text{ meters}.$$

Other integration formulas obtained at once from differentiation formulas are:

> **Theorem 4.2.12.** *As functions on the entire real line* $(-\infty, \infty)$, *we have*
>
> *1.*
> $$\int \cos x \, dx \;=\; \sin x + C;$$
>
> *2.*
> $$\int \sin x \, dx \;=\; -\cos x + C; \text{ and}$$
>
> *3.*
> $$\int e^x \, dx \;=\; e^x + C.$$

> Note that the integration formulas for sin and cos have the negative sign in the "opposite" place from the differentiation formulas. This frequently leads to confusion. It shouldn't. Remember: you are finding anti-derivatives. This means that the derivative of what you end up with should be what you started with (i.e., the integrand).

Let's look at another example in which you are given the acceleration of an object, and are asked to find the velocity and position, but, this time, we have an acceleration which is not constant.

Example 4.2.13. Suppose that the acceleration, in m/s^2, of an object moving in a straight line is $a = \sin t$, where t is the time in seconds. Find the velocity and position of the object, as functions of time, in terms of the initial velocity and initial position.

Solution:

We find that
$$v \;=\; \int a \, dt \;=\; \int \sin t \, dt \;=\; -\cos t + C,$$
and, plugging in the tautological initial data, we find $v_0 = -\cos(0) + C = -1 + C$. Thus, $C = v_0 + 1$, and
$$v \;=\; -\cos t + v_0 + 1 \ \ \text{m/s}.$$
Now, we integrate the velocity to find the position:

$$p \;=\; \int v \, dt \;=\; \int (-\cos t + v_0 + 1) \, dt \;=\; -\int \cos t \, dt + (v_0 + 1) \int 1 \, dt \;=\;$$

$$-\sin t + (v_0 + 1)t + C.$$

Finally, we plug in the tautological initial data, in order to give physical meaning to this last C:

$$p_0 = -\sin(0) + (v_0 + 1)(0) + C.$$

Therefore, $C = p_0$, and we find

$$p = -\sin t + (v_0 + 1)t + p_0 \quad \text{meters.}$$

We shall not rewrite every one of our differentiation formulas from Appendix B as integration formulas; however, we'll go ahead and give two more, before looking at the Chain Rule and the Product Rule in their indefinite integral forms.

Theorem 4.2.14. *As functions on the open interval* $(-1, 1)$,

$$\int \frac{1}{\sqrt{1 - x^2}}\, dx = \sin^{-1} x + C.$$

As functions on the open interval $(-\infty, \infty)$,

$$\int \frac{1}{1 + x^2}\, dx = \tan^{-1} x + C.$$

Substitution: the Chain Rule in anti-derivative form:

The Chain Rule as an anti-derivative formula is

Theorem 4.2.15. (Integration by Substitution) *If f and g are differentiable functions, then*

$$\int f'(g(x))g'(x)\,dx \;=\; f(g(x)) + C,$$

or, letting $u = g(x)$,

$$\int f'(u)\frac{du}{dx}\,dx \;=\; \int f'(u)\,du \;=\; f(u) + C.$$

The second formula for substitution is particularly easy to use; it looks as though the dx's cancel, as in multiplying fractions. This is **not** what's happening, but it does make substitution easier to remember. It also means that the differential notation

$$du = \frac{du}{dx}\,dx,$$

from Section 3.4, yields correct formulas in integrals, and, hence, we use it extensively.

Example 4.2.16. Calculate the integrals

$$\int \cos(e^x + 7) \cdot e^x\,dx \qquad \text{and} \qquad \int \frac{t}{1+t^2}\,dt.$$

Solution:

How should you approach an integral like

$$\int \cos(e^x + 7) \cdot e^x\,dx?$$

First, you should realize that it's not just a linear combination of integrals of specific functions that you've memorized. You should then think "Well...$\cos(e^x + 7)$ is a composition of functions. What if I make a substitution for the "inside" function? I'll let $u = e^x + 7$, so that $\cos(e^x + 7)$ becomes $\cos u$, and see if the remaining part of the integrand looks like du."

In fact, for easy cases, you can do this in your head. If $u = e^x + 7$, then $du = e^x\,dx$, and you see that this is the remaining "factor" in the integral. Thus, by substitution, our original integral is transformed into an integral in terms of u that is very simple:

$$\int \cos(e^x + 7) \cdot e^x\,dx \;=\; \int \cos u\,du \;=\; \sin u + C \;=\; \sin(e^x + 7) + C.$$

Nice.

What about

$$\int \frac{t}{1 + t^2}\,dt?$$

It's not some linear combination of integrals you should immediately know, and there's no obvious composition of functions this time. You might think that $\tan^{-1} t$ is involved somehow, since $\int 1/(1 + t^2)\,dt = \tan^{-1} t + C$, but the t in the numerator causes a problem. You might think that you can factor the integrand and use that in some way:

$$\int \frac{t}{1 + t^2}\,dt \;=\; \int t \cdot \frac{1}{1 + t^2}\,dt \;\overset{?}{\underset{?}{=}}\; \int t\,dt \cdot \int \frac{1}{1 + t^2}\,dt \;=\; \frac{t^2}{2} \cdot \tan^{-1} t + C.$$

THIS IS COMPLETELY WRONG. You can't just integrate each factor in a product and then multiply the results; dealing with products in an integral is more complicated than that. We will get to the integral form of the Product Rule shortly, but, for now, you should differentiate $t^2 \tan^{-1} t/2$ (using the Product Rule) and verify that you don't get anything close to our integrand $t/(1 + t^2)$.

Great. Now we know one way **NOT** to find the integral $\int t/(1 + t^2)\,dt$. We also know that we're discussing substitution here, and so you should suspect that a substitution is involved.

With practice, you should actually see relatively quickly that if you let $w = 1 + t^2$, then $dw = 2t\,dt$, and we have a $t\,dt$ in the integral, so this substitution might be good. We can always "fix" multiplying by a constant, like the 2 in $dw = 2t\,dt$. We divide by 2 to get

$$\frac{1}{2}\,dw \;=\; t\,dt,$$

and

$$\int \frac{t}{1 + t^2}\,dt \;=\; \int \frac{1}{1 + t^2} \cdot t\,dt \;=\; \int \frac{1}{w} \cdot \frac{1}{2}\,dw \;=$$

$$\frac{1}{2}\int \frac{1}{w}\,dw \;=\; \frac{1}{2}\ln|w| + C \;=\; \frac{1}{2}\ln|1+t^2| + C \;=\; \frac{1}{2}\ln(1+t^2) + C,$$

where the last equality follows from the fact that $1 + t^2 \geq 0$ (in fact, $1 + t^2 \geq 1$).

It would be a good exercise for you to differentiate our final answer above, and see how the Chain Rule comes into play to produce our initial integrand.

Integration by Parts: the Product Rule in anti-derivative form:

The Product Rule as an anti-derivative formula is

Theorem 4.2.17. (Integration by Parts) *If f and g are differentiable functions, then*

$$\int f(x)g'(x)\,dx + \int g(x)f'(x)\,dx \;=\; f(x)g(x) + C,$$

or, letting $u = f(x)$ and $v = g(x)$,

$$\int u\,dv + \int v\,du \;=\; uv + C$$

or

$$\int u\,dv \;=\; uv - \int v\,du.$$

It is the last formula above that most people memorize as THE formula for integration by parts.

Here's how a basic integration by parts attack on a problem goes. You look at your integral, and make a choice that lets you write the integral in the form $\int u\,dv$. Then, you apply the integration by parts formula to obtain that your integral equals $uv - \int v\,du$. Then, you hope that the new integral $\int v\,du$ is easier (or, at least, no harder) to integrate than the integral that you started with.

Example 4.2.18. Calculate

$$\int ze^z\, dz \qquad \text{and} \qquad \int z^2 e^z\, dz.$$

Solution:

You should look at $\int ze^z\, dz$, and realize quickly that ze^z does not result from one of the basic derivative formulas that you should have memorized, and so this integration is not a basic one that you should know immediately. In addition, if you think for a minute or so, you should be able to convince yourself that there's no substitution that will help. However, we do see that the integrand is the product of two very different-looking functions, z and e^z; this is a hint that integration by parts may be good to use.

Now what do you do? You identify some factor in the integrand that will be u; that factor should **not** contain the differential (here, dz). The remaining part of the integrand, together with the differential, should be dv. With u and dv determined, you calculate du and $v = \int dv$. You do not need to include an arbitrary constant in your calculation of v; we need **some** v, not all possible v's. Then, you write $uv - \int v\, du$, look at your new integral, and hope that it's easier.

Let's see how this works for $\int ze^z\, dz$. There are two obvious choices for u: either $u = z$ or $u = e^z$. Either one of these will lead to a formula that is **true**, but only one will lead to something **useful**.

Let's look at the **bad** choice first, so that you can see how you can tell when you've made a bad choice. Let's try $u = e^z$. That leaves $z\, dz$ to be dv. Now, if $u = e^z$ and $dv = z\, dz$, then $du = e^z\, dz$ and $v = \int dv = \int z\, dz = z^2/2$, where, as we discussed, we don't include a $+C$ in our calculation of v. Applying the integration by parts formula, we find

$$\int ze^z\, dz \;=\; \int e^z z\, dz \;=\; \int u\, dv \;=\; uv - \int v\, du \;=$$

$$(e^z)(z^2/2) - \int (z^2/2)e^z\, dz. \tag{4.5}$$

This is true, but not particularly helpful for calculating $\int ze^z\, dz$. The power of z went up, and, after moving the constant $1/2$ outside the integral, the rest of the integral, the e^z, is the same as what we started with. You should realize that the new integral is harder to deal with than the original. We could try to integrate by parts again, but, we leave it as an exercise for you to verify, depending on your new choice of u, that either the power of z goes up again (and the e^z remains), or the power of z goes back down to 1, but you end up with exactly the integral that we started with.

So, let's make the other choice for u in integrating by parts to calculate $\int ze^z\,dz$. Let $u = z$, which means that $dv = e^z\,dz$. Then, we find that $du = dz$ and $v = \int dv = \int e^z\,dz = e^z$. Thus, we obtain

$$\int ze^z\,dz \;=\; \int u\,dv \;=\; uv - \int v\,du \;=\;$$

$$ze^z - \int e^z\,dz \;=\; ze^z - e^z + C.$$

Why did this choice of u work better than our earlier one? Before, when we picked u to not include the power of z, the power of z was left in dv; when we integrated dv to get v, the power of z went up. Now, when we pick u to be the power of z, namely z^1, the power of z goes down in the calculation of du. For this reason, it is frequently (but not always – see the next example) a good idea in integration by parts problems which include powers of the variable to let the power of the variable be u.

With this in mind, how do you integrate $\int z^2 e^z\,dz$? You integrate by parts, letting $u = z^2$, which means $dv = e^z\,dz$. What you should get is actually exactly what we got in Formula 4.5, except you need to multiply Formula 4.5 by 2, and rearrange, to obtain:

$$\int z^2 e^z\,dz \;=\; z^2 e^z - 2\int ze^z\,dz.$$

Now, even if we had not already calculated $\int ze^z\,dz$, you should realize that the new integral on the right above is easier than the one you started with, and you would calculate it by a second integration by parts (if we had not already done so). What we find is

$$\int z^2 e^z\,dz \;=\; z^2 e^z - 2\int ze^z\,dz \;=\; z^2 e^z - 2\left(ze^z - e^z + C\right) \;=\; z^2 e^z - 2ze^z + 2e^z + C.$$

It might be instructive to differentiate the final result above, on the right, and verify that you obtain $z^2 e^z$.

Example 4.2.19. Calculate

$$\int t^5 \ln t\,dt \qquad \text{and} \qquad \int \ln t\,dt.$$

Solution:

You look at $\int t^5 \ln t \, dt$, and you realize immediately that this doesn't come from one of our basic derivative/integral formulas. You might think about a substitution, like $w = \ln t$, for a minute or so, but then realize that it doesn't get you anywhere. Then, you decide that, since the integrand is the product of two different kinds of functions, maybe integration by parts would be a good thing to try.

If you look at our previous example, you'd probably be tempted to let u be the power of t, i.e., let $u = t^5$, which would lead to $dv = \ln t \, dt$. However, this is one of those times when picking the power of the variable to be u is a bad idea. Perhaps the most obvious reason why this is bad is because we don't know how to calculate $v = \int dv = \int \ln t \, dt$. You may think that we've discussed this integral, and that it equals $1/t + C$. This is **very wrong**. We know that $(\ln t)' = 1/t$ or, what's the same thing, $\int 1/t \, dt = \ln t + C$ (for $t > 0$), but we don't know $\int \ln t \, dt$ (it's possible that you do, but it hasn't been discussed in the book up to this point). In fact, calculating $\int \ln t \, dt$ is the second part of this example.

So, to calculate

$$\int t^5 \ln t \, dt,$$

we'll try integration by parts with $u = \ln t$, which means that $dv = t^5 \, dt$.

We find that $du = (1/t) \, dt$ and $v = \int dv = \int t^5 \, dt = t^6/6$ (remember, we don't need a $+C$ here). Applying the integration by parts formula, we find

$$\int t^5 \ln t \, dt \;=\; \int u \, dv \;=\; uv - \int v \, du \;=\; (\ln t)(t^6/6) - \int (t^6/6)(1/t) \, dt \;=\;$$

$$\frac{t^6 \ln t}{6} - \frac{1}{6} \int t^5 \, dt \;=\; \frac{t^6 \ln t}{6} - \frac{1}{6} \cdot \frac{t^6}{6} + C \;=\; \frac{t^6 \ln t}{6} - \frac{t^6}{36} + C.$$

Now, let's look at $\int \ln t \, dt$. How could this possibly be an integration by parts problem? There's no product in the integrand! Admittedly, this does not look like an integration by parts problem. Nonetheless, it is. Let $u = \ln t$, which means that $dv = dt$. Then, $du = (1/t) \, dt$, and $v = \int dv = \int dt = t$. Applying the integration by parts formula, we obtain

$$\int \ln t \, dt \;=\; \int u \, dv \;=\; uv - \int v \, du \;=\; (\ln t)(t) - \int t(1/t) \, dt \;=\;$$

$$t \ln t - \int 1 \, dt \;=\; t \ln t - t + C.$$

This agrees with our calculation in Example 2.5.4.

We wish to look at one more integration by parts example, a complicated example.

Example 4.2.20. Calculate

$$\int e^x \cos x \, dx.$$

Solution:

We will calculate this integral via integration by parts. We could use either $u = e^x$ or $u = \cos x$; this time, it actually makes little difference. We will pick $u = e^x$, which means that $dv = \cos x \, dx$. (As an exercise, you should try starting with $u = \cos x$ instead.) We find, then, that $du = e^x \, dx$ and $v = \int dv = \int \cos x \, dx = \sin x$. The integration by parts formula tells us that we have

$$\int e^x \cos x \, dx \;=\; \int u \, dv \;=\; uv - \int v \, du \;=\; e^x \sin x - \int (\sin x) e^x \, dx. \qquad (4.6)$$

But our new integral, $\int (\sin x) e^x \, dx$ is clearly just as difficult to integrate as the original integral. Are we getting anywhere? The answer is "yes", but it's not obvious yet. We will integrate $\int (\sin x) e^x \, dx$ by parts also.

You have to be careful. If you make the choice $u = \sin x$ here, you will obtain that $\int (\sin x) e^x \, dx$ equals $e^x \sin x - \int e^x \cos x \, dx$. If you substitute this into Formula 4.6, you will find that you have undone our original integration by parts, and you will come to the stunning conclusion that $\int e^x \cos x \, dx = \int e^x \cos x \, dx$. It's true, but not very helpful.

When integrating $\int (\sin x) e^x \, dx$ by parts, you need to make the choice of u that is analogous to your choice of u for the original integration by parts; you let $u = e^x$ again, and so $dv = \sin x \, dx$. Then, $du = e^x \, dx$ and $v = \int dv = \int \sin x \, dx = -\cos x$. Applying integration by parts, we find

$$\int (\sin x) e^x \, dx \;=\; \int u \, dv \;=\; uv - \int v \, du \;=\; e^x(-\cos x) - \int (-\cos x) e^x \, dx \;=$$

$$-e^x \cos x + \int e^x \cos x \, dx.$$

If you insert this result into Formula 4.6, you obtain

$$\int e^x \cos x \, dx \;=\; e^x \sin x + e^x \cos x - \int e^x \cos x \, dx.$$

At this point, you may be thinking to yourself "Aaaaagggghhhhhh! We spent all that time integrating by parts, only to end up with the same integral that we started with!" However, the fact that the new occurrence of the original integral has a negative sign in front of it saves us. If you simply add $\int e^x \cos x\, dx$ to each side of this last equation (and fix the loss of the arbitrary constant on the right), you obtain

$$2 \cdot \int e^x \cos x\, dx \;=\; e^x \sin x + e^x \cos x + C,$$

and so

$$\int e^x \cos x\, dx \;=\; \frac{e^x \sin x + e^x \cos x}{2} + C.$$

We end this section with a possibly surprising complication that exists for anti-differentiation; a type of complication which does not occur for differentiation.

Remark 4.2.21. From our derivative formulas in Chapter 2, we see that the derivative of any elementary function is again an elementary function. You might hope that anti-derivatives/integrals would behave equally as well. **They do not.** It is easy to give elementary functions $f(x)$ for which it is possible to prove that there is no elementary function $F(x)$ such that $F'(x) = f(x)$, i.e., $f(x)$ has no elementary anti-derivative. Such functions $f(x)$ include e^{x^2}, $\sin(x^2)$, and $\cos(x^2)$. This was first proved by Liouville in 1835.

The Fundamental Theorem of Calculus guarantees that the functions e^{x^2}, $\sin(x^2)$, and $\cos(x^2)$, and, in fact, all continuous functions, have **some** anti-derivative, but those anti-derivatives need not be elementary functions.

4.2.1   Exercises

Calculate the general anti-derivatives in Exercises 1 through 21.

1. $\displaystyle\int (4x^2 + 4x + 9)\, dx$

2. $\displaystyle\int \left(\frac{5}{w} - 7e^w + 6\sqrt[3]{w} \right) dw$

3. $\displaystyle\int \left(5\sin t - \frac{3}{\sqrt{1-t^2}}\right) dt$

4. $\displaystyle\int \frac{1+v+\sqrt{v}}{v^2}\, dv$

5. $\displaystyle\int \left(\frac{1}{y} + \frac{1}{y^2+1}\right) dy$

6. $\displaystyle\int \frac{5}{3z-7}\, dz$

7. $\displaystyle\int \cos(2\theta - 1)\, d\theta$

8. $\displaystyle\int e^{p+4}\, dp$

9. $\displaystyle\int \frac{r}{r^2-4}\, dr$

10. $\displaystyle\int \left(\frac{x}{\sqrt{x^2-1}} + \frac{1}{|x|\sqrt{x^2-1}}\right) dx$

11. $\displaystyle\int (5\omega - 3)^{100}\, d\omega$

12. $\displaystyle\int \big(2\cos(2t+5) + 3\sin(9t)\big)\, dt$

13. $\displaystyle\int \ln\left[(x+2)^{x+5}\right] dx$

14. $\displaystyle\int \frac{e^{1/x}}{x^3}\, dx$

15. $\displaystyle\int \frac{5}{x\ln x}\, dx$

16. $\displaystyle\int \frac{e^{1/x}}{6x^2}\, dx$

17. $\displaystyle\int \tan\theta\, d\theta$

18. $\displaystyle\int \frac{5}{4+x^2}\, dx$

19. $\displaystyle\int t^4\sqrt{t^5+6}\, dt$

20. $\displaystyle\int \frac{1}{x^2+4x+5}\, dx$ Hint: $x^2 + 4x + 5 = (x+2)^2 + 1$.

21. $\displaystyle\int \frac{1}{\sqrt{-x^2 + 6x - 8}}\, dx$

In each of Exercises 22 through 31, find the anti-derivative of the given function which satisfies the given initial condition. The anti-derivative of each function with a lower-case name is denoted by the upper-case version of the same letter.

22. $h(x) = 4x^2 + 4x + 9$, such that $H(-1) = 2$.

23. $p(w) = \dfrac{5}{w} - 7e^w + 6\sqrt[3]{w}$, such that $P(-1) = 0$.

24. $q(t) = 5\sin t - \dfrac{3}{\sqrt{1 - t^2}}$, such that $Q(0) = 7$.

25. $k(v) = \dfrac{1 + v + \sqrt{v}}{v^2}$, such that $K(1) = -2$.

26. $b(y) = \dfrac{1}{y} + \dfrac{1}{y^2 + 1}$, such that $B(1) = 0$.

27. $f(x) = x^2 + x\sin x$, such that $F(\pi) = 2\pi$.

28. $s(t) = \dfrac{2}{t(\ln t)^2}$, such that $S(e^2) = 5$.

29. $g(x) = x\sqrt{x + 1}$, such that $G(0) = 1$.

30. $w(y) = \dfrac{\tan^{-1}\left(\frac{y}{2}\right)}{4 + y^2}$, such that $W(2) = \pi$.

31. $r(t) = te^{1-t^2} - t$, such that $R(1) = \sqrt{2}$.

In each of Exercises 32 through 41, use integration by parts to find the indicated anti-derivative.

32. $\int xe^{3x}\, dx$

33. $\int (x - 5)^2 e^x\, dx$

34. $\int t\sin(2t)\, dt$

35. $\int t^2 \cos t\, dt$

36. $\int \sqrt{p}\, \ln p\, dp$

37. $\displaystyle\int \frac{\ln t}{t^2}\, dt$

38. $\int e^x \sin x\, dx$

39. $\int e^{2x} \sin(5x)\, dx$

40. $\int \tan^{-1} w\, dw$

41. $\int w \tan^{-1} w\, dw$ Hint: At some point, you may want to use that $w^2 = (1 + w^2) - 1$.

42. Suppose that the net force F, acting on an object of mass m, pushes the object along the x-axis with an acceleration function, in m/s^2, of

$$a(t) = \sin(2t),$$

where $t \geq 0$ is measured in seconds.

  a. Recall that $F = ma$ and that momentum $p = mv$. Find the momentum of the object, as a function of time, if the mass of the object is 10 kilograms, and momentum at time $t = 0$ is 20 kilogram meters per second.

  b. What is the momentum of the object at time $t = 4$ seconds?

43. Repeat the preceding problem with the new acceleration function $a(t) = t \cdot \sin(2t)$.

44. For each positive integer n, define $f_n(\theta) = \sin^n \theta \cos \theta$ and $g_n(\theta) = \cos^n \theta \sin \theta$

  a. Find $\displaystyle\int f_n(\theta)\, d\theta$ and $\displaystyle\int g_n(\theta)\, d\theta$.

  b. Find the specific anti-derivatives $F_n(\theta)$ and $G_n(\theta)$ that satisfy the initial conditions $F_n(0) = 5$ and $G_n\left(\dfrac{\pi}{2}\right) = 4$.

In Exercises 45 through 48, you are given the velocity of a particle at time t, and the position $p(t_0)$ of the particle at a specific time t_0. Find the position function.

45. $v(t) = 3t^2 - 4t + 3$, $p(1) = 4$.

46. $v(t) = t + \cos(t)$, $p(0) = 0$.

47. $v(t) = 2t\sqrt{18 + 7t^2}$, $p(1) = 8$.

48. $v(t) = at + b$, $p(2) = 5$. Leave your answer in terms of a and b.

49. In the following steps, you will calculate the general anti-derivatives for $\sin^2 x$ and $\cos^2 x$.

  a. Apply integration by parts to $\displaystyle\int \sin x \cdot \sin x\, dx$ (written suggestively).

b. Integration by parts yields a new anti-derivative. Use a trigonometric identity to write this new anti-derivative in terms of $\sin^2 x$, and solve your integration by parts equation for $\int \sin^2 x \, dx$.

c. What is $\int \cos^2 x \, dx$? Hint: Use your answer to part (b).

d. From the cosine double angle formula, $\cos(2x) = 2\cos^2 x - 1$. Use this to integrate $\cos^2 x$, and explain why the different-looking answer that you obtain is, in fact, the same as your answer from part (c).

Exercises 50 and 51 show that the argument in Exercise 49 can be generalized to calculate anti-derivatives of higher powers of sin and cos.

50.   a. Use integration by parts to prove that

$$\int \sin^n(t) \, dt = -\frac{1}{n} \cos t \sin^{n-1} t + \frac{n-1}{n} \int \sin^{n-2} t \, dt.$$

Assume $n \geq 2$.

b. Use this formula to calculate $\int \sin^2 t \, dt$. Check your answer by comparing to the previous problem.

51.   a. Use integration by parts to prove that

$$\int \cos^n t \, dt = \frac{1}{n} \cos^{n-1} t \sin t + \frac{n-1}{n} \int \cos^{n-2} t \, dt.$$

b. Use the formula in part (a) to determine $\int \cos^2 t \, dt$.

52. Suppose that instantaneous rate of change, with respect to time, of a population of an island at time t, measured in years, where $t = 0$ corresponds to the year 2000, is given by

$$\frac{1}{2\sqrt{6,250,000 + t}} - \frac{500}{(t+1)^2}.$$

The population of the island in 2000 is 3000.

a. Find an explicit formula for the population at time t.

b. What is the predicted population in 2050?

53. Prove that the argument used to calculate $\int \ln t \, dt$ can be generalized. Assume that $f(t)$ is differentiable and prove that

$$\int f(t) \, dt = t f(t) - \int t f'(t) \, dt.$$

54. Calculate $\int e^x \sinh x \, dx$. Hint: do *not* use integration by parts.

55. Consider the following logic in calculating $\int e^x \sinh x \, dx$ using integration by parts.

$$\int e^x \sinh x \, dx = e^x \sinh x - \int e^x \cosh x \, dx$$
$$= e^x \sinh x - \left(e^x \cosh x - \int e^x \sinh x \, dx \right) \Rightarrow$$
$$0 = e^x \sinh x - e^x \cosh x \Rightarrow$$
$$e^x \sinh x = e^x \cosh x.$$

Since $e^x > 0$, we can divide and conclude that $\sinh x = \cosh x$. What is the flaw in this argument?

56. Prove the formula
$$\int t^n e^t \, dt = t^n e^t - n \int t^{n-1} e^t \, dt.$$

Assume that $n \geq 1$.

57. Calculate $\int \dfrac{1}{x^2 + a^2} \, dx$, where $a \neq 0$ is a constant. Hint: use Theorem 4.2.14 and an appropriate substitution.

58. Prove that $\int \dfrac{dx}{\sqrt{x^2 + a^2}} = \sinh^{-1}\left(\dfrac{x}{a}\right) + C.$

59. Calculate $\int \dfrac{1}{\sqrt{a^2 - x^2}} \, dx$, where $a > 0$ is a constant.

60. Calculate $\int \dfrac{1}{|x|\sqrt{x^2 - 1}} \, dx$. Hint: consider $\sec^{-1} x$.

61. Calculate $\int e^x e^{e^x} \, dx$. Hint: use substitution. ▶

62. Let $f_1(x) = e^x$ and, for all integers $n \geq 2$, let $f_n(x) = e^{f_{n-1}(x)}$. Prove that

$$\int f_1(x) \cdot f_2(x) \cdot \ldots \cdot f_n(x)\, dx = f_n(x) + C.$$

63. Calculate $\displaystyle\int 2^x\, dx$. Hint: rewrite 2^x as an exponential expression with base e.

64. Calculate $\displaystyle\int \frac{x+4}{\sqrt{x+2}}\, dx$.

65. Calculate $\displaystyle\int \ln(1 + x^2)\, dx$.

66. Calculate $\displaystyle\int \frac{e^{3x} - e^{2x}}{e^{2x} - 1}\, dx$.   ▶

Consider a simple electric circuit with an inductor, but no resistor or capacitor. A battery supplies voltage $V(t)$. If inductance is constantly L, in henrys, then Kirchoff's Law from Example 2.7.18 tells us that the current i at time t satisfies the differential equation $L\dfrac{di}{dt} = V(t)$. In Exercises 67 through 70, find an explicit formula for $i(t)$, given the condition $i(t_0) = i_0$.

67. $L = 12$, $V(t) = \sin t$, $i(0) = 0$.

68. $L = 9$, $V(t) = 12$, $i(3) = 12$.

69. $L = 3$, $V(t) = \dfrac{t}{t^2 + 1}$, $i(0) = 0$.

70. $L = 13$, $V(t) = \dfrac{\ln(1/t)}{t^2}$, $i(1) = 2$.

For each of the functions in Exercises 71 through 74, verify that (a) $\dfrac{d}{dx}\left[\displaystyle\int f(x)\, dx\right] = f(x)$ and (b) $\displaystyle\int \left[\dfrac{d}{dx} f(x)\right]\, dx = f(x) + C$. Assume an appropriate domain for $f(x)$.

71. $f(x) = x^4$.

72. $f(x) = 3\cos(2x)$.

73. $f(x) = \ln x$.

74. $f(x) = \dfrac{1}{1 + x^2}$.

75. In the following steps, you will find the general anti-derivatives for functions of the form

$$t^n \ln t.$$

    a. Suppose that $n \neq -1$. Apply integration by parts to find $\int t^n \ln t \, dt$.

    b. Find $\int t^{-1} \ln t \, dt$.

76. Suppose that water flows out of a hole 0.1 square meters in area from the bottom of a cylindrical tank with a base radius of 2 meters and an initial height of 10 meters at a rate

$$\frac{dV}{dt} = 0.007798t - 1.3999 \quad \text{cubic meters per second.}$$

    a. If the tank starts out full, what is the function $V(t)$ for the volume in the tank at time t?

    b. Calculate the amount of water remaining in the tank one minute after the leak starts.

    c. Verify that $\frac{dV}{dt} = 0$ precisely when the tank is empty.

77. A particle is traveling along the curve $y = x^2$, so that its x-coordinate is a function of time t, measured in minutes. Suppose that the horizontal velocity (i.e., velocity in only the horizontal direction) is given by $dx/dt = 0.5 \cos^3 t$ miles per minute.

    a. What is the maximum horizontal *speed* (absolute value of velocity) on the time interval $0 \leq t \leq 20\pi$?

    b. Find the function $x(t)$, the x-coordinate of the particle at time t, subject to the condition that the particle is at the origin at time $t = 0$.

    c. Find the function $y(t)$, the y-coordinate of the particle at time t.

    d. What is the vertical velocity of the particle at time $t = \pi$ minutes?

78. An enclosed room is built in order to experiment with the effects of pressure changes on objects. Suppose that it is capable of decreasing the pressure in the room at a rate of $-\frac{2}{t+1} - 0.04t$ kilopascals, kPa, per second, and it can run for up to one minute before overheating.

    a. If the internal pressure in the room is 101.325 kPa (equal to the standard atmospheric pressure or atm), find an expression for $P(t)$.

    b. How long will it take to reach 50 kPa (this is approximately the atmospheric pressure five and a half kilometers above the surface of the Earth)?

c. What is the lowest pressure that can be reached in the room?

In Exercises 79 through 82, you are given the acceleration, $a = a(t)$ in m/s^2, of an object moving in a straight line, where t is the time in seconds. Find the velocity v and position p of the object, as functions of time, in terms of the initial velocity v_0 and initial position p_0.

79. $a = t + 1$

80. $a = \sin t + \cos t$

81. $a = e^{-3t}$

82. $a = te^{-t}$

4.3 Separable Differential Equations

In the last section, we looked at the simplest type of differential equation: differential equations of the form $dy/dx = f(x)$, where $f(x)$ is given. The general solution to such an equation is just the general anti-derivative of $f(x)$. Of course, producing the general anti-derivative (in a nice form) may not be so simple.

In this section, we will look at the next simplest type of differential equations: *separable differential equations.*

> **Definition 4.3.1.** *A differential equation is called* **separable** *if it can be written in the form*
> $$\frac{dy}{dx} = f(x)g(y),$$
> *for some functions f and g.*

We will state a theorem which tells you how to solve separable equations, but an example first will make things more clear.

Example 4.3.2. Let's look at the differential equation that we discussed at length in Section 4.1:

$$\frac{dA}{dt} = tA^2.$$

This equation is separable; it is of the form: the derivative of A, with respect to t, equals a product of a function of t and a function of A.

How do you solve such an equation? We will write what most people say and do, and afterwards explain why it makes sense.

Divide both sides of the differential equation by A^2 and multiply both sides by dt to obtain $A^{-2}\, dA = t\, dt$. This is called *separation of variables*. Now, integrate both sides:

$$\int A^{-2}\, dA = \int t\, dt$$

and so

$$A^{-1}/(-1) \;=\; t^2/2 + C.$$

Multiply both sides by -1 (and replace $-C$ with simply C) to obtain

$$A^{-1} \;=\; -t^2/2 + C,$$

and, hence,

$$A \;=\; \frac{1}{-t^2/2 + C} \;=\; \frac{2}{C - t^2},$$

where, in the last step, we multiplied the numerator and denominator by 2, and replaced $2C$ with just C.

Note that, in Section 4.1, we also included the constant function $A = 0$ as a solution to this differential equation. Where did we lose this possible solution? In our first step, we divided by A^2, which we can't do when $A = 0$; that's why we have to check as a separate case whether or not $A = 0$ is a solution (which it obviously is).

We are finished with finding the general solution to the separable differential equation, but we still need to discuss why it's okay to think of "multiplying both sides of an equation by dt" and using that to "cancel" and "denominator" in dA/dt. We also need to explain why we wrote only a $+C$ after integrating, instead of writing arbitrary constants on each side of the equation.

As long as $A \neq 0$, it does make mathematical sense to go from the differential equation $dA/dt = tA^2$ to

$$A^{-2} \frac{dA}{dt} \;=\; t.$$

We may then anti-differentiate both sides, with respect to t, to obtain

$$\int A^{-2} \frac{dA}{dt} \, dt \;=\; \int t \, dt.$$

Now, substitution, Theorem 4.2.15, which is the Chain Rule in an anti-derivative form, tells us that the left-hand integral above is equal to $\int A^{-2} \, dA$. Therefore, we obtain

$$\int A^{-2} \, dA \;=\; \int t \, dt.$$

This means that we get precisely what we got by informally multiplying both sides of $A^{-2} \dfrac{dA}{dt} = t$

by dt, then using our differential notation to write $A^{-2}\,dA = t\,dt$, and then integrating both sides.

Thus, in the future, we will stick with this very useful, informal way of discussing the method of separation of variables.

What about the single $+C$? When we had

$$\int A^{-2}\,dA \;=\; \int t\,dt,$$

we should have obtained

$$A^{-1}/(-1) + C_1 \;=\; \frac{t^2}{2} + C_2,$$

where C_1 and C_2 could be any constants. But then, we can immediately put the two constants on the same side of the equation, and group them together, calling the difference C. That is, we write

$$A^{-1}/(-1) = \frac{t^2}{2} + (C_2 - C_1),$$

and simply let C denote $C_2 - C_1$. Thus, we end up with

$$A^{-1}/(-1) = \frac{t^2}{2} + C,$$

where C can be anything.

The point is: when you have an equality of two indefinite integrals, you put a $+C$ on one side or the other, but not both; it's not wrong to put arbitrary constants on both sides of the equation, but it's a waste of time.

In the example above, we had two distinct types of solutions to consider: one where we assumed that A^2 was never (on some interval) equal to 0, so that we could divide by A^2, and one where we assumed that A^2 was always 0. This is the general situation for $dy/dx = f(x)g(y)$, when $f(x)$ and $g(y)$ are continuous.

Before we state a general theorem, we should point out that, in Example 4.3.2, we could easily solve explicitly for A as a function of t. For general separable equations, obtaining such an explicit solution is frequently **not** possible. You always arrive at $\int \left(1/g(y)\right) dy = \int f(x)\,dx$, and the y integral can produce an almost arbitrarily ugly function, such that there is no algebraic method for solving explicitly for y as a function of x. In this case, you simply leave your solution in *implicit* form (recall the definition of an implicitly defined function from Definition 2.10.3). That is, you leave your answer as some function of y equals some function of x.

The theorem which generalizes what we have seen in Example 4.3.2 is:

> **Theorem 4.3.3.** *Suppose that f and g are continuous functions. Then, every solution $y = H(x)$, on an open interval I, of*
>
> $$\frac{dy}{dx} \; = \; f(x)g(y)$$
>
> *is obtained by considering two cases and, possibly, extending solutions from smaller open intervals:*
>
> **Case 1**: *Find all values y_0 such that $g(y_0) = 0$. Then, each constant function $y = y_0$ is a solution to the differential equation (an equilibrium solution).*
>
> **Case 2**: *Assume that $g(y) \neq 0$, and let $y = H(x)$ be implicitly defined by*
>
> $$\int \frac{1}{g(y)} \, dy \; = \; \int f(x) \, dx.$$

Proof. Since f and g are continuous, the anti-derivatives in the final equation exist. See [1], [3], or [2]. That Case 1 produces solutions is trivial. That Case 2 produces solutions near (x_0, y_0), where $g(y_0) \neq 0$, follows from the existence of the anti-derivatives, the Chain Rule, and the Implicit Function Theorem, Theorem 2.A.4.

There remains the question of why you need to worry only about the two extreme cases: where $y = y_0$ and $g(y) = g(y_0)$ is **always** 0, and where $g(y)$ is **never** 0.

First, if $y = H(x)$ is a solution to the differential equation on an open interval I, and $g(H(x)) = 0$ for all x in I, then the differential equation tells us that $dy/dx = 0$ for all x in I; thus, $H(x)$ is a constant.

Suppose, instead, that $y = H(x)$ is a solution to the differential equation on an open interval I and that there exists x_0 in I such that $g(H(x_0)) \neq 0$. Let $y_0 = H(x_0)$. As $g(y_0) \neq 0$ and g is continuous, there exists an open interval J around y_0 such that, for all y in J, $g(y) \neq 0$. As

H is continuous, there exists an open interval K around x_0 such that, for all x in K, $H(x)$ is in J, i.e., $g(H(x))$ is never 0 for x in K. Thus, the restriction of $y = H(x)$ to the interval K is a solution that can be found via the method in Case 2. The original solution on the interval I is therefore an extension of the solution on the interval K. $\qquad\square$

Remark 4.3.4. We need to make a couple of remarks about Theorem 4.3.3.

First, note that there is no claim in the statement or proof of the above theorem that extensions of solutions to bigger open intervals are unique. In fact, if a solution extends through a point y_1 such that $g(y_1) = 0$, then we can "piece together" different-looking solutions from either side of y_1. See the example below.

Second, our definition of an implicitly defined function in Definition 2.10.3 required that we have a point in the xy-plane (x_0, y_0) (or (a, b)), and that the implicitly defined function need be defined only **near** (x_0, y_0). Therefore, what Case 2 in Theorem 4.3.3 really means is the following:

You assume that $g(y) \neq 0$. The equation $\int \frac{1}{g(y)} \, dy = \int f(x) \, dx$ means that you have an anti-derivative (with respect to y) $L(y)$ of $1/g(y)$, an anti-derivative (with respect to x) $F(x)$ of $f(x)$, an arbitrary constant C, and an equality $L(y) = F(x) + C$. For every initial condition of the form $y(x_0) = y_0$, where $g(y_0) \neq 0$, C is determined by $L(y_0) = F(x_0) + C$, and the implicitly defined function referred to in the theorem is the function implicitly defined, near (x_0, y_0), by $L(y) = F(x) + (K(y_0) - F(x_0))$.

Example 4.3.5. Recall the differential equation from Example 4.1.1:

$$\frac{dy}{dx} = y^{2/3}.$$

By Theorem 4.3.3, we have two kinds of solutions. One will be found by setting $y^{2/3}$ equal to 0, and solving for y. We find the equilibrium solution, the function

$$y = y(x) = 0.$$

Now, we assume that $y \neq 0$, and separate the variables to obtain

$$\int y^{-2/3}\, dy \;=\; \int 1\, dx,$$

which yields

$$\frac{y^{1/3}}{1/3} \;=\; x + C.$$

We can solve this explicitly for y. We obtain

$$y \;=\; \left(\frac{1}{3}x + C\right)^3,$$

(where this final C is $1/3$ of the previous C). Note that, while we assumed that $y \neq 0$ to obtain this solution, the solution that we obtain extends to a solution on all of $(-\infty, \infty)$.

Let's look at this as an initial value problem, as we did in Example 4.1.1. Start with the initial value problem

$$\frac{dy}{dx} \;=\; y^{2/3} \qquad \text{and} \qquad y(0) = y_0.$$

If $y_0 = 0$, then there are two distinct solutions to the initial value problem; namely, the equilibrium solution $y = y(x) = 0$, and $y = (x/3 + C)^3$, where C is chosen so that $0 = (0 + C)^3$, i.e., $C = 0$ and so $y = x^3/27$. These two distinct solutions exist on any open interval which contains $x = 0$.

On the other hand, if $y(0) = y_0 \neq 0$, then there is a unique solution to the initial value problem in an open interval around $x = 0$.

To be specific, let's suppose that $y_0 = 1$. Then, the unique solution near 0 is given by $y = (x/3 + C)^3$, where C is chosen so that $1 = (0 + C)^3$. Thus, $C = 1$ and $y = (x/3 + 1)^3$. This solution is unique on any open interval I containing 0, as long as I does not contain an x value which makes $y = (x/3 + 1)^3$ a place where $g(y) = y^{2/3}$ equals 0. This means that we need to avoid $x = -3$.

Therefore, $y = (x/3 + 1)^3$ is the unique solution to $\int \frac{1}{g(y)}\, dy \;=\; \int f(x)\, dx$ and $y(0) = 1$, for x in the open interval $(-3, \infty)$.

However, there is not a **unique** solution to this initial value problem on any larger open interval. If $a < -3$, then two solutions to the initial value problem on the interval (a, ∞) are

1.

$$y = (x/3 + 1)^3 \text{ ; and}$$

2.

$$y = \begin{cases} 0 & \text{if } a < x \leq -3; \\ (x/3+1)^3 & \text{if } x \geq -3. \end{cases}$$

Of course, the second solution above looks sort of silly; we just "messed up" the "real solution". That, however, is exactly the point: it is possible to alter what looks like a nice, unique solution, if you let the solution extend so far that you pass through a place where $g(y)$ becomes 0.

Example 4.3.6. Find the general solution to

$$\frac{dy}{dx} = y \cos x.$$

Solution:

First, we find the equilibrium solution $y = y(x) = 0$.

Now, assuming $y \neq 0$, we divide both sides of the equation by y and multiply by dx to obtain

$$\frac{1}{y} \, dy = \cos x \, dx.$$

We integrate

$$\int \frac{1}{y} \, dy = \int \cos x \, dx,$$

and find

$$\ln |y| = \sin x + C.$$

The solutions $y = 0$ and the implicit solutions $\ln |y| = \sin x + C$ together form the general solution to the separable equation. However, since it is easy to solve for y explicitly, we will.

If we raise e to each side of $\ln |y| = \sin x + C$, we find

$$|y| = e^{\ln |y|} = e^{\sin x + C} = e^C \cdot e^{\sin x}.$$

Therefore,

$$y = \pm e^C \cdot e^{\sin x},$$

where $\pm e^C$ could be any positive or negative constant; call it B. Then, we find that the general solution is given by $y = 0$ or $y = Be^{\sin x}$, where $B \neq 0$. But, if B were 0, then $y = Be^{\sin x}$ would equal 0, and give us our equilibrium solution $y = 0$.

Hence, the nicest form of the general solution is $y = Be^{\sin x}$, where B is an arbitrary constant.

Example 4.3.7. In Example 2.2.9 and Example 2.4.24, we discussed the logistic model for population growth. This model uses the differential equation:

$$\frac{dP}{dt} = kP(M - P),$$

where M and k are non-zero constants (which are positive, in population applications). In this example, we will derive the general solution to this differential equation.

There are two equilibrium solutions $P(t) = 0$ and $P(t) = M$.

Now, assume that $P(M - P) \neq 0$. Then, we separate the variables and find

$$\int \frac{1}{P(M - P)}\, dP \;=\; \int k\, dt. \tag{4.7}$$

The integral on the right is certainly easy, but the integral on the left is complicated. We must figure out how to anti-differentiate $1/[P(M - P)]$.

We use here an algebraic technique, which is normally referred to as *partial fractions*. You have certainly added two fractions before by first obtaining a common denominator. Partial fractions is essentially the reverse of this; we have the fraction with the common denominator, and we wish to write it as a sum of two fractions with simpler denominators.

While a complete discussion of the technique of partial fractions would be quite lengthy, in our current situation, what we want to do is find constants A and B such that we have an equality of functions

$$\frac{1}{P(M - P)} \;=\; \frac{A}{P} \;+\; \frac{B}{M - P}. \tag{4.8}$$

Assuming that such A and B exist, we multiply both sides of the equation by $P(M - P)$ and obtain

$$1 \;=\; A(M - P) \;+\; BP, \tag{4.9}$$

which is an equality of functions of P, though our domains in Formula 4.8 omitted places where the denominators were zero, namely, $P = 0$ and $P = M$. However, this last equality in Formula 4.9 is an equality of continuous functions, which are defined when $P = 0$ and $P = M$. Therefore, the equality of functions in Formula 4.8, for $P \neq 0$ and $P \neq M$, implies that the equality in Formula 4.9 holds for **all** values of P, because we may take limits of both sides as P approaches 0 and M.

It is important that Formula 4.9 holds when $P = 0$ and $P = M$, for this allows us to find A and B quickly. By substituting $P = 0$ into Formula 4.9, we obtain $1 = AM + 0$; so, $A = 1/M$. By substituting $P = M$ into Formula 4.9, we obtain $1 = 0 + BM$; so, $B = 1/M$. Therefore, we obtain that

$$\frac{1}{P(M-P)} = \frac{1/M}{P} + \frac{1/M}{M-P} = \frac{1}{M}\left[\frac{1}{P} - \frac{1}{P-M}\right]. \tag{4.10}$$

(Technically, what we have shown is that, **if there are A and B such that Formula 4.8 holds**, then both A and B must be $1/M$; we have not actually shown that $A = B = 1/M$ produces the equality in Formula 4.10. We leave it as an exercise for you to verify that Formula 4.10 is correct.)

Thus, we have that

$$\int \frac{1}{P(M-P)}\, dP = \int \frac{1}{M}\left[\frac{1}{P} - \frac{1}{P-M}\right] dP =$$

$$\frac{1}{M}\Big[\ln|P| - \ln|M-P|\Big] + C_1 = \frac{1}{M}\ln\left|\frac{P}{M-P}\right| + C_1,$$

where we used the substitution $u = P - M$ to integrate $1/(P-M)$. We also used that $|P-M| = |M-P|$, that the quotient of absolute values is the absolute value of the quotient, and an algebraic property of logarithms in the final step.

Substituting this anti-derivative back into Formula 4.7 (and omitting/absorbing the $+C_1$), we obtain that

$$\frac{1}{M}\ln\left|\frac{P}{M-P}\right| = kt + C.$$

Now, we multiply both sides by M, let C denote the old CM, and raise e to both sides to obtain

$$\left|\frac{P}{M-P}\right| = e^{Mkt+C} = e^C \cdot e^{Mkt}.$$

Thus,

$$\frac{P}{M-P} = Re^{Mkt},$$

where R is the constant $\pm e^C$.

Now is a good time to insert the tautological initial data that when $t = 0$, $P = P_0$. We find that

$$\frac{P_0}{M - P_0} = R,$$

and so

$$\frac{P}{M - P} = \frac{P_0}{M - P_0} e^{Mkt}.$$

Taking reciprocals, we find

$$\frac{M}{P} - 1 = \frac{M - P}{P} = \frac{(M - P_0)e^{-Mkt}}{P_0}.$$

Hence,

$$\frac{M}{P} = \frac{(M - P_0)e^{-Mkt} + P_0}{P_0},$$

and so

$$P = \frac{MP_0}{(M - P_0)e^{-Mkt} + P_0}.$$

This is the solution that we gave in Example 2.4.24. We leave it to you to show that the solution, written in this form, also includes the two equilibrium solutions.

Example 4.3.8. Consider the *family of curves* defined by $y^4 = kx$, where k can be any constant. Fixing a value of k determines a member of the family.

A second family of curves consists of *orthogonal trajectories* to a given family of curves if, whenever a curve from the first family intersects a member of the second family, the tangent lines are orthogonal. This, of course, means that the slopes of the tangent lines are negative reciprocals of each other.

How can we find the orthogonal trajectories of a given family? In particular, how can we find the orthogonal trajectories of $y^4 = kx$?

You proceed as follows.

1. Write the original family of curves in the form $H(x, y) = k$.

2. Differentiate, implicitly, both sides of $H(x, y) = k$, with respect to x. The reference to k drops out, and you are left with a differential equation that is satisfied by all members of your original family.

3. Write the differential equation from Step 2 in the form $\dfrac{dy}{dx} = f(x, y)$.

4. Write a differential equation which is satisfied by the orthogonal trajectories. The orthogonal trajectories to the first family of curves have negative reciprocal slopes, and so must satisfy the differential equation

$$\frac{dy}{dx} = -\frac{1}{f(x, y)}.$$

5. Solve $\dfrac{dy}{dx} = -\dfrac{1}{f(x, y)}$ to find the orthogonal trajectories.

Let's go through this process with the family $y^4 = kx$.

Step 1: We rewrite the family as $y^4/x = k$.

Step 2: We differentiate both sides of $y^4/x = k$, with respect to x, and obtain

$$\frac{x4y^3y' - y^4}{x^2} = 0,$$

so that we have

$$4xy^3 \frac{dy}{dx} - y^4 = 0.$$

Thus, either $y = 0$ or $4x\dfrac{dy}{dx} - y = 0$. As $y = 0$ is a solution to $4x\dfrac{dy}{dx} - y = 0$, we don't need to consider $y = 0$ as a special case.

Step 3: We find that the members of our original family of curves all satisfy

$$\frac{dy}{dx} = \frac{y}{4x}.$$

Step 4: So the orthogonal trajectories to our original family of curves satisfy the differential equation which is obtained from the above by taking the negative reciprocal of the right-hand side, i.e., the orthogonal trajectories satisfy

$$\frac{dy}{dx} = -\frac{4x}{y}.$$

Step 5: All that remains is for us to solve

$$\frac{dy}{dx} = -\frac{4x}{y}.$$

We separate the variables to find

$$\int y \, dy = \int -4x \, dx,$$

and so

$$\frac{y^2}{2} = -2x^2 + C.$$

Rearranging, dividing by 2, and calling $C/2$ just C again, we obtain the family of orthogonal trajectories

$$x^2 + \frac{y^2}{4} = C,$$

which you should recognize as a family of ellipses.

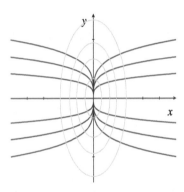

Figure 4.1: $y^4 = kx$ in blue. $x^2 + y^2/4 = C$ in green.

4.3.1   Exercises

In each of Exercises 1 through 8, find the general solution to the differential equation. Give explicit solutions where possible, implicit solutions where necessary.

1. $\dfrac{dy}{dx} = xy^2$.

2. $\dfrac{dy}{dx} = (x^2 - x + 1) \cdot (1 + y^2)$.

3. $\dfrac{1}{t} \cdot \dfrac{dy}{dt} = \dfrac{\cos t}{y}$.

4. $\dfrac{dr}{dt} = \dfrac{2}{\sin r + t^2 \sin r}$.

5. $\dfrac{dw}{dt} = (\tan w) t e^{-t^2}$.

6. $\dfrac{dy}{dx} = xy + y + x + 1$.

7. $\dfrac{dx}{dt} = \dfrac{1 + x^2}{tx + x}$.

8. $\dfrac{dp}{dt} = \dfrac{\sqrt{p^3 - 1}}{p^2}$

In each of Exercises **9** and **11**, solve the differential equation, subject to the given initial condition. Give explicit solutions where possible, implicit solutions where necessary.

9. $\dfrac{dy}{dx} = xe^{x+y}$, and $y(1) = -1$.

10. $\dfrac{dx}{dt} = \dfrac{t \sin t}{3x^2}$, and $x(\frac{\pi}{2}) = 2$.

11. $\dfrac{dr}{dt} = \dfrac{t}{2r + 1}$, and $r(0) = 2$.

Separable differential equations are sometimes written in the form $F(x)\,dx + G(y)\,dy = 0$. By manipulating the symbols algebraically, this equation is "equivalent to"

$$\frac{dy}{dx} = -\frac{F(x)}{G(y)},$$

and the method of solution is identical to the one outlined in this chapter. In Exercises **12** through **14**, solve the differential equation written in this format.

12. $\sin(3x)\,dx + \cos(5y)\,dy = 0$.

13. $-x^2\,dx + (1 + y^2)\,dy = 0$.

14. $-y\cos(x)\,dx + (1 + 2y^2)\,dy = 0$.

15. Show that any differential equation of the form $\dfrac{dy}{dx} = f(x)$, considered in the previous chapter, is separable.

16. Recall that an autonomous differential equation is one in the form $\dfrac{dy}{dx} = F(y)$. Show that this equation is separable. Is the logistic differential equation autonomous?

Recall that the current $i(t)$ in a simple circuit with a resistor, an inductor, but no capacitor is given by $L\dfrac{di}{dt} + Ri = V(t)$. R is the resistance, L the inductance and V the voltage. Use this information in Exercises 17 through 20.

17. Find $i(t)$ if $V(t) = 40$, $L = 5$, $R = 8$, $i(0) = 0$.

18. Find $i(t)$ if $V(t) = 60$, $L = 12$, $R = 2$, $i(0) = 0$.

19. Show that if the voltage is constant, then the differential equation is separable.

20. Find the general solution to the differential equation in terms of L, R and V when $i(0) = 0$ and $V(t) = V$ is constant.

Use the method of Example 4.3.8 to find the orthogonal trajectories of the families of curves in Exercises 21 through 25.

21. $x^2 - y^2 = k$.

22. $y = k/x$.

23. $y = kx^3$.

24. $y = kx$.

25. $y = kx + 3$.

In Exercises 26 through 30, use the method of partial fractions outlined in this chapter to solve the separable equations.

26. $y' = \dfrac{21x + 14}{x^2 + 2x}$.

27. $\dfrac{dy}{dx} = \dfrac{-2y}{x^2 - 1}$.

28. $y' = \dfrac{(\csc y)\,(7x + 6)}{x^2 + 5x + 6}$.

29. $\left(x^4 + 3x^2 + 2\right) y' = x^2 + 4$. Use the fact that $\displaystyle\int \frac{dx}{x^2 + a^2} = (1/a) \tan^{-1}(x/a) + C$.

30. $y'(t) = \dfrac{t^2 + t + 1}{t^3 + t^2} \cdot \sec y$.

In Exercises 31 through 34, identify the equilibrium solutions. You need not solve the differential equation.

31. $\dfrac{dy}{dx} = \left(4 - y^2\right) e^x$.

32. $\dfrac{dy}{dx} = e^x \sqrt{y^2 - 4}$.

33. $\dfrac{dy}{dx} = (y^2 + 4)e^x$.

34. $\dfrac{dy}{dx} = 5x \left(\dfrac{1}{y^2} + \dfrac{3}{y} + 2\right)$.

35. Show that the equation $\dfrac{dy}{dx} = e^x \cos y$ has infinitely many equilibrium solutions.

A differential equation $\frac{dy}{dx} = f(x, y)$ is said to be *homogeneous*, when $f(x,y)$ is a function of the ratio y/x. Note that the term "homogeneous" means different things in different contexts. It can also refer to a linear differential equation equal to zero. Exercises 36 through 41 show how the techniques in this chapter can be extended and used to solve homogeneous equations.

36. Suppose $\dfrac{dy}{dx} = \dfrac{y^2 + xy}{x^2}$. Note that this equation is not separable.

   a. Let $w = y/x$ and rewrite the right hand side of the differential equation in terms of w.

   b. View w as a function of x. Use the Product Rule to show that $\dfrac{dy}{dx} = x\dfrac{dw}{dx} + w$.

   c. Use the results of parts (a) and (b) to rewrite the differential equation as

$$x\frac{dw}{dx} + w = w^2 + w,$$

   and show that this new equation is separable.

   d. Solve the separable equation and show that $y = \dfrac{-x}{\ln |x| + C}$ is a family of solutions to the original equation.

Use the technique outlined in the previous problem to solve the homogeneous equations in Exercises 37 through 39.

37. $\dfrac{dy}{dx} = \dfrac{x^2 + y^2}{xy}$.

38. $\dfrac{dy}{dx} = e^{-y/x} + (y/x)$.

39. $\dfrac{dy}{dx} = \sec(y/x) + (y/x)$.

40. Solve the initial value problem $\dfrac{dy}{dx} = \dfrac{y^2 + xy}{x^2}$, where $y(1) = 1$.

41. Show that, if you have the homogeneous equation $\dfrac{dy}{dx} = f(y/x)$, and you let $w = y/x$, then the differential equation can be rewritten as the separable equation:

$$\frac{dx}{x} = \frac{dw}{f(w) - w},$$

assuming that $f(w) \neq w$.

42. Torricelli's Law gives the rate of flow of water out of a hole in the bottom of a container as a function of the area a, in square meters, of the hole, and the height y, in meters, of water above the hole. Namely:

$$\frac{dV}{dt} = -a\sqrt{2gy},$$

where V is the volume, in cubic meters, of the water remaining in the container, and $g = 9.8$ meters per second squared is the gravitational acceleration.

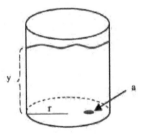

a. If the container is a right circular cylinder, of radius 2 meters and $a = 0.1$ square meters, what is y as a function of V?

b. Substitute your answer from part (a) for y in Torricelli's law to yield a separable differential equation.

c. If $a = 0.1$ square meters, solve the differential equation subject to the initial condition that $V(0) = 100$ cubic meters.

d. At what time t will the tank be empty under the condition from (c)?

43. Find the general solution to $dy/dx = ky$ by solving the separable equation. Your solution should, of course, agree with what we gave in Theorem 2.4.14.

4.4   Applications of Differential Equations

In this section, we will introduce no new mathematics, but instead just present some word/modeling problems that yield differential equations and initial value problems, which we can solve using the techniques of the previous two sections.

Example 4.4.1. A surprising array of applied problems yield a separable differential equation of the form

$$\frac{dy}{dx} \; = \; a + by, \tag{4.11}$$

where a and b are constants, $b \neq 0$. This includes exponential growth and decay problems, Newton's Law of Cooling problems, velocity and acceleration problems, and population problems, just to name a few.

Let's find the general solution to this equation.

First, we have the equilibrium solution where $a + by$ is **always** 0. That is, the solution

$$y = y(x) = -a/b.$$

Now, we look on intervals where $a + by \neq 0$, and separate the variables to find

$$\int \frac{1}{a + by}\, dy \; = \; \int 1 \, dx. \tag{4.12}$$

To evaluate the integral on the left, we make the substitution $u = a + by$, so that $du = b\, dy$ or, equivalently, $(1/b)\, du = dy$. We obtain

$$\int \frac{1}{a + by}\, dy \; = \; \int \frac{1}{u} \cdot \frac{1}{b}\, du \; = \; \frac{1}{b} \ln |u| \; = \; \frac{1}{b} \ln |a + by|,$$

where we have omitted the arbitrary constant, as we shall include it on the other side of our equation below.

Thus, Formula 4.12 yields

$$\frac{1}{b} \ln |a + by| \; = \; x + C.$$

We multiply by b, simply write C again in place of bC, and raise e to each side of the above equation to obtain

$$|a + by| = e^{bx+C} = e^C \cdot e^{bx}.$$

We remove the absolute value signs from the left, and then must insert a $\pm$ on the right; however, we then write A for the constant $\pm e^C$ to find

$$a + by = Ae^{bx},$$

and so, letting B denote A/b, we arrive at

$$y = -\frac{a}{b} + Be^{bx}.$$

As $\pm e^C \neq 0$, our method above yields that $B \neq 0$, which implies that y is never equal to $-a/b$, i.e., that $a + by$ is never 0; thus, this solution is valid for all x. Note also that our equilibrium solution is precisely what you get if you let B equal 0.

Therefore the general solution to $dy/dx = a + by$ is

$$y = -\frac{a}{b} + Be^{bx},$$

where B can be any real number.

If we also have initial data $y(0) = y_0$, then we can solve for B in terms of y_0. We find that $y_0 = -a/b + B$, so that $B = y_0 + a/b$. Thus, the solution to the initial value problem is

$$y = -\frac{a}{b} + \left(y_0 + \frac{a}{b}\right)e^{bx}. \tag{4.13}$$

Note that, if $a = 0$, our solution collapses to what we gave in Theorem 2.4.14.

Example 4.4.2. Suppose that we are considering a population of people, animals, bacteria, etc. Let $P = P(t)$ denote the number of individuals/organisms in the population at time t. (We assume that having fractional values for P gives a reasonable approximation to an actual integer value of P.) The population can change for four different reasons: organisms are born, organisms die, organisms are added (from the outside), and organisms are removed (to the outside).

The birth and death rates are normally expressed as percentages, or fractions, of the current population. Thus, a birth rate β of 3% per year means that the rate of change of the population, due to births, is $0.03P$ organisms per year. Similarly, a birth rate δ of 4% per year means that the rate of change of the population, due to deaths, is $0.04P$ organisms per year.

The rate at which organisms are added to, and subtracted from, the population is frequently given either as a fraction of the current population, or as a number per time interval.

Finally, as a reasonable approximation, all of these rates of change in the population are assumed to be instantaneous.

Let's look at a specific problem. Suppose that, in the year 2000, a certain city had a population of $500,000$, and that this city has an annual birth rate of 3%, and an annual death rate of 4%. Finally, suppose that, each year, the city has a net gain of 1000 people, due to people moving into and out of the city.

Find the population of the city at time t years, measured from the year 2000.

Solution:

The rate of change of the population, with respect to time, in people per year, is dP/dt. This rate of change is the sum of the rates of change in the population for all reasons. The rates of change in the population are:

1. the rate of change per year due to births: $\beta P = 0.03P$;

2. the rate of change per year due to deaths: $-\delta P = -0.04P$;

3. the net change per year due to moving: 1000.

Therefore, we obtain the initial value problem

$$\frac{dP}{dt} = 0.03P - 0.04P + 1000 = 1000 + (-0.01)P; \quad P(0) = 500,000.$$

This differential equation is of the form given in Formula 4.11. If we look at the solution to the initial value problem given in Formula 4.13, we conclude that the solution to our current initial value problem is

$$P = -\left(\frac{1000}{-0.01}\right) + \left(500,000 + \frac{1000}{-0.01}\right)e^{-0.01t},$$

That is

$$P = 100,000 + 400,000e^{-0.01t}.$$

Notice that, as $t \to \infty$, $e^{-0.01t} \to 0$, and thus, as time goes on, the population $P = 100,000 + 400,000e^{-0.01t}$ approaches the equilibrium solution of $P = 100,000$.

Example 4.4.3. Suppose that an object of mass m is moving through a fluid, in a direction that has been designated as positive (so its velocity at time 0 is $v_0 > 0$). We assume that the only force which acts on the object (at least, in the direction we care about) is the fluid resistance, which always acts in the direction opposite that of the velocity. It is common to approximate the magnitude of the resistance force by stating that it is proportional to some (positive) power of the velocity.

Thus, we assume that the total force acting on the object is $F = -kv^p$, where k is a positive constant, $p > 0$ is constant, and the minus sign is present to make the force act in the opposite direction from the velocity, whose direction we take as positive.

Assume that $p = 1$, i.e., that the force of resistance is proportional to the velocity. Find the velocity of the object and its position at time t.

Solution:

We are not given units, so we will write formulas assuming that all the units are consistent.

We use Newton's 2nd Law of Motion, which, since F is the only force and the mass m is constant, tells us that $F = ma$, where a is the acceleration of the object. Now, $a = dv/dt$, where v is the velocity of the object. Therefore, as we are in the $p = 1$ case, we obtain the differential equation

$$-kv \;=\; m\frac{dv}{dt} \qquad \text{or, equivalently,} \qquad \frac{dv}{dt} \;=\; bv,$$

where $b = -k/m$.

Now, we use either Theorem 2.4.14, or Example 4.4.1 with $a = 0$, to conclude that

$$v \;=\; v_0 e^{-kt/m},$$

where v_0 is the velocity to $t = 0$.

The velocity v is the derivative dx/dt, where x is the position. Therefore, we have

$$\frac{dx}{dt} \ = \ v_0 e^{-kt/m} \quad \text{or, equivalently,} \quad x \ = \ \int v_0 \, e^{-kt/m} \, dt.$$

To calculate this anti-derivative, we "pull out" the constant v_0, and make the substitution $u = -kt/m$. Then, $du = -k \, dt/m$, or $dt = (-m/k) \, du$. Thus, we obtain

$$x \ = \ \int v_0 \, e^{-kt/m} \, dt \ = \ v_0 \int e^u \, (-m/k) \, du \ = \ -m v_0 e^u / k + C.$$

Putting back in that $u = -kt/m$, we find

$$x \ = \ -\frac{m v_0}{k} \, e^{-kt/m} + C.$$

Now we plug in the tautological initial data $x(0) = x_0$ to determine C in terms of physical data:

$$x_0 \ = \ -\frac{m v_0}{k} \, e^0 + C,$$

and so $C = x_0 + m v_0 / k$.

Finally, we conclude that

$$x \ = \ -\frac{m v_0}{k} \, e^{-kt/m} + \left(x_0 + \frac{m v_0}{k} \right) \ = \ x_0 + \frac{m v_0}{k} \left(1 - e^{-kt/m} \right).$$

It is interesting to note that, in theory, the velocity of the object approaches 0, but it never actually stops, and its position approaches $x_0 + m v_0 / k$, but it never quite gets there. Of course, in real life, once the object is moving at a small velocity, like a nanometer per millennium, other effects, which were negligible at higher velocities would come into play to actually stop the object.

Example 4.4.4. Assume that you have the situation of the previous example, but that now the force of resistance is proportional to the **square** of the velocity, so that $p = 2$, and the force acting on the object is $F = -kv^2$, where k is a positive constant.

Find the velocity of the object and its position at time t.

Solution:

We gave the solution to this in Example 2.5.7, but we simply handed you the solution; here, we will derive it.

Using Newton's 2nd Law of Motion, we obtain

$$m \frac{dv}{dt} = -kv^2.$$

There is, of course, the equilibrium solution $v = v(t) = 0$.

Assume that $v \neq 0$. Divide both sides by v^2 and m, multiply by dt, and integrate:

$$\int v^{-2} \, dv = \int -\frac{k}{m} \, dt,$$

and so

$$\frac{v^{-1}}{-1} = -\frac{kt}{m} + C, \quad \text{and so} \quad v^{-1} = \frac{kt}{m} + C,$$

where we have written simply C in place of $-C$. Now is the best time to plug in the tautological initial data $v(0) = v_0 > 0$; we immediately conclude that this (last) $C = v_0^{-1}$. Therefore, we have

$$v = \frac{1}{\frac{k}{m} t + \frac{1}{v_0}} = \frac{mv_0}{v_0 kt + m}.$$

Note that, in this final form, we can also include the equilibrium solution, where $v_0 = 0$ and $v = v(t) = 0$. Note, also, that, as k and m are positive, and $v_0 \geq 0$, this solution for v is the unique solution for t in the open interval $(-m/(v_0 k), \infty)$, which includes $t = 0$.

Now, let's find the position $x = x(t)$.

If $v_0 = 0$, so that $v = v(t) = 0$, then, clearly, $x = x(t) = x_0$, where x_0 is the position of the object at time 0.

Suppose that $v_0 \neq 0$. Then, we have, for $t > -m/(v_0 k)$,

$$x = \int v \, dt = \int \frac{mv_0}{v_0 kt + m} \, dt = \frac{m}{k} \ln \left(\frac{v_0 kt}{m} + 1 \right) + C,$$

where we leave it as an exercise for you to verify this last equality, by making the substitution

$u = (v_0 kt/m) + 1$, and we have omitted absolute value signs, since $(v_0 kt/m) + 1 > 0$ when $t > -m/(v_0 k)$. Plugging in that $x(0) = x_0$, we find that $C = x_0$, and obtain

$$x = \frac{m}{k} \ln \left(\frac{v_0 kt}{m} + 1 \right) + x_0.$$

Note that this includes the special solution $x = x(t) = x_0$, if $v_0 = 0$.

It is interesting to compare the result of the previous example with this one. When the force of resistance was proportional to the velocity, we found that the position of the object approached some finite limit; this, of course, means that the object does not travel arbitrarily far as time goes on. However, when the force of resistance is proportional to the square of the velocity, we have that

$$\lim_{t \to \infty} x = \lim_{t \to \infty} \left[\frac{m}{k} \ln \left(\frac{v_0 kt}{m} + 1 \right) + x_0 \right] = \infty.$$

This may seem contradictory; we frequently think of v^2 of being bigger than v, which would leave us with the question: why, when there's **more** resistance, does the object travel farther? However, eventually, the velocity becomes less than 1, and then it remains less than 1 for the rest of time. When $0 < v < 1$, v^2 is **less than** v, which resolves our "contradiction".

Example 4.4.5. We will now consider a skydiver of mass m, dropping through the atmosphere. There are two forces which act on the skydiver: the force of gravity F_g, which pulls down, and air resistance F_r, which pushes up. As a general rule, it is best to pick the direction of the motion as the positive direction; hence, we select the downward direction as the positive direction and assume that $v_0 \geq 0$.

Gravity provides a constant (to a reasonable approximation) acceleration, g, which is independent of the mass of the object. The value of g is approximately 32 (ft/s)/s or 9.8 (m/s)/s. We shall just write g throughout. Hence, $F_g = mg$.

The force of air resistance on a skydiver is (approximately) proportional to the velocity v; so, we assume that $F_r = -kv$, where $k > 0$ and the minus sign indicates that air resistance acts in the negative direction (up), assuming that, in fact $v > 0$. If $v < 0$, so that (strangely) the skydiver was moving upward, then both gravity and air resistance would act downward.

Newton's 2nd Law of Motion gives us

$$mg - kv \;=\; F_g + F_r \;=\; \text{net force} \;=\; ma \;=\; m\frac{dv}{dt}.$$

Dividing by m, we obtain

$$\frac{dv}{dt} \;=\; g - \frac{k}{m}v,$$

which is in the form of Formula 4.11.

Therefore, we use Formula 4.13 to conclude that

$$v \;=\; -\left(\frac{g}{-k/m}\right) + \left(v_0 + \frac{g}{-k/m}\right)e^{-kt/m},$$

that is

$$v \;=\; \frac{mg}{k} + \left(v_0 - \frac{mg}{k}\right)e^{-kt/m}.$$

Note that, as k and m are positive, $\lim_{t\to\infty} e^{-kt/m} = 0$, and so $\lim_{t\to\infty} v = mg/k$; this is called the *terminal velocity*. Of course, it is interesting/important to know if the skydiver is accelerating towards the ground, decelerating, or sometimes accelerating and sometimes decelerating as the velocity approaches mg/k. We could take the derivative of our final answer to get $a = dv/dt$, and check when its positive and when its negative, but we don't need to.

Remember how we began the problem: our differential equation (after a little factoring) was $dv/dt = -(k/m)(v - mg/k)$. Thus, if we differentiate our final $v = v(t)$, we will find that we get $-(k/m)(v - mg/k)$. As k and m are positive, this last quantity, which equals the acceleration a, is positive if $v < mg/k$, and is negative if $v > mg/k$. So, there are three cases: where $v_0 = mg/k$, where $v_0 < mg/k$, and where $v_0 > mg/k$. Recall that downward is our positive direction.

What we have found is:

1. If the initial velocity is less than the terminal velocity (the usual case for a skydiver), then the skydiver will always being accelerating towards the ground and the velocity will approach the terminal velocity. Note that this case includes the (bizarre) case in which the skydiver is shot upward out of a plane, so that the initial (downward) velocity is actually negative; the acceleration is still always downward, even though the velocity is upward for a time.

2. If the skydiver is shot downward at exactly the terminal velocity, then the velocity will always be the terminal velocity.

3. If the skydiver is shot downward with an initial velocity greater than the terminal velocity, then the diver's downward velocity will always decrease, and approach the terminal velocity.

Example 4.4.6. In Exercise 52 in Section 2.4, we stated Newton's Law of Cooling, which says: if $T = T(t)$ is the temperature of an object, which is placed in surroundings which are kept at a constant (ambient) temperature, T_{amb}, then the rate of change of the temperature of the object, with respect to time, is proportional to the difference between the ambient temperature and the temperature of the object.

Thus, the relevant differential equation is

$$\frac{dT}{dt} \;=\; k(T_{\text{amb}} - T) \;=\; kT_{\text{amb}} - kT,$$

where $k > 0$ is a constant.

This equation is used as the basis for one method of estimating how long a corpse has been dead. Under "normal conditions", the value of k for the human liver is approximately 0.08 per hour, and the initial temperature is approximately 98.6°F. Once a person dies, and the liver temperature is not maintained by the body, Newton's Law of Cooling applies.

If the ambient temperature is (and has been for a long time) 72°F, and the liver temperature of a body found in "normal conditions" is 80°F, estimate how long the corpse has been dead.

Solution:

We are given that $T_{\text{amb}} = 72$°F, $k = 0.08$ per hour, and we have initial data that $T(0) = 98.6$°F. We will suppress the units until the final answer, and also not insert the values for the constants until the end.

The differential equation is, once again, in the form of Formula 4.11. Therefore, we use Formula 4.13 to conclude that

$$T \;=\; -\left(\frac{kT_{\text{amb}}}{-k}\right) + \left(T_0 + \frac{kT_{\text{amb}}}{-k}\right)e^{-kt} \;=\; T_{\text{amb}} + (T_0 - T_{\text{amb}})\,e^{-kt}.$$

Substituting the given values for the constants, we have

$$T = 72 + 26.6\, e^{-0.08t}.$$

We want to find t when $T = 80$. We find

$$t = \frac{\ln(8/26.6)}{-0.08} \approx 15 \text{ hr.}$$

Example 4.4.7. A company begins selling a new style of heavy wool coat on January 1, 2009 in a small city. Market research indicates that, in this city, there is a maximum number of 1000 possible buyers of this coat, and that, at any given point in time, the instantaneous rate of change, per month, in the number of coats sold is proportional to the number of possible buyers who had not yet purchased one of these coats.

On the other hand, research indicates that, due to the seasonal temperature variations, for a fixed number of coats already sold, the instantaneous rate of change, per month, in the number of coats sold is proportional to $1 + \cos(\pi t/6)$, where t is the number of months since January 1, 2009 (note that the period of this function is 12 months, i.e., it repeats each year).

Therefore, let $W = W(t)$ be the number of wool coats sold since January 1, 2009, and assume we have the differential equation

$$\frac{dW}{dt} = k\big(1 + \cos(\pi t/6)\big)(1000 - W). \tag{4.14}$$

Suppose that, at the beginning of January 1, 2009, no coats have been sold, but that the rate of change in the number of coats sold, with respect to time, is 20 coats per month.

How many of their new coats will the company sell in the first 6 months? in the first 9 months? in the first year?

Solution:

We will omit the units until the end of the problem.

We are given that, when $t = 0$, $W = 0$ and $dW/dt = 20$. This enables us to immediately calculate k from the differential equation in Formula 4.14; we find

$$20 \;=\; k(1 + \cos(0))(1000 - 0).$$

Therefore, $k = 1/100 = 0.01$.

Formula 4.14 is separable. Separating the variables, we obtain

$$\int \frac{1}{1000 - W}\, dW \;=\; \int 0.01\big(1 + \cos(\pi t/6)\big)\, dt \;=\; 0.01 \int 1\, dt \;+\; 0.01 \int \cos(\pi t/6)\, dt.$$

If you make the substitution $u = 1000 - W$ on the left, and $z = \pi t/6$ on the right, you should obtain:

$$-\ln|1000 - W| \;=\; 0.01t + \frac{0.06}{\pi}\, \sin(\pi t/6) + C,$$

or, equivalently (with a new C),

$$\ln|1000 - W| \;=\; -0.01\left(t + \frac{6}{\pi}\, \sin(\pi t/6)\right) + C.$$

We solve for W by raising e to both sides, factoring off the e^{C}, removing the absolute value signs, and letting B denote $\pm e^{C}$:

$$W \;=\; 1000 + B\, e^{-0.01\left(t + \frac{6}{\pi}\, \sin(\pi t/6)\right)}.$$

We now use the initial data (again) that $W(0) = 0$, and find that $B = -1000$.

Therefore, the number of coats sold t months after January 1, 2009 is

$$W \;=\; 1000\left(1 - e^{-0.01\left(t + \frac{6}{\pi}\, \sin(\pi t/6)\right)}\right).$$

Using a calculator, we find

$$W(6) \;=\; 1000(1 - e^{-0.06}) \;\approx\; 58.23546641575 \text{ coats,}$$

$$W(9) = 1000(1 - e^{-0.01(9 - 6/\pi)}) \approx 68.44626756343 \text{ coats},$$

and

$$W(12) = 1000(1 - e^{-0.12}) \approx 113.07956328284 \text{ coats}.$$

You should be asking: "Why did you go out to such a ridiculous number of decimal places, when we're trying to estimate the **number** of coats sold, which would be a positive integer?!?"

The answer is that we would like to compare the "actual" answers above with what you get from linear approximation (recall Definition 3.4.3). For the given t values, the exponent, $x = -0.01 \left(t + \frac{6}{\pi} \sin(\pi t/6) \right)$, of e in the solution, is fairly close to 0, and the linear approximation for e^x at 0 should give us a reasonably close approximation for that portion of the solution.

We leave it as an exercise for you to show that the linear approximation of e^x at 0 is that $e^x \approx 1 + x$. If we continue to let $x = -0.01 \left(t + \frac{6}{\pi} \sin(\pi t/6) \right)$, then, for the t values in which we're interested

$$W = 1000(1 - e^x) \approx 1000 \big(1 - (1 + x) \big) = 10 \left(t + \frac{6}{\pi} \sin(\pi t/6) \right).$$

Using this linear approximation, we find that

$$W(6) \approx 60 \text{ coats},$$

$$W(9) \approx 10(9 - 6/\pi) \approx 70.90140682897 \text{ coats},$$

and

$$W(12) \approx 120 \text{ coats}.$$

So, what we see is that the linear approximation yields a fairly good estimate, but not a great estimate; on the other hand, it has the advantage that we could estimate $W(6)$ and $W(12)$ **by hand**.

One way to look at the linear approximation here is that, when t is close to 0, W will be close to 0, and so $1000 - W$ will be close to the constant value of 1000. We would expect then that, using the same initial data, the solution to

$$\frac{dW}{dt} = k\big(1 + \cos(\pi t/6)\big)(1000)$$

would be a good approximation, for values of t close to 0, of the solution of

$$\frac{dW}{dt} = k\big(1 + \cos(\pi t/6)\big)(1000 - W).$$

The solution to this former equation is precisely what we got by using linear approximation to estimate the solution to the latter equation.

As our final example, we will look at a 2nd-order differential equation, a differential equation in which the highest-order derivative is the 2nd derivative.

Example 4.4.8. In Corollary 2.7.15, we considered the differential equation $y'' = -y$, where the primes denote derivatives with respect to x.

We reduced this 2nd order differential equation to a 1st-order equation by noting that

$$\big[y^2 + (y')^2\big]' = 2yy' + 2y'y'' = 2y'(y + y'').$$

Therefore, if $y'' = -y$, then $\big[y^2 + (y')^2\big]' = 0$, and so, there exists a constant C such that

$$y^2 + (y')^2 = C. \tag{4.15}$$

As it equals the sum of squares, C must be greater than or equal to 0; so, let $a \geq 0$ be the square root of C, so that $C = a^2$. Note that, if $a = 0$, we must have $y = y(x) = 0$.

Now, unlike earlier, we have a systematic way of solving this resultant 1st-order equation; it's separable. We find that

$$y' = \frac{dy}{dx} = \pm\sqrt{a^2 - y^2}.$$

First, there is the equilibrium solution $y = y(x) = a$. If we insert this into the original 2nd order equation, $y'' = -y$, we find that a must be 0. Therefore, $y = y(x) = 0$ is our one equilibrium solution.

Now suppose that $a > 0$ and that $y \neq a$. Separating the variables, we obtain

$$\int \frac{1}{\sqrt{a^2 - y^2}}\, dy = \int \pm 1\, dx. \tag{4.16}$$

The integral on the left looks like it should involve $\sin^{-1} y$, but there is an a where we would like to have a 1. We can factor out $\sqrt{a^2}$, which equals a since $a \geq 0$, but then the integral still isn't exactly the one we know:

$$\int \frac{1}{\sqrt{a^2 - y^2}}\, dy = \int \frac{1}{a \cdot \sqrt{1 - (y/a)^2}}\, dy = \int \frac{1}{\sqrt{1 - (y/a)^2}} \left(\frac{1}{a}\right) dy.$$

However, this makes it clear that the substitution $u = y/a$ will be helpful. We find $du = (1/a)\, dy$, and so

$$\int \frac{1}{\sqrt{1 - (y/a)^2}} \left(\frac{1}{a}\right) dy = \int \frac{1}{\sqrt{1 - u^2}}\, du = \sin^{-1} u = \sin^{-1}\left(\frac{y}{a}\right),$$

where, as usual, we omit the arbitrary constant on the indefinite integral, since we will put it on the other side of our equation momentarily.

Thus, we find that Formula 4.16 becomes

$$\sin^{-1}\left(\frac{y}{a}\right) = \pm x + B.$$

Hence, we obtain that

$$y = a\sin(\pm x + B) = \pm a\sin(x \mp B) = A\sin(x + E),$$

where the next-to-last equality follows from the fact that sine is an odd function, and the last equality results from letting $A = \pm a$ and $E = \mp B$. Note that, when $A = 0$, we obtain the equilibrium solution that we found earlier.

You can easily verify that $y = A\sin(x + E)$ is a solution to the initial differential equation $y'' = -y$, for all x, even though the solution will include places where y has the value a, i.e., our solution on an open interval on which $y \neq a$ extends to the entire real line.

There remains the question of why this solution agrees with what we found in Corollary 2.7.15. We leave it to you, as an exercise, to use the angle addition formula for sine, in order to make the two answers agree.

4.4.1   Exercises

1. Complete the proof of Example 4.4.4. Namely, show that

$$\int \frac{mv_0}{v_0 kt + m}\, dt = \frac{m}{k} \ln \left(\frac{v_0 kt}{m} + 1 \right) + C.$$

2. Suppose a particle is moving through a fluid, and that the force of resistance is given by $F = -kv^p$ Newtons, where v is in m/s. What are the units of k in the cases where $p = 1$ and $p = 2$?

In Exercises 3 through 5, assume the setup in Example 4.4.5. Specifically, you are given the mass m, initial velocity v_0, and constant k of a skydiver experiencing air resistance with force $F_r = -kv$ Newtons. For each exercise, (a) write a formula for the velocity v at time t, (b) determine the terminal velocity, (c) determine which of the three relationships between the initial and terminal velocity holds.

3. $m = 60$ kg, $v_0 = 294$ m/s, $k = 2$ N/(m/s).

4. $m = 75$ kg, $v_0 = 50$ m/s, $k = 2.5$ N/(m/s).

5. $m = 40$ kg, $v_0 = 100$ m/s, $k = 4$ N/(m/s).

In Exercises 6 through 9, you are given the annual birth rate, β, death rate, δ, the net change per year due to immigration, I, and the initial population P_0 for some country. For each problem, (a) determine the differential equation describing the population change, (b) solve the differential equation, and (c) calculate the limiting population.

6. $\beta = 0.07$, $\delta = 0.10$, $I = 2000$, $P_0 = 400,000$.

7. $\beta = 0.04$, $\delta = 0.06$, $I = 3000$, $P_0 = 1,000,000$.

8. $\beta = 0.06$, $\delta = 0.04$, $I = 5000$, $P_0 = 200,000$.

9. $\beta = 0.04$, $\delta = 0.02$, $I = -3000$, $P_0 = 100,000$.

10. More generally, what is the population $P(t)$ in terms of β, δ, P_0 and I according to this model?

11. What limiting value does the population approach if $I > 0$ and $\beta - \delta > 0$? Show that the population explodes in the sense that $P(t) \to \infty$ as $t \to \infty$. What adjustment could be made to the model to prevent this from happening?

12. What limiting value does the population approach if $I < 0$ and $\beta - \delta < 0$? Does this make sense?

13. Suppose a corpse is found and, referring to Example 4.4.6, $k = 0.08$ per hour. If, as is the usual case, $T_0 = 98.6°F$, then what is the predicted liver temperature 7 hours after death, if the ambient temperature is 75°F?

14. Suppose a corpse is found and, referring to Example 4.4.6, $k = 0.075$ per hour. Assume $T_0 = 98.6°F$, that the current temperature of the liver is 77°F and that the corpse has been dead for 6 hours. What is the ambient temperature? Assume the ambient temperature has been constant for some time.

15. In Example 4.4.7, what is the limit as $t \to \infty$ of $W(t)$, the number of wool coats sold since January 1, 2009?

16. Verify the claim made in Example 4.4.7 that the linear approximation of e^x at $x = 0$ is $e^x \approx 1 + x$.

17. What property of the cosine function makes it appealing for modeling wool coat purchases in Example 4.4.7?

18. Assume that the city in which wool coats are sold experiences particularly cold winters every three years. How would you modify the differential equation in Example 4.4.7?

19. Suppose an object with mass m is moving through a fluid, and that the force of fluid resistance is proportional to the square-root of the velocity. What is the velocity of the object at time t? Leave your answer in terms of the initial velocity v_0, proportionality constant k, r, and mass m. Assume the velocity is always positive. Assume the only force experienced by the object is fluid resistance.

20. Suppose an object with mass m is moving through a fluid, and that the force of fluid resistance is proportional to v^r, where r is a positive constant and $r \neq 1$. What is the velocity of the object at time t? Leave your answer in terms of the initial velocity v_0, proportionality constant k, r, and mass m. Assume the velocity is always positive. Assume the only force experienced by the object is fluid resistance.

21. Suppose a particle is moving vertically downward through a fluid with resistance proportional to the square of the velocity.

    a. Write down the differential equation that describes this situation.

    b. Solve this differential equation for $v(t)$. Hint: use partial fractions to facilitate the integration.

22. Given the setup in the previous problem, what is the limit as $t \to \infty$ of the velocity?

23. Generalize the argument in Example 4.4.8 to find solutions of the differential equation $y'' = -k^2 y$, where $k \neq 0$.

24. Verify the conclusion of Example 4.4.8. that the solution calculated in this chapter concurs with that given in Corollary 2.7.15.

25. Consider a chicken farm where the birth rate is proportional to the population with growth rate 10%. There are two causes of death: each year 200 chickens are slaughtered and sold. Also, the death rate due to natural causes is 0.5%. Suppose that the current ($t = 0$) population is 2200.

    a. Write an equation for the population P of chickens at time t.

    b. What is the limit as $t \to \infty$ of the population?

26. Using the results of Example 4.4.5, find the position function of a skydiver of mass m subject to the force of gravity $F_g = mg$ and air resistance $F_r = -kv$, with initial velocity v_0 and initial height h_0.

27. Suppose a rocket of mass m is projected upwards (perpendicular to the surface of the earth) with an initial velocity v_0. For simplicity, assume there is no resistance and that the only force acting on the rocket is gravity. This force is given by

$$F_g = -\frac{mgR^2}{(R+h)^2}$$

where R is the radius of the earth and h is the height of the rocket. This formula agrees with our intuition that the greater heights correspond to lower forces of gravity, or increased weightlessness. This gives us the differential equation

$$ma = m\frac{dv}{dt} = -\frac{mgR^2}{(R+h)^2}.$$

    a. Use the Chain Rule to rewrite the differential equation as $v\dfrac{dv}{dh} = -\dfrac{gR^2}{(R+h)^2}$. That is, express the differential equation with height as the independent variable rather than time.

    b. Separate the variables to arrive at the equation

$$\frac{v^2}{2} = \frac{gR^2}{R+h} + C.$$

Assume an initial velocity of v_0 at height $h = 0$. Solve for C.

c. Use the fact that the velocity of the rocket when it reaches its maximum height, h_{max}, is zero to show that

$$h_{max} = \frac{v_0^2 R}{2gR - v_0^2}.$$

d. Reverse part (c); find the initial velocity v_0 required to lift the rocket to a height of h_{max}.

e. Show that the limit as $h_{max} \to \infty$ of the initial velocity required to lift the rocket to a height of h_{max} is $\sqrt{2gR}$. The is called the *escape velocity* and is the velocity required to put a rocket into orbit.

28. The table below shows the surface gravities and radii (in km) of objects in our solar system. Use the information in the previous problem and the data in the table below to determine the escape velocities. Note that surface gravities are expressed in "Earth units." That is, surface gravity of 1.0 is 9.8 meters/sec^2.

Object	Surface Gravity	Radius	Escape Velocity
Mercury	0.380	2439	
Venus	0.903	6051	
Earth	1.0	6378	
Mars	0.380	3397	
Jupiter	2.53	71,492	
Saturn	1.07	60,268	
Uranus	0.92	25,559	
Neptune	1.12	24,764	
Pluto	0.06	1160	

29. A superhero of mass 55 kilograms leaps from a plane 3000 meters above the Earth's surface. Being well-prepared, she has a jet pack, capable of exerting 535 Newtons downwards, which she turns on as soon as she leaves the plane. Furthermore, assume that the force, in Newtons, acting upon her due to air resistance is proportional to her velocity, in m/s, with proportionality constant 1 N/(m/s).

a. Use Newton's 2nd Law of Motion to find a differential equation which you could solve to find the velocity of the superhero.

b. Noting that she starts with no vertical velocity, solve the differential equation from part (a) to find $v(t)$ the velocity of the superhero.

c. What is the terminal or equilibrium velocity of the superhero in the fall?

d. What is the velocity of the skydiver when she reaches the ground? (Note: You will need the position function $y(t)$.)

30. A grove of trees is being logged for paper. Each year, the company that owns the grove removes 10% of the adult trees, and plants 200 nascent trees. Suppose further that it takes 5 years for the nascent trees to become adult trees, and let $N(t)$ be the number of *adult* trees in the grove t years after logging starts.

   a. Find a differential equation for $N(t)$ that describes the growth of N during the first five years of the logging.

   b. Solve this differential equation under the assumption that the grove starts with 1000 adult trees.

   c. Set up and solve a differential equation for $N(t)$ that describes the growth of N after the first five years. Use your result from part (b) to determine the initial conditions for this differential equation.

   d. What limiting value does $N(t)$ approach as $t \to \infty$?

31. A tank is being slowly filled with drainage water, and simultaneously water is evaporating from the tank. Water is streaming in at a rate of 0.1 cubic centimeters per minute, and 5% evaporates each minute.

   a. Find a differential equation for the volume of water in the tank, with respect to time.

   b. Solve the differential equation from part (a), assuming that, at time $t = 0$ minutes, the volume of water in the tank is 100 cubic centimeters.

   c. Assuming the reaction continues as described, find the limit of the volume of water in the tank as $t \to \infty$.

32. To make a salt water solution, a tank containing 2000 liters of pure water is infused with 10 liters per minute of a solution containing 11 kilograms of salt per liter and, each minute, 10 liters are siphoned off.

   a. Give an initial value problem whose solution is the number of kilograms of salt in the tank at time t minutes.

   b. Solve the initial value problem.

   c. What is the salt concentration in the tank after 10 minutes?

   d. What value does the salt concentration approach as time goes on? Does this make intuitive sense?

33. As the ecology expert in a small town, you are put in charge of a fishing pond. The pond starts out stocked with $P_0 = 1000$ fish. Furthermore, the annual birthrate for the specific breed of fish in the pond is 10% (that is, the total increases by 10% each year due to births), the annual death rate is 5%, and, annually, 7% of the population is fished out.

    a. Write a differential equation for $P(t)$, the population of fish in the pond at year t.

    b. Solve the differential equation you found in part (a). After how many years does the fish population dwindle to 0?

    c. Suppose you decide to add a fixed amount of 100 fish into the pond each year to avoid extinction. How will this effect the differential equation that you found in part (a)?

    d. Solve the differential equation from part (c).

    e. What limiting value does the population approach as t goes to infinity?

34. Chicken cutlets at an initial temperature of $45\,°F$ are placed in a $350\,°F$ oven.

    a. Assuming that it takes 20 minutes for the chicken to reach $165\,°F$ (minimum safe internal temperature), find the function $T(t)$, the temperature of the chicken t minutes after being put into the oven.

    b. How long should the chicken be left in the oven if, to be safe, you want it to reach $170\,°F$?

35. Suppose a right, conical icicle melts in such a way that the dripping water refreezes at the tip so that, while the icicle remains a cone with fixed height h, the radius of its base is decreasing. Suppose further that the rate at which the volume is decreasing is proportional to the total surface area of the shape (including the base) with proportionality constant $k > 0$.

    a. Find a differential equation relating the volume and the surface area.

    b. Use the formulas for surface area and volume for a cone (see, for instance, Exercise 40 in Section 3.3) to find a differential equation relating the rate of change of the radius to the radius itself.

    c. Solve the differential equation in part (b) under the assumption that the radius of the base at time $t = 0$ is precisely the height.

    d. If the dynamics of the system remain unchanged, how long will it take for the icicle to melt entirely?

36. Assume that the graph $y = f(x)$ satisfies the following property: the tangent line to the curve at the point $(x, f(x))$ also passes through the point $\left(-\dfrac{1}{x}, -\dfrac{1}{f(x)}\right)$. First, find a differential equation relating $\dfrac{dy}{dx}$ and y, and then solve it under the assumption that the $f(1) = 2$.

37. In Exercise 48 in Section 2.8, you showed that the tangent line to a point (x, y) on a circle with center (x_c, y_c) is orthogonal to the line that passes through both the point and the center. Show that if a curve has this property, i.e., that there is a point (a, b) so that the

line orthogonal to the curve at every point passes through (a, b), then the curve is a circle with center (a, b).

38. Assume that the graph $y = f(x)$ satisfies the following property: for each point $(x, f(x))$ on the curve, the right triangle constructed as given in the diagram has area 2 square units. Find a differential equation with $f(x)$ as its solution and solve for this function.

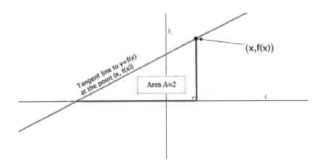

39. A differential equation of the form

$$\frac{dP}{dt} = kP^{1+r},$$

where k and r are positive, is called a *doomsday equation*. It is called this because there is a (finite) value t_0 such that $\lim\limits_{t \to t_0^-} P(t) = \infty$. (Thus, if P represents population and t represents time, then the differential equation would predict that the population explodes.)

a. Suppose that the population function of a certain prolific organism fits a doomsday differential equation with $r = 0.01$. At time $t = 0$, there are 1000 animals and, after two years, there are 2000 animals. Find $P(t)$, the population function.

b. Find t_0, the so-called doomsday, for this population.

40. A college student racks up a $8,000 credit card bill on a credit card that accrues 6.0% yearly interest, compounded continuously. She decides to set up automatic payments of A_0 dollars per month from her checking account towards her credit card bill.

a. What should the payment A_0 be if she wants her debt to stay steady?

b. Suppose that she can afford to pay $50 per month. How long will it take her to pay off her debt?

41. The gravitational acceleration near the surface of Mars is approximately 3.8 meters per second squared. A 1000 kilogram satellite that is 2000 meters above the surface of Mars is falling at a rate of 100 meters per second toward Mars and exerts 3700 Newtons of force upwards using its rockets. Furthermore, due to the thin atmosphere, the force, in Newtons, of air resistance is proportional to the velocity, in m/s, of the satellite with constant of proportionality is 0.2 N/(m/s). ▶

    a. Find the velocity and position functions for the satellite by applying Newton's 2nd Law of Motion.

    b. How long after the start of the problem does the satellite reach the surface of the planet? (*This is not easily done by hand, so it is better to either employ Newton's Method or a calculator with solving capabilities.*)

    c. Using your answer from part (b), calculate the speed of the satellite when it crashes into Mars (use the approximation of 5 seconds for part (b), if necessary).

42. As we have discussed, Newton's Law of Cooling is normally stated as: if $T = T(t)$ is the temperature of an object, which is placed in surroundings which are kept at a constant (ambient) temperature, T_{amb}, then the rate of change of the temperature of the object, with respect to time, is proportional to the difference between the ambient temperature and the temperature of the object. Hence,

$$\frac{dT}{dt} = k(T_{\text{amb}} - T), \tag{4.17}$$

where $k > 0$ is a constant.

However, this equation **still** applies if T_{amb} is **not** constant. If the ambient temperature is changing solely from being in contact with the object, then the rate of change of the ambient temperature is governed by the analogous differential equation, in which the roles of T_{amb} and T are reversed:

$$\frac{dT_{\text{amb}}}{dt} = k_a(T - T_{\text{amb}}), \tag{4.18}$$

and $k_a \geq 0$ is a constant. This situation, with $k_a \neq 0$, occurs, for instance, when warm drinks are placed inside a cold (well-insulated) cooler. Note that we recover the standard situation by letting $k_a = 0$, for then dT_{amb}/dt is 0, and so T_{amb} is constant.

Thus, in a situation in which the ambient temperature changes from being in contact with the object, the temperature of the object and the ambient temperature must simultaneously satisfy the pair of differential equations given in Formula 4.17 and Formula 4.18.

Show first that

$$T_{\text{amb}} \;=\; -\frac{k_a}{k}\,T + E,$$

for some constant E, and then show that

$$T \;=\; \frac{kE}{k_a + k} + B\,e^{-(k_a+k)t},$$

where B is a constant, and E is the same constant as in the previous equation.

Conclude that, as $t \to \infty$, both T and T_{amb} approach the common temperature $kE/(k_a + k)$. Finally, conclude, in the standard case, where $k_a = 0$, that

$$T \;=\; T_{\text{amb}} + B\,e^{-kt}.$$

4.5 Approximating Solutions of $y' = m(x, y)$

In the previous sections of this chapter, we looked at solving differential equations that have fairly simple forms: either ones of the form $y' = f(x)$, which are solved by anti-differentiating, or separable equations, which are of the more general form $y' = f(x)g(y)$. However, it is easy for real-world applications to yield more-complicated differential equations/initial value problems, which are very difficult to solve, or for which no method of explicitly solving is known. In such cases, solutions of the initial value problems are frequently approximated by numerical and/or graphical techniques.

In this section, we will look at differential equations of the form $dy/dx = m(x, y)$, where $m(x, y)$ is some function which may depend on both x and y, e.g., $dy/dx = x - y$, and discuss how you can analyze/approximate solutions of initial value problems that have such underlying differential equations. The numerical method that we will discuss is known as *Euler's Method*, and the graphical manner of analyzing solutions is via *slope fields*.

We should remark that our treatment in this section is relatively informal; we will put no restrictions, beyond continuity, on the function $m(x, y)$, restrictions that would guarantee "nice" behavior of Euler's Method and slope fields. We will also not analyze the amount of error that might be involved in Euler's Method. These are topics for courses in differential equations or numerical analysis.

Euler's Method:

The differential equation $dy/dx = x/y$ is separable and easy to solve, but what if we changed the division on the right-hand side to subtraction? Would the differential equation really become substantially more difficult? **Yes**.

The differential equation $dy/dx = x - y$ is **not** separable, and a method of solution is beyond the scope of this textbook. Nonetheless, it is not TOO difficult to show that the general solution is $y = Ae^{-x} + x - 1$, where A could be any constant.

But, what if you don't know how to produce the general solution that we gave (and, presumably, you don't)? Or what if the equation had been $dy/dx = x - y^2$? Or what if it had been something truly awful, like $dy/dx = xe^y - \sin(xy)$? What can you do if you have an initial value problem that involves one of these differential equations?

More generally, we want to be able to handle the situation in which we are given a 1st-order initial value problem (IVP) of the form

$$\frac{dy}{dx} \; = \; m(x,y); \qquad y(a) = y_0, \tag{4.19}$$

where $m(x,y)$ is a continuous function of x and y, and we know initial data: when $x = a$, $y = y_0$. While we will not give the technical definition of a continuous function of two (or more) variables (see [3] or [2]), intuitively, it means what you think it means: very small changes in the pair (x,y) do not cause big jumps in the value of $m(x,y)$. In particular, recall from the discussion just before Theorem 2.10.5 that all elementary functions of two variables are continuous.

Now, ideally, we could explicitly solve the IVP in Formula 4.19, and find a unique solution function $y = y(x)$. But, frequently, this is problem is too hard, and so, instead, we assume that there a unique solution $y = y(x)$ and ask: given an arbitrary x value, b, can we estimate, in some reasonable fashion, the numerical value of $y(b)$?

If b is "close to" a, the answer is easy; we use the linear approximation from Definition 3.4.3, and putting in b for x:

$$y(b) \; \approx \; y(a) \; + \; \frac{dy}{dx}\bigg|_{a} \cdot (b - a) \; = \; y_0 \; + \; \frac{dy}{dx}\bigg|_{a} \cdot (b - a)$$

Since we're assuming that $y = y(x)$ is a solution to the IVP in Formula 4.19, we have that

$$\frac{dy}{dx}\bigg|_{a} \; = \; m(a, y(a)) \; = \; m(a, y_0)$$

and the linear approximation above becomes

$$y(b) \; \approx \; y_0 \; + \; m(a, y_0) \cdot (b - a).$$

As easy as it is, this last approximation, that, if $y = y(x)$ is a solution to the IVP in Formula 4.19 and $\Delta x = b - a$ is close to 0, then

$$y(b) \; \approx \; y_0 \; + \; m(a, y_0) \cdot \Delta x, \tag{4.20}$$

is the entire basis of Euler's Method.

But what if b is not close to a? What do we do then? We chop up the interval from a to b (or from b to a) into smaller subintervals, and then use our linear approximation above to estimate the corresponding y values as we move from one x value to a nearby one, and we keep going until we reach $x = b$. We'll discuss this now in general, and then go through a couple of examples.

Suppose a and b are "far" apart. We subdivide the interval $[a, b]$ or $[b, a]$ into smaller subintervals, all of the same "small" (signed) length Δx. Call the number of subintervals n, and let $\Delta x = (b - a)/n$. This Δx is usually referred to as the *step size* (which could be negative) and n is the *number of steps*. We let $x_0, \ldots, x_n$ denote the endpoints of the subintervals; thus,

$$x_0 = a, \qquad x_1 = x_0 + \Delta x, \qquad x_2 = x_1 + \Delta x = x_0 + 2\Delta x, \qquad \ldots$$

$$x_k = x_{k-1} + \Delta x = x_0 + k\Delta x, \qquad \ldots \qquad x_n = x_{n-1} + \Delta x = x_0 + n\Delta x = a + (b - a) = b.$$

Now, if n is big, then Δx is close to 0 and, for each k, the linear approximation

$$y(x_{k+1}) \approx y(x_k) + m(x_k, y(x_k)) \cdot \Delta x$$

would give us a good approximation for $y(x_{k+1})$, **provided** that we know x_k and $y(x_k)$. Of course, we know x_k, but how do we determine $y(x_k)$? In fact, we **don't** determine $y(x_k)$ **exactly**, we approximate it from earlier data, and call the approximation y_k. And, since we know y_0, we have a way to get started.

So, the entire Euler's Method for approximating the value of $y(b)$, using n steps, goes like this:

Definition 4.5.1. (Euler's Method) *Suppose that you have the initial value problem*

$$\frac{dy}{dx} = m(x, y); \quad y(a) = y_0.$$

Let n be a positive integer. Let $\Delta x = (b - a)/n$ and let $x_0 = a$. For all k between 0 and n, let $x_k = x_0 + k\Delta x$, so that $x_n = b$. For all k from 0 to $n - 1$, define y_{k+1} iteratively by

$$y_{k+1} = y_k + m(x_k, y_k)\Delta x.$$

Then, y_n **is the approximation of** $y(b)$**, obtained from Euler's method, using** n **steps**.

Of course, we expect/hope that, when n is bigger, the approximation is better. Let's look at some examples.

Example 4.5.2. Consider the initial value problem:

$$\frac{dy}{dx} = x - y; \qquad y(1) = -1.$$

This means that we are considering the case where $m(x, y) = x - y$, $a = x_0 = 1$, and $y_0 = -1$.

As we stated earlier, the general solution to the differential equation in the IVP is $y = Ae^{-x} + x - 1$. If we require that $y(1) = -1$, then we find that the unique solution to the IVP is

$$y = -e \cdot e^{-x} + x - 1 = -e^{-x+1} + x - 1. \tag{4.21}$$

Having the explicit solution to the IVP is good in this example, because it will enable us to look at the error in the approximation that we will obtain via Euler's Method. However, you should understand:

> The whole point of Euler's Method is that it enables you to numerically approximate solutions to initial value problems **in cases where finding an explicit solution is difficult/impossible.**

Let's temporarily ignore the fact that we have the explicit solution to the IVP. How would we use Euler's Method, Definition 4.5.1, to estimate $y(2)$?

We could simply think of 2 as being "close to" 1, and take use Euler's Method with 1 step, i.e., take $n = 1$, so that $x_1 = 2$. When $n = 1$, Euler's Method just amounts to using linear approximation to estimate $y(2)$. We find

$$y(2) \approx y_0 + m(x_0, y_0) \cdot (x_1 - x_0) = -1 + (1 - (-1)) \cdot (2 - 1) = 1.$$

How good is the approximation $y(2) \approx 1$? Well, since we have the explicit solution, we know that, in fact,

$$y(2) = -e^{-2+1} + 2 - 1 = -e^{-1} + 1 \text{ "} = \text{" } 0.632120558829,$$

where this last number is to the full accuracy of our calculator.

You can see that our approximation by Euler's Method with one step is not so good; the error, the absolute value of the difference between the approximation and the actual value, is around 0.37, a large percentage of the actual value.

What if we use 2 steps? Let's see what happens when we let $n = 2$, so that $\Delta x = (2-1)/2 = 0.5$, and $x_0 = a = 1$, $x_1 = x_0 + \Delta x = 1.5$, and $x_2 = x_1 + \Delta x = 2$. Of course, y_0 still equals -1.

We expect the Euler's Method approximation of $y(x_1) = y(1.5)$ to be better than our approximation of $y(2)$ because 1.5 is closer to 1 than 2 is. Of course, then our estimate y_1 for $y(1.5)$ is not perfect before we once again apply the linear approximation, the Euler's Method formula, again. Still, we have some hope that the compounded error from using $\Delta x = 0.5$ twice is smaller than the error from using $\Delta x = 1$ once. Let's check.

We find

$$y_1 \; = \; y_0 \; + \; m(x_0, y_0)\Delta x \; = \; y_0 \; + \; (x_0 - y_0)\Delta x \; = \; -1 \; + \; (1 - (-1)) \cdot 0.5 \; = \; 0,$$

and then

$$y_2 \; = \; y_1 \; + \; m(x_1, y_1)\Delta x \; = 0 \; + \; (x_1 - y_1)\Delta x \; = \; 0 + (1.5 - 0) \cdot 0.5 \; = \; 0.75.$$

This is significantly closer to the actual value than when we used a single step.

We would like to see how much better the approximation is when we take the number of steps to be $n = 4$. In this case, $\Delta x = (2 - 1)/4 = 0.25$ and, in order, x_0, x_1, x_2, x_3, and x_4 are 1, 1.25, 1.5, 1.75, and 2. We're trying to find the value of y_4, our new approximation of $y = y(x_4) = y(2)$. We go through Euler's Method a step at a time:

$$y_1 \; = \; y_0 \; + \; m(x_0, y_0)\Delta x \; = \; y_0 \; + \; (x_0 - y_0)\Delta x \; = \; -1 \; + \; (1 - (-1)) \cdot 0.25 \; = \; -0.5.$$

$$y_2 \; = \; y_1 \; + \; m(x_1, y_1)\Delta x \; = \; y_1 \; + \; (x_1 - y_1)\Delta x \; = \; -0.5 \; + \; (1.25 - (-0.5)) \cdot 0.25 \; = \; -0.0625.$$

$$y_3 \; = \; y_2 + m(x_2, y_2)\Delta x \; = \; y_2 + (x_2 - y_2)\Delta x \; = \; -0.0625 + (1.5 - (-0.0625)) \cdot 0.25 \; = \; 0.328125.$$

And, finally,

$$y_4 \; = \; y_3 + m(x_3, y_3)\Delta x \; = \; y_3 + (x_3 - y_3)\Delta x \; = \; 0.328125 + (1.75 - 0.328125) \cdot 0.25 \; = \; 0.68359375.$$

As you can see, by taking $n = 4$, we have gotten much closer to the "actual" value of $y(2)$, 0.632120558829.

In the above example, we calculated everything "by hand". To get better accuracy in our approximation via Euler's Method, we want to take even larger values of n. We'd like to use $n = 100$ or $n = 500$, or something even larger. But, nobody in their right mind would actually calculate Euler's Method by hand for $n = 100$ or 500. So, instead, we typically use a spreadsheet program to implement Euler's Method. What you put in the cells is actually pretty simple (we assume that you some minor experience with spreadsheet software).

◇	A	B	C	D
1	k	x_k	y_k	m(x_k, y_k)
2				
3	0	1	-1	2
4	1	1.25	-0.5	1.75
5	2	1.5	-0.0625	1.5625
6	3	1.75	0.328125	1.421875
7	4	2	0.68359375	1.31640625

Figure 4.2: Euler's Method spreadsheet for $y' = x - y$; $y(1) = -1$; $n = 4$.

In Figure 4.2, we have the $n = 4$ data from Example 4.5.2. Column A contains the k values, 0 through 4, Column B contains the corresponding x_k values, Column C contains the corresponding y_k approximations of $y(x_k)$, and Column D is an auxiliary column containing the corresponding values $m(x_k, y_k)$, which, in this example, is given by $m(x_k, y_k) = x_k - y_k$. You see our approximation of $y(2) = y(x_4)$ in cell C7; this is the value of y_4, namely 0.68359375.

What formulas did we enter into the cells? First, either in another portion of the spreadsheet, or by hand, we calculated the value of $\Delta x = (b - a)/n = (2 - 1)/4 = 0.25$. Now, in Column A, we put a 0 in cell A3. Then, in cell A4, we entered "$= A3 + 1$" (without the quotes), and then filled down the A column, beginning with cell A4. This caused each value to be 1 plus the value above it. In cells B3 and C3, we entered the initial data x_0 and y_0, respectively; in this example, 1 and -1. In cell B4, we entered "$= B3 + 0.25$", and then filled down Column B, beginning with cell B4. This caused each entry in Column B to be the entry above it plus $\Delta x = 0.25$.

In Column D, the formula for $m(x, y)$ entered into the calculation; in cell D3, we put "$= B3 - C3$", corresponding to $m(x_k, y_k) = x_k - y_k$. In other problems, in which $m(x, y)$ is some other function, its formula, translated into spreadsheet language, would go in cell D3. Then, we filled down Column D, beginning with cell D3. Finally, we dealt with Column C, the y_k column, the column we're after. This column contains the linear approximation that's at the heart of Euler's Method. Recall that we already placed the initial y value, here -1, in cell C3. In cell

C4, we entered "$= C3 + D3 * 0.25$"; this is the spreadsheet version of $y_1 = y_0 + m(x_0, y_0) \cdot \Delta x$. We then filled down Column C, so that $y_2 = y_1 + m(x_1, y_1) \cdot \Delta x$, $y_3 = y_2 + m(x_2, y_2) \cdot \Delta x$, and $y_4 = y_3 + m(x_3, y_3) \cdot \Delta x$, i.e., $y_{k+1} = y_k + m(x_k, y_k) \cdot \Delta x$.

Of course, once everything is in a spreadsheet, it's easy to pick a larger n, calculate the new Δx, and fill down n rows, even if n is large. In addition, in an example in which we can actually find an explicit solution to the IVP (like the problem we've been discussing), it's interesting to add a column for the exact value of the solution $y = y(x)$ at each x_k value, another column for the error, that is, the absolute value of the difference between y_k and $y(x_k)$, and another column for what percent of the actual value the error is.

For instance, here's how the beginning and end of the $n = 100$ spreadsheet looks for our IVP $y' = x - y$; $y(1) = -1$:

	A	B	C	D	E	F	G
1	k	x_k	y_k	m(x_k, y_k)	y(x_k)	error	% error
2							
3	0	1	-1	2	-1	0	0
4	1	1.01	-0.98	1.99	-0.98004983	4.9834E-05	0.00508482
5	2	1.02	-0.9601	1.9801	-0.96019867	9.8673E-05	0.01027634
6	3	1.03	-0.940299	1.970299	-0.94044553	0.00014653	0.01558129

Figure 4.3: Start of spreadsheet for $y' = x - y$; $y(1) = -1$; $n = 100$.

	A	B	C	D	E	F	G
101	98	1.98	0.60653572	1.37346428	0.6046889	0.00184682	0.30541629
102	99	1.99	0.620270362	1.369729638	0.61842331	0.00184705	0.29867137
103	100	2	0.633967659	1.366032341	0.63212056	0.0018471	0.2922069

Figure 4.4: End of spreadsheet for $y' = x - y$; $y(1) = -1$; $n = 100$.

How did we create Columns E, F, and G? In Column E, we had to put in the formula for the explicit solution $y = -e^{-x+1} + x - 1$, in spreadsheet terms, of course. So, in cell E3, we entered "$= -EXP(-B3+1) + B3 - 1$", and then filled down from there. In cell F3, we entered "$= ABS(C3 - E3)$", and filled down. Finally, in cell G3, we entered "$= 100 * ABS(F3/E3)$", and filled down.

As you can see, the error is much smaller for $n = 100$ than it was when we had $n = 4$. We could get even less error by taking $n = 500$, 1000, or $10,000$. After all, it's easy to fill down in a spreadsheet!

We should point out: in a **real** Euler's Method problem, where you can't find an explicit solution to the initial value problem, you wouldn't be able to include the last three columns that we included in the spreadsheet clips above.

Example 4.5.3. In this example, we'll consider an IVP which we cannot solve explicitly. Consider

$$\frac{dy}{dx} = \sqrt{x} + \cos y; \quad y(2) = 0,$$

and let's approximate $y(2.4)$.

Thus, we have $m(x, y) = \sqrt{x} + \cos y$, $a = x_0 = 2$, and $y_0 = 0$, and $b = 2.4$.

First, we'll use Euler's Method with $n = 2$ steps, and do this "by hand". We put "by hand" in quotes, since we'll use a calculator to give us values in the calculation. As before, we shall use an equals sign, in quotes, to denote the result to the full accuracy of our calculator.

After we go through Euler's Method by hand, for $n = 2$, we'll use a spreadsheet to see what we get with $n = 100$ and $n = 10,000$.

- $n = 2$, by hand and calculator:

We find that $\Delta x = (b - a)/n = (2.4 - 2)/2 = 0.2$, $x_0 = 2$, $x_1 = 2.2$, $x_2 = 2.4$ and

$$y_1 = y_0 + m(x_0, y_0)\Delta x = 0 + m(2, 0) \cdot 0.2 = \left(\sqrt{2} + \cos 0\right)(0.2) \text{ "} = \text{" } 0.482842712475.$$

$$y_2 = y_1 + m(x_1, y_1)\Delta x \text{ "} = \text{" } 0.482842712475 + m(2.2, 0.482842712475) \cdot 0.2 \text{ "} = \text{"}$$

$$\left(\sqrt{2.2} + \cos(0.482842712475)\right)(0.2) \text{ "} = \text{" } 0.956626379002.$$

So, our approximation of $y(2.4)$, using $n = 2$, is 0.956626379002.

- $n = 100$, by spreadsheet:

◇	A	B	C	D
1	k	x_k	y_k	m(x_k, y_k)
2				
3	0	2	0	2.414213562
4	1	2.004	0.009656854	2.415580442
5	2	2.008	0.019319176	2.416852557

Figure 4.5: Start of spreadsheet when $n = 100$.

From Figure 4.6, we see that the Euler's Method approximation of $y(2.4)$, using $n = 100$, is 0.935932321.

◇	A	B	C	D
101	98	2.392	0.918727094	2.153441586
102	99	2.396	0.92734086	2.147865292
103	100	2.4	0.935932321	2.142261362

Figure 4.6: End of spreadsheet when $n = 100$.

- $n = 10,000$, by spreadsheet:

◇	A	B	C	D
1	k	x_k	y_k	m(x_k, y_k)
2				
3	0	2	0	2.414213562
4	1	2.00004	9.65685E-05	2.4142277
5	2	2.00008	0.000193138	2.414241828

Figure 4.7: Start of spreadsheet when $n = 10,000$.

◇	A	B	C	D
10001	9998	2.39992	0.935263985	2.142773522
10002	9999	2.39996	0.935349695	2.142717454
10003	10000	2.4	0.935435404	2.142661383

Figure 4.8: End of spreadsheet when $n = 10,000$.

So, our approximation of $y(2.4)$, using $n = 10,000$, is 0.935435404.

What do we conclude from all of the above data? Well...we can't **prove** anything from the data, but it certainly appears that $y(2.4)$ is approximately equal to 0.935.

Before we leave our discussion of Euler's Method, we should mention that there are other numerical methods for approximating solutions to IVP's. Two common ones are the *Improved Euler's Method* and the *Runge-Kutta Method*. Either one of these typically yields much better accuracy with far fewer steps than Euler's Method. However, the trade-off is that the formulas that go into the methods/spreadsheets for the Improved Euler's Method or the Runge-Kutta Method are substantially, or dramatically, more complicated than the formulas in Euler's Method. You can remember that Euler's Method simply uses linear approximation, and produce the appropriate spreadsheet in minutes, without having to consult references for the formulas.

In contrast, the Runge-Kutta Method is accurate with stunningly few steps, but Runge-Kutta's formulas look incredibly complicated and essentially everyone would need to consult a reference to get them right. Of course, if you're programming a computer to deal with a variety of IVP's, over a period of time, you look up the formulas once and use the Runge-Kutta Method, or some variation of it; this is what numerical routines for solving IVP's use in most sophisticated mathematics software. However, if you just have one or two IVP's to approximate solutions for, and you're entering the formulas into a spreadsheet yourself, then you probably want to use Euler's Method; you could enter Euler's Method into a spreadsheet, and fill down 1000 rows a lot faster than you could enter the Runge-Kutta Method and fill down 10 rows.

Slope fields:

In our discussion of Euler's Method, we looked at a technique for numerically approximating solutions to initial value problems of the form:

$$\frac{dy}{dx} = m(x, y); \quad y(x_0) = y_0,$$

where $m(x, y)$ is continuous.

The question we address now is: Is there a graphical method for estimating, or analyzing qualitatively, solutions to such IVP's?

Let's temporarily ignore the initial data, and concentrate on the differential equation $y' = m(x, y)$. We will call the graph of a solution to this differential equation a *solution curve*.

If we have a solution $y = y(x)$ to our differential equation, then the slope of the tangent line to the solution curve at any point (x, y) on the curve is just y', but this would equal $m(x, y)$, i.e., at any point on the solution curve, the slope of the tangent line to the solution curve at that point is given by the value of m.

This means that, if we select a whole bunch of points (x, y) in the xy-plane and draw little line segments through each point, with slope $m(x, y)$, then any solution curve which passes through some of our selected points must have those little line segments as tangent lines (actually, tangent line segments). A collection of little line segments, at various points (x, y), with slope given by $m(x, y)$, is called a *slope field* for the differential equation $y' = m(x, y)$. See Figure 4.9. It is easy for computers to draw **lots** of little line segments.

How does this help you sketch solution curves and, in particular, a solution curve which passes through our initial data point (x_0, y_0)? It's like connecting-the-dots from when you were a kid, except that now the dots aren't numbered, but the dots tell you in what directions your curve goes. You start at (x_0, y_0), and you try to draw a smooth curve which is tangent to any

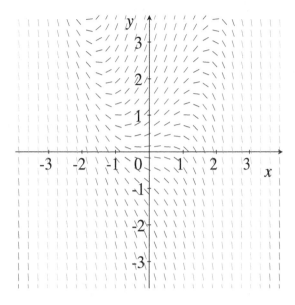

Figure 4.9: A slope field for $y' = y - x^2$.

line segment in the slope field that it passes through. Of course, you don't have little slope field line segments **everywhere**, so you have to interpolate in-between line segments, that is, you estimate from the slopes of nearby line segments what the slopes would be at each point on the curve you're drawing (this implicitly uses our assumption that $m(x, y)$ is continuous). See Figure 4.10, where we have used the slope field for $y' = y - x^2$ from Figure 4.9, but have had a computer draw in the solution curve $y = y(x)$ which passes through the point given by the initial data $y(0) = 1$, i.e., passes through $(0, 1)$. You may wonder how a computer determines the points on the solution curve. The answer: it uses a modified form of the Runge-Kutta Method.

Once you have a solution curve, drawn by hand and calculator, or by computer, you can use it to estimate the value of y and different x values. For instance, from the solution curve in Figure 4.10, you might estimate that $y(2)$ is approximately 2.6.

However, the real value of having a slope field is not that it lets you estimate the values of the solution to a single initial value problem, but rather that it lets you see the qualitative nature of how the solution curves for a single differential equation $y' = m(x, y)$ change as you change the initial data, so that the slope field lets you compare a family of solution curves to IVP's associated to a single differential equation. In Figure 4.11, we continue to look at a slope field for $y' = y - x^2$, but have now included solution curves which correspond to the initial data $y(0) = -2, 0, 1, 2$.

Perhaps, at this point, you're convinced of the value of having a slope field for a differential

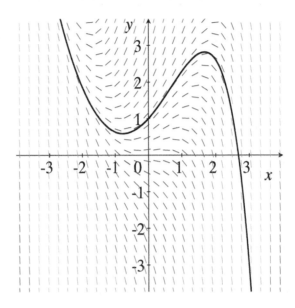

Figure 4.10: A slope field and solution curve for $y' = y - x^2$; $y(0) = 1$.

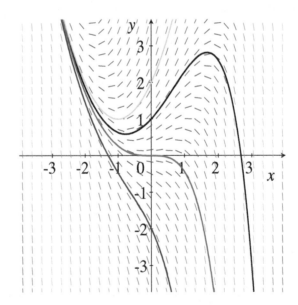

Figure 4.11: A slope field and solution curve for $y' = y - x^2$; $y(0) = -2, 0, 1, 2$.

equation of the form $y' = m(x, y)$, but you probably want to know how you obtain a slope field, right?

Well...by hand, with a calculator, you **could** just plug a bunch of points into $m(x, y)$, and then try to carefully draw little line segments with the appropriate slopes. This is pretty painful. There is a faster way to do this by hand, provided that $m(x, y)$ is something fairly familiar. Actually, what we need, for various constants c, is for the graph of $m(x, y) = c$ to be something familiar.

How does this work? Consider trying to draw by hand a slope field for $y' = y - x^2$. Then, $m(x, y) = y - x^2$, and the graph of $m(x, y) = 0$ is just the graph of $y - x^2 = 0$, i.e., the parabola given by $y = x^2$. This means that the points on the parabola $y = x^2$ are precisely the places where $m(x, y) = 0$; that is, the places where the line segments in the slope field have slope 0. So, now, you draw lots of little line segments of slope 0 at lots of points on the parabola $y = x^2$. What about the points where $m(x, y) = y - x^2 = 1$? These all lie along the parabola $y = x^2 + 1$, which is the same as the parabola $y = x^2$, except shifted up one unit. Thus, along the curve $y = x^2 + 1$, you should draw a bunch of little line segments of slope 1. You can continue in this manner by looking at $m(x, y) = c$ for other values of the constant c, and quickly draw quite a few line segments in the slope field. See Figure 4.12.

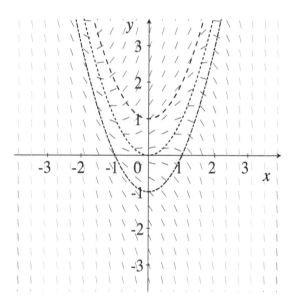

Figure 4.12: A slope field for $y' = y - x^2$, with isoclines where $m(x, y) = -1, 0, 1$.

Any curve along which the slope is constant is called an *isocline*: "iso-" for "same", and "-cline" for "inclination" or "slope".

Do not confuse the isoclines with solution curves. Isoclines are simply auxiliary curves that help you draw the slope field.

However, even by using isoclines, drawing slope fields by hand/calculator is very tedious. Ideally, you would use a computer or a graphing calculator.

Example 4.5.4. Looking back at the IVP from Example 4.5.2,

$$\frac{dy}{dx} = x - y; \qquad y(1) = -1,$$

it's interesting to draw a slope field, and solution curve, and get the big picture of what's going on.

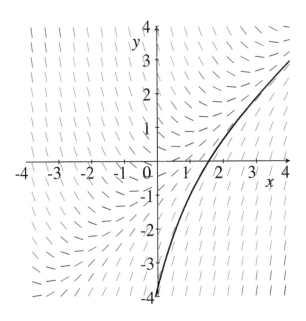

Figure 4.13: A slope field for $y' = x - y$, and solution curve where $y(1) = -1$.

Example 4.5.5. Now, looking back at the IVP from Example 4.5.3,

$$\frac{dy}{dx} = \sqrt{x} + \cos y; \qquad y(2) = 0,$$

it's once again interesting to draw a slope field, and solution curve, and get the big picture of what's going on.

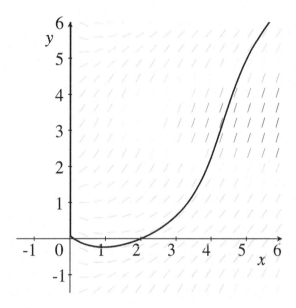

Figure 4.14: A slope field for $y' = \sqrt{x} + \cos y$, and solution curve where $y(2) = 0$.

Example 4.5.6. As our last example on slope fields, let's consider the differential equation

$$\frac{dA}{dt} = (A+1)(A-0.5)(A-1.5), \qquad (4.22)$$

where we are thinking of A as the value of some physical quantity, that can be positive or negative, that is changing with time t, and we care about the long-term qualitative nature of solutions, as time increases, and how the nature of the solutions depend on the initial value $A(0)$. When we write "qualitative", we mean things such as "is A always increasing?", "is A always decreasing", "in the long run, does A approach some specific value?", "in the long run, does A get arbitrary large?", etc.

Note that the function $m(t,A) = (A+1)(A-0.5)(A-1.5)$, in fact, depends only on A. Recall that, in Section 4.1, Exercises 20-22, we referred to a differential equation of the form $dy/dx = F(y)$ or, here, $dA/dt = F(A)$, as *autonomous*. At a fixed value of the dependent/function variable (here, A), a slope field for an autonomous differential equation will have line segments all of which have the same slope; that is, horizontal lines are isoclines for autonomous differential equations.

You should also recall the notion, from Section 4.1, of an *equilibrium solution*. An equilibrium solution is a solution which is constant; here, a solution of the form $A = A(t) = C$ for some constant C. This, of course, means that $dA/dt = 0$ for an equilibrium solution. For an autonomous equation, this means that the roots, or zeroes, of the right-hand side correspond to equilibrium solutions. Thus, we quickly see that there are three equilibrium solutions for Formula 4.22: $A(t) = -1$, $A(t) = 0.5$, and $A(t) = 1.5$.

You should realize that autonomous differential equations are always separable; just divide both sides by the entire quantity on the right, and multiply both sides by dt (or, in another problem, whatever the differential is on the bottom of the derivative). However, the resulting integral involving A is a fairly painful integration by partial fractions, and the result would not enable you to (easily) determine how the solutions depend on the initial data.

Instead, let's look at a slope field for the differential equation, in which we have included the equilibrium solution curves, in red, and other typical solution curves, in black; see Figure 4.15.

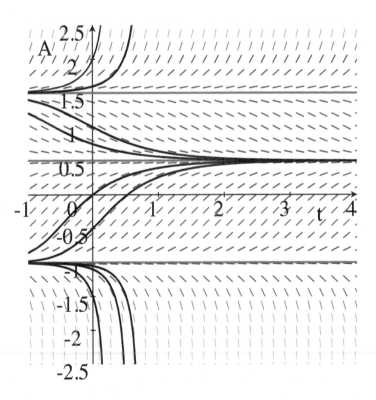

Figure 4.15: A slope field for $A' = (A+1)(A-0.5)(A-1.5)$, with equilibrium solution curves.

For an autonomous 1st-order differential equation, a diagram like that in Figure 4.15, in which you've included the equilibrium solution curves, and typical solution curves on each side of each equilibrium solution curve, is called a *phase diagram*, regardless of whether or not the slope field is present.

Remember that we are thinking of some physical process, which starts at time 0, and we're measuring the value of some physical quantity as time increases. This means that we really don't care what happens to solutions when $t < 0$, and that you want to think about what the solutions are doing as t increase, i.e., as you move from left to right along the solution curves.

What do we see in the phase diagram? Well, certainly, if $A(0)$ equals one of the equilibrium values, then the solution $A(t)$ to the corresponding IVP remains at that equilibrium value for all time. If $A(0) > 1.5$, then the solution to the corresponding IVP always increases and becomes unboundedly large, i.e., goes to ∞. If $A(0)$ is strictly between the equilibrium values 0.5 and 1.5, then the solution to the corresponding IVP always decreases and approaches the equilibrium solution $A(t) = 0.5$. If $A(0)$ is strictly between the equilibrium values -1 and 0.5, then the solution to the corresponding IVP always increases and approaches the equilibrium solution $A(t) = 0.5$. And, if $A(0)$ is less than -1, then the solution to the corresponding IVP always decreases and approaches $-\infty$.

In fact, even without the phase diagram, it is not hard to reach the above conclusions about the solutions increasing and decreasing; you simply consider for which A values the product $(A + 1)(A - 0.5)(A - 1.5)$ is positive and for which it's negative. However, the conclusions about the solutions approaching the equilibrium solutions, or heading off to $\pm\infty$, despite being geometrically apparent from the slope field, are difficult to prove, and are beyond the scope of this textbook.

4.5.1 Exercises

In each of Exercises 1 through 4, you are given an initial value problem, involving a separable differential equation. Find an explicit solution to the IVP, and then use Euler's Method, with 4 steps, by hand and calculator, to approximate the y value indicated. Give the error in the final approximation and the percentage that the error is of the actual value.

1. $\dfrac{dy}{dx} = xy^2$; $y(0) = -1$. Approximate $y(1)$.

2. $\dfrac{dy}{dx} = xy^2$; $y(1) = -1$. Approximate $y(3)$.

3. $\dfrac{dy}{dx} = y \cos x$; $y(0) = 1$. Approximate $y(2)$.

4. $\dfrac{dy}{dx} = y(3 - y)$; $y(-1) = 1$. Approximate $y(0)$.

In Exercises 5 through 8, you are to consider the same four IVP's as in Exercises 1 through 4, but now use a spreadsheet program to approximate the y value indicated by using Euler's Method with $n = 10$, 100, and 1000 steps. Include columns in your spreadsheets for the actual y values, the error, and the percentage that the error is of the actual value.

5. Use the IVP from Exercise 1.

6. Use the IVP from Exercise 2.

7. Use the IVP from Exercise 3.

8. Use the IVP from Exercise 4.

In each of Exercises 9 through 12, you are given an initial value problem, involving a non-separable differential equation. Use Euler's Method, with 4 steps, by hand and calculator, to approximate the y value indicated.

9. $\dfrac{dy}{dx} = x + y$; $y(2) = -1$. Approximate $y(1)$.

10. $\dfrac{dy}{dx} = x - \cos y$; $y(1) = 0$. Approximate $y(1.4)$.

11. $\dfrac{dy}{dx} = 1 - ye^x$; $y(-0.5) = -1$. Approximate $y(0.5)$.

12. $\dfrac{dy}{dx} = y^2 - y + x$; $y(0) = 1$. Approximate $y(1)$.

In Exercises 13 through 16, you are to consider the same four IVP's as in Exercises 9 through 12, but now use a spreadsheet program to approximate the y value indicated by using Euler's Method with $n = 10$, 100, and 1000 steps. Include columns in your spreadsheets for the actual y values, the error, and the percentage that the error is of the actual value.

13. Use the IVP from Exercise 9.

14. Use the IVP from Exercise 10.

15. Use the IVP from Exercise 11.

16. Use the IVP from Exercise 12.

In each of Exercises 17 through 20, you are given an initial value problem and a slope field for the underlying differential equation. Print or copy each exercise, and then sketch a solution curve to the IVP on top of the slope field. Use your solution curve to approximate the indicated y value.

17. $\dfrac{dy}{dx} = xy^2$; $y(1) = -1$. Approximate $y(3)$. Use Figure 4.16

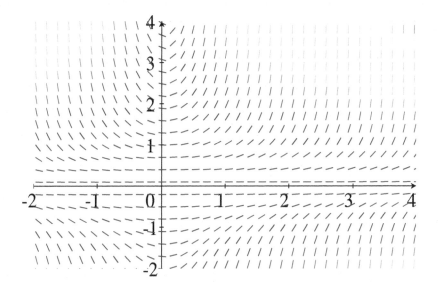

Figure 4.16: A slope field for $y' = xy^2$.

18. $\dfrac{dy}{dx} = x - \cos y$; $y(1) = 0$. Approximate $y(1.4)$. Use Figure 4.17

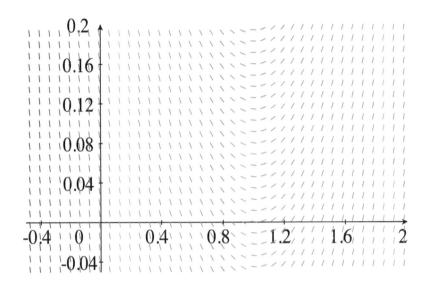

Figure 4.17: A slope field for $y' = x - \cos y$.

19. $\dfrac{dy}{dx} = 1 - ye^x$; $y(-0.5) = -1$. Approximate $y(0.5)$. Use Figure 4.18

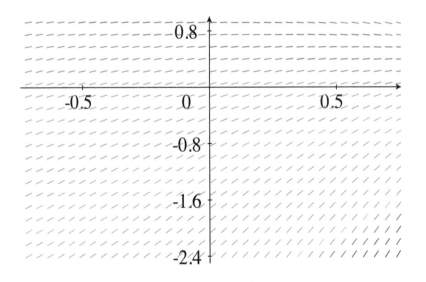

Figure 4.18: A slope field for $y' = 1 - ye^x$.

20. $\dfrac{dy}{dx} = y^2 - y + x$; $y(0) = 1$. Approximate $y(1)$. Use Figure 4.19

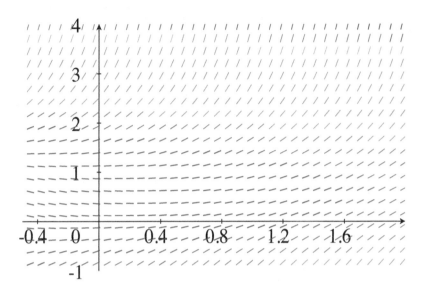

Figure 4.19: A slope field for $y' = y^2 - y + x$.

In each of Exercises 21 through 24, you are given an autonomous differential equation. Determine the equilibrium solution(s) of the differential equation. Sketch a slope field for the differential equation by using isoclines of the form $A = c$ for various constants c. Use your slope field to sketch various solution curves to form a phase diagram. For each of your solution curves, indicate the approximate value of $A(0)$.

21. $\dfrac{dA}{dt} = A(A^2 + 1)$.

22. $\dfrac{dA}{dt} = A(A^2 - 1)$.

23. $\dfrac{dA}{dt} = A^2(A + 1)$.

24. $\dfrac{dA}{dt} = (A - 1)^2(A + 1)$.

25. Recall Example 4.4.7 in which a company was selling a new style of heavy wool coat in a small city. We assumed that there was a maximum number of 1000 possible buyers of this coat, and that, at any given point in time, the instantaneous rate of change, per month, in the number of coats sold was proportional to the number of possible buyers who had not yet purchased one of these coats.

We also assumed that, due to the seasonal temperature variations, for a fixed number of coats already sold, the instantaneous rate of change, per month, in the number of coats sold was proportional to $1 + \cos(\pi t/6)$, where t was the number of months since January 1, 2009.

We let $W = W(t)$ be the number of wool coats sold since January 1, 2009, and were given that $W(0) = 0$ and $W'(0) = 20$. All of this data allowed us to produce a corresponding IVP:

$$\frac{dW}{dt} = 0.01\big(1 + \cos(\pi t/6)\big)(1000 - W); \quad W(0) = 0.$$

In Figure 4.20, we give a slope field for this differential equation. Use this slope field to respond to the parts of this exercise.

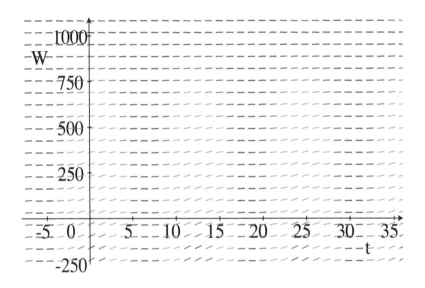

Figure 4.20: Wool coat slope field.

a. Sketch a solution curve that corresponds to the initial data $W(0) = 0$. What approximation do you get for $W(12)$? How does this compare with what we found in Example 4.4.7?

b. Use your solution from part a) to approximate $W(24)$ and $W(36)$.

c. Suppose that $W(0) = 250$. Now what approximations do you get from a solution curve for $W(12)$, $W(24)$, and $W(36)$?

d. Sketch a solution curve corresponding to the initial data $W(0) = 1000$. Explain what you get.

e. There are t values for which the slope field seems to be horizontal, regardless of the W value. Estimate such t values by looking at the slope field, and then use the differential equation to determine their precise values.

Appendix A

Parameterized Curves and Motion

A.1 Parameterized Curves

Throughout most of this book, when we deal with motion, we deal with an object (e.g., a car, a particle, a block) which is moving in a straight line. However, many/most types of problems in mathematics and physics deal with objects moving in a plane or in space, and, in a few problems in this book, we look at such motion without presenting it as anything new. We can get by with this because, in a sense, such problems can be analyzed by breaking the problem up into two or three separate motions "in a straight line".

For instance, suppose that a particle is moving in $\mathbb{R}^2$ (the xy-plane) and, at time t, the x-coordinate of the particle is $x = x(t) = 2t - 1$ and the y-coordinate is $y = y(t) = (2t - 1)^2$. If you simply look at each coordinate separately, then the motion breaks up into what looks like motion along the x-axis and motion along the y-axis. Of course, the particle is definitely **not** actually moving along the x- or y-axis. In fact, in this example, we see that the y-coordinate of the particle is always the square of the x-coordinate. In other words, the x- and y-coordinates of the particle always satisfy $y = x^2$. Thus, the particle is moving along the parabola given by $y = x^2$; we also say that the particle traces out the parabola (or, at least, part of it).

Of course, specifying the curve $y = x^2$ provides far less information than knowing where, along the parabola, the particle is at each time. When we specify a curve, like $y = x^2$, by specifying each coordinate in terms of a new variable, like t, we say that we are given the curve *parametrically*, where t is the *parameter*. Note that, when we are given a curve (in $\mathbb{R}^2$) parametrically, we specify one real number, the value of the parameter t, and that parameter value determines two real numbers $x(t)$ and $y(t)$, or what's the same thing, t determines a point

$(x(t), y(t))$ in $\mathbb{R}^2$.

The most general definition of a parameterized curve is:

> **Definition A.1.1.** *A* **parameterized curve** *in* $\mathbb{R}^2$ *(respectively,* $\mathbb{R}^3$*) is a function* $(x(t), y(t))$ *(respectively,* $(x(t), y(t), z(t))$*), whose domain is contained in the real line, and whose codomain is contained in* $\mathbb{R}^2$ *(respectively,* $\mathbb{R}^3$*).*
>
> *The variable* t *(or whatever the independent variable is) is the* **parameter***, and the individual functions* $x(t)$*,* $y(t)$*, and* $z(t)$ *are the* **component functions***.*
>
> *Parameterized curves are also known as* **vector functions***, and are frequently denoted by letters with little arrows over them, e.g.,* $\vec{p}(t)$*.*

Remark A.1.2. Recall that the range of a function is the set of values that the function actually attains. Thus, the curve described by $y = x^2$ is precisely the range of the parameterized curve $(x(t), y(t)) = (2t - 1, (2t - 1)^2)$.

Parameterized curves are usually defined by specifying the individual component functions, rather than actually writing the ordered pair $(x(t), y(t))$ or ordered triple $(x(t), y(t), z(t))$ of functions. That is, we usually write something like: consider the parameterized curve $x(t) = 2t - 1$ and $y(t) = (2t - 1)^2$.

We should also comment that Definition A.1.1 is very broad, and allows for parameterized curves that don't "look like" curves. For instance, by our definition, the range of the parameterized curve could be a single point. There are other "degenerate cases" as well, which many people would not want to call "curves". In most problems, the domain of a parameterized curve would be an interval containing more than a single point, and the component functions would be required to be at least continuous.

There are conditions that can be put on the derivatives of the component functions, such as requiring that the derivatives are never all zero at the same time, which guarantee that the range of a parameterized curve actually looks a curve or a line.

Finally, we should mention that, for mathematicians, a line is considered to be a curve; a line is just a curve that doesn't curve (even if that seems like silly terminology).

Below, we define some important concepts, related to an object moving in space. The same definitions apply for motion in the plane; you just need to omit the references to the z-coordinate. We assume at this point that you have read Chapter 1.

Definition A.1.3. *Suppose that an object is moving in space, and its position at time t is given by the parameterized curve $(x(t), y(t), z(t))$. This ordered triple is referred to as the* **position vector** *of the object, and is frequently denoted by $\vec{p}(t)$.*

Assuming that it exists, the ordered triple $\vec{v}(t) = \vec{p}\,'(t) = (x'(t), y'(t), z'(t))$ is the **velocity** *(vector) of the object. The* **magnitude** *of the velocity is*

$$\sqrt{(x'(t))^2 + (y'(t))^2 + (z'(t))^2};$$

the magnitude of the velocity is called **speed**, *and is denoted by $|\vec{v}(t)|$.*

Assuming that it exists, the ordered triple $\vec{a}(t) = \vec{v}\,'(t) = \vec{p}\,''(t) = (x''(t), y''(t), z''(t))$ is the **acceleration** *(vector) of the object. The* **magnitude** *of the acceleration is*

$$\sqrt{(x''(t))^2 + (y''(t))^2 + (z''(t))^2},$$

which is denoted by $|\vec{a}(t)|$.

In fact, we define the magnitude of any vector quantity $\vec{\beta} = (b_1, b_2, b_3)$ to be the square root of the sum of the squares of the components; the magnitude is denoted by "absolute value" signs, i.e.,

$$|\vec{\beta}| = \sqrt{b_1^2 + b_2^2 + b_3^2}.$$

Remark A.1.4. Note that the analogous concept of speed for an object moving along the x-axis, with position $x(t)$, at time t, would be that speed is $\sqrt{(x'(t))^2} = |x'(t)|$, which agrees with our definition in Definition 1.2.8.

Let's look at an example, an example that should, essentially, contain everything that you need to know about parameterized curves and motion.

Example A.1.5. Suppose that a particle is moving in the xy-plane, which is marked off in feet on each axis. The particle moves in such a way that, at time t seconds (from some initial starting time), the x-coordinate of the particle is $x = x(t) = 1 - t^2$, and the y-coordinate is $y = y(t) = t^4 + 2$.

There are many questions that we can ask.

Where is the particle at times $t = 0$, $t = 1$, and $t = 2$ seconds? Can we describe the path that the particle is moving along, just in terms of x and y, without reference to t? Can we determine which way the particle is moving along the path, and whether or not the particle ever "backs up"? What is the average rate of change of the x-coordinate of the particle, with respect to time, between times $t = 0$ and $t = 2$ seconds? What is the average rate of change of the y-coordinate of the particle, with respect to time, between times $t = 0$ and $t = 2$ seconds?

Assuming that you've read Chapter 1, there are more questions that we should look at.

What are the velocity, speed, and acceleration of the particle at time t? What are the x-components of the velocity and acceleration at time $t = 2$ seconds? What is the slope of the tangent line to the path of the particle at time $t = 2$ seconds?

Solution:

To determine the location of the particle at any given time, you simply plug in the given time. So, at times $t = 0$, $t = 1$, and $t = 2$ seconds, the particle is located at the points

$$(x, y) \;=\; (x(0), y(0)) \;=\; (1 - 0^2, 0^4 + 2) \;=\; (1, 2),$$

$$(x, y) \;=\; (x(1), y(1)) \;=\; (1 - 1^2, 1^4 + 2) \;=\; (0, 3), \text{and}$$

$$(x, y) \;=\; (x(2), y(2)) \;=\; (1 - 2^2, 2^4 + 2) \;=\; (-3, 18),$$

respectively.

To describe the path of the particle, without reference to t, we need to take the two equations $x = 1 - t^2$ and $y = t^4 + 2$, and eliminate the t. It is **not** always possible to do this, given two arbitrary functions of t. However, for the functions in this example, it's easy. We find that $t^2 = 1 - x$, which means that $t^4 = (1 - x)^2$. Substituting this into the equation for y, we find that, at any time, the x- and y-coordinates of the particle satisfy the relation $y = (1 - x)^2 + 2$, i.e., the particle is moving along the parabolic path $y = x^2 - 2x + 3$. We say that $x(t)$ and $y(t)$ *parameterize* the curve $y = x^2 - 2x + 3$.

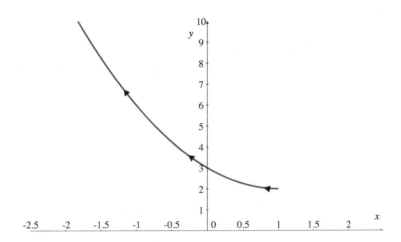

Figure A.1: The path of the particle.

If we assume that the particle begins its journey at $t = 0$, and continues for all $t > 0$, then, as $x = 1 - t^2$, x will assume every value ≤ 1, and the particle traces out the portion of the parabola shown in Figure A.1.

Note that, if we forget the given parameterization, and just consider the parabola defined by $y = x^2 - 2x + 3$, then we have lost a great deal of information; we would have no idea where the particle is at any particular time. In addition, we wouldn't even know how much of the parabola is traced out as the particle moves along.

Since $x = 1 - t^2$, as t increases from time 0, the x-coordinate of the particle decreases. Thus, the particle moves from right to left, and never "backs up". In Figure A.1, we have indicated with arrows the direction in which the particle is moving.

The average rates of changes, with respect to time, of the x- and y-coordinates of the particle, between $t = 0$ and $t = 2$ seconds are easy to calculate:

$$\frac{\Delta x}{\Delta t} = \frac{x(2) - x(0)}{2 - 0} = \frac{-3 - 1}{2} = -2 \text{ ft/s, and}$$

$$\frac{\Delta y}{\Delta t} = \frac{y(2) - y(0)}{2 - 0} = \frac{18 - 2}{2} = 8 \text{ ft/s.}$$

We now need to assume knowledge of Chapter 1 and Section 2.1.

The position vector of the particle is $\vec{p}(t) = (1 - t^2, t^4 + 2)$. Hence, the velocity is

$$\vec{v}(t) \;=\; \vec{p}\,'(t) \;=\; (x'(t), y'(t)) \;=\; ((1 - t^2)', (t^4 + 2)') \;=\; (-2t, 4t^3) \text{ ft/s},$$

and the acceleration is

$$\vec{a}(t) \;=\; \vec{v}\,'(t) \;=\; ((-2t)', (4t^3)') \;=\; (-2, 12t^2) \text{ (ft/s)/s}.$$

The speed of the particle is

$$|\vec{v}(t)| \;=\; \sqrt{(-2t)^2 + (4t^3)^2} \;=\; 2t\sqrt{1 + 4t^4} \text{ ft/s},$$

where we assumed that $t \geq 0$.

At time $t = 2$ seconds, the x-components of the velocity and acceleration, $\vec{v}_x$ and $\vec{a}_x$, respectively, are $(-2t)_{|_2} = -4$ ft/s and -2 (ft/s)/s, respectively. Note that we did not need the entire velocity and acceleration vectors to determine this. These results depend on only the x-coordinate data $x(t) = 1 - t^2$, and thus would be the same if the particle were actually moving along the x-axis, with its position given by $x(t) = 1 - t^2$.

Finally, we come to the slope of the tangent line question. Since we know that the path of the particle is given by $y = x^2 - 2x + 3$, we can simply calculate $dy/dx = 2x - 2$ at the x-coordinate at time $t = 2$ seconds, i.e., at $x = 1 - 2^2 = -3$. Thus, at $t = 2$ seconds, the slope of the tangent line to the path of the particle is $2(-3) - 2 = -8$.

Could we have found this slope without explicitly solving for y in terms of x? Yes, we need to use results from Section 2.3; specifically, we need the Chain Rule, Theorem 2.3.1, and Theorem 2.3.8. Combining those two theorems we obtain that, if dy/dt and dx/dt exist, and $dx/dt \neq 0$, then

$$\frac{dy}{dx} \;=\; \frac{dy/dt}{dx/dt}, \tag{A.1}$$

i.e., it looks like the dt's "cancel". Therefore, we find

$$\left(\frac{dy}{dx}\right)_{|t=2} \;=\; \left(\frac{dy/dt}{dx/dt}\right)_{|t=2} \;=\; \left(\frac{4t^3}{-2t}\right)_{|t=2} \;=\; \frac{32}{-4} \;=\; -8,$$

as we found before.

It may seem strange, but it is sometimes useful to start out with a curve C in the plane, defined by some equation involving x and y, and then produce **some** *parameterization* of the curve, i.e., produce a parameterized curve $(x(t), y(t))$ whose range is C. Moreover, we usually would like for $(x(t), y(t))$ to be one-to-one or, maybe, one-to-one when restricted to some open interval around each t value. In terms of the motion of an object, this would be a bit absurd; we would be "making up" where the object is at each time, after being given its path. Nonetheless, parameterizations of curves can be very useful.

Suppose the curve C is the graph of a function $y = f(x)$. Then, there is always the trivial parameterization: $x = t$, and $y = f(t)$. For instance, the graph of $y = 7 + x^3$ can be parameterized by $x = t$ and $y = 7 + t^3$. However, this parameterization is in no way unique; we could let $x = 2t$ and $y = 7 + (2t)^3$, or, as another example, we could let $y = t$ and $x = (t - 7)^{1/3}$. There are an infinite number of choices of parameterizations.

Parameterizing curves which are not the graphs of functions is more useful. Consider a standard ellipse, centered at (h, k), with semi-axes of lengths $a, b > 0$, i.e., the set of points satisfying

$$\frac{(x - h)^2}{a^2} + \frac{(y - k)^2}{b^2} = 1.$$

This curve is not the graph of a function; it fails the vertical line test. However, Calculus applies to **functions**, and so, for dealing with many problems about this curve, the question is: can we describe this curve in terms of a function? The answer is "yes"; we parameterize the curve.

Assuming that you are familiar with sine and cosine as periodic functions (see Section 2.7), there is a standard parameterization of this ellipse (or, circle, if $a = b$); let $x = h + a \cos t$ and $y = k + b \sin t$, for $0 \le t < 2\pi$. You should verify that these x- and y-coordinates yield points on the given ellipse (this is easy), and that you actually obtain every point on the ellipse exactly once (this is not so easy). Also, as t begins increasing from 0, $\cos t$ decreases, while $\sin t$ increases. Such considerations, or plotting points for different t values, should convince you that, if this parameterization describes the position of a particle at time t, then the particle starts at the right-hand vertex of the ellipse, moves counterclockwise around the ellipse, and approaches the right-hand vertex again as t approaches 2π.

A.1.1   Exercises

While some of these exercises use only polynomial functions, others use functions that are discussed in Chapter 1 and Chapter 2.

In each of Exercises 1 through 4, you are given a parameterized curve. Assume that the parameterization gives the position, in meters, of a particle at time t seconds. For each component of the position, find the average velocity in that "direction", between times $t = 0$ and $t = 2$ seconds.

For curves in the plane, eliminate the parameter t to find an equation, involving only x and y, which describes the range of the curve (or, possibly, describes a bigger set of points). Sketch these plane curves, indicating the points corresponding to $t = -1$, $t = 0$, $t = 1$, and $t = 2$.

1. $x(t) = t^2 - 7$, $y(t) = 5t$.

2. $x(t) = t^2 - 7$, $y(t) = 5t$, $z(t) = t^2$.

3. $x(t) = e^t$, $y(t) = t$.

4. $x = 3\cos(\pi t/2)$, $y = 3\sin(\pi t/2)$, $z = t$.

In each of Exercises 5 through 8, consider a particle, whose position is given in Exercises 1 through 4 above. Find all of the components of the velocity and acceleration at $t = 2$ seconds or, equivalently, find the velocity and acceleration vectors at $t = 2$ seconds. What is the speed of the particle at time $t = 2$ seconds?

5. Use the parameterization from Exercise 1.

6. Use the parameterization from Exercise 2.

7. Use the parameterization from Exercise 3.

8. Use the parameterization from Exercise 4.

In each of Exercises 9 through 12, you are given an equation in terms of x and y, which describes a curve C in the xy-plane. Find a parameterized curve, $(x(t), y(t))$, whose range is precisely C, and such that $x(t)$ and $y(t)$ are each differentiable. There are an infinite number of possible correct answers here, though some are far easier than others. Use a calculator or computer to graph these curves.

9. $x - 5 = y^4$.

10. $y^2 = x^3$.

11. $(x + 2)^2 + (y - 3)^2 = 1$.

12. $y^2 = x^3 + x^2$. (Hint: For $x \neq 0$, let $t = y/x$, and go from there.)

In Exercises 13 through 16, calculate the indicated derivative at the point Q using the chain rule. Assume that, as usual, the first component is $x(t)$, the second $y(t)$, and the third, if it exists, is $z(t)$.

13. $\vec{\alpha}(t) = (\cos t, \sin t, t)$.

     a. dy/dx, $Q = (0, 1, \pi/2)$.

     b. dz/dx, $Q = (0, 1, \pi/2)$.

     c. dz/dy, $Q = (1, 0, 0)$.

14. $\vec{\beta}(t) = (t, t, t^2)$, dz/dx, $Q = (1, 1, 1)$.

15. $\vec{g}(t) = (t^3 + 2, t^3 - 5, t^3 + 17)$, dy/dx, $Q = (10, 3, 25)$.

16. $\vec{h}(t) = (a\cos t, b\sin t)$ where a and b are positive, $t \in [0, 2\pi]$, dy/dx, $Q = (0, b)$. Note that this is a parameterization of an ellipse, as mentioned in the chapter.

17. Consider the curve $\vec{r}(t) = (3e^{-t}\cos t, 3e^{-t}\sin t)$.

     a. Show that $|\vec{r}(t)| \to 0$ as $t \to \infty$.

     b. Show that $\vec{r}'(t) \to (0, 0)$ as $t \to \infty$.

Assume that each component of the position vector $\vec{p}(t) = (x(t), y(t), z(t))$ is differentiable and that the speed is never zero. Then we define the *unit tangent vector* at a given time as:

$$\vec{T}(t) = \frac{\vec{p}'(t)}{|\vec{p}'(t)|} = \frac{\vec{v}(t)}{|\vec{v}(t)|}.$$

The unit tangent vector has a physical interpretation. By the Law of Inertia, a particle must be experiencing a force if it is not moving in a straight line or at rest. If all forces are removed at time t_0, then the particle will move along a straight line in the direction of the unit tangent vector. Calculate the unit tangent vector in Exercises 18 through 21.

18. $\vec{p}(t) = (9\cos t, 9\sin t, 0)$.

19. $\vec{p}(t) = (3t + 9, 5t - 7, 4t)$.

20. $\vec{p}(t) = (3\cos t, 3\sin t, 4t)$.

21. $\vec{p}(t) = (\cosh t, \sinh t, t)$.

22. The curve $\vec{p}(t) = (\cos(kt), \sin(kt))$, $k > 0$ is a parameterization of the unit circle. The bigger k is, the faster the particle moves around the circle. Show that $\vec{p}'(0)$ depends on k, but that the unit tangent vector $\vec{T}(0)$ does not.

23. Recall the definition of the dot product from Chapter 2.2, Exercise 46. In that exercise, you showed that

$$\frac{d}{dt}\left[\vec{p}(t) \cdot \vec{q}(t)\right] = \vec{p}\,'(t) \cdot \vec{q}(t) + \vec{p}(t) \cdot \vec{q}\,'(t).$$

   a. Show that $|\vec{p}(t)|^2 = \vec{p}(t) \cdot \vec{p}(t)$. This result holds more generally for any point (x, y, z) in space: the self dot product gives the square of the distance.

   b. Show that $\dfrac{d}{dt}|\vec{p}(t)| = \dfrac{\vec{p}(t) \cdot \vec{p}\,'(t)}{|\vec{p}(t)|}$.

24.   a. Show that $\vec{T}(t) \cdot \vec{T}(t) = 1$.

   b. Differentiate the equation in part (a) to show that $\vec{T}\,'(t) \cdot \vec{T}(t) = 0$.

25. The height of a projectile, at time t, launched from an initial height h_0, with initial velocity v_0, is given by $h(t) = -\frac{1}{2}gt^2 + v_0 t + h_0$. The height can be expressed parametrically as $\alpha(t) = (t, h(t))$. Take the second derivative of this parameterization to show that the magnitude of the acceleration is g.

26. Show that, if a particle moves so that its position at time t is $\vec{p}(t) = (t, f(t))$, where $f(t)$ is twice differentiable, then the speed is $|\vec{v}(t)| = \sqrt{1 + f'(t)^2}$ and the magnitude of the acceleration is $|f''(t)|$.

27. The *unit normal vector* for a given position vector is defined as $\vec{N}(t) = \dfrac{\vec{T}\,'(t)}{|\vec{T}\,'(t)|}$, for any t such that $\vec{T}\,'(t)$ exists and $|\vec{T}\,'(t)| \neq 0$. Calculate the unit normal vector for the curves in Exercises 18 - 21. You need not recalculate $\vec{T}(t)$. ▶

28. Show that the dot product of the unit tangent vector and unit normal vector is zero, i.e., $\vec{T}(t) \cdot \vec{N}(t) = 0$. ▶

29. Suppose $\vec{\alpha}(t) = (x(t), y(t), z(t))$ is a differentiable space curve where $t \in I$, I some open interval, and that (x_0, y_0, z_0) is a point in space not on the curve. Assume further that $|\vec{\alpha}\,'(t)|$ is never zero. If $(x_1, y_1, z_1) = \vec{\alpha}(t_1)$ is the closest point on the curve to (x_0, y_0, z_0), show that

$$(x_0 - x_1, y_0 - y_1, z_0 - z_1) \cdot (x'(t_1), y'(t_2), z'(t_3)) = 0.$$

If each component of the position vector $\vec{p}(t)$ is twice differentiable and the speed is never zero, then the *curvature* of the curve at the point $\vec{p}(t)$ is

$$\kappa(t) = \frac{|\vec{T}\,'(t)|}{|\vec{p}\,'(t)|} = \frac{|\vec{T}\,'(t)|}{|\vec{v}(t)|},$$

provided that $|\vec{v}(t)| \neq 0$. **The curvature measures how quickly the curve is changing directions. Calculate** $\kappa(t)$ **in Exercises 30 through 32.**

30. $\vec{p}(t) = (2 + 4\cos t, 3 + 4\sin t, 0)$.

31. $\vec{p}(t) = (4 + 3t, 2 + 7t, 1 - t)$.

32. $\vec{p}(t) = (4\sin t, 3t, 4\cos t)$.

33. Show that in the case where the curve is planar and the graph of a twice differentiable function, $\vec{p}(t) = (t, y(t))$, the curvature is given by

$$\kappa(t) = \frac{|y''(t)|}{\left(1 + [y'(t)]^2\right)^{3/2}}.$$

Note that this means that the notion of curvature which we are using here is actually the absolute value of what we used in Section 2.2, Exercise 54 and in a series of exercises beginning with Exercise 46 in Section 2.10.

34. The reciprocal of the curvature is called the *radius of curvature*. What is the radius of curvature of a circle of radius R with parameterization $x(t) = R\cos t$, $y(t) = R\sin t$?

35. Any straight line in three dimensional space can be parameterized as

$$x(t) = x_0 + at$$
$$y(t) = y_0 + bt$$
$$z(t) = z_0 + ct.$$

Show that the curvature of a line is zero (as we would hope).

36. Let $\vec{\alpha}(t)$ be a differentiable parameterized curve in either two or three dimensions. The tangent line to the curve at the point t_0 is given by the equation $\vec{L}(t) = \vec{\alpha}(t_0) + t\vec{\alpha}'(t_0)$.

a. Recall that the graph of a differentiable function $f(t)$ is the range of the parameterized curve $\vec{\alpha}(t) = (t, f(t))$. Show that the tangent line at $t = t_0$ is given by $\vec{L}(t) = (t_0 + t, f(t_0) + tf'(t_0))$.

b. Let $x_0 = t_0$, $y_0 = f(t_0)$ and $x = x_0 + t$. Show that the tangent line coincides with our more familiar representation of the tangent line: the graph of $y = f'(x_0)(x - x_0) + y_0$.

In Exercises 37 through 39, give a parameterization of the tangent line to the curve at the given point(s).

37. $\vec{p}(t) = (\cos t, 0, \sin t)$, $t = \pi/2$.

38. $\vec{p}(t) = (3 \sin t, 3 \cos t, t)$, $\pi/2 + 2\pi n$, n an integer.

39. $\vec{\alpha}(t) = (e^{-t} \cos t, e^{-t} \sin t)$, $t = \pi$.

40. Show that the tangent line to a straight line in three dimensional space is the line itself.

41. Let $x(t) = \dfrac{k^2}{2}(t - \sin t)$ and $y(t) = \dfrac{k^2}{2}(1 - \cos t)$ and assume $0 < t < \pi$. This is a parameterization of a *cycloid*.

    a. Calculate dy/dx.

    b. Show that $\left[1 + \left(\dfrac{dy}{dx}\right)^2\right] y = k^2$.

This famous curve has many applications. It is the solution of the famous *Brachistochrone Problem*: to find the curve along which a particle will slide from one point to another the most quickly (assuming no friction).

42. Let

$$\vec{p}(t) = \begin{cases} (t, 0, e^{-1/t^2}) & t > 0; \\ (t, e^{-1/t^2}, 0) & t < 0; \\ (0, 0, 0) & t = 0. \end{cases}$$

Show that $\vec{p}(t)$ is differentiable for all t. Is $|\vec{p}'(t)|$ ever zero?

Appendix B

Tables of Derivative Formulas

In the tables below, a, b, and p denote arbitrary real constants; in the formulas for b^x and $\log_b$, we assume that $b > 0$. We use f and g to denote differentiable real functions, and u, v, and x to denote variable names, either independent variables, or dependent variables, i.e., the values of functions. In any formulas involving divisions, we assume that we are in a situation that does **not** lead to division by zero. More generally, we assume that we are in a situation where all of the functions involved in the formulas are defined.

You may also want to try WolframAlpha.com, as a way of obtaining the derivative of a complicated function.

Formulas for Reducing Complicated Derivatives to Easier Ones	
Linearity:	$(af \pm bg)'(x) = af'(x) \pm bg'(x).$
Product Rule:	$(f \cdot g)'(x) = f(x)g'(x) + g(x)f'(x).$
Quotient Rule:	$(f/g)'(x) = \big(g(x)f'(x) - f(x)g'(x)\big)/[g(x)]^2.$
Chain Rule:	$(f \circ g)'(x) = f'(g(x)) \cdot g'(x)$ or $\dfrac{dv}{dx} = \dfrac{dv}{du} \cdot \dfrac{du}{dx}.$
Inverse Function Derivative:	$(f^{-1})'(x) = \dfrac{1}{f'(f^{-1}(x))}$ or $\dfrac{dx}{du} = \dfrac{1}{du/dx}.$

Algebraic Derivatives	
$a' = 0.$	$(b^x)' = b^x \ln b.$
$(x^p)' = p x^{p-1}.$	$\ln' |x| = 1/x.$
$\exp'(x) = (e^x)' = e^x.$	$\log_b' |x| = 1/(x \ln b).$

Trigonometric Derivatives	
$\sin' x = \cos x.$	$(\sin^{-1})'(x) = \dfrac{1}{\sqrt{1 - x^2}}.$
$\cos' x = -\sin x.$	$(\cos^{-1})'(x) = \dfrac{-1}{\sqrt{1 - x^2}}.$
$\tan' x = \sec^2 x.$	$(\tan^{-1})'(x) = \dfrac{1}{1 + x^2}.$
$\cot' x = -\csc^2 x.$	$(\cot^{-1})'(x) = \dfrac{-1}{1 + x^2}.$
$\sec' x = \sec x \tan x.$	$(\sec^{-1})'(x) = \dfrac{1}{|x|\sqrt{x^2 - 1}}.$
$\csc' x = -\csc x \cot x.$	$(\csc^{-1})'(x) = \dfrac{-1}{|x|\sqrt{x^2 - 1}}.$

Appendix C

Answers to Odd-Numbered Problems

Chapter 1

Section 1.1

1. 30.005

3. 333

5. 1

7. $-0.04\overline{44}$

9. ≈ 5

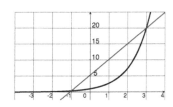

11. units of output/units of input

13. real function

15. not a real function

17. real function

19. $y = t$

21. $y = 0$

23. $w = (x + z)y + xz$

25.

   a. $T(r) = \pi r^2 D(r)$.

   b. $23,012.16619$ people per mile.

   c. Approximately $30,866.14782$ people. Over-estimate.

27. Approx. 7.47 % spam/yr

29. a. $364\pi/3$ in^3/in   b. $19/30$ in^3/in^2 c. $15/19$ in^2/in^3

31. a. 0.681 citizen/yr   b. 0.523 citizen/yr   c. decreasing

33. a. 0   b. m

35. $a = 3, b = 18$

37. $b = 59/23$

39. The statement is false.

41. a. -1282.61 \$/%   b. -3420.97 \$/yr

43.

   a. 30 mph/min= 1800 mi/hr^2

   c. $[2, 3]$

   d. Yes, by the Intermediate Value Theorem.

45. a. $66.6\overline{6}$ mph   b. cannot be done

Section 1.2

1. AROC: always 7. IROC: 7

3. AROCs: 7, 6.1, 6.01, 6.001, 6.0001. IROC: 6

5.   AROCs: 8,   21.68,   23.7608,   23.976008, 23.997960008. IROC: 24

7. AROCs: undefined when $h = 1$, 1 for other values of h. IROC: 1

9. AROCs: $3a + b$, $2.1a + b$, $2.01a + b$, $2.001a + b$, $2.0001a + b$. IROC: $2a + b$

11. $v = -2/t^3$ ft/s

13. $v = m$ ft/s

15. $6/t^4$ ft/s^2

17. 0 ft/s^2

19. 1/14

21. 3

23. a. 0   b. does not exist

31. IROC ≈ 0.02335 kg/(km/s)

33.

    a. 0.267, 0.217, 0.171

    b. 0.267, 0.167, 0.133

    c. 0.217, 0.167, 0.111

    d. decreasing

35. Both are 0.

37. Both are 3.

39. undefined vs. 0

41.   b. -0.00003, -0.00002, -0.000015, and $-0.0000166\overline{6}$ c. Approx. -0.000025

43. a. 2, 3, 1 °F/min   b. 3, 2, 1 °F/min

45. horiz. $v = 1$, vert. $v = -t/\sqrt{9 - t^2}$

Section 1.3

1. 3

3. 1

5. 0

7. $\delta = 0.2$ works.

9. $M = 40$ works.

11. $f(x) = \dfrac{(x - 4)(x + 2)}{x + 2}$ is one possibility.

13. a. 0   b. ∞

15. $\sqrt{v}$ isn't defined to the left of 0. The limit from the right is 0.

17.   The limit from the right does not exist, so neither does the two-sided limit. The limit from the left is 0.

19. $(-\infty, -2) \cup (-2, 2) \cup (2, \infty)$

21. $[-1, 3) \cup (3, \infty)$

23. Points to check: $x = -10, 0, 2$.

$x = -10$: left and right limits are both -10, not continuous, but removable lack of continuity (in fact, removable discontinuity)

$x = 0$: left and right limits are both -20, continuous

$x = 2$: limit from left is $-\infty$ (which does not exist), limit from right is ∞ (which does not exist), not continuous (not even defined), but the lack of continuity is also not removable.

29. a. No.   b. a/m   c. a_n/b_n

33.   The Extreme Value Theorem does not generalize to the case of an infinite union of closed intervals.

35.   a. For $x \le 500$, $P(x) = 0.8$; for $500 < x \le 1000$, $P(x) = 0.72$; for $x > 1000$, $P(x) = 0.65$.

b. $C(x) = xP(x)$.

c. $\lim_{x \to 1000+} C(x) = 650$, $\lim_{x \to 1000-} C(x) = 720$.

37.   The argument is invalid, since $x = -1$ is not in the domain of f.

39.

    a. $J = (0, 1]$

    b. $f^{-1}(y) = \sqrt{\dfrac{1}{y} - 1}$

    c. That $f^{-1}(y)$ is continuous.

41. f is continuous only at 0.

45. -6

47. $-1/4$

49. $-2/125$

Section 1.4

1. $-1/16$

3. 0

5. 1

7. 6

9. $a = -6, b = 19$.

11. $6z + 4$

13. $14u + \sqrt{2}$

15. $2/3$

17. $y = -9(x+5)$ and $y = 9(x-4)$.

19. At $x = 0, y - 1 = -1(x - 0)$; at $x = 1, y - 1/2 = -\dfrac{1}{4}(x-1)$; at $x = 2$, $y - 1/3 = -\dfrac{1}{9}(x-2)$.

21. $y - (7a + 3) = 7(x - a)$, i.e., $y = 7x + 3$.

23. $y - \dfrac{a+9}{a-4} = \dfrac{-13}{(a-4)^2}(x - a)$, $a \neq 4$.

25. a. $5t$ b. 5 m/s

27. a.

i. 8   ii. 3   iii. 1.25   iv. 0.5625

b. 0

29. $v = 5$ m/s, $a = 0$ (m/s)/s, $F = 0$ N.

31. $v = t/\sqrt{t^2 + 1}$ m/s, $a = 1/(t^2+1)^{3/2}$ (m/s)/s, $F = 5/(t^2 + 1)^{3/2}$ N.

33. $v = 10t + 4t/\sqrt{t^2 + 1}$ m/s, $a = 10 + 4/(t^2 + 1)^{3/2}$ (m/s)/s, $F = 50 + 20/(t^2 + 1)^{3/2}$ N.

35. $2x$

37. $1/(2\sqrt{x})$

39. Show that $g'(x) = 3x^2 + 2ax + b$, and find where $g'(x) = 0$.

41. $3x^2 + 2y$

43. $x = 4$.

45. c. When $a = b = c$.

47. a. positive   b. negative   c. $c(t)$ would need to equal a or b.

49. The points occur where $x = 4$ and $x = -1$.

55.

a. $t = 4$ seconds

b. 1 cubic centimeter per second

c. 9 cubic centimeters per second

57.

a.

$$s(t) = \begin{cases} 2t^2 - 42t + 180 & \text{if } t < 6 \text{ or } t > 15 \\ -2t^2 + 42t - 180 & \text{if } 6 \leq t \leq 15. \end{cases}$$

b. $s'(1) = -38$ meters per minute squared

c. $t = 6$ and $t = 15$ minutes.

Section 1.5

1. $(x, f(x)) = (1/27, 2/27), (-1/27, -2/27)$

3. $\left(\dfrac{-4+\sqrt{19}}{3}, -4.06067\right)$;

$\left(\dfrac{-4-\sqrt{19}}{3}, 8.20882\right)$

5. $(x, g(x)) = (h, k)$

7. $(x, p(x)) = (0, 1), (1, 0), (-1, 0)$

9. increasing on $(-\infty, -1), [0, 1)$. decreasing on $(-1, 0], (1, \infty)$

11. increasing on $(-\infty, -1), (1, \infty)$. decreasing on $(-1, 1)$.

13. $c = 3/2$

15. $c = (-1 \pm \sqrt{19})/3$

17. a. 120 mm HG   c. $c = 25$

27. False

29. a. $p(t) = -4.9t^2 + 5t + 10$ m.   b. $t \approx 2.03$ s. c. incr. on $(0, 0.51)$, decr. on $(0.51, 2.03)$ d. $t \approx 0.51$ s.

31. Their sum is constant.

35. a. $x = 0$ b. $x = 0, \pm 1$

39. a.

$$P(x) = \begin{cases} x & \text{if } x \leq D; \\ D & \text{if } D \leq x \leq D + L; \\ x - L & \text{if } x \geq D + L \end{cases}$$

b.

$$I(x) = \begin{cases} 0 & \text{if } x \leq D; \\ x - D & \text{if } D \leq x \leq D + L; \\ L & \text{if } x \geq D + L \end{cases}$$

c. $x = 0$ and all x in $[D, D + L]$

45. $(0, 2), (6, 12)$

Section 1.6

1. $f'(x) = 2x + 1$; $f''(x) = 2$

3. $f'(x) = 1 + 5/2\sqrt{x}$; $f''(x) = (-5/4)(x^{-3/2})$

5. $f'(x) = 2x + 6$; $f''(x) = 2$

7.

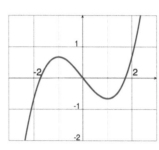

9. moving left and losing speed on $(-\infty, 6)$; moving right and gaining speed on $(6, \infty)$

11. moving right and losing speed on $(-\infty, -0.026225)$ and $(0, 0.0334476)$; moving right and gaining speed on $(-0.026225, 0)$; moving left and gaining speed on $(0.0334476, \infty)$

13. $v = -gt + v_0$;　$a = -g$.

15. $v = -3t^2 - 10t - 14$;　$a = -6t - 10$

17. $f''(t) = 2 + (2/t^3)$

19. $x''(c) = -2a/[(b^2 - 4ac)^{3/2}]$.

21. $f' > 0$ for all x;　$f'' > 0$ for $x < 7$;　$f'' < 0$ for $x > 7$

23. $f' > 0$ on $(-\infty, 2.75)$ and $(3.25, \infty)$; $f' < 0$ on $(2.75, 3.25)$;　$f'' < 0$ for $x < 3$; $f'' > 0$ for $x > 3$

25. $f'(x) > 0$ on $(0, 0.5)$, $(2, 6)$, $(12.5, 20)$; $f'(x) < 0$ on $(0.5, 2)$, $(6, 12.5)$, and $(20, 24)$; $f''(x) < 0$ on $(0, 1)$, $(4, 9)$, $(16, 24)$; $f''(x) > 0$ on $(1, 4)$, $(9, 16)$

27. $f^{(n)}(x) = (-1)^n n! \, x^{-n-1}$

29.
$$f^{(n)}(x) = \begin{cases} \frac{k!}{(k-n)!} x^{k-n} & n \le k; \\ 0 & n > k; \end{cases}.$$

33. local minima at $x = 0, 1$; local maximum at $x = 1/2$

35. local minima at $x = -1, 1$;　local maxima at $x = -2, 0, 2$.

37. $P_2(x) = 1 + x + x^2$

39.
$$P_2(x) = 1 + \frac{1}{2}(x - 1) - \frac{1}{8}(x - 1)^2$$

41. $x = a$ or $x = b$.

43.

a.
$$g'(x) = 3x^2 y + \frac{y^2}{2\sqrt{x}}.$$

b.
$$\frac{d}{dy}\left(3x^2 y + \frac{y^2}{2\sqrt{x}}\right) = 3x^2 + \frac{y}{\sqrt{x}}.$$

c.
$$h'(y) = x^3 + 2y\sqrt{x}.$$

d.
$$\frac{d}{dx}\left(x^3 + 2y\sqrt{x}\right) = 3x^2 + \frac{y}{\sqrt{x}}.$$

45. a. $U(0) = 9,400,000$ people. b. $U'(1) = 500,000$ people/month. c. $U'(9) > 0$ and $U''(9) < 0$

47. Yes, at $x = -a$

Chapter 2

Section 2.1

1. all domains: $(-\infty, \infty)$; $f'(x) = 36x^3 + 6x + 2$; $f''(x) = 108x^2 + 6$

3. all domains: $(-\infty, \infty)$; $f'(x) = 1$; $f''(x) = 0$

5. all domains: $x \neq 3$; $f'(x) = 1$; $f''(x) = 0$

7. all domains: $(-\infty, \infty)$; $f'(x) = 4x^3 + 12ax^2 + 12a^2 x + 4a^3$; $f''(x) = 12x^2 + 24ax + 12a^2$

9. all domains: $x \neq 2, -3$; $f'(x) = 1$; $f''(x) = 0$

11. domain of f: $(-\infty, \infty)$; domain of f' and f'': $x \neq -2$; $f'(x) = -1$ if $x < -2$; $f'(x) = 1$ if $x > -2$; $f''(x) = 0$

13. domain of f: $(-\infty, \infty)$; domain of f' and f'': $x \neq -5, 2$.

$$f'(x) = \begin{cases} 2x + 3 & \text{if } x < -5 \text{ or } x > 2 \\ -2x - 3 & \text{if } -5 < x < 2 \end{cases}$$

and

$$f''(x) = \begin{cases} 2 & \text{if } x < -5 \text{ or } x > 2 \\ -2 & \text{if } -5 < x < 2 \end{cases}.$$

15. $f'(x) = 3x^2 + 14x + 10 \neq 2x + 5$

17. $f'(x) = 2x - 4 \neq 1$

19. $f(x) = 5.5x^2 + C$

21. $m(x) = x^3 + x + C$

23. $F'(x) = 2x + 3$

25. $F'(x) = 10(x^5 - x^4 + x^2)/3$

27. $-47/60$

29. 708

33. $f'(x) = (e^x + e^{-x})/2$

35. $f'(x) = 18/(1 + x^2)$

37. a. $dA/dh = 2\pi r$; b. $dA/dr = 2\pi h + 4\pi r$; c. 40π in^2/in

43. $k''(1) = 0$; not an inflection point

47.

  a. $V = x(8 - 2x)(6 - 2x)$

  b. $dV/dx = 12x^2 - 56x + 48$

  c. $x \approx 1.131482908$

  d. $A = (8 - 2x)(6 - 2x) + 2x(8 - 2x) + 2x(6 - 2x)$

  e. $x = 0$, i.e., remove no paper

49.

  a. $x'(t) = 12t^3 - 60t^2 - 48t + 240$ meters per second

  b. $t = -2, 2, 5$

  c. increasing $(-2, 2)$, $(5, 10)$ and decreasing on $(-3, -2)$, $(2, 5)$

  d. 11,177 meters

  e. $t = 10$ the particle is farthest right; $t = -2$ the particle is farthest left

Section 2.2

1. $f'(x) = \dfrac{-1}{(x-1)^2}$; $f''(x) = \dfrac{2}{(x-1)^3}$

3. $f'(x) = \dfrac{-6x^4 + -9x^3 + 22x^2 + 4}{(3x^3 + 2x)^2}$.

$$f''(x) = \frac{1}{(9x^6 + 12x^4 + 4x^2)^2} \cdot \Big(108x^9 +$$
$$243x^8 - 792x^7 + 108x^6 - 792x^5 -$$
$$36x^4 - 192x^3 - 32x\Big).$$

5.

$$f'(x) = \frac{-6\pi x^{13} - 4x^{11} - 9x^{10} + 18x + 3}{(x^{10} + 3)^2};$$

$$f''(x) = \frac{1}{(x^{10} + 3)^4} \cdot \big\{42\pi x^{32} + 36x^{30} +$$
$$90x^{29} + (360\pi - 468)x^{22} -$$
$$366x^{20} - 90x^{19} - 2070x^{10} - 990x^9 + 162\big\}$$

7. $f'(x) = (-8x)/(x^4 + 4x^2 + 4)$. $f''(x) = (-24x^4 - 32x^2 - 32)/(x^4 + 4x^2 + 4)^2$.

9. Critical point at $x = -1$. Increasing on $(-\infty, -1]$. Decreasing on $[-1, \infty)$.

11. Critical point at $x = -3/2$. Increasing on $[-3/2, \infty)$. Decreasing on $(-\infty, -3/2]$.

13. False.

15. True.

17. $f' = f_1 f_2 f_3' + f_1 f_2' f_3 + f_1' f_2 f_3$

21. $y - \frac{b}{d} = \frac{ad - bc}{d^2} x$.

23. $y - 1/2 = -\frac{n}{4}(x - 1)$.

25. $(fg)''' = f'''g + 3f''g' + 3f'g'' + g'''f$.

27. $F'(t) = -1000Gm(t^{-2} + 4t^{-5})$.

29. $y' = (x - a)[2f(x) + (x - a)f'(x)]$.

31. $y' = [xf'(x) - f(x)]/x^2$.

33. $\left(\frac{fg}{h}\right)' = \frac{f'gh + fg'h - fgh'}{h^2}$.

35. $a(t) = (2t + 3)(t^5 - 1) + (5t^4)(t^2 + 3t + 1)$. $j(t) = 42t^5 + 90t^4 + 20t^3 - 2$.

37. $a(t) = \frac{t^2 + 6t - 1}{t^2 + 6t + 9}$. $j(t) = \frac{20}{(t+3)^3}$.

43.

  a. $I' = (RV' - R'V)/R^2$.

b. $I'(1) = 0$.

c. $t = 1$ second.

d. $V(1) = 10$ volts. $R(1) = 2$ ohms.

45.

    a. $dP/dt = 1.6P(25 - P)$.

    b. $0 < P < 25$

    c. $P > 25$

    d. $P'' = 2.56P(25 - P)(25 - 2P)$.

    e. $0 \leq P \leq 12.5$

    f. $P \geq 12.5$

Section 2.3

1. $f'(x) = 6(2x + 1)^2$

3. $h'(x) = \frac{7}{3}\left(\frac{1}{3}x + 12\right)^6$

5. $f'(s) = 5s^4 + 2$

7.
$$g'(p) = (54p + 18)(3p^2 + 2p)^8$$
$$+132p^2(4p^3 + 72)^{10}$$

9. $f'(r) = (56r^6 + 60r^4)(4r^7 + 6r^5)^{99}$

11. $t'(x) = -\dfrac{300}{(5x + 1)^3}$

13.
$$f'(t) = \frac{1}{(t^2 + t + 2)^2} \cdot \{12(3t + 2)^3(t^2 + t + 2) -$$
$$(2t + 1)(3t + 2)^4 - 14t - 7\}$$

15. $m'(x) = 4\left[(x + 5)^3 + (2x + 5)^5\right]^3 \cdot \left(3(x + 5)^2 + 10(2x + 5)^4\right)$

17. $g'(x) = 4((x^2 + 1)^5 + x^6 + x^3)^3 \cdot \left(10x(x^2 + 1)^4 + 6x^5 + 3x^2\right)$

19. $h'(x) = \frac{1}{2}\left[x + (x + x^{1/2})^{1/2}\right]^{-1/2} \cdot \left[1 + \frac{1}{2}\left(x + x^{1/2}\right)^{-1/2}\left(1 + \frac{1}{2}x^{-1/2}\right)\right]$.

21. $n = 1$

23. no solution.

25. $C_2 = (50C_{total})/(50 - C_{total})$

27. c

29. a. $[f(g(x))]' = f(g(x))g'(x)$ b. $[g(f(x))]' = g'(f(x))f(x)$

31. $1/27$

33. $1/36$

35. -9

37. f and g crit. pt.: $x = 3$; h crit. pts.: $x = 3 \pm \sqrt{6}$

39. 880 watts per sec.

41. 294 joules per sec.

43. $53,958$ joules per sec.

45.

    a. $\dfrac{dA}{dt} = 4\pi r\dfrac{dr}{dt} + 2\pi h\dfrac{dr}{dt} + 2\pi r\dfrac{dh}{dt}$

    b. 228π in^3/hr

    c. $dA/dh = 2\pi + 3\pi h^{1/2}$; $dV/dh = 2\pi h$

    d. always decreasing

47.

    a. $y = -3x/4 + 25/2$

    b. $y = 4x/3 + 16$

    c. $y = x^2/36 + x + 1$

49.

    a. approx. 0.996 ft/s

    b. $t \approx 0.7368$ sec.

    c. yes

    d. $x \approx 0.57$ feet

53. $-(135)^{-3} \cdot 450$

Section 2.4

1. $f'(x) = 5e^x$

3. $h'(x) = 6xe^{x^2+1}$

5. $q'(t) = (12t + 87)e^t$

7. $k'(x) = 14xe^{x^2}$

9. $f'(x) = -\frac{1}{12}e^{-x}$

11. $f'(r) = 30r^4 + 15e^r - 35r^{2/5}$

13. $g'(t) = -5.52e^{-0.3t}$

15. $q'(t) = e^t - 3e^{-2t}$

17.

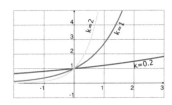

19. $b_1 = 1! = 1$, $b_2 = 2! = 2$, $b_3 = 3! = 6$, $b_4 = 4! = 24$.

21. $a_4 = 3$, $a_5 = 5$, $a_6 = 8$.

23.

   a. $\sigma_5(0.8) \approx 3.36$, $\sigma_{10}(0.8) \approx 4.46$, $\sigma_{50}(0.8) \approx 4.999$.

   b. 5

   c. $\sigma_5(1.4) \approx 10.95$, $\sigma_{10}(1.4) \approx 69.8$, $\sigma_{50}(1.4) \approx 50,622,288$.

   d. -2.5

   e. For some values of r (actually, those such that $|r| < 1$), $\sigma_n(r)$ converges to $1/(1 - r)$ as n goes to ∞. For some values of r (those such that $|r| \geq 1$), this convergence does not occur.

25.

   a. $g'(x) = e^{f(x)} \cdot f'(x)$.

   b. $g'(x) = e^{f(e^x)} \cdot f'(e^x) \cdot e^x$.

   c. $g'(x) = f'\left(e^{\left(\frac{1}{x}\right)}\right) \cdot e^{\left(\frac{1}{x}\right)} \cdot \frac{-1}{x^2}$.

   d. $g'(x) = f'(x) \cdot e^x + f(x)e^x = (f'(x) + f(x))e^x$.

27. a. $(-\infty, \infty)$ b. $[0, \infty)$ c. $x \neq 1$

29. a. λ b. $\lambda^2 + \lambda$

33. $e - P_3(1) \leq 5/(4! \cdot 4)$, $e - P_4(1) \leq 6/(5! \cdot 5)$, $e - P_5(1) \leq 7/(6! \cdot 6)$.

37. 4

39.

   a. $f'(x) = -(x - \mu)f(x)/\sigma^2$.

   b. $f''(x) = [(x - \mu)^2 - \sigma^2]f(x)/\sigma^4$.

   c. Yes, at $x = \mu \pm \sigma$.

43. a. There are none. b. Inflection pt. at $x = 0$.

47. 1/2

51.

a. $R(p) = p \cdot D(p)$

b. $R'(p) = Ce^{-kp}(1 - pk)$ dollars (of revenue) per dollar (of price)

c. $p = 1/k$ dollars

53. $y = ax^2 - x/k + 1$

Section 2.5

1. $f'(x) = 3x + 3/x$

3. $h'(x) = \frac{11(2x+1)}{x^2+x+1}$

5. $p'(x) = 12x$

7. $f'(t) = \frac{t-1}{t^2+t}$

9. $f'(x) = -\frac{2\ln x - 1}{x^3(\ln x)^2}$

11. $h'(x) = -\frac{\ln x + x}{x^2 \ln x}$

13. $r'(x) = \frac{x}{x^2-1}$

15. $dy/dx = 1.8y = 1.8(75/(12e^{-1.8x}))$

17. $dy/dx = y[6x + 2 + 1/(2x + 2) - 4x/(2x^2 + 5)]$

19. $y' = (y/3)[1/(x - a) + 1/(x - b) + 1/(x - c)]$

21. $g'(x) = x^x(1 + \ln x)$

23. $j(t) = 2/t^3 \text{ m/s}^3$

25. $t = -1$; no inflection points

27. $t = 2$ and $t = 3$; both are inflection points

29. a. Only $x = 12$ is in domain. b. $x = 12$ yields a minimum.

31. $b = 1/54$

33. $v(t) = 5/(15t + 1)$; $a(t) = -75/(15t + 1)^2$

37. $y = x$

39. $y = 10ex - 9e$

41. $y' = (\ln k)^2 k^x k^{k^x}$

47. $f^{(k)}(x) = (-1)^{k-1}(k - 1)!x^{-k}$

49. $\chi_{n,m} = (m - n)/(e^m - e^n)$; c is unique $c = (e^m - e^n)/(m - n)$

Section 2.6

1. $f'(x) = (\ln 32)(2^x)$

3. $h'(x) = 2\ln 3$

5. $r'(t) = 5^t\left(\ln 5 \ln t + \frac{1}{t}\right)$

7. $f'(x) = \dfrac{2}{x \ln 3} \log_3 x$

9. $r'(x) = 2x(\ln 6) 6^{x^2+1}$

11. $f'(t) = \dfrac{t 7^t \ln(t+7) - 1.1 \cdot 7^t}{t(\ln t)^2}$

13. $h'(x) = (x^2 + 5x + 3)^{7x-3} \{(7x - 2) \cdot (2x + 5) + 7(x^2 + 5x + 3) \ln(x^2 + 5x + 3)\}$

15. $f'(t) = -325.5 e^{-3.5t} - 284 \ln(13) 13^{-4t}$

17. $g'(x) = \dfrac{1}{x \log_b x (\ln b)^2}$

19. $h'(x) = \dfrac{\sqrt{1+x^2}+x}{(x\sqrt{1+x^2}+(1+x^2)) \ln 2}$

21. $g'(x) = (x^{2x})(\ln x^2 + 2)$

27. $g'(x) = (1/x)(\ln x)^{(\ln x)}(1 + \ln(\ln x)), x \neq 0$

29. Reflection over the line $y = x$.

31. $f'(x) = \dfrac{\ln(\ln x) - 1}{x(\ln(\ln x))^2}$

35. $y = (\ln a)x + 1$

37.

    a. $A(t) = 1000 \cdot 1.0075^{\lfloor 4t \rfloor}$

    b. $E(t) = 1000 \cdot (1.030339)^t$

    c. $A'(1/12) = 0$ dollars per year

    d. $E'(1/12) = 29.962408$ dollars per year

    e. $E(25) \approx 2111.074074$, $A(25) \approx 211.08384$.

39.

    a. $f(x)$ and $g(x)$ intersect at $(0, 1)$ for each $b \neq 1$ (they are the same graph otherwise)

    b. $f'(x) = b^x \ln b$ and $g'(x) = -b^{-x} \ln b$. So the slopes of the tangent lines to the graphs of $f(x)$ and $g(x)$ at $x = 0$ are $f'(0) = \ln b$ and $g'(0) = -\ln b$, respectively. In order for these to be orthogonal, we would need

$$\ln b = -\dfrac{1}{-\ln b}$$
$$(\ln b)^2 = 1$$
$$\ln b = \pm 1.$$

    i.e., $b = e$ or $b = e^{-1}$.

49.

    a. $A(t) = 100 e^{-0.0693147t}$

    b. 10 years.

    c. $A(t) = 100(1/2)^{t/10}$

51.

    a. $E'(x) = \dfrac{1}{x \ln 10}$

    b. $E'(1000) \approx 0.00043429$

    c. $E(100) = 3, E(10) = 2$

    d. $c = \dfrac{90}{\ln 10} \approx 30.0865$

Section 2.7

1. $f'(t) = 10 \cos(2t)$

3. $h'(x) = \left(6 \sin x + \dfrac{2}{5}\right) \cos x$

5. $q'(\theta) = 4\pi \sin^2 \theta + 2\pi \cos 2\theta + 8\pi\theta \sin \theta \cos \theta - 4\pi\theta \sin 2\theta$

7. $g'(x) = \cos\left(\dfrac{\pi}{6} + x\right)$

9. $f'(x) = 2x \sin x + x^2 \cos x$

11. $h'(x) = \dfrac{-\sqrt{x} \sin \sqrt{x} - 2 \cos \sqrt{x}}{2x^2}$

13. $f'(\theta) = \dfrac{1}{\cos^2 \theta}$

15. $q'(x) = \dfrac{-1}{1 + \sin x}$

17. $g'(x) = 4 \cos 2x - 4x \sin 2x$

19. $p'(\theta) = \cos\left(\dfrac{\theta}{4}\right) + 3 \sin 3\theta$

21.

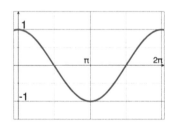

    a. Increasing on $(\pi, 2\pi)$, and decreasing on $(0, \pi)$

    b. $g'(x) = \sin(-x)$, critical points at $x = n\pi$ for every integer n, so in the interval $[0, 2\pi]$, we have critical points $x = 0, \pi, 2\pi$, interior critical point $x = \pi$

23. $f^{(n)}(x) = \begin{cases} (-1)^{\frac{n}{2}} ab^n \sin(bx) & \text{if } n \text{ is even} \\ (-1)^{\frac{n-1}{2}} ab^n \cos(bx) & \text{if } n \text{ is odd} \end{cases}$

25. $g^{(33)}(x) = g'(x) = \cos(x) - \sin(x)$

27. $1/2$

29. 0

31. $j(x) = \dfrac{d-c}{2}\left[\sin\left(\dfrac{2\pi}{b-a}(x-a)\right) + 1\right] + c$

33. $i(t) = 10\cos(0.5t) + 12\sin(0.5t)$

35. $i(t) = 3\cos(4t) + 5\sin(4t)$

37. a. $x = k\pi/3$, k is integer. b. $x = j\pi/3 + 2n\pi$, $j = 1$ or 2, and n is integer

41. a. $0.524; 0.4997; 0.50000$. b. $O_2(\pi/6) \to \sin(\pi/6) = 0.5$

43. $y' = \frac{-y\sin(xy)}{2y + x\sin(xy)}$

45. $y' = \frac{-1 + y(x+y)\cos(xy)}{1 - x(x+y)\cos(xy)}$

53.

    a. 220 is the maximum voltage; first achieved after 0.005 seconds

    b. $t = 0.02$ seconds.

57. $f'(x) = \cos x \cos x - \sin x \sin x = \cos^2 x - \sin^2 x = \cos 2x$

59.

    a. $t = \pi/2$, $t = 3\pi/2$

    b. $x(t)$ decreases from 0 to $\pi/2$, increases until $3\pi/2$ then decreases until 2π

    c. $t = 3\pi/2$

    d. $t = \pi/2$

    e. Inflection point at $t = \pi$

61. $x(t) = v_0\sqrt{m/k}\sin(\sqrt{k/m}t) + (x_0 - F_0/k)\cos(\sqrt{k/m}t) + F_0/k$

Section 2.8

1. By Theorem 2.8.1, $y'(\theta) = -\csc^2 x$

3. $h'(u) = -21\csc^2(3u - 9)$

5. $f'(\theta) = -e^{-2\theta}\csc\theta(2 + \cot\theta)$

7. $g'(x) = \frac{1}{2\sqrt{1+x}}\sec^2\left(\sqrt{1+x}\right)$

9.

$$p'(x) = 28x^3\csc x - 7x^4\cot x\csc x$$
$$- 6x^2\sin x - 2x^3\cos x$$

11. $f'(x) = \frac{1}{2}\tan x\sqrt{\sec x}$

13. $p'(x) = 3(\sec^2 2x)\sqrt{\tan 2x}$

15. $r'(x) =$
$3^{-x}\left(-(\ln 3)\tan(2x^3) + 6x\sec^2(2x^3)\right)$

17.

a. Increasing on $\left(0, \frac{\pi}{2}\right)$; decreasing on $\left(-\frac{\pi}{2}, 0\right)$

b. $g'(\theta) = \frac{\sin\theta}{\cos^3\theta|\tan\theta|}$; critical point at $x = 0$

19. $g''(t) = \sec^3 t + \tan^2 t\sec t$

21. $p'' = 2\csc^2 u\cot u$

25. No local differentiable inverse function exists.

29. $1/2$

31. $a(t) = (\csc^3 t + \cot^2 t\csc t)(\cos(\csc t)) - \cot^2 t\csc^2 t\sin(\csc t)$

33. a. 2π. b. 2π c. π d. π e. 2π f. 2π

35. $-\coth x\operatorname{csch} x = -\frac{2(e^x + e^{-x})}{(e^x - e^{-x})^2}$

39. $x = \pi/2 + k\pi$ where k is any integer

41. j has no inflection points

45. a. Odd. b. Even. c. Odd. d. Odd. e. Even. f. Odd

47.

    a. $x = n\pi + \frac{\pi}{2}$ for all integer n

    b. $x = n\pi + \frac{\pi}{2}$ for odd values of n

    c. $x = n\pi + \frac{\pi}{2}$ for even values of n

    d.

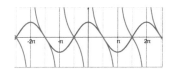

49.

    a. $s(\theta) = \tan\theta$, and $t(\theta) = \sec\theta$

    b. $s'(\pi/4) = 2$

    c. $t'(\pi/3) = 2\sqrt{3}$

Section 2.9

1. $f'(x) = \frac{2}{\sqrt{1 - x^2}} + 2x$

3. $h'(u) = \frac{3}{(\sin^{-1} u)\sqrt{1 - u^2}}$

5. $g'(x) = -1 - \frac{x\cos^{-1}x}{\sqrt{1 - x^2}}$

7. $r'(t) = \frac{1}{2\sqrt{t}(1 + t)}$

9. $q'(u) = 0$.

11. $h'(\theta) =$

$$7^\theta \left(\ln 7 \csc^{-1}\theta - \frac{1}{|\theta|\sqrt{\theta^2-1}} \right)$$

13. $r'(t) = \frac{3(\sec^{-1} t)^2 + 12}{|t|\sqrt{t^2-1}}$

15. $h'(x) = 0$

17. $q'(u) = \frac{-e^u}{e^{2u} - 2e^u + 2}$

19. $g'(z) = \frac{a - 2a^2 z \tan^{-1}(az)}{(1+a^2 z^2)^2}$

21. $x = 0$ is the only inflection point.

23. $x = 0$ is the only inflection point of h.

27. a. false. b. true. c. false. d. true.

39. The formula holds for all x and y such that $y \ne 1/x$

43. Maximum at $(1, \pi/2)$, minimum at $(-1, -\pi/2)$.

47.

b. $\theta(s) = \sec^{-1}\left(\frac{s}{r_1 + r_2}\right)$

c. $\lim_{s\to\infty}\theta(s) = \pi/2$

d.

$$\frac{d\theta}{ds} = \frac{1}{s}\sqrt{\frac{(r_1+r_2)^2}{s^2 - (r_1+r_2)^2}}$$

49.

a. $\theta(x) = \tan^{-1} f'(x)$.

b. $\frac{d\theta}{dx} = \frac{f''(x)}{1 + f'(x)^2}$

Section 2.10

1. $\frac{dy}{dx} = -x/y$

3. $\frac{dy}{dx} = -2/y$

5. $\frac{dy}{dx} = \frac{-6x + 3y}{4y - 3x + 1}$

7. $\frac{dy}{dx} = \frac{e^{x+y}}{1 - e^{x+y}} = \frac{y}{1-y}$

9. $\frac{dy}{dx} = \frac{y\sin(xy)}{2y - x\sin(xy)}$

11. $\frac{dy}{dx} = \frac{2(x-y) - 2xy^3}{2(x-y) + 3y^2 x^2}$

13. $\frac{dy}{dx} = \frac{ye^{xy} - e^x}{e^y - xe^{ey}}$

15. $\frac{dy}{dx} = \frac{x + \sqrt{y^2 - x^2}}{y} = \frac{2x}{y}$

17. $\frac{dy}{dx} = \frac{2x\tan(x^2)\sec^2(x^2)}{1 - 2y\sec^2(y^2)}$

19. $-\frac{\partial F/\partial x}{\partial F/\partial y} = y' = \frac{1 - y\cos(xy)}{1 + x\cos(xy)}$

21. $-\frac{\partial F/\partial x}{\partial F/\partial y} = y' = \frac{1 - e^{x+y}}{e^{x+y} + 1}$

23. a. $y' = \frac{1 - \sqrt{\frac{y}{x}}}{\sqrt{\frac{x}{y}} - 1}$. b. $x = y$

25.

a. $\frac{xb^2}{ya^2} = \frac{dy}{dx}$

b. $y - y_0 = \frac{x_0 b^2}{y_0 a^2}(x - x_0)$

c. i. $y_0^2 = \frac{b^2}{a^2}x_0^2$ ii. no tangent through origin.

35. a. y' is defined for $x \in (-8, 0) \cup (0, 8)$. b. $y' = \pm\frac{\sqrt{4 - x^2}}{\sqrt[3]{x}}$, y' is undefined at $x = 0, x = \pm 8$.

37. $p' = \frac{\sin(t)\sin(p)}{\cos(t)\cos(p)} = \tan(t)\tan(p)$

41.

a. $\frac{dy}{dx} = \frac{-2 - 2y}{10y + 2x - 5}$

b. There are no horizontal tangent lines to the graph

c. $y = -\frac{1}{5}x + \frac{3}{5} \pm \frac{1}{5}\sqrt{x^2 - 16x - 9 - 5y}$

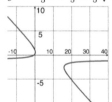

d. $\frac{dy}{dx} = \frac{-4}{5}$

43. a. $\frac{dy}{dx} = -2$; $\frac{d^2 y}{dx^2} = 0$, b. $y + 2x = \pm 1$

45.

a. $\cos y = -\sin(5x)$

b. $-\sin(5\pi/4) = \sqrt{2}/2$

c. $y - \frac{\pi}{4} = \frac{1}{5}\left(x - \frac{\pi}{4}\right)$

47. curvature is zero; straight lines.

49. $\kappa(x) = \frac{1}{(x^2 + y^2)^{3/2}}$

51. $\kappa'(x) = 0$

53. a. x, x^2 both constant. b. $f_0(x) = b = 9$. c. $f_1(x) = x^2$. d. $f_2(x) = x^2$

Chapter 3

Section 3.1

1. $2\sqrt{3}$

3. $-9/2$

5. 45

7. $-12e^5$

9. $y'(t_0) = 0$

11. $\frac{dz}{dt} = 10$

13. $\frac{dz}{dt} = 18$

15. $x = 1/4$

17. -2

19. $\frac{dr}{dt} = 5/32\pi \approx 0.0497$

21. $h'(t) = \sqrt{3}/15 \approx 0.1155$

23. a. $V = S^{3/2}/(6\sqrt{\pi})$. b. $dV/dS = S^{1/2}/(4\sqrt{\pi})$ c. $dV/dt = 3\sqrt{2}/(2\sqrt{\pi})$

27. $v \approx 1047$ mi/h

29. $\omega = \frac{t^2+2t-1}{15(t+1)^2}$

31. $dV/dt = 22,500$ m^3/s

33. $dy/dt = 60$

35. $dE/dt = 1500$ J/s

37.

    a. $x(t)^2 + y(t)^2 = 10^2 = 100$

    b. $x'(t) = -0.1333$ m/s, $x''(t) = -0.004294$ m/s^2

    c. model allows $v \to -\infty$

39. approx. 188 mi/h

41. a. $t \approx 2617.994$ seconds. b. 0.0101859 m/s. c. 0.0025465 m/s

43. a. $d = [(100 + 20\cos(\theta_2) - 10\cos(\theta_1))^2 + (20\sin(\theta_2) - 10\sin(\theta_1))^2]^{1/2}$. b. $\theta_1(t) = 0.4t$ rad; $\theta_2(t) = 0.2t$ rad. c. $d'(t) \approx 1.5017165$ m/s.

45. a. $h = 1$. b. $h'(t) = 0.15$ m/s. c. $h'(t) = -0.15306$ m/s

Section 3.2

1.

3.

5.

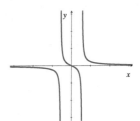

7.

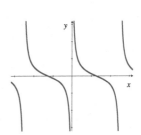

9.

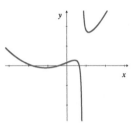

11.

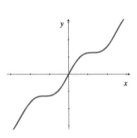

13.

a. The graph must be concave up, since if the function is decreasing and positive, with a limit of 0, then it cannot be concave down (it must approach 0 less and less quickly as x tends towards infinity. Some sort of shifted exponential-type curve is often used:

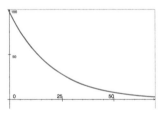

b. In the long run, just as above, the function should be concave up, so if there is one inflection point, it must be where the function turns from concave down to concave up. The logistics curve can be modeled to fit this:

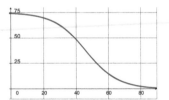

15. (a)

17. (e)

19. (h)

21. (f)

Section 3.3

1. g attains a maximum of 32 at $x = 2$, and a minimum of -32 at $x = -2$

3. $r_{\max}$ at $\theta = -\pi$; $r_{\min}$ at $\theta = 3\pi/4$

5. $g_{\max}$ at $p = 3/2$; g_{min} at $p = 0$

7. $a = b = 100$

9.

   a. $y'(t) = e^{-t} - te^{-t}$

   b. $y(t)$ is smallest at $t = 0$ and largest at $t = 1$

   c. Swiftest motion upward is at time $t = 0$ and swiftest motion downwards $t = 2$

13. Max resistance when $R_1 = R_2 = \Omega/2$. $R_{\text{total}} = \Omega/4$ ohms

15. $R_{\text{total}} \approx 12.32$

17. $h = r = \sqrt[3]{1000/\pi} \approx 6.2878$

19. $2r^2$

21. a. $A(25) = 1250$. b. $y = 25, x = 50$

23. $A(F/4) = F(F/4) - 2(F/4)^2 = F^2/4 - F^2/8 = F^2/8$

25. $t = 1$

27. Rectangle base $2\sqrt{2}$, height $\sqrt{0.5}$

29. $4\sqrt{a^2 + b^2}$

31. $\sqrt{f\left(\sqrt{\frac{17}{324}}\right)} \approx 0.329$

37. $f(x) = x$ does not attain a maximum on its domain

39. a. 2000. b. $n^2 + 2 + 2000/n, n > 0$. c. 302

41. a. $10000pe^{-2p}$. b. maximum revenue of $1839.40 when $p = \$0.50$. c. $20000e^{-2p} + 5000$. d. $2.5/hat

43. a. 28.8675 miles from the end of the waterway. b. $109,282$. c. Yes.

Section 3.4

1. $L_f(x; -1) = 1 - 8(x + 1)$

3. $L_f(x; 1/2) = 0 + 2(x - 1/2)$

5. $L_f(x; -\sqrt{2}) = 3\pi/4 + (1/\sqrt{2})(x + \sqrt{2})$

7. $dy = \frac{-2}{(x+1)^3} dx$; $dy = 0.025$

9. $dy = \frac{1}{x} dx$; $dy = 0.01$

11. $x_2 = 1.2358078$

13. $x_2 = 1.179747$

15. $x_2 = -0.3516893316$

17. ≈ 0.05152

19. $\approx -7 \cdot 10^{-5}$

21. $R_f(b; a) \approx 0.0024$

23. ≈ 0.62 cents/unit

25. $|\Delta| = 1.55 \cdot 10^{-19} \cdot e^{-0.465} \approx 9.74 \cdot 10^{-20}$ kg

29. 2.94 m

33. $x_1 = -1$; $x_2 = 1$

35.
   a. $x_1 = 4/3$; $x_2 \approx 1.23581$
   b. $x_1 = 1 - (-1)/13 \approx 1.03226$; $x_2 \approx 1.06119$
   c. Newton's Method

37.
   a. $M = f'(0) = 2.25$
   b. $x_1 = 5/9$; $x_2 = 5/9 - f(5/9)/2.25 \approx 0.148$

39. $x_2 = 134/37$; $x_3 = 3.6789$

43. $x_1 = 4.25$; $x_2 = 4.345625$

45.
   a. $dS = 0.01759291886$
   b.
   $$\frac{dS}{S} = \frac{h + 2r}{1 + r} \cdot \frac{dr}{r}$$
   c. 0.000875%
   d. (a) $dV \approx 0.0301592895$
   d. (b) $dV/V = 2(dr/r)$
   d. (c) $0.05\ \%$

47.
   a. $dh \approx 0.0015915$
   b. 0.16666%

49. $\sqrt[3]{10} \approx 2.51450362$

51.

a. $L_f(x; \pi/6) = (1/2) + (\sqrt{3}/2)(x - \pi/6)$

b. 0.9534498411

c. -0.0874244373

d. 0.11871289

Section 3.5

1. p/q

3. Does not exist

5. 0

7. $-1/2$

9. $1/2$

11. 0

13. Does not exist (approaches $-\infty$)

15. e^{-6}

17. 5

19. 0

21. $1/2$

23. 0

25. Does not exist

27. $p(0)$

29. na^{n-1}

31. 1

33. e^3

35. $e^{-1/2}$

37. $e^{3\pi/2}$

39. 40

41. $f''(x)$

51. a. 0. c. 5

53.
   a. $V(2) = -2e^{-4} - e^{-4} - 1/4e^{-4} + 1/4 \approx 0.19047417$.
   b. $1/4$

55.
   a. $T(\theta) = -\frac{\sin\theta}{\cos\theta - 1}$
   b. $T'\left(\frac{\pi}{2}\right) = \frac{1}{\cos\pi/2 - 1} = -1$
   c. $\frac{d^2 T}{d\theta^2} = \frac{\sin\theta}{(\cos\theta - 1)^2}$. Has inflection point.
   d. $\lim_{\theta \to 0^+} T(\theta) = \lim_{\theta \to 0^+} \cot\theta = \infty$

Chapter 4

Section 4.1

1. $y = \sin x$

3. $y = x + e^x$

7. (e)

9. (c)

11. (f)

13. 1

15. 2

17. $A(t) = \frac{6}{1 - 3t^2}$

21. The differential equation is autonomous.

23. 2

35. $y(t) = (10/3)e^{-5t} - (10/3)e^{-8t}$

37. $y(t) = 3e^{6t} - 18te^{6t}$

39.

    a. $P = 400 - 370e^{-3t/100}$

    b. $t = -100/3 \ln 35/37 \approx 1.852$ sec

    c. 400

41.

    a. $T(t) = 70 - 38e^{-0.5t}$

    b. $t = -2\ln 10/19 \approx 1.2837$

    c. $T(\infty) \to 70$. Over time the ice cube will approach the ambient temperature.

43.

    a. $v = 0.5dv/dt$

    b. $v(t) = e^{0.5t}$

    c. $10e^5 \approx 1484.131591$ kg·m/s

45.

    a. $3500 - 0.06P(t)$

    b. $P(t) = Ce^{-0.06t} + 58333.3333$

    c. At $t = -66.7889$, hence the population will not reach this level.

    a. $\frac{m}{2}\left(\frac{dh}{dt}\right)^2 + mgh(t) = c$

    b. $m\frac{dh}{dt}\frac{d^2h}{dt^2} + mg\frac{dh}{dt} = 0$

    $m\frac{dh}{dt}\left(\frac{d^2h}{dt^2} + g\right) = 0$.

    Therefore, if $dh/dt \neq 0$, we must have

$$\frac{d^2h}{dt^2} = -g.$$

    c. $d^2h/dt^2 = -g$

    d. $T_E = -m\left(\frac{v_0{}^2}{2} + gh_0\right)$

Section 4.2

1. $\frac{4}{3}x^3 + 2x^2 + 9x + C$

3. $-5\cos t - 3\sin^{-1} t + C$

5. $\ln|y| + \tan^{-1} y + C$

7. $\sin(2\theta - 1)/2 + C$

9. $\ln|r^2 - 4|/2 + C$

11. $(5w - 3)^{101}/505 + C$

13. $-(x+2)(x+5) + (x+7)(x+2)\ln|x+2| + C$

15. $5\ln|\ln x| + C$

17. $-\ln|\cos\theta| + C$

19. $\frac{2}{15}(t^5 + 6)^{3/2} + C$

21. $\sin^{-1}(x - 3) + C$

23. $P(w) = 5\ln|w| - 7e^w + (9/2)w^{4/3} + 7e^{-1} - 9/2$

25. $K(v) = -v^{-1} + \ln|v| - 2v^{-1/2} + 1$

27. $F(x) = 1/3x^3 - x\cos x + \sin x + \pi - 1/3\pi^3$

29. $G(x) = -2/3(x+1)^{3/2} + 2/5(x+1)^{5/2} + 19/15$

31. $R(t) = -1/2e^{1-t^2} - 1/2t^2 + \sqrt{2} + 1$

33. $(x-5)^2e^x - 2[(x-5)e^x - e^x] + C = (x^2 - 12x + 37)e^x + C$

35. $t^2\sin t + 2t\cos t - 2\sin t + C$

37. $-\dfrac{\ln t}{t} - \dfrac{1}{t} + C$

39. $\dfrac{1}{29}e^{2x}[2\sin(5x) - 5\cos(5x)] + C$

41. $\dfrac{1}{2}(w^2\tan^{-1} w + \tan^{-1} w - w) + C$

43.

a. $p(t) = -5t\cos(2t) + +2.5\sin(2t) + 20$

b. $p(4) \approx 25.3833963$

45. $p(t) = t^3 - 2t^2 + 3t + 2$

47. $p(t) = 2/21(18 + 7t^2)^{3/2} - 82/21$

49.

   a. $-\sin x \cos x + \int \cos x \cos x\, dx$

   b. $-1/2\sin x \cos x + 1/2x + C$.

   c. $1/2\sin x \cos x + 1/2x + C$

   d. $1/2\sin x \cos x + 1/2x + C$

57. $(\tan^{-1}(x/a))/a + C$.

59. $\sin^{-1}(x/a) + C$

61. $e^{e^x} + C$

63. $\frac{2^x}{\ln 2} + C$

65. $x\ln(1 + x^2) - 2x + 2\tan^{-1}(x) + C$

67. $(1 - \cos t)/12$

69. $-(\ln(t^2 + 1))/6$

75.

   a. $\int t^n \ln t\, dt = \frac{t^{n+1}}{n+1}\ln t - \frac{t^{n+1}}{(n+1)^2} + C$

   b. $(\ln t)^2/2 + C$

77.

   a. maximum speed is 0.5 mi/h

   b. $x(t) = 1/2\sin t - 1/6\sin^3 t$.

   c. $y(t) = (x(t))^2 = \left(\frac{1}{2}\sin t - \frac{1}{6}\sin^3 t\right)^2$

   d. 0 mi/h

79. $v = t^2/2 + t + v_0$, $p = t^3/6 + t^2/2 + v_0 t + p_0$

81. $v = -e^{-3t}/3 + v_0 + 1/3$, $p = e^{-3t}/9 + (v_0 + 1/3)t + p_0 - 1/9$

Section 4.3

1. $y = \frac{-1}{0.5x^2 + c}$

3. $y = \pm\sqrt{2t\sin t + 2\cos t + c}$

5. $w = \sin^{-1}\left(Ce^{-0.5e^{-t^2}}\right)$

7. $x = \pm\sqrt{-1 + C(t + 1)^2}$

9. $y = -\ln(-xe^x + e^x + e)$

11. $r^2 + r = 0.5t^2 + 6$

13. $x^3 = 3y + y^3 + C$

17. $i(t) = 5 - 5e^{-8t/5}$

19. If V is constant, then $\frac{di}{dt} = \frac{V - Ri}{L}$.

21. $xy = C$

23. $(x^2/3) + y^2 = C$

25. $x^2 + (y - 3)^2 = C$

27. $\ln|y| = \ln|x + 1| - \ln|x - 1| + C$

29. $y = -\sqrt{2}\tan^{-1}(x/\sqrt{2}) + 3\tan^{-1}x + C$

31. $y = 2$, $y = -2$

33. no equilibrium solutions

35. $y = \pi/2 + k\pi$, where k is an integer

37. $y^2 = 2x^2(\ln|x| + C)$

39. $y = x\sin^{-1}(\ln|x| + C)$

43. $y = Ce^{kx}$

Section 4.4

3.

   a. $v(t) = 294$

   b. Terminal velocity $\lim_{t\to\infty} v(t) = 294$ m/s

   c. Velocity is constant, thus terminal velocity is equal to initial velocity

5.

   a. $v(t) = 98 + 2e^{-t/10}$

   b. Terminal velocity $\lim_{t\to\infty} v(t) = 98$ m/s

   c. The skydiver is decelerating from 100 to 98 m/s

7.

   a. $dP/dt = -0.02P + 3000$

   b. $P(t) = 150,000 + 850,000e^{-0.02t}$

   c. 2nd term goes to zero as $t \to \infty$; limiting population is 150,000

9.

   a. $dP/dt = 0.02P - 3000$

   b. $P(t) = 150,000 - 50,000e^{0.02t}$

   c. The second term goes to infinity as $t \to \infty$. Eventually the population will vanish.

11. There is no limiting value, as the population will increase without bound. The model can be adjusted by including a *carrying capacity* variable.

13. $T(7) \approx 88.48$

15. $\lim_{t\to\infty} W(t) = 1000$

17. It is cyclic, and thus can capture cyclic seasonal effects

19.
$$v(t) = \frac{k^2 t^2}{4m^2} - \sqrt{v_0}\frac{kt}{m} + v_0$$

21.

a.
$$m\frac{dv}{dt} = mg - kv^2$$

b.
$$v(t) = \sqrt{\frac{gm}{k}} \frac{e^{2t\sqrt{g/mk}} - 1}{e^{2t\sqrt{g/mk}} + 1}$$

23. $y = J\sin(kx + F)$

25. a. $P(t) \approx 2105 + 95e^{0.095t}$. b. Pop. grows w/o bound.

27.

b. $C = (v_0^2/2) - gR$

d. $v_0 = \sqrt{\frac{2gRh_{\max}}{R + h_{\max}}}$

29.

a. $F(t) = 55\frac{dv}{dt} = -4 - v$

b. $v(t) = 4e^{-t/55} - 4$.

c. Terminal velocity when $dv/dt = 0$, when $v = 4$ m/s

d. $v(805)$=-3.999999824 m/s

31.

a. $\frac{dV}{dt} = 0.1 - 0.05V$

b. $V(t) = 98e^{-0.05t} + 2$

c. 2

33.

a. $\frac{dP}{dt} = -0.02P$

b. $P(t) = 1000e^{-0.02t}$; it takes approximately 264.915 years

c. $\frac{dP}{dt} = 100 - 0.02P$

d. $P(t) = -4000e^{-0.02t} + 5000$

e. 5000 fish

35.

a. $\frac{dV}{dt} = -k\frac{S}{r}$

b. $\frac{dr}{dt} = \frac{-3k}{2h} \cdot \frac{\sqrt{r^2 + h^2}}{r}$

c.
$$r(t) = \sqrt{\left(\frac{-6k}{h}t + h\sqrt{2}\right)^2 - h^2}$$

d.
$$t = \frac{h^2(\sqrt{2} - 1)}{6k}$$

37. Differential equation
$$\frac{dy}{dx} = -\frac{x - a}{y - b};$$

Solving this differential equation, one arrives at
$$(y - b)^2 = -(x - a)^2 + c$$
$$(y - b)^2 + (x - a)^2 = c.$$

39.

a. $P(t) =$
$$\frac{1}{(-0.00322325t + 0.9332543)^{100}}$$

b. $t = 289.5383$

41.

a. $v(t) = 400e^{-0.0002t} - 500$
$p(t) = -2,000,000e^{-0.0002t} - 500t + 2,002,000$

b. Approximately 19.842715 seconds from the start of the fall

c. $v(19.842715) = -101.5842715$

Section 4.5

1. $y = -2/(x^2 + 2)$; $y(1) \approx -0.6992025$; error$\approx$ 0.03253583; % error$\approx$ 4.88037393

3. $y = e^{\sin x}$; $y(2) \approx 2.83817685$; error$\approx$ 0.35559912; % error$\approx$ 14.323786

5.

$n = 10$: $y(1) \approx -0.6782133$; error$\approx$ 0.01154661; % error$\approx$ 1.73199214

$n = 100$: $y(1) \approx -0.6677451$; error$\approx$ 0.00107845; % error$\approx$ 0.16176791

$n = 1000$: $y(1) \approx -0.6667738$; error≈ 0.00010712; % error≈ 0.016068

7.

$n = 10$: $y(2) \approx 2.6307403$; error≈ 0.14816259; % error≈ 5.96809466

$n = 100$: $y(2) \approx 2.49758864$; error≈ 0.01501091; % error≈ 0.60465035

$n = 1000$: $y(2) \approx 2.48408038$; error≈ 0.00150265; % error≈ 0.06052789

9. $y(1) \approx -1.1815025$

11. $y(0.5) \approx 0.30167222$

13. $n = 10$: $y(1) \approx -1.0311002$; $n = 100$: ≈ -0.9439698; $n = 1000$: ≈ -0.9353762

15. $n = 10$: $y(0.5) \approx 0.2495403$; $n = 100$: ≈ 0.22107094; $n = 1000$: ≈ 0.21833261

17. $y(3) \approx -0.2$

19. $y(0.5) \approx 0.2$

21. Equilibrium sol. $A = 0$

23. Equilibrium sol. $A = 0, -1$

25. d. The horizontal solution curve at $W = 1000$ is an equilibrium solution. e. $t = 6, 18, 30$.

Appendix A

1. avg. x-vel. $= 2$ m/s; avg. y-vel. $= 5$ m/s; $x = (y/5)^2 - 7$

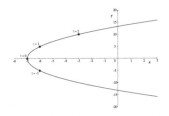

3. avg. x-vel. $= \dfrac{e^2 - 1}{2}$ m/s; avg. y-vel. $= 1$ m/s; $y = \ln x$

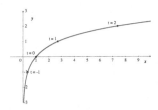

5. $\vec{v}(2) = (4, 5)$ m/s

$\vec{a}(2) = (2, 0)$ m/s

speed at $t = 2$ sec. $= \sqrt{41}$ m/s

7. $\vec{v}(2) = (e^2, 1)$ m/s

$\vec{a}(2) = (e^2, 0)$ m/s

speed at $t = 2$ sec. $= \sqrt{e^4 + 1}$ m/s

9. One possibility: $x = 5 + t^4$, $y = t$

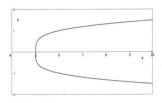

11. One possibility: $x = -2 + \cos t$, $y = 3 + \sin t$

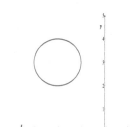

13. a. 0 b. -1 c. 1

15. 1

19. $\sqrt{2}(3, 5, 4)/10$

21. $\sqrt{2}(\tanh t, 1, 1/\cosh t)/2$

27.

a. $(-\cos t, -\sin t, 0)$

b. undefined

c. $(-\sin t, -\cos t, 0)$

d. $(\operatorname{sech} t, 0, -\tanh t)$

31. 0

37. $\vec{L}(t) = (-t, 0, 1)$

39. $\vec{L}(t) = (-e^\pi, 0) + te^{-\pi/2}(1, -1)$

41. a. $dy/dx = \sqrt{(k^2 - y)/y}$

Bibliography

[1] Massey, D. *Worldwide Integral Calculus, with infinite series.* Worldwide Center of Math., 1st edition, 2010.

[2] Rudin, W. *Principles of Mathematical Analysis.* International Series in Pure and Applied Mathematics. McGraw-Hill, 1953.

[3] Trench, William F. *Introduction to Real Analysis.* Prentice Hall, 2003.

Index

1-to-1, 143
RLC circuit, simple, 274
Δ notation, 3
e, value of, 215
y-intercept, 350

acceleration, 1
acceleration vector, 525
acceleration, average, 11
acceleration, instantaneous, 31
alternating current, 275
amplitude, 268
Angle Addition Formulas, 259
angular speed, 343
anti-derivative, 115, 230, 435
arccos, 298
arcsin, 298
AROC, 1, 2, 4
astroid, 324
asymptote, horizontal, 68, 350
asymptote, vertical, 72, 350
asymptotes, 350
autonomous differential equation, 430, 515

basket, 327
bell curve, 224
Bernoulli, Johann, 401
bijection, 143
binomial expansion, 160
blackbody, 224
blackbody radiation, 224
bound, lower, 142

bound, upper, 142
bounded, 142
bounded above, 142
bounded below, 142
brachistochrone problem, 534

capacitance, 274
capacitor, 274
carrying capacity, 255
Cauchy, Augustin-Louis, 43
Chain Rule, 186, 188
Chain Rule, multivariable, 344
characteristic equation, 431
charge across a capacitor, 274
circuit elements, 274
circuit, electrical, 192
closed, 141
codomain, 3, 142
compact set, 142
complementary angles, 286
component functions, 524
composition, 144, 186
concave down, 129, 130
concave up, 129, 130
concavity, 129, 349
connected set, 142
Conservation of Energy, 433
constraint, 370
constructive proof, 329
continuous, 25, 51
continuous function, 41, 502
continuous, uniformly, 79

contraction mapping, 396

converges, 205

converges uniformly, 207

correspondence, 1-to-1, 143

cosecant, 286

cosine, 259

cost, 30, 367

cotangent, 286

Coulomb's Law, 106

critical point, 111

critical value, 111

critically damped, 401

current, 274

curvature, 204, 327, 532

cycle, 267

cycloid, 534

decreasing function, 115

decreasing, strictly, 115

deductible, 124

deleted open interval, 42

deleted open left interval, 46

deleted open right interval, 46

demand, 39, 122, 226, 361, 367, 377

density function, 138, 167

derivative, 25, 82

derivative, 0-th, 127

derivative, n-th, 126

derivative, higher-order, 126

derivative, second, 126

difference quotient, 4

differentiable, 82

differentiable, continuously, 118

differential, 387

differential approximation, 387

Differential Calculus, 2

differential equation, 115, 177, 219, 225, 226, 264, 419, 420

differential equation, homogeneous first-order, 475

differential equation, homogeneous linear, 430

differential equation, linear, 430

differentials, relative, 387

differentiate, 82

differentiation, 82

differentiation, implicit, 198, 280, 312

disconnected set, 142

discontinuity, 10

discontinuity, jump, 52

discontinuity, removable, 52, 72

discontinuous, 51

distribution, 167

distribution function, cumulative, 138

distribution, cumulative, 167

diverges, 205

domain, 142

doomsday equation, 498

dot product, 184, 532

Double Angle Formulas, 260

drugs, exponential decay of, 247

dummy variable, 27, 163

e, 213

e, approximate value of, 214

Einstein, Albert, 17, 202

electromotive force, 274

elementary function, 41, 59, 318

energy, kinetic, 201

equilibrium position, 266

equilibrium solution, 421, 516

error analysis, 388

Euler's Method, 501

Euler, Leonhard, 83

even function, 19, 204, 259, 350

exponential decay, 211, 225

exponential function, 207

exponential function, base b, 243

exponential function, general, 243
exponential growth, 211, 225
extrema, 109
Extreme Value Theorem, 55, 152
extreme values, 109

factorial, 206
family of curves, 470
Fibonacci sequence, 222
First Derivative Test, 119
fixed cost, 366
folium of Descartes, 324
fractional part, 77
free-fall, 122, 309
frequency, 267
function, elementary, 41, 59, 318
function, inverse, 56, 143
function, rational, 49, 65
function, real, 3, 143
Fundamental Theorem of Calculus, 436, 452
Fundamental Trig. Identity, 259

general solution, 421
geometric finite sum, 214
Gompertz equation, 241
graphing, 349
gravity, force of, 181

half-life, 246
Hermite Equation, 430
higher-order derivative, 126
homeomorphism, 79
Hooke's Law, 273, 284
horizontal line test, 56
hyperbolic cosecant, 294
hyperbolic cosine, 225, 294
hyperbolic cotangent, 294
hyperbolic secant, 294
hyperbolic sine, 225, 294
hyperbolic tangent, 225, 294

hyperbolic trigonometric functions, 307

ideal gas, 178
Ideal Gas Law, 178
identity function, 144
image, 144
implicit differentiation, 198, 280, 312
Implicit Function Theorem, 332
Implicit Function Theorem, informal, 318
implicit solution, 464
implicitly defined function, 198, 312, 316
Improved Euler's Method, 509
increases without bound, 30
increasing function, 115
increasing, strictly, 115
indeterminate form, 400
indexing set, 144
indifference curve, 328
inductance, 274
induction, 49
inductor, 274
infinite series, 207
infinity, negative, 140
infinity, positive, 140
inflection point, 130, 349, 355
initial data, 419, 422
initial value problem, 115, 419, 422
injection, 143
Integral Calculus, 436
integral, definite, 436
integral, indefinite, 436
integrand, 437
integration, 436
integration by parts, 447
integration by substitution, 445
interest, compound, 236
interest, simple, 236
interference, 281
interior, 81

interior point, 81, 318

Intermediate Value Theorem, 55, 151

Intermediate Value Theorem for Derivatives, 155

interval, 140

interval, closed, 140

interval, half-closed, 140

interval, half-open, 140

interval, open, 140

inverse cosecant, 302

inverse cosine, 298

inverse cotangent, 300

Inverse Function Theorem, 194, 332

inverse image, 144

inverse secant, 302

inverse sine, 298

inverse tangent, 300

IROC, 2, 24

isocline, 513

isotherm, 326

jerk, 127

jump discontinuity, 52

kinetic energy, 344, 433

Kirchoff's Law, 274, 458

l'Hôpital's Rule, 401, 415

l'Hôpital, Guillaume, 401

Lagrange Form, 415

Lagrange, Joseph-Louis, 83, 415

Law of Cosines, 257, 343

Law of Inertia, 74, 531

Law of Sines, 257

least upper bound, 142

Leibniz, 436

Leibniz, Gottfried Wilhelm, 83

limit, 24, 43

limit, from left, 46

limit, from right, 46

limit, one-sided, 46

limit, two-sided, 46

linear approximation, 381

linear combination, 440

linear function, 380

linear operation, 161

linear speed, 343

linearization, 380

Liouville, Joseph, 452

local extrema, 349

logarithm, base b, 248

logarithm, natural, 228

logarithmic differentiation, 233

logistic model, 177, 219, 468

Maclaurin polynomial, 414

Maclaurin, Colin, 414

magnitude, 525

magnitude of a vector, 525

map, 142

maps into, 142

maps onto, 143

maps to, 142

marginal analysis, 389

marginal cost, 30

marginal price, 30

marginal profit, 30, 368

marginal rate of substitution, 328

marginal revenue, 30

marginal value, 30

maxima, local, 109

maximum value, 109

maximum value, absolute, 109

maximum value, global, 109

maximum value, local, 109

maximum value, relative, 109

Mean Value Theorem, 113

Mean Value Theorem, Cauchy's, 415

minima, local, 109

minimum value, 109

minimum value, absolute, 109

minimum value, global, 109

minimum value, local, 109

minimum value, relative, 109

mixing problem, 424

modeling, 363

momentum, 92

monotonic, 115, 349

monotonic, strictly, 115

multiplicity, 182

Newton's 1st Law of Motion, 74

Newton's 2nd Law of Motion, 92, 179, 231, 285

Newton's Law of Cooling, 226, 433, 486, 499

Newton's Method, 390

Newton, Isaac, 83

nodes, 281

number of steps in Euler's Method, 503

odd function, 19, 204, 259, 350

Ohm's Law, 183

one-to-one, 56, 143

onto, 56, 143

open ball, 315

open mapping, 379

open set, 141

optimization, 363

order of a derivative, 126

order of a differential equation, 420

orthogonal, 203, 227

orthogonal trajectory, 470

parameter, 523, 524

parameter error, 395

parameterization, 529

parameterize, 526

parameterized curve, 524

parametrically, 523

partial derivative, 322, 332, 344

partial fractions, 468

particular solution, 423

period, 257, 267

periodic function, 257

periodic functions, 257

phase diagram, 517

Pinching Theorem, 45

Planck's Law, 224

Poisson distribution, 223, 395

position, 11

position vector, 525

potential energy, 433

power loss, 192

Power Rule, 174

Power Rule for Integration, 437

Power Rule for Rational Exponents, 195

Power Rule, for Natural Exponents, 160

Power Rule, with general exponents, 249

Power-Exponential Rule, 250

present value, 19

price, 30

probability mass function, 395

Product Rule, 171

Product Rule for dot product, 184

profit, 30, 367, 377

Quotient Rule, 171

radians, 257, 258

radioactive decay, 246

radius of curvature, 533

range, 3, 143

rate of change, average, 1

rate of change, fractional, 164

rate of change, instantaneous, 2, 24, 25, 82

rate of change, relative, 164

rate of decrease, 84

rate of increase, 83

rational function, 49

real numbers, 140

real numbers, extended, 140

related rates, 335

Relativity, Special Theory of, 17, 37, 202

remainder, 383

removable discontinuity, 52, 72

removable lack of continuity, 52, 72

resistance, 274

resistor, 274

resonance, 282

restriction, 144

revenue, 30, 122, 226, 367, 377

Rolle's Theorem, 112

Rolle, Michel, 112

root, 58

Runge-Kutta Method, 509

secant, 286

secant line, 7

secant line method, 397

Second Derivative Test, 132

separable differential equation, 461

separation of variables, 461

sequence, 205

sequence of real functions, 206

sigma notation, 162

signed point, 351

simple harmonic motion, 267

sinc function, 411

sine, 259

sine approximation, 278

slope field, 501, 510

slope of f, 86

slope of graph, 86

solution curve, 510

sound pressure, 254

speed of light, 17, 37

speed, average, 11

speed, instantaneous, 31, 525

spline, cubic, 137

spring, 266

spring constant, 273

Squeeze Theorem, 45

standing wave, 281

step size, 503

stereographic projection, 412

sublimation, 345

substitution in integrals, 445

substitution in limits, 53, 54

surjection, 143

symmetric about the y-axis, 350

symmetric about the origin, 350

symmetry, 350

tangent, 286

tangent line, 29, 86

tangent line approximation, 381

tangent line, vertical, 195

tautological initial data, 442

Taylor polynomial, 414

Taylor remainder, 415

Taylor's Theorem, 413

Taylor, Brook, 414

terminal velocity, 485

Torricelli's Law, 476

transcendental, 214

triangle inequality, 145

triangular numbers, 168

underdamped, 400

uniform convergence, 207

uniformly continuous, 79

unit circle, 258

unit normal vector, 532

unit tangent vector, 531

Universal Gravitation, Law of, 176

unsigned point, 351

utility, 327